POLITICAL ECONOMY, OLIGOPOLY AND EXPERIMENTAL GAMES

ECONOMISTS OF THE TWENTIETH CENTURY

General Editors: David Colander, *Christian A. Johnson Distinguished Professor of Economics, Middlebury College, Vermont, USA* and Mark Blaug, *Professor Emeritus, University of London, UK, Professor Emeritus, University of Buckingham, UK and Visiting Professor, University of Amsterdam, The Netherlands*

This innovative series comprises specially invited collections of articles and papers by economists whose work has made an important contribution to economics in the late twentieth century.

The proliferation of new journals and the ever-increasing number of new articles make it difficult for even the most assiduous economist to keep track of all the important recent advances. By focusing on those economists whose work is generally recognized to be at the forefront of the discipline, the series will be an essential reference point for the different specialisms included.

A list of published and future titles in this series is printed at the end of this volume.

Political Economy, Oligopoly and Experimental Games

The Selected Essays of Martin Shubik Volume One

Martin Shubik

Seymour H. Knox Professor of Mathematical Institutional Economics, Cowles Foundation for Research in Economics, Yale University, USA

ECONOMISTS OF THE TWENTIETH CENTURY

Edward Elgar
Cheltenham, UK • Northampton, MA, USA

© Martin Shubik 1999

All rights reserved. No part of this publication may be reproduced, stored in a retrieval system or transmitted in any form or by any means, electronic, mechanical or photocopying, recording, or otherwise without the prior permission of the publisher.

Published by
Edward Elgar Publishing Limited
Glensanda House
Montpellier Parade
Cheltenham
Glos GL50 1UA
UK

Edward Elgar Publishing, Inc.
136 West Street
Suite 202
Northampton
Massachusetts 01060
USA

A catalogue record for this book
is available from the British Library

Library of Congress Cataloguing in Publication Data

Shubik, Martin.
   Political economy, Oligopoly and experimental games / Martin
Shubik.
     p. cm. – (Economists of the twentieth century series)
(Selected essays of Martin Shubik ; v. 1)
   Includes index.
   1. Game theory. 2. Oligopolies. 3. Operations research.
4. Shubik, Martin. I. Title. II. Series. III. Series: Shubik,
Martin. Essays. Selections ; v. 1.
HB144.S584   1999
338.8′2–dc21
                                                                                     99–16257
                                                                                        CIP

ISBN 1 85898 241 3

Printed and bound in Great Britain by MPG Books Ltd, Bodmin, Cornwall

Contents

Acknowledgements ix
Introduction xiii

PART I POLITICAL ECONOMY

1 'A Business Cycle Model with Organized Labor Considered', *Econometrica*, **20**(2), April 1952, pp. 284–94   3
2 'Information, Theories of Competition, and the Theory of Games', *Journal of Political Economy*, **60**, February–December 1952, pp. 145–50 (reset)   14
3 'Information, Risk, Ignorance and Indeterminacy', *The Quarterly Journal of Economics*, **68**, November, 1954, pp. 629–40   21
4 'Market Form, Intent of the Firm and Market Behavior', *Zeitschrift für Nationalökonomie*, **17**(2), 1957, pp. 186–96   33
5 'Simulation of the Industry and the Firm', *American Economic Review*, **50**(5), December 1960, pp. 908–19   44
6 'Approaches to the Study of Decision-Making Relevant to the Firm', *The Journal of Business of the University of Chicago*, **34**(2), April 1961, pp. 101–18   56
7 'Objective Functions and Models of Corporate Optimization', *The Quarterly Journal of Economics*, **75**(3), August 1961, pp. 345–75   74
8 'Incentives, Decentralized Control, the Assignment of Joint Costs and Internal Pricing', *Management Science*, **8**(3), April 1962, pp. 325–43   105
9 'Ownership and the Production Function' (with L.S. Shapley), *The Quarterly Journal of Economics*, **81**, February 1967, pp. 88–111   124
10 'A Curmudgeon's Guide to Microeconomics', *The Journal of Economic Literature*, **8**(2), June 1970, pp. 405–34   148
11 'On Different Methods for Allocating Resources', *Kyklos*, **23**(2), 1970, pp. 332–7   178
12 'The "Bridge Game" Economy: An Example of Indivisibilities', *Journal of Political Economy*, **79**(4), July–August 1971, pp. 909–12   184
13 'A Note on the Shape of the Pareto Optimal Surface' (with G. Bradley), *Journal of Economic Theory*, **8**(4), August 1974, pp. 530–38   188
14 'On the Role of Numbers and Information in Competition', *Revue Economique*, **26**(4), 1975, pp. 605–21   197
15 'An Example of a Trading Economy with Three Competitive Equilibria' (with L.S. Shapley), *Journal of Political Economy*, **85**(4), August 1977, pp. 873–5   214
16 'On Concepts of Efficiency', *Policy Sciences*, **9**, 1978, pp. 121–6 (reset)   217

17 'Information Conditions, Communication and General Equilibrium' (with P. Dubey), *Mathematics of Operations Research*, **6**(2), May 1981, pp. 186–9 223
18 'The Many Approaches to the Study of Monopolistic Competition', *European Economic Review*, **27**, 1985, pp. 97–114 227

PART II OLIGOPOLY

19 'A Comparison of Treatments of a Duopoly Situation' (with J.P. Mayberry and J.F. Nash), *Econometrica*, **21**, January 1953, pp. 141–54 247
20 'A Comparison of Treatments of a Duopoly Problem (Part II)', *Econometrica*, **23**(4), October 1955, pp. 417–31 261
21 'Games of Economic Survival' (with G.L. Thompson), *Naval Research Logistics Quarterly*, **6**(2), June 1959, pp. 111–23 276
22 'Price Strategy Oligopoly with Product Variation' (with L.S. Shapley), *Kyklos*, **22**(1), 1969, pp. 30–44 289
23 'Price Strategy Oligopoly: Limiting Behavior with Product Differentiation', *Western Economic Journal*, **8**(3), September 1970, pp. 226–32 304
24 'Price Variation Duopoly with Differentiated Products and Random Demand' (with R. Levitan), *Journal of Economic Theory*, **3**(1), March 1971, pp. 23–39 311
25 'Price Duopoly and Capacity Constraints' (with R. Levitan), *International Economic Review*, **13**(1), February 1972, pp. 111–22 . . 328
26 'Duopoly with Price and Quantity as Strategic Variables' (with R. Levitan), *International Journal of Game Theory*, **7**(1), 1978, pp. 1–11 340
27 'Noncooperative Oligopoly with Entry' (with K. Nti), *Journal of Economic Theory*, **24**(2), April 1981, pp. 187–204 351

PART III GAMING

28 'Some Experimental Non–Zero Sum Games with Lack of Information about the Rules', *Management Science*, **8**(2), January 1962, pp. 215–34 371
29 '"So Long Sucker" – A Four Person Game' (with M. Hausner, J. Nash and L.S. Shapley), in *Game Theory and Related Approaches to Social Behavior*, M. Shubik (ed.), New York: John Wiley & Sons, 1964, pp. 359–61 391
30 'A Note on a Simulated Stock Market', *Decision Sciences*, **1**, January–April 1970, pp. 129–41 394
31 'An Artificial Player for a Business Market Game' (with G. Wolf and S. Lockhart), *Simulation and Games*, March 1971, pp. 27–43 407
32 'The Dollar Auction Game: A Paradox in Noncooperative Behavior and Escalation', *The Journal of Conflict Resolution*, **15**(1), 1971, pp. 109–11 424
33 'On the Scope of Gaming', *Management Science*, **18**(5), January Part 2, 1972, pp. 20–36 427

34	'Some Experiences with an Experimental Oligopoly Business Game' (with G. Wolf and H.B. Eisenberg), *General Systems*, **17**, 1972, pp. 61–75 (reset)	444
35	'Perception of Payoff Structure and Opponent's Behavior in Related Matrix Games' (with G. Wolf and B. Poon), *Journal of Conflict Resolution*, **18**(4), December 1974, pp. 646–55	468
36	'Teams Compared to Individuals in Duopoly Games with an Artificial Player' (with G. Wolf), *Southern Economic Journal*, **41**(4), April 1975, pp. 635–48	478
37	'Cooperative Game Solutions: Australian, Indian, and U.S. Opinions', *Journal of Conflict Resolution*, **30**(1), March 1986, pp. 63–76	492

PART IV   GAME THEORY AND OPERATIONS RESEARCH

38	'Solutions of N-Person Games with Ordinal Utilities' (with L.S. Shapley), *Econometrica*, **21**(2), April 1953, p. 348	509
39	'A Method for Evaluating the Distribution of Power in a Committee System' (with L.S. Shapley), *The American Political Science Review*, **43**(3), September 1954, pp. 787–92	510
40	'Does the Fittest Necessarily Survive?', in *Readings in Game Theory and Political Behavior*, M. Shubik (ed.), New York: Doubleday, 1954, pp. 43–6	516
41	'The Assignment Game I: The Core' (with L.S. Shapley), *International Journal of Game Theory*, **1**(2), 1971, pp. 111–30 (reset from RAND paper R-874-RC, October 1971, pp. 1–41)	520
42	'What is an Application and When is Theory a Waste of Time?', *Management Science*, **33**(12), December 1987, pp. 1511–22	539

Name index   551

Acknowledgements

The publishers wish to thank the following who have kindly given permission for the use of copyright material.

Academic Press for articles: 'A Note on the Shape of the Pareto Optimal Surface' (with G. Bradley), *Journal of Economic Theory*, **8**(4), August 1974, pp. 530–38; 'Price Variation Duopoly with Differentiated Products and Random Demand' (with R. Levitan), *Journal of Economic Theory*, **3**(1), March 1971, pp. 23–39; 'Noncooperative Oligopoly with Entry' (with K. Nti), *Journal of Economic Theory*, **24**(2), April 1981, pp. 187–204.

American Economic Association for articles: 'Simulation of the Industry and the Firm', *American Economic Review*, **50**(5), December 1960, pp. 908–19; 'A Curmudgeon's Guide to Microeconomics', *The Journal of Economic Literature*, **8**(2), June 1970, pp. 405–34.

The Decision Sciences Institute for article: 'A Note on a Simulated Stock Market', *Decision Sciences*, **1**, January–April 1970, pp. 129–41.

The Econometric Society for articles: 'A Business Cycle Model with Organized Labor Considered', *Econometrica*, **20**(2), April 1952, pp. 284–94; 'A Comparison of Treatments of a Duopoly Situation' (with J.P. Mayberry and J.F. Nash), *Econometrica*, **21**, January 1953, pp. 141–54; 'A Comparison of Treatments of a Duopoly Problem (Part II)', *Econometrica*, **23**(4), October 1955, pp. 417–31; 'Solutions of N-Person Games with Ordinal Utilities' (with L.S. Shapley), *Econometrica*, **21**(2), April 1953, p. 348.

Elsevier Science for article: 'The Many Approaches to the Study of Monopolistic Competition', *European Economic Review*, **27**, 1985, pp. 97–114.

Helbing & Lichtenhahn Verlag for articles: 'On Different Methods for Allocating Resources', *Kyklos*, **23**(2), 1970, pp. 332–7; 'Price Strategy Oligopoly with Product Variation' (with L.S. Shapley), *Kyklos*, **22**(1), 1969, pp. 30–44.

The Institute of Management Sciences for articles: 'Incentives, Decentralized Control, the Assignment of Joint Costs and Internal Pricing', *Management Science*, **8**(3), April 1962, pp. 325–43; 'Information Conditions, Communication and General Equilibrium' (with P. Dubey), *Mathematics of Operations Research*, **6**(2), May 1981, pp. 186–9; 'On the Scope of Gaming', *Management Science*, **18**(5), January Part 2, 1972, pp. 20–36; 'Some Experimental Non–Zero Sum Games with Lack of Information about the Rules', *Management Science*, **8**(2), January 1962, pp. 215–34; 'What is an

Application and When is Theory a Waste of Time?', *Management Science*, **33**(12), December 1987, pp. 1511–22.

International Economic Review for article: 'Price Duopoly and Capacity Contraints' (with R. Levitan), *International Economic Review*, **13**(1), February 1972, pp. 111–22.

International Society for the Systems Sciences for article: 'Some Experiences with an Experimental Oligopoly Business Game' (with G. Wolf and H.B. Eisenberg), *General Systems*, **17**, 1972, pp. 61–75.

John Wiley & Sons, Inc. for articles: 'Games of Economic Survival' (with G.L. Thompson), *Naval Research Logistics Quarterly*, **6**(2), June 1959, pp. 111–23; '"So Long Sucker" – A Four Person Game' (with M. Hausner, J. Nash and L.S. Shapley), in *Game Theory and Related Approaches to Social Behavior*, M. Shubik (ed.), 1964, pp. 359–61.

Kluwer Academic Publishers for article: 'On Concepts of Efficiency', *Political Sciences*, **9**, 1978, pp. 121–6.

MIT Press for articles: 'Information, Risk, Ignorance and Indeterminacy', *The Quarterly Journal of Economics*, **68**, November 1954, pp. 629–40; 'Objective Functions and Models of Corporate Optimization', *The Quarterly Journal of Economics*, **75**(3), August 1961, pp. 345–75; 'Ownership and the Production Function' (with L.S. Shapley), *The Quarterly Journal of Economics*, **81**, February 1967, pp. 88–111.

Physica-Verlag for articles: 'Duopoly with Price and Quantity as Strategic Variables' (with R. Levitan), *International Journal of Game Theory*, **7**(1), 1978, pp. 1–11; 'The Assignment Game I: The Core' (with L.S. Shapley), *International Journal of Game Theory*, **1**(2), 1971, pp. 111–30.

Presses de Sciences PO for article: 'On the Role of Numbers and Information in Competition', *Revue Economique*, **26**(4), 1975, pp. 605–21.

Sage Publications for articles: 'An Artificial Player for a Business Market Game' (with G. Wolf and S. Lockhart), *Simulation and Games*, March 1971, pp. 27–43; 'The Dollar Auction Game: A Paradox in Noncooperative Behavior and Escalation', *The Journal of Conflict Resolution*, **15**(1), 1971, pp. 109–11; 'Perception of Payoff Structure and Opponent's Behavior in Related Matrix Games' (with G. Wolf and B. Poon), *Journal of Conflict Resolution*, **18**(4), December 1974, pp. 646–55; 'Cooperative Game Solutions: Australian, Indian and U.S. Opinions', *Journal of Conflict Resolution*, **30**(1), March 1986, pp. 63–76.

Southern Economic Journal for article: 'Teams Compared to Individuals in Duopoly Games with an Artificial Player' (with G. Wolf), *Southern Economic Journal*, **41**(4), April 1975, pp. 635–48.

Springer Verlag for article: 'Market Form, Intent of the Firm and Market Behavior', *Zeitschrift für Nationalökonomie*, **17**(2), 1957, pp. 186–96.

The University of Chicago for articles: 'Information, Theories of Competition, and the Theory of Games', *Journal of Political Economy*, **60**, February–December 1952, pp. 145–50; 'Approaches to the Study of Decision-Making Relevant to the Firm', *The Journal of Business of the University of Chicago*, **34**(2), April 1961, pp. 101–18; 'The "Bridge Game" Economy: An Example of Indivisibilities', *Journal of Political Economy*, **79**(4), July–August 1971, pp. 909–12; 'An Example of a Trading Economy with Three Competitive Equilibria' (with L.S. Shapley), *Journal of Political Economy*, **85**(4), August 1977, pp. 873–5.

Western Economic Journal for article: 'Price Strategy Oligopoly: Limiting Behavior with Product Differentiation', *Western Economic Journal*, **8**(3), September 1970, pp. 226–32.

Every effort has been made to trace all the copyright holders but if any have been inadvertently overlooked the publishers will be pleased to make the necessary arrangements at the first opportunity.

Introduction

The occasion of publishing a large set of one's articles provides an opportunity for reflection and a consideration of the author's *apologia pro sua vita*. The test of one's own work lies with time and with others. Nevertheless the exercise of introspection provides an attempt to explain to oneself and to others what one was trying to do, is doing and still hopes to do. This might offer some insight on the gap between the wish and the reality as well as the self-assessment and the assessment of others.

In trying to classify my publications I find that they fall into several categories: general economics and political economy; oligopoly theory; the theory of money and financial institutions; pure game theory; game theory applied to economics, law, political science, management, experimental gaming, simulation, defense studies and operations research. I have concentrated my selection primarily on my economics publications, thus I have left out most of my work in defense studies, in operations research and in other behavioral sciences.

The period covered is from 1952 to 1992. Although I have worked on many topics, my basic long-term concerns have been with the understanding of several fundamental features of the economy, in particular the functioning of the price system and the purpose and evolution of money and financial institutions.

My early interests by the end of high school were in history, politics, society and the economy. Although by then I suspected that I had at best a mediocre talent for mathematics, I chose to go into the Mathematics and Physics program at the University of Toronto, not because I thought that I might become a mathematician, but because I thought that at the undergraduate level one could learn some techniques and ways of thought which would be of considerable value no matter what one did. In contrast, undergraduate studies in the Social Sciences at the time struck me as hardly worth while. I felt that the 'great books' could be read by oneself and, if one went to graduate school and were careful in course selection, there might be a course or two which would explain the great books.

I learned a great deal as an undergraduate in Mathematics and Physics. In particular I learned that I did not have the type of mind or talent to become either a decent mathematician or a physicist. But even though I was both a poor mathematician and a sloppy experimentalist in physics I learned a great amount about the precision of mathematics and the care needed to perform even a trivial experiment in physics. I was impressed by several of my teachers at Toronto. While I sat as an undergraduate in Professor Richard Brauer's class in algebra, I kept thinking that it was a waste of such obvious talent on a bunch of ordinary Ontario undergraduates. I was fascinated by the enthusiasm of Professor Coxeter, teaching the geometry of perfect solids. But at the same time it struck me that there were many forms of intelligence and that a great mathematician could be a totally naive citizen. As I was politically active as an undergraduate I was impressed by the different form of intelligence possessed by local politicians as contrasted with the processors of mathematics.

On graduating from Mathematics and Physics I thought for a while that I would take a higher degree in Applied Mathematics, but it was evident to me that I had neither the real interest, insight nor talent. I decided instead to take an MA in economics at Toronto. There were several positive reasons for doing so. I was involved with the CCF Party and contemplated running for parliament. But as I examined the platform concerning the electrification of North Ontario I kept asking myself, 'Where is the money going to come from?' I found the politicians to be willing to make dogmatic statements concerning items about which they knew next to nothing. This contrasted with the mathematicians who required a definition and precision for virtually everything. I decided that some formal economic theory might help. I took a course in Great Books in the Social Sciences from Harold Innis and found it to be fascinating and certainly mind stretching. Yet at the same time, while listening to his major thesis about the relationship between democracy and the oral tradition and written language and communication and the authoritarian state, I did some research on the Inca empire and found that it did not fit into the Innis thesis. I had the uneasy feeling that in the Social Sciences one merely refitted the facts to make sure that they fitted appropriately onto the Bed of Procrustes.

In my course in Economic Theory, we were required to write a book review of a major book in economics. While browsing the University of Toronto library I came across, in the newly purchased book display, in 1948, the *Theory of Games and Economic Behavior*. I read parts of it and did not really understand it. I probably wrote a relatively bad essay on it. I have no recollection whatsoever about what I said, but within a few days of finding the book, I felt that this *had to be* the way to go. Within a few weeks I decided that I wanted to go to Princeton (although I had originally contemplated MIT as well). I wrote to Princeton, saying that I wanted to study with Professor Morgenstern. I applied to both the Economics and Mathematics Departments at Princeton and the Mathematics Department had the good sense to reject my application, thereby saving all parties from potential grief.

Princeton was a sheer pleasure, far better than I suspect any drug high might offer. On the faculty I grew to like and admire Oskar Morgenstern as a great intellectual entrepreneur with great common sense and a marvelous sense of purpose. William Baumol became a friend for life and I was impressed by both the fine scholarship and the unpleasant competitive debating attitude of Jacob Viner; it seemed to me that winning the debate appeared to be more important to him than trying to grope towards whatever the truth might be. Von Neumann's intellect staggered me and gave me a feeling for what major talent looks like. When Irwin Panovsky, the great art historian, gave a lecture which I attended, I had the feeling of having seen a completely different, but nevertheless major, talent.

My intellectual stimulus came from the mathematics graduate students far more than from the economics graduate students. In particular, John McCarthy, Marvin Minsky, John Nash and Lloyd Shapley provided considerable intellectual stimulation. Nash, Shapley and I interacted on our mutual interest in game theory. A possible distinction between Nash, Shapley and myself is that they were clearly both interested in and capable of making basic contributions to the underlying mathematical theory, as is evinced by the Nash noncooperative equilibrium and the Shapley value, whereas I was primarily concerned with application to economics in particular, and the social sciences in general.

My first papers came out in 1952. I published 'A Business Cycle Model with Organized Labor Considered' in *Econometrica* (1952) where my concern was to catch the idea that the threat power of the union varied with the position of the business cycle. The next paper was both less technical and far more important. It seemed to me that the information requirements in much of economic theory were far too high. An elementary consideration of the combinatorics and costs has to limit the amount of information utilized. 'Information, Theories of Competition, and the Theory of Games' appeared in *The Journal of Political Economy* (1952).

In my discussions with Nash, who had finished his thesis on the existence of noncooperative equilibria, it was clear to me that the Cournot model of oligopoly utilized an equilibrium concept which was in essence a noncooperative equilibrium. Nash and I agreed to apply both his noncooperative and cooperative papers to a duopoly model. After my making many errors in lengthy and boring calculations, we enlisted Jim Mayberry to join us in this enterprise.

Shapley and I had worked previously on two-person cooperative games with variable threats, but stopped after Nash produced his solution. I am still not satisfied with the economic meaning of how to smooth away a continuum of equilibria to yield a unique outcome.

Although my interest in cooperative game theory was, and remains, high, it seemed to me that there was much to be done, especially in oligopoly theory, utilizing noncooperative theory. I decided that this would be the main concept employed in my thesis. I had read both Cournot and Edgeworth's *Mathematical Psychics* and they both, in different ways, suggested to me the idea of studying the behavior of solutions to games under replication. Intuitively the idea that the influence of a single individual becomes small as numbers of like, or almost like, individuals increases seemed to be commonsensical.

My thesis, which eventually became a book entitled *Strategy and Market Structure* (1959) utilized John Nash's noncooperative equilibrium, which I regarded as a considerable generalization of Cournot's equilibrium concept (although my impression in conversations with Nash was that he was not aware of Cournot's work). While working on my thesis it seemed to me that the most satisfactory link between oligopoly models and perfect competition had to involve the appropriate model of very small agents, so small that a single individual would not influence price. Shapley pointed out that to do this right would require measure theory, where one could consider a single agent as a set of measure zero, but as I knew no formal measure theory I settled for examining replicated finite sets of agents.

In the second part of my 1959 book, I wanted to take some steps towards dynamics. I concluded that an economic dynamics had to place stress on information conditions and on some basic aspects of institutional structure. In order to stress this, I entitled the second part of the book 'Mathematical Institutional Economics'.

My goal was to understand economic behavior and, to a lesser extent, political and bureaucratic strategic voting behavior. Thus, when Shapley showed me his newly developed value solution and described the class of simple games as a formal way to represent voting structures, it appeared to me that this offered an opportunity for a basic application to Political Science. We jointly wrote an article entitled 'A Method for Evaluating the Distribution of Power in a Committee System' (1954). Much to my surprise, it was accepted almost immediately by *The American Political Science Review*; to this day I do not know how or why this major journal in political science took a more or less mathematical article from two unknown graduate students, neither of whom was in Political Science.

Even as a graduate student, I was struck by the contrast between cooperative game theory, the seeds of which I regarded as already present in Edgeworth, and noncooperative theory, which was present in Cournot. I wanted to see if game theoretic methods applied to both of these works would yield limiting results related to the concept of economic competition. In the early 1950s, Shapley, Gillies and I had discussed the core of an n-person game in cooperative form. It seemed to me that Edgeworth's work on two-sided markets could be regarded as a precursor of a cooperative game approach to markets. While at the Center for Advanced Study in the Behavioral Sciences, I wrote an article entitled 'Edgeworth Market Games' (1959) which was finally published four years later. At the time of writing, except for some assistance by Howard Raiffa, I had no one to check my poor mathematics, and the article appeared with several false theorems. By the time it was finally published, I had realized most of my errors and had conjectured on the full significance of the core and its convergence to the competitive equilibria under broad conditions, but I knew that I was incapable of proving it. Most of my earlier errors were due to a confusion of the relationship between the sidepayment and no-sidepayment games (this linkage was eventually resolved with the analysis of the λ-transfer games). I corrected most of my errors in a publication which took several years to get published, entitled 'Extended Edgeworth Bargaining Games and Competitive Equilibrium' (1968). I posed my general conjecture to Lloyd Shapley and then to Herbert Scarf in a walk back from a seminar which he had given at Columbia University (Shapley proved, but I believe did not publish, the sidepayment results) and Scarf was able to prove the general conjecture for games without sidepayments.

Another major challenge which I perceived as a graduate student was to construct satisfactory models of the firm in a dynamic context which included the possibility of bankruptcy. I constructed a class of games which I called 'Games of Economic Survival' (1959) and managed to enlist the mathematical collaboration of Gerald Thompson in analyzing them as stochastic games with an infinite horizon and with bankruptcy represented by an absorbing barrier for the controlled random walk. At the time I did not yet see the critical importance of bankruptcy as a necessary feature in constructing any satisfactory viable dynamic theory of competition.

The few years I spent at General Electric gave me an appreciation for both macroeconomic detail and managerial practice which would have been difficult to obtain in academia. While there I became interested in the possibilities for gaming as a teaching and planning device; George Feeney was building a business game and I had talked with Richard Bellman about his game for the AMA. Even earlier, a chance meeting with Sidney Siegal in 1956 in the High Sierras had convinced me that experimental gaming was going to be of considerable importance. We began what we hoped was going to be a long collaboration; I commuted every now and then from General Electric in New York to Pennsylvania State University, but distance and Sid's early death cut our plans short. In 1960, while visiting Yale, I ran an experiment which was published as 'Some Experimental Non-Zero Sum Games with Lack of Information about the Rules' (1962). It is my belief that, to this day, the full importance of experimental gaming and games for teaching in economics has not been appreciated.

While I was at General Electric, the importance of the relationship between the assignment of joint costs and incentive systems within the firm seemed to me to be critical. I had talked about this with Harlan Mills, whom I regarded as a fine operations research practitioner. It appeared to me that one way of looking at this problem was via the Shapley value which assigned a combinatoric marginal worth to the players. I wrote up these observations in 'Incentives, Decentralized Control, the Assignment of Joint Costs and Internal Pricing' (1962).

Another of my concerns has been defense studies. In particular there is a central problem which has certainly been the object of formal study for around 40 years, but which still is by no means adequately solved. It is the meaning and credibility of threat in international relations, in economic competition and in social life.

My debt to Lloyd Shapley cannot be overstated. There are two major books on game theory published under my name which, as far as I am concerned, are joint work. But when one works with a close colleague and friend for many decades, individual choice should be respected, even if it is not understood. He chose to not appear as a coauthor, but consented to have me publish what I regard as essentially joint work. In order to fully explain our collaboration to myself, I have said, partly in jest, partly quite seriously, that I could usually tell when a paper with Lloyd was finished – I could no longer read it.

My strengths have been in conjecturing and model building, not in proof. Much of my time has been spent in finding an individual with the mathematical talent who was willing and able to understand my models and conjectures and interested enough to collaborate. The great benefits of working with Lloyd were that, not only does he have great talent in game theory, but his ability to find both logical and substantive flaws in models is considerable.

Volume One

The collection of reprints presented here covers three areas of interest. The first is devoted to my general writings in political economy and the others to two more specific topics which interested me early in my career: oligopoly theory and experimental gaming.

I was convinced from the start of my professional career that the treatment of information was critical to the understanding of economic process. This has run as a constant theme through several of my papers on theory and experimentation and has culminated in my concern for the information processing aspects of money and financial institutions and for what I regard as one of the key challenges for the behavioral sciences of the future: that is, the development of adequate models to describe decision making among individuals with limited abilities in perception, memory, computation and prediction.

Much of economic activity involves fiduciary decision making. The spenders of large sums of money are usually playing primarily with other people's money. Thus, in order to understand some basic aspects of economic decision making, it is desirable to understand how hierarchies and bureaucracies differ from individuals in the production of economic decisions.

About 25 years ago I offered my opinions on the problems and future of microeconomics in a paper entitled 'A Curmudgeon's Guide to Microeconomics' (1970). On rereading it,[1] although some of my more sanguine predictions may be running late, I believe that for the most part the development has been along the lines suggested, although the importance of computational methods and the computer were probably underestimated, even though they were advocated.

Oligopoly
For at least one brand of mathematical microeconomist, oligopoly theory, especially duopoly theory, plays the same role as the five-finger exercises for the pianist, or even worse the double acrostic for those who have nothing better to do with their time. When I saw the connection between John Nash's thesis and Cournot, I was convinced that games in strategic or extensive form provided a fruitful way to extend our knowledge of competition among the few. I believe now that I was simultaneously right in some aspects and wrong in others. The formal process models of game theory have permitted us to begin to obtain an appreciation of the importance of production, price, product variation, location, capacity, entry, inventories and other basic economic features of the firm in competition. We have also been able to gain better insight into the role of numbers of competitors. Most of my papers are addressed to these features. But it has become more and more apparent that we do not have an adequate multiperiod dynamic theory for competition among the few. The three-firm problem in economics is probably far harder analytically and more difficult to model than the three-body problem in physics. In recent years the 'New Industrial Organization' has taken over, highly laced with game theory and based on a noncooperative equilibrium theory which does not appear to be adequate to capture the dynamics. Mathematical economics and game theory are undoubtedly helpful in illustrating interesting possibilities and ways of thinking about oligopolistic competition. They are probably of value in providing expert witnesses in antitrust litigation with intellectual ammunition for their cannons. But I suspect that the future development in industrial organization will swing back more and more to better models of administrative behavior and more detailed institutional understanding of specific industries, perhaps bolstered by simulations and specific game models. An exception may be had in the construction of unified engineering, marketing and financial models which will blend these diverse aspects of the control of the firm into far richer (but computationally feasible) models than is currently the practice.

Experimental gaming
In my discussions with Sidney Siegal, Merrill Flood and others dating back to the 1950s I was (and am still) convinced of the enormous importance of gaming in economics and in business for teaching, training, planning and experimentation. Furthermore experimental games provide a natural forum for cooperative work among behavioral scientists with different backgrounds: in particular, economists, psychologists and social psychologists. Much of my own work was done in collaboration with Siegal and Gerrit Wolf. Economists working alone, especially with highly computerized games, may easily fail to appreciate the many artifacts and constraints guiding their results.

Although one can hardly classify as a formal experiment asking individuals their opinions about how they would play certain games, I have found that considerable insight concerning several basic aspects of game theory can be gleaned from doing so. In 'Cooperative Game Solutions: Australian, Indian, and U.S. Opinions' (1986) the attractiveness of the core as a solution concept is examined as its size and 'fairness' are varied.

I have been concerned with the design of simple games to illustrate problems and principles in multiperson decision making. Two games are presented here. '"So Long Sucker"' (1964) was concocted by a group of us when we were graduate students at Princeton in the early 1950s. We wanted to design a game where informal cooperation was necessary, but not sufficient to win. At some point there would appear great pressures to doublecross one's partner. 'The Dollar Auction Game' (1971) struck me as a nice illustration of the potential for mutually destructive escalation.

Except in highly structured games where the players interact through anonymous markets, experience with gaming indicates (at least to some of us) the inadequacy of the model of the well informed rational decision maker of *homo oeconomicus*. The actual beast is far more complex and subtle. The power of general equilibrium theory and the writings on the price system derives for the most part from the fortunate feature that some aspects of economics can be treated as an *asocial science*: the large numbers of agents, to a great extent, wipe out much of the sociopsychological interaction. The wise economist who wishes to experiment is well advised to appreciate how differently the social psychologist, historian, anthropologist, zoologist or biologist will look at the same phenomenon.

Volume Two

The price system
My specific goal for many years (essentially from 1952 to 1984) was to understand the relationship between the competitive equilibrium price system and the various game theoretic solutions applied to economic exchange and production modeled as a game. Without the deep collaboration with Shapley, I would not have had a hope of understanding the deep relationships that are present. My first attempts involved the understanding of the core, which could be regarded as reflecting countervailing power. My program was to examine and contrast the competitive price system with the core, the von Neumann–Morgenstern stable set, the Shapley value (the nucleolus, kernel and bargaining set some years later) and the noncooperative equilibrium. Already in my two articles on the core it was clear to me that, although the price equilibrium would be in the stable set, there was no indication that the stable set would converge towards the price system. This was further illustrated in work with Shapley in 'Concepts and Theories of Pure Competition' (1967). In this publication we also were able to present a nonsymmetric closed strategic model of exchange with one group of agents as price takers, but the others as strategic Cournot quantity strategy players.

In 1967, William Lucas published his counterexample to a conjecture of von Neumann that all coalitional games would have a stable set solution. This result intrigued me, especially when, in a chance conversation, Michael Maschler mentioned that this game had a core. In particular, I wanted to know if this could possibly apply to a model of economic exchange or production formulated as a game in coalitional form. I spent several weeks trying to cook up an example. I finally patched up an example which I suspected was wrong, but I sent it to Shapley anyhow, hoping that what did happen, would happen. He found the errors, brooded over the problem and constructed an accurate example which established my hunch that the Lucas counterexample could arise from an innocent-looking economic background. We then talked further and made an extremely simple but fundamental observation. Any subgroup of agents in an n-person exchange economy can also form an exchange economy. But as we knew that the game representation of an exchange economy has a core, this implied that every one of the 2^n games which can be formed from the n players also has a core. From this observation in the paper, 'On Market Games' (1969), we developed and explored the relationship between exchange economies and totally balanced games and showed that all exchange economies could be represented as totally balanced games. Some years later, in 'Competitive Outcomes in the Cores of Market Games' (1976), we went in the opposite direction and showed that any totally balanced game would have at least one representation as an exchange economy.

While at IBM in 1962, I had conjectured that the Shapley value of an exchange economy did not necessarily lie in the core. I suggested this to Scarf and Shapley and actually was able to construct a simple and correct example. I then guessed that, even though it was not in the core, on replication it might converge to the limit point of the core. I interested Shapley in this idea and he was able to establish my conjecture. We published 'Pure Competition, Coalitional Power and Fair Division' (1969) showing that the value also converged to the same limit point as the core which could be interpreted as the competitive price system.

There are three further solution concepts, little known to the broad public, but regarded as of interest by the faithful in game theory. They are the nucleolus, the kernel and the bargaining set. As the nucleolus solution is always in the core, if the core is nonempty, the proof of convergence came as a free good, given the core convergence. Shapley and I worked on the convergence of the bargaining set and kernel solutions, but we never published a paper on it, although I provided a painstakingly calculated (and possibly inaccurate?)[2] example in my book, *A Game-Theoretic Approach to Political Economy* (1944, pp. 343–9).

Missing from the solutions noted above are the convergence results for the non-cooperative equilibrium models of the economy which should reflect individual strategic power. They are covered in the next section, for reasons which are indicated below.

A brief summary of the investigation of the functioning of the price system in an exchange economy is as follows:

- the competitive equilibrium – decentralization;
- the core – the limitations imposed by countervailing power;
- the value – an *a priori* concept of fair division;[3]
- the nucleolus – the minimization of the maximum complaint any group might have against the outcome;

- the noncooperative equilibrium – the equilibrium outcome resulting from the use of individual strategic power in the presence of consistent (self-fulfilling) expectations.

The theory of money and financial institutions

In 1961, while at the IBM Watson Laboratories, I had begun to think as deeply as I could about the role of money in the economy. I tried to obtain a satisfactory model and failed miserably. This problem dogged me for years. In the late 1960s, I made some small headway, writing and publishing 'Pecuniary Externalities: A Game Theoretic Analysis' (1971). But it did not seem to go anywhere.

In 1970–71, I took a leave from Yale and spent over a year at RAND. In late 1970, I reviewed my work of the previous years and came to the conclusion that there were two major problems of deep interest to me which were still outstanding. They were the symmetric embedding of the Cournot oligopoly model in a closed exchange model and a macroeconomic theory of money. I decided that, as I was already 44, and had been working unsuccessfully for around 10 years on the theory of money, even though it was my number one problem I should finally abandon it and concentrate on the last piece of the package of game theoretic solutions applied to understanding models of a closed exchange economy. This was the noncooperative equilibrium.

The reason why I had failed to obtain the correct formulation for the Cournot and Bertrand–Edgeworth models of competition, although Shapley and I had marched through essentially all of the cooperative solutions, was that I had failed to note a fundamental conceptual difference in the basic models required. The characteristic function or coalitional form representation of a game provides a list of how much any grouping of individuals can obtain if they all coordinate their actions. It leaves out all transactions costs; communication is implicitly free; all the problems of coalition formation, cooperation and contract are costless. The coalitions do not exist, they are 'in being'; they are mathematical artifacts which enable us to consider the combinatorics of finite *n*-person games at a high level of abstraction, using only the characteristic function. In essence, it is a context and institution free representation of multiperson collaboration. The primitive concepts utilized are players, coalitions and payoffs to coalitions.

The general equilibrium representation of economic competition is also presented at a high degree of abstraction, but not quite as high as the characteristic function. The primitive concepts behind the general equilibrium are individually owned, traded and consumed goods (and services), production processes and prices. Beyond those features there is no institutional specification and no context is supplied.

Neither the characteristic function representation of a game nor the general equilibrium model of the economy provides a specification of process. The mathematical specification of the mechanisms which serve to carry process amount to adding (albeit at a high level of abstraction) an institutional description. Essentially *the rules of the game are the carriers of process* and, in the economy, *the carriers of process are the elementary institutions*. It was an eventual understanding that the Cournot description of competition and Bertrand's diatribe against Cournot could be recast as arguments about the appropriate *extensive form* description of the game which enabled me to see why the noncooperative equilibrium models did not fit in with our previous analysis. The models had to be utterly different and required, at the least, even if at a high level of abstraction, an extensive form description of the game. An extensive form description spells out the nature of moves and information conditions in detail. It is far more engineering and physics oriented than most mathematical economic models built solely for equilibrium analysis. It deals with mechanisms. Markets are mechanisms where market prices are formed. In general, equilibrium theory market price is a primitive concept and the existence of a set of efficient market prices is proved. In a process description of the economy prices are formed in a market mechanism.

Von Neumann and Morgenstern, in their great book on the theory of games, developed three separate representations of a game, any one of which could be treated as a primitive concept: (1) the game in extensive form (2) the game in strategic form and (3) the coalitional or cooperative or characteristic function form representation. The first is the most detailed: it provides an explicit description of all potential paths of play and all information conditions. The second, the strategic form, provides a great reduction which obliterates the fine structure of moves and information and reduces the portrayal to one dealing only with sets of strategies for each individual. The formal description uses, as its primitives, players, strategies and payoffs. The cooperative form of the game goes a step further in the abstraction. It obliterates the details on strategic interaction and deals only with players' coalitions and their associated joint payoffs.

The work of Shapley and myself on the price system had utilized games in coalitional or cooperative form and the various solutions applied to this form. In order to encompass the noncooperative equilibrium solution, the model needed to be cast in either the extensive or the strategic form. Fortunately these forms coincide in the simplest case of interest, that is games with a single move for each agent, each made without information about the actions of others. For pure simplicity, I started to try to construct a model in strategic form in order to produce an intrinsically symmetric model of an exchange economy as a game in strategic form.

At this time, I reread Chamberlin's *Monopolistic Competition* and Mrs Robinson's *Imperfect Competition* to see if I might get any modeling ideas from them. A chapter heading entitled 'A World of Monopolies' from Mrs Robinson's book gave me an idea. I would try to construct a model with n traders each owning a different commodity. Thus there would be n traders and n commodities. I began by considering a strategic game in which each trader would announce price. But at this point I hit a snag. It appeared to me that a price was an exchange ratio, some quantity divided by another quantity. I could build a well-defined strategic game by selecting one of the commodities and using it as numeraire and have all other traders quote prices in terms of an exchange for this good. But if I did this, the game constructed was not intrinsically symmetric. The act of selecting the numeraire would take away the strategy set of the individual owning the commodity selected as numeraire. I fiddled with this for a while and then decided that, if I wanted to get a fully symmetric game, I seemed to be short of one degree of freedom. There appeared to be an easy way to patch up my model. Consider the model with n traders, each one a monopolist, but where there are $n + 1$ commodities and each trader holds a supply of his own good and a supply of the $n + 1$st commodity. For various technical reasons, I decided at this point that the quantity strategy model was easier to formulate and analyze than the price strategy model, so I switched to the Cournot-style strategic model where each trader offered a quantity of his good for sale and offered a quantity of the $n + 1$st good to bid for the other n goods. I then considered the replication of this model with k traders of each type. I took my half-baked model to Shapley who constructively tore it to bits in many different ways. But after each tearing up we were able to repatch and strengthen it. In particular Lloyd was quick in pointing out that, if no one offered a commodity for sale, but the $n + 1$st commodity had been bid for it, as price of the jth commodity was defined as the quantity of $n + 1$st good offered for it divided by the quantity of the jth commodity put up for sale, there would be a division by zero. We discussed this and observed that the specialist's role on the New York exchange requires that he make an 'orderly market', that is, that he has a small inventory available for sale. This can be mathematically approximated by relating an 'epsilon-defined' game to the game under consideration where we assume that there is always some small amount of 'epsilon' available in each market. This would guarantee compact payoff sets, and Lloyd thought that this might work.

I worked at a simple example and the convergence seemed to be going through. We apparently had the model that was going to complete the package for the study of the price system.

Some four to six weeks later, somewhere in early/mid-January of 1971, while walking on the beach near the end of Sunset Boulevard, it dawned on me that the $n + 1$st commodity in the model could be interpreted as money. I had been looking at the solution to my second most important problem, the Cournot convergence, and it turned out to provide the key insight into the problem I had abandoned. To this day, I am amazed that I could have worked for about two months without even considering the connection. My concerns were with symmetry and an extra degree of freedom and when I introduced the $n + 1$st commodity I did not understand what I was doing in the context of introducing the role of money in the economy.

My first published paper containing the new model was 'Commodity Money, Oligopoly, Credit and Bankruptcy in a General Equilibrium Model' (1972). I would have preferred to publish a mathematically more sophisticated joint paper, but I felt that it was important not to wait much more than a year or two, given my experience with the core. Some years later, Shapley and I published 'Trade Using One Commodity as a Means of Payment' (1977). In this paper, primarily through Lloyd's ingenuity, we were able to milk the Edgeworth Box diagram for basic exposition of a model in which all of one good was offered for sale. Pradeep Dubey and I, in a paper entitled 'The Noncooperative Equilibria of a Closed Trading Economy with Market Supply and Bidding Strategies' (1978), published the basic proof for the existence and convergence of the model where individuals can both buy and sell all commodities. A key lemma was supplied by Shapley. We originally had an elaborate example of 'wash sales' where the markets were made thicker by all traders buying and selling in the same markets, but the editor cut it out and I published it separately in my book, *A Game-Theoretic Approach to Political Economy* (1944).

During my stay at RAND, I suspected that the eventual development of the models I had begun to construct would call for modeling the economy as a set of parallel dynamic programs with a continuum of agents. I spent a certain amount of time trying to persuade Richard Bellman, who was a friend, to consider working with me on this. It took me considerable time to interest him, but before we began a serious collaboration he was tragically stricken with a brain tumor and never fully recovered from the ensuing operation. Subsequently I obtained the collaboration of Ward Whitt and he and I published a lengthy paper analyzing an infinite horizon two-person game with fiat money (Shubik and Whitt, 1973)[4]. There was a hiatus of 20 more years before I could find two colleagues to work on the infinite horizon dynamic model with stochastic elements. The paper by Karatzas, Shubik and Sudderth (1994)[5] which has appeared in *Mathematics of Operations Research*, and several subsequent papers, are based on several unpublished papers of Whitt preliminary to a projected joint paper which was never completed.

Some time in the late 1970s, I decided that the class of games being considered was sufficiently important and different from the class of market games that Shapley and I had developed and used to study the price system for these to need a name to distinguish them from the others. 'Strategic Market Games' seemed to be appropriate, as it emphasized that the agents have a strategic role.

Possibly through a combination of my interest in gaming and the poorness of my mathematics, my approach to the theory of money has been to break the analysis into many small simple parts and to try to digest them one at a time prior to reassembling them to obtain an integrated overview. Thus the first attack was concentrated on an exchange economy with a commodity money. Having managed to formulate and analyze the Cournot model, the Bertrand–Edgeworth model remained to be investigated. The article 'The Noncooperative Equilibria of a Closed Trading Economy with Market Supply and Bidding Strategies' (1978), by Pradeep Dubey and myself, contained my basic model. Dubey published the full technical version with formal proofs somewhat later in *Econometrica*.

A key concern of mine, dating back to my book, *Strategy and Market Structure* (1959), has been (and remains) the understanding of the role of bankruptcy and reorganization in an ongoing economy. In an economy with a credit structure and incomplete markets, rules covering the possibility of bankruptcy are required. For many years bankruptcy has been ignored in macroeconomic theory, or regarded as a minor institutional nuisance. My belief is that it is central to the understanding of equilibrium with complete or incomplete markets. Beginning with a model of complete markets, Charles Wilson and I were able to explore the requirement that a bankruptcy rule in an economy using fiat money be harsh enough to prevent strategic bankruptcy. In a paper entitled 'The Optimal Bankruptcy Rule in a Trading Economy using Fiat Money' (1977) we explored the relationship between the money supply and the bankruptcy penalty.

Over a decade later, Pradeep Dubey, John Geanakoplos and I examined the equilibrium analysis of an economy with bankruptcy and incomplete markets; this paper is not yet published. The fundamental contrast between the economies with complete and incomplete markets is that, in the former, equilibrium requires bankruptcy conditions harsh enough to prevent strategic election of bankruptcy, but in the latter the bankruptcy rules emerge as an important public good which set the standards of risktaking acceptable to society as a whole.

It appears to me that there is a deep analogy between the bankruptcy rules of an economy and the role of mutation among biological organisms. The more lenient the bankruptcy laws, the higher is the level of innovation and the greater the failure rate. The higher the mutation rate, the greater is the potential for adaptability of the organism, but the higher is the level of nonviable mutants.

A general criticism leveled against my strategic market game models from the start was that they were too unrealistic. My attitude is fairly simple. Actual economic life is a highly complex, but playable, game. Economic models must be simpler, but they still should be cast as playable games. The essay form or the pure equilibrium model both run the danger of containing subtle omissions which prevent them from being adequate representations to be used in the study of process. In the analysis of both the Cournot and the Bertrand–Edgeworth models, I was looking for the simplest mechanisms to form competitive prices. My idea was to look for *minimal institutions*. After one had achieved that goal and had looked at enough special cases, the next step would be to generalize by trying to axiomatize the concept of price formation and a market. I had discussed this with Pradeep Dubey and we found out that Andreu Mas-Colell had been thinking somewhat along the same lines. We got together and wrote 'Efficiency Properties of Strategic Market Games: An Axiomatic Approach' (1980).

I have described my first approach to embedding the Cournot model in a closed economic system; several years later I realized that there was a general problem concerning the number of markets which remained open. I had considered a world with $n + 1$ commodities and n markets where the $n + 1$st commodity played a special role. However, in contrasting this type of model with a general equilibrium exchange economy, one could conceive of an economy which had trading posts for every pair of goods. In this instance this would require $n(n + 1)/2$ markets. In such an economy, every commodity would serve as a 'money' in the sense that it would be a means of exchange which could be utilized in the purchase of every other commodity. In my model, individuals would offer goods in all markets or 'trading posts' and each market would clear independently. Shapley suggested a somewhat different model where the strategies were the same, but with a clearing house which summed together all the goods of the same type and then a central mechanism would calculate the prices to clear the markets for all goods. Two excellent graduate students, Sahi at Yale and Yao at UCLA, worked on the different models as part of their PhD theses. Amir, Sahi, Shubik and Yao (1990) published a paper showing the existence of equilibria in the decentralized 'trading posts' model. This paper also established the possibility that the distribution of resources could be so skewed that the usual price consistency conditions brought about by an arbitrage chain would not hold without the introduction of credit. The concept of enough money in an economy depends, not only on the market and clearing structure and the institutional needs to cover transactions, but on both the quantity of the money and its distribution.

In fact, the world does not have active markets for each pair of commodities, but the models with all markets provide a natural format to consider the endogenous selection of a money based on consideration of transactions costs and the thickness of markets.

The macroeconomic theory of money and financial institutions, after many years, is finally beginning to emerge, not merely in the work noted but also in the works of

Grandmont, Geanakoplos, Cass, Shell and others. The various individuals involved have each approached this topic from their own bents, be it temporary general equilibrium theory, or enlarged general equilibrium models with incomplete markets, or strategic market games.

Methodologically I suspect that the mathematics involved in the different approaches is coming together. Although some of us have spent a great amount of time investigating one-period models using game theory methods for a finite number of agents, many of the broad interesting questions concerning money and financial institutions lie with infinite horizon dynamic models and with overlapping generations. In my opinion, there is no unique adequate game theoretic solution concept for these models with a finite number of agents. The system dynamics is based on local optimization.

In the actual world of business and finance, some individuals have considerable strategic power in some markets. Face-to-face bargaining and contract and agency problems are of concern. Furthermore, a well-written contract may be more productive than 30 years of hard work. Thus attention to microdetail is critical. The applied game theorist, along with the lawyers and practicing financial analysts may be able to offer some insights into some of the basic features of the drawing up of contracts, but both the conceptual and mathematical difficulties to be faced in the application of n-person dynamic games are such that they are limited to special *ad hoc* problems.

If, however, one limits one's attention to models with considerable economic structure and with a continuum of agents, then the mathematical methods required for the study of the noncooperative equilibrium solution applied to strategic market games appear to be the same as or closely related to those employed in variants of competitive equilibrium analysis. This means that a whole body of methods is available with relatively easy modification both for proofs of the existence of equilibrium and for computational methods. The basic difference in approach, however, remains. By adhering to the playable game test of a model, the stress is on carriers of process, or minimal institutions, not on equilibrium. As the models are fully defined process models, they can be simulated or otherwise studied for behavioral solutions other than the noncooperative equilibrium.

Even if one adopts a strategic player view of the individual, I believe that both the competitive equilibrium and the perfect noncooperative equilibrium solutions with so-called 'individualistic economic maximizers' are useful but merely crude behavioral approximations for the type of socialized, capacity-constrained, limited information individuals we are. The next major advance in economic understanding will come in being able to model agents with different levels of expertise and ability to process information. The key concerns will be with expectations, inference and prediction. All of these are critical to the understanding of economic dynamics, but are essentially finessed or grossly simplified in the emphasis on equilibrium studies.

But for now, the next steps in the creation of a theory of money and financial institutions call for the modeling of the economy as a set of parallel dynamic programs with a continuum of agents, with stochastic elements, with given rules for the creation and destruction of money and credit and bankruptcy and reorganization rules specified. The process models describe the nature of the institutions. From there the next steps will involve the study of limited rationality and the use of solution concepts stressing the methods of inference used in forecasting. Details such as the financing of the float and why fiat money is the only financial instrument which causes the model of exchange to be a set of nonhomogeneous equations must be explained. The reasons why the intrinsic lack of consumer value of fiat money fixes the price level and removes the indeterminacy in the general equilibrium price level have to be spelled out. The importance of the relationship between human error in decision making and the length of time used to make the decisions remains to be explored. The thrust of my current work is in this direction.

Notes

1. I did Hicks an injustice, underestimating the importance of his comments on dynamics.
2. I suspect that no one except myself has read it and checked it, and I do not trust my own checking.
3. This is not unique for nosidepayment games. Further exposition would require a lengthy discourse on λ-transfer.
4. 'Fiat Money in an Economy with One Nondurable Good and No Credit (A Noncooperative Sequential Game)' (with W. Whitt) in *Topics in Differential Games* (ed. A. Blaquiere), Amsterdam: North Holland, 1973, pp. 401–48.
5. 'Construction of Stationary Markov Equilibria on a Strategic Market Game' (with Ioannis Karatzas and William D. Sudderth), *Mathematics of Operations Research*, **19**(4), November 1994.

PART I

POLITICAL ECONOMY

[1]

A BUSINESS CYCLE MODEL WITH ORGANIZED LABOR CONSIDERED

By Martin Shubik[1]

A model is constructed of the business cycle that is adjusted to include union versus firm wage bargaining. Stress is laid on the role of incomplete information in the economy and a method of taking this into account when analyzing the possible outcomes of a union and firm wage bargain is suggested. The use of the concepts of game theory is stressed in the construction of this model.

I. INTRODUCTION

This paper will discuss an aspect of the bargaining of organized labor with large industry during the business cycle. In much of present business cycle theory some sort of labor supply function is postulated in order to determine the level of employment. It is suggested here that the concept of a labor supply function should be abandoned when dealing with organized labor and that the data available in the economy should be used to set up a bilateral monopoly situation or a "labor-entrepreneur game" (two-person nonzero-sum)[2] to determine the final level of employment.

Until recently, the use of a supply curve for labor might have been justified, for labor did not have any appreciable amount of organized power and hence could be treated as an inanimate commodity subject to certain commodity peculiarities. Now, by postulating a bilateral monopoly situation, a somewhat different and to some extent indeterminate solution is suggested.

The plan of this paper is as follows: A cycle equation is taken. It is shown that, from this, the position of the union vis-a-vis the industry can be determined. This can be regarded as a normalized two-person nonzero-sum game during each period. A modification introduces incomplete information.[3] This gives rise to a modified two-person nonzero-sum game. The solution of the game depends to some extent on random factors. An extra term is added to the original cycle equation. This term enables the effect of the bargaining situation to be accounted for in the new cycle equation.

[1] The author is deeply grateful to Mr. L. S. Shapley, of Princeton University and The RAND Corporation, for his help in clarifying ideas in many general discussions on this topic. Part of the work here was done with the support of the Office of Naval Research Contract N6onr-27009.

[2] J. von Neumann and O. Morgenstern, *Theory of Games and Economic Behavior*, Princeton: Princeton University Press, 2nd ed., 1947, p. 47.

[3] Incomplete information in the normalized form means lack of knowledge of some of the payoffs. This is not the same as the lack of information in a game in extensive form, such as poker. In poker the payoffs are still known.

II. BARGAINING POSITION AND THE CYCLE

The first step in building up the model will be to consider a set of cycle equations and to solve them, regarding labor as an exogenous variable. We will take a standard example that has already been fully discussed elsewhere. Consider the Hansen-Samuelson model:[4]

$$Y_t = C_t + I_t + G_t,$$
$$C_t = \alpha Y_{t-1},$$
$$I_t = \beta(C_t - C_{t-1}),$$
$$G_t = 1,$$

where Y_t represents national income; C_t, consumer expenditure; I_t, investment; and G_t, government spending. This gives

$$Y_t = \alpha(1 - \beta) Y_{t-1} + \alpha\beta Y_{t-2} + 1.$$

We can examine the characteristic equation and get a solution. For complex roots, $Y_t = K(\alpha\beta)^{t/2} \sin(t - \tan^{-1} a/b - \varphi)$, where K and φ are constants and $a = \sqrt{4\alpha\beta - \alpha^2(1 + \beta)^2}/2$ and $b = \alpha(1 + \beta)/2$.

By regarding G as a constant we can have a cycle for some of the values of α, β, and G. By varying G, an anticyclical policy may be indulged in by the government.

In order to bring in a bilateral monopoly or a game concept of labor supply, we will enlarge the cycle model slightly. We postulate that $Y_t = Y_{1t} + Y_{2t}$, where Y_{1t} is the income of the workers (hence the spending of it is controlled by them) and Y_{2t} is the income of the entrepreneurs. We replace $C_t = \alpha Y_{t-1}$ by $C_{1t} = \alpha_1 Y_{1,t-1}$ and $C_{2t} = \alpha_2 Y_{2,t-1}$, where $C_t = C_{1t} + C_{2t}$ and the α's represent the different propensities of the different segments of the community.

In general, the game suggested by an industry would be of the three-person nonzero-sum[5] type, with the players being the consumers, the wage earners, and the entrepreneurs. In most cases, however the consumers are without much organized power, and if we regard the problem as a zero-sum game with an extra but "fictitious" player, Nature, then we can consider the consumers as part of Nature.[6] It will be seen that any large group of wage earners, such as the steel workers, the coal miners, or the auto workers, will be enough to formulate a two-person game that can have a noticeable effect on the cycle.

A production function is introduced that is a function of the quantities of factors, including labor as a factor. We consider the factors used

---

[4] P. A. Samuelson, "Interactions between the Multiplier Analysis and the Principle of Acceleration," *Review of Economic Statistics*, Vol. 21, May, 1939, pp. 75–78.

[5] Von Neumann and Morgenstern, *op. cit.*, p. 550.

[6] *Ibid.*, Chapter XI; especially pp. 505–513.

in period t as variables, while the factors used in periods $t-1$, $t-2$, $\cdots$ can be regarded as giving restraint conditions on the production function (this is to account for goods that require more than one period to complete). We write the function as

$$n_t = \psi(q_{1t}, q_{1,t-1}, \cdots, q_{2t}, q_{2t-1}, \cdots, q_{kt}q_{kt-1}, \cdots),$$

where n_t are the number of finished goods in period t and q_{it} ($i = 1, 2, \cdots, k$) are factors of production put in period t. We have here a technological function which tells the entrepreneurs how much labor they need to produce any output. We assume that, in the short run, fairly fixed proportions of labor are needed in relation to other factors of production. The firm is now in a position to draw up a schedule of desired amounts of labor and the range of prices for labor in which it can still operate without a loss or without having to close down. The firm wishes to maximize the function $\Pi_t = n_t p_t(n_t) - \sum_{i=2}^{k-1} q_{it} p_{it} + q_{1t} p_{1t}$, where p_t, p_{it}, p_{1t} are the prices of the finished goods, the input of materials, and the input of labor. We can assume that the union wishes to maximize either $q_{1t}p_{1t}$ or $\overset{*}{q}_{1t}p_{1t}$, where $\overset{*}{q}_{1t}$ is a specific number of workers. The p_{1t} is taken to include not only the nominal rates per hour but also the money value of company, social security, pension plans, etc.[7]

Suppose that we are given the initial conditions of our cycle, namely the values Y_{10}, Y_{20}, C_{10}, C_{20}, I_0, and N_0, where N_0 is the number of people employed in the initial period. Given the constants α_1, α_2, and β, we can get the value of all the variables in the first period except N_1. We are now in a position to set up a game for the first period. Consider first the position of the firm in the bargain. The following assumptions are made: The variation in the cost of maintenance of a closed plant, though varying in some manner directly as the cycle, is of an order of magnitude small enough in comparison to the firm's operation to be regarded as a constant. We can now regard the cost of maintenance of a closed plant as the zero point of the firm's bargain. We assume that the firm has unused capacity which is called into use as the cycle proceeds on the upswing. Profits of the firm increase monotonically with output and of course depend in magnitude on the wages paid to labor. Let Y_{At} be the income accruing to the firm and union as a whole, where $Y_{At} = Y_{1At} + Y_{2At}$, where Y_{2At} is the firm's share and Y_{1At} is the union's share. The absolute lower limit for Y_{1At} would be relief payment, although, of course, if one took into consideration the possibility of alternative employment, etc., it would be higher. At any rate, the maximal point of the bargain for the firm in any period

[7] This is similar to the wage-bill assumption used by John Dunlop.

would be at most $Y_{At} - \bar{Y}_{1At}$, where $\bar{Y}_{1At}$ is relief payment. The period-to-period bargaining range of the firm can be seen to vary from 0 to $Y_{At} - \bar{Y}_{1At}$.

The monetary factors to the union would be at least the financial strength of the union, the personal savings of its members, and the institutional setup of relief payments. We see that the minimum point of the union's bargain will vary from period to period. The lowest minimum will be at a time of no work, no money left in the union treasury, and no personal savings left to its members. We have a function depending upon the previous Y_{1At}, $Z_t = (Y_{1At}, \cdots, Y_{1A0})$ to set the period-to-period lower limit. The maximum point of the union's bargain would be such that the firms in the industry are just operating above the point where it would pay them to close down. In other words, if Y_{2At} is zero, then $Y_{At} = Y_{1At}$ and the union attains its maximal position. The range of the union's period-to-period bargain is then from Z_t to Y_{At}.

Consider the bargaining position in the first period. We have the range of both sides. The value to each side of any settlement can be worked out. Enough is known to set this up as a two-person nonzero-sum game.

An assumption about the dynamic aspects of the situation must be made explicit before proceeding. Neither the union nor the firm it is fighting worry about their effect on the business cycle. Each period during which there is negotiation or renegotiation can be treated as giving rise to a "static" game. The results of any bargain in one period will affect the initial conditions of the bargain in the next period, but it is beyond the ability and the time horizons of both sides to take into account these future effects.

During any period there is a great deal of common interest in striking a bargain, for otherwise both parties lose. However, some of the bargains are much more advantageous to one side than the other. Two aspects of the situation will be noted, the von Neumann-Morgenstern solution and the Nash game solution.

Von Neumann and Morgenstern show that, for the two-person non-zero-sum game, "There exists precisely one solution. It consists of all those imputations where each player gets individually at least that amount which he can secure for himself, while the two get together precisely the maximum amount which they can secure together."[8] This emphasizes the collusive duopoly aspects of this bilateral monopoly situation. The first action is to maximize against the outside, or the market, and then to argue over the proceeds. (Often the rise in price

---

[8] *Ibid.*, p. 555.

of the good being sold by the firm may have feedback, in the sense that the union may use the goods it produces and hence will suffer from the rise in price caused by its gain in wages; but in the case of most single industries this feedback can be ignored.) Some unions have occasionally come out with apparently altruistic statements to the effect that they do not want to see the price of the good they produce rise; this is usually followed by the remark that the firm or industry could afford to pay higher wages without raising prices. However, if we work on the assumption that the producers are maximizing, the effect of the wage rise may be to change the marginal value of the loss of good will, as compared to the increment of gain from a price rise, in such a manner that a price rise becomes inevitable if the producers have any restrictive control on the market.

The von Neumann-Morgenstern solution brings out the duopoly aspect of the situation inasmuch as it emphasizes the two competitive elements present rather than the one usually stressed; they are the union and firm against the market and the union against the firm. It does not offer any light on the possible side-payment settlement between the union and the firm. However, Nash's work on the two-person cooperative game gives a method for evaluating a threat[9] which decides the size of the side payment (and has the additional advantage of not assuming comparable utilities).

Returning to the cycle aspect, suppose we consider a steady sine function generated by our cycle conditions. We can discuss the shape of the fluctuation in the bargaining range (Figures 1 and 2). The minimal point for the industry remains fixed. The maximal point varies from period to period in a sinusoidal fashion in phase with the cycle (some distortion would arise if the percentage of national income accruing to the industry differed greatly during the cycle). A suggested shape for Z_t, the union's minimal point, is a complex sinusoidal function which will lag behind the cycle at least at the peaks. It has a long and fairly level minimum that lasts from before the trough of the cycle until after it; then there is slow growth, followed by quickening, until the maximum is reached past the maximum of the cycle; a quick decline brings the minimum back again.

The following observations can be made: The value of the firm's threat varies inversely with the period of the cycle, whereas that of the union varies directly. The period during which the union should expect to get the largest payoff occurs slightly after the down-turn of the cycle. Given the Nash bargaining method, this model as constructed would never lead to a strike or lockout since this action can only bring loss to

---

[9] In this case, the method outlined in "The Bargaining Problem," ECONOMETRICA, Vol. 18, April, 1950, pp. 155–162, will give rise to the same situation.

both sides. However, we have assumed complete information. It has been pointed out above that although the shape of the Z_t curve suggests itself, the actual evaluation of a point on it depends upon a complex of factors that the union itself would find difficult to evaluate accurately. It is most probable that the firm would miscalculate the threat position of the union. Whenever this happens there appears a "danger zone" in which strikes may occur. Both sides have been assumed to be equal in such psychological aspects as toughness and bargaining ability. Differences in these qualities could be introduced explicitly and would impart a certain amount of asymmetry to the outcome of the bargains but would not change the solutions essentially. The assumption of fairly complete information in a union-firm or union-industry bilateral monopoly situation is not too unreasonable in some cases at the present time since both unions and firms have become increasingly interested in having accurate and complete data on the state of each other. However, this is not always the case.

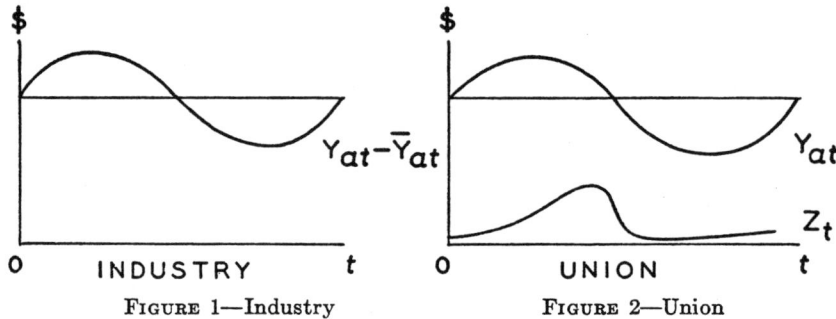

FIGURE 1—Industry   FIGURE 2—Union

In this model if the information of both sides were correct, there would be no strikes, but the division of profits would vary with the position in the cycle. When the propensities of the two groups are appreciably different, this changing distribution could have a feedback effect large enough that it could change the main cycle slightly due to fluctuations in the aggregate propensity to consume.

In our economy "danger zones" due to wrong estimations of the opponent's position are to be expected. Irrational actions by industrial or union "captains" would tend to reinforce this view. If a strike occurs, then the feedback into the general economy is at a maximum, and given a sufficiently large union and industry the result on the economy could be a noticeable change in the cycle.

III. INFORMATION ASPECTS OF THE SITUATION

It is quite apparent that the amount of information available to an organization varies with the time available to gather it. If the object or institution is not static, then the problem of accurate and complete information gathering becomes more and more difficult with the rate of change of growth of the organization. We are now in a position to write as a first and very crude approximation,

$$S_u = F(\Delta Y_t), \qquad S_I = G(\Delta Y_t),$$

where F and G are decreasing functions of $|\Delta Y_t|$. S_u is the set of information about the industry available to the union and S_I is the set of information about the union available to the industry. Naturally these functions should include many more variables that bear on the availability of information; however, the variable used here is certainly a major factor.

We have noted that the completeness of the sets of information available to the industry and union decreases directly as the rate of change of the income level increases. This means that in periods of moderate stability one can expect the union and industry to indulge in negotiations that in effect could be predicted by the solution of the appropriate two-person cooperative game with complete information. When there is a period of change, the following interpretations can be made: Consider a two-person nonzero-sum game in normalized form. It can be characterized by two payoff matrices, (a_{ij}) and (b_{ij}). Complete information would mean that each side knew the value for every a_{ij} and b_{ij}. Now we can weaken this condition. As a first approximation we can regard player A as being in possession of the matrix (a_{ij}) and part of the matrix (b_{ij}), with some of the b_{ij} values filled in by question marks; similarly for B. Next we may consider that the referee appears and tells A and B separately the values for the blanks in their respective matrices. We assume that they believe him and that he may not have given them correct information. This gives rise to two nonzero-sum bargaining games being played. A plays a game using matrices (a_{ij}) and $(b_{ij})'$ and B plays a game with $(a_{ij})'$ and (b_{ij}). Depending upon the state of the misinformation, two points will be obtained between which will lie either a range of very great danger of strike or a range of indeterminacy in which reconciliation will take place. The same sort of result would have been obtained if the referee had given out probability distributions over the unknown values.

A second and better approximation would be still to consider two separate bargaining games being played, but trying to make allowance for A's knowledge of B's lack of information, and vice versa. Consider the following: A obtains his own matrix in this manner; there is a deck of A matrices. B knows the frequency of the matrices in the deck. A is given a matrix at random from this deck. This means that B has a probability distribution on the state of A. At the same time, A knows that B has a probability distribution on his state of information, but does not know what it is. The same sort of situation holds for B. This gives us A knowing (a_{ij}) and a series of $p_k(b_{ij})_k$; and B knowing (b_{ij}) and a series of $p_1(a_{ij})_1$. A conservative action that suggests itself is that both sides minimize "regret." This will still give a range as before.

If we had considered in this second example that A knew the probability distribution that B had on A's state, and vice versa, then in effect we once more have a game with complete information.[10,11]

In the cooperative game models with incomplete information the

[10] Hans Brems has voiced an objection to a solution of a game problem that involves using a mixed strategy. It should be noted that the playing of a mixed strategy does not necessarily mean that each player tosses a die to decide what he shall finally do. Any random device, such as the mood of the union leader after breakfast, may be used. What is meant by the playing of a mixed strategy, however, is that the other side has no way of determining exactly what move is going to be made.

[11] The problems of dynamics and incomplete information appear at a very early stage in any consideration of oligopolistic competition. The Cournot solution may be interpreted to involve both of these aspects. We can regard each move of the players as taking place in one time period. The moves are made simultaneously. Each man operates as if the other were going to produce a certain fixed amount. This can be regarded as acting under misinformation, in a manner that would seem rather foolish to the spectator. The state of information assumes a vital role in the action of each player. Here Brems points out that:

"In order to reap the profits promised by action, the entrepreneur has got to act at once. Thus he cannot escape uncertainty. But since profits—and losses—depend decidedly upon what the rival is going to do, the entrepreneur will try to figure out, the best he is able, what the rival is going to do. Uncertainty implies that there are several alternatives. The entrepreneur, when thinking this over, is likely to attach weights to these alternatives. Such weights might be called 'probabilities' if one is aware that the latter word is used in its everyday sense, not in the sense that von Neumann and Morgenstern use it. According to von Neumann and Morgenstern the probability approach cannot be applied in case the variable to be expected is governed by rational principles. The weather may be regarded as a purely statistical phenomenon but the responses of a rival cannot be. Thus von Neumann and Morgenstern would apply probability theory to the former but not to the latter case. Other theorists, however, do not refrain from using the probability distribution as a tool. And it is hard to find any substitution for it. Von Neumann and Morgenstern hardly offer any." ("Some Notes on the Structure of the Duopoly Problem," *Nordisk Tidsskrift for Teknisk Økonomi*, Vol. 37, No. 1, 1948, pp. 50–51.)

This, I feel, is a somewhat confused interpretation of the theory of games. Brems, in the above quote, actually offers a procedure quite in keeping with the theory. The entrepreneur, in trying to figure out what his rival will do, will attach probabilities to several of the alternatives. This does not mean that the "free will" of the opponent has been removed and replaced by predicted action. On the contrary, what it means is that, as the amount of information to a player grows, he is more accurately able to ascribe the physical bounds within which his opponent will exercise his free will.

actual cause of a strike or a settlement in the zone between the two points will depend on random factors such as personalities or toughness, while the width of the zone depends heavily on the information functions. This means that in order to take into account the bilateral monopoly aspects in the economy we must add a stochastic term to the cycle model. This is the term $\sum_{i=1}^{k} L_i(d)$.

The individual component $L_i(d)$ is formed in the following manner: Consider the difference $d = x_{(i)} - x_{(u)}$, where $x_{(i)}$ and $x_{(u)}$ are the settlement points computed by both sides. If $d > 0$, there is a strike; if $d \leqslant 0$, there is reconciliation. If $Y_{t-1} - Y_{t-2} = \Delta y = 0$, then eventually a state of complete information on both sides is reached; hence $d = 0$. If $|\Delta Y| > 0$, there will be incomplete information and $d \gtreqless 0$ randomly.

The new model is of the form,

$$Y_t = \alpha(1 - \beta) Y_{t-1} + \alpha\beta Y_{t-2} - \sum_{i=1}^{k} L_i(d) + G_t,$$

where L_i is near zero if $d \leqslant 0$ and L_i is positive if $d > 0$. Each component $L_i(d)$ represents one bilateral monopoly wage bargain. When the extra term is added to the cycle equation, we put it in with a summation to signify that all bilateral monopoly wage bargaining taking place in the economy is accounted for.

In brief, the term $L_i(d)$ can be interpreted thus: The rate of change of the cycle plays a great part in determining change in both the firms and unions. The amount of accurate information available to each side varies inversely as the rate of change of the object being observed. When the state of information is very high on both sides, a recognized bargaining range will be reached in which there will be a settlement. The effect on the cycle will be quite small and will be due only to possible slight shifts of spending power between groups with different propensities to consume. If the state of information is low but both sides overestimate each other's power, then there will be no strike and the effect on the cycle will be small. When the state of information is low and both sides underestimate each other's power, the possibility of a strike or lockout becomes great and the effect of such an event will be fed back into the economy with some force.

The role of pessimism and optimism in the interpretation of incomplete information will be a factor of great effect upon the determination of the width of the "danger zone." An assumption on this will yield an idea of the effect of $L_i(d)$. If we assume, for instance, that both the firm and the union were pessimists during the down-swing and optimists during the up-swing, then a reasonable form for $L_i(d)$ would be

$$\begin{aligned} L_i(d) &= A(Y_{t-1} - Y_{t-2}) \text{ when } Y_{t-1} > Y_{t-2}, \\ &= 0 \hspace{2.5cm} \text{when } Y_{t-1} \leqslant Y_{t-2}. \end{aligned}$$

This gives a difference equation for the corrected cycle which is quite amenable to analysis. A in the equation is a constant. This assumption, however, does not seem to be too well borne out by past experience, and another and more complicated one is probably needed before it becomes worthwhile to solve the adjusted equation.

The information aspect of the union-firm bargain suggests that, regardless of the stage of the cycle, the entry of a new firm or union should always produce a period during which the danger of a strike or lockout is very high owing to bad information estimates.

In the general model, we see that there is always the possibility of a random disturbance to the cycle unless a steady state has existed for sometime. The problem of government anticyclical budgeting becomes, in effect, a two-person zero-sum game, with the authorities having to play some sort of mixed strategy to take care of the possible perturbation. As the policy becomes effective this perturbation factor becomes smaller and smaller until, in the steady state, the government can resort to a pure strategy and the random effect will have gone with the vanishing of the "danger zone."

The model developed here has been based upon the Nash solution. This particular solution of the two-person nonzero-sum game puts a very strong set of assumptions upon the concept of "rational man." Even with this, the barest consideration of the states of information introduced a "strike zone" without having to consider any sociological aspects of the problem.

If the von Neumann-Morgenstern solution is used, then under complete information we can say that the joint union-firm action against the market can be decided on, but the only comment feasible about the splitting of income is that any one of a set of imputations will be possible, depending upon the bargainers. In this model, unless one wishes to specify some characteristics for the rest of society apart from the union and the firm, a strike or lockout will never take place; they will be threatened, but not carried out. The weakness, and possibly the strength, of the von Neumann-Morgenstern solution is that it makes no comment on how long it takes the players to agree upon accepting a given imputation as the final division of spoils in any game; it merely delineates the zone of agreement. The introduction of incomplete information here may lead both sides to believe that certain imputations lie within the solution set when they do not; when this happens, the possibility of a strike grows.

This paper has attempted to stress two points. First, "homo oeconomicus" is a cooperative man in the sense that he will cooperate with others to gain the nonzero-sum aspects out of any situation even though he competes for his share of the nonzero-sum increment. Second, his ability to maximize successfully both with and against his fellow man depends heavily upon his state of information and communication. Kickbacks, tributes, strikes, and many other similar phenomena are not necessarily examples of the noneconomic aspects of society. Game models with incomplete information concerning payoffs appear to be the type of model that merits much study in the eventual development of economic theory. One very simple model of this sort has been offered here.

Princeton University

[2]

Information, theories of competition, and the theory of games[1]

Economic theorists have dealt with monopolistic and oligopolistic competition, on the one hand, and pure competition, on the other; yet little effort has been made to unify these various theories of competition. This short note suggests a method of unifying these theories by means of game and information theory.

Oligopoly theory has been treated in two ways. One is essentially static and assumes complete information to be present, as in the theory of games.[2] The other assumes lack of complete information. The latter treatment has led to ideas of conjectural interdependence. The mathematical models of Cournot and Bertrand are of this variety. When dealing with the problem formulated in this manner, it is difficult to distinguish between statics and dynamics. The literature is not clear, even in the Cournot duopoly case, as to whether or not each production adjustment is meant to represent one period. Is it to be regarded as taking place instantaneously or possibly only taking place in the entrepreneurs' minds until they both produce at the equilibrium rate?[3]

In pure competition theory, the problem of information is dealt with by saying that each competitor knows that he cannot influence the market; hence he engages in a simple maximization problem. Professor Viner has suggested that pure competition does not imply complete information but implies equal ignorance on the part of all competitors. Professor Knight suggests: 'Chief among the simplifications of reality prerequisite to the achievement of perfect competition is, as has been emphasized all along, the assumption of practical omniscience on the part of every member of the competitive system.'[4] Professor Morgenstern has pointed out the error in Knight's statement,[5] yet very little formal analysis of the information aspects of pure competition has been done.

The theory of monopoly deals with one individual maximizing against either an inanimate object in the market case or with a machine with a given law of motion in the long-run equilibrium case. For the market situation, the monopolist is assumed to know some sort of demand curve or function between price and quantity taken. In the long-run situation, sociological, psychological, or economic reasons can be thrown in; and the monopolist's reaction to a shift in the demand curves caused by any one of these reasons can be studied. A degree of uncertainty can be introduced by setting up the problem in a manner amenable to period analysis and then giving the monopolist only a probability distribution over the possible shapes of the future-demand curves [145] (how this probability distribution is arrived at is another problem that has not been looked at very closely). Essentially complete information is implicit in most treatments of monopoly.

A very abstract formal scheme or model of firms in competition is sketched here in order to examine the role played by information. A firm is regarded as an organization designed to obtain, process, store, and act on information. Each firm is assumed to be maximizing some quantity which is in a monotonic relationship to money. For the sake of simplicity the firms can be taken to be maximizing money.[6] This assumption permits the use of the von Neumann–Morgenstern type of game theory, which, although it is a highly simplified model, will serve to examine the basic structure of competition.

In order to simplify the discussion here, the game models will be discussed only in normalized form.[7] This means that the games are of the sort in which each player makes a move simultaneously, then, when all the moves have been made, a pay-off is given to each player. An example of this is where the move consists of offering a certain number of goods to the market, and the pay-off is the amount of money received for the goods.

Competition under the von Neumann–Morgenstern conditions of complete information about the pay-offs must be examined first; then information modifications will be introduced.

Monopoly as treated in current theory is easily reformulated in game-theory terms, and the old results remain. It can be regarded as a one-person nonzero sum game.[8] There is just one pay-off matrix (a_i) which tells the monopolist what profit he would make for any production rate. These numbers can be obtained by considering that the monopolist knows what his marginal cost and demand curves look like. He picks a strategy which will maximize his pay-off. This is the production rate at which marginal cost equals marginal revenue. Strictly speaking, since game theory is a general equilibrium theory, monopoly is actually an $(N+1)$-person nonzero sum game, where the players are N buyers and 1 seller. However, it has been implicitly assumed here that the N buyers can be replaced by a demand curve that is fixed and known to the seller. The reason for this will be discussed when information is dealt with.

Duopoly can be treated as a two-person nonzero sum game. Here, if the players have complete information about the pay-offs but play a completely nonco-operative game and do not even communicate with each other, then the best that either player can *enforce* for himself is an equilibrium point.[9] This would yield a situation somewhat like the Cournot solution, depending upon what the competitor's variables were regarded to be. This solution implies one possible state of society which is logically possible but does not seem to be very relevant to societies as we know them. We can consider the possibility of co-operation. Game theory suggests that joint maximization will be striven for, and there will be a side-payment. Limits can be placed on the size of the side-payment by this economic theory, but the actual side-payment made can be determined only by additional assumptions.

Three players in competition in a non-zero [146] sum game bring in many complications, and the number of solutions[10] becomes great. Such situations as a discriminatory solution in which two of the players gang up against the third can exist and obey the economic conditions that each firm strives to enforce for itself an outcome as favorable as possible. The multiplicity of solutions in this case indicates that economic assumptions made here are not enough to determine a result that will

16 *Political Economy, Oligopoly and Experimental Games*

be a single solution consisting of a single imputation of the money and goods involved. To close the model in the sense of obtaining a one-imputation solution, sociological and psychological assumptions must be added. The assumption as to how the two members of a coalition are going to split the gains obtained by squeezing the third man is a psychological one, for it requires postulates about each individual's bargaining strength and other similar factors. The selection of a discriminatory solution to start out with requires a sociological assumption. A whimsical example here would be to consider a society consisting of two Nazis and a Jew.

Games with four, five, and more players appear to become more and more complex as the number of players increases. All indicate the need for extra-economic assumptions to get a single-imputation solution. As N becomes large, there seems to be no indication that the situation will simplify; on the contrary, it would appear that the reverse is true. Qualitatively new phenomena appear in the six- and seven-person games; and, although the analysis has not yet been carried out for many games, the number of players seems very important to the structure of the game. An addition of one player may cause a totally new phenomenon to appear. Given a large N-person game with complete knowledge of the pay-off values to all players, then game theory gives no indication of ever reaching a state analogous to that of pure competition.

The von Neumann-Morgenstern theory is one in which it is assumed that the players all know the pay-off values for every possible outcome to every player in the game. Discussion and argument between the players takes place before they settle down actually to play the game. The aspects of negotiation and information-gathering are left out of the game model. In economic life these play an important role; hence it is desirable to look at the theory of games in the light of some modifications that are suggested.

The processes of obtaining, processing, and storing information and the processes of evaluating and negotiating coalitions cost money and take time. On examination it will be seen that these are not minor 'frictions', as has been suggested sometimes, but are major economic limitational factors.

Information storage is examined first. It is reasonable to assume that any firm has a given finite upper bound to its capacity for storing information. This storage can be regarded as an ability to keep a certain amount of numbers. Let this amount for a particular firm be I_i.

Let the ith player have σ_i strategies open to him (these can be regarded as decisions to produce 10, 11, 12 ... etc. ... goods to bring to market). Then the total amount of numbers needed for complete knowledge of the pay-offs in an n-person normalized game is

$$n \prod_{i=1}^{n} \sigma_i.$$

If we consider, just for the purposes of illustration, that each player has the same number of strategies k, then this becomes nk^n. As the number of people in the game increases, the proportion of information that a firm can store in comparison to the amount that it needs for complete knowledge is

$$\frac{I_i}{n \prod_{i=1}^{n} \sigma_i}$$

or in the simpler case I_i/nk^n. This proportion goes to zero asymptotically at a very rapid rate. If there were 2 players in the game and each had 10 strategies, then 200 numbers would have to be stored. Increase the number of players to 10, then 100,000,000,000 numbers are required for complete information.

The firm will have an upper limit on the amount of information it can obtain. Each number represents an observation. If we assume that there is no trouble involved in gathering information other than making a phone call to the bureau of statistics, noting down the information, and handing it to the information-storage section, then we have a problem of the same order of magnitude as that in the storage section. Actually it will be much worse because the acquisition of information will rarely be as simple as the method outlined above. If the upper bound to the firm's ability to obtain information is P_i, the proportion it can handle to the amount it needs is

$$\frac{P_i}{n \prod_{i=1}^{n} \sigma_i}$$

or in the simpler case P_i/nk^n.

Once the firm has obtained and stored all the numbers it needs, then it must process them. If the firm is to take advantage of the fine structure of coalitions that exist in the von Neumann–Morgenstern game, it must evaluate the power of every possible coalition that can be formed. If there are n players, each firm must evaluate all possible combinations which give

$$\sum_{r=1}^{n} nC_r$$

or $(2^n - 1)$ numbers that must be processed and evaluated from the data on hand.

If we assume that all the firms have worked out everything that is necessary, we are still left with the problem of communication and discussion among the players prior to agreement. Regard each player as being situated in a room having telephone connections with all other players via a central switchboard. There are conference rooms available for any groups who desire to meet. The problem is, how do the number of telephone calls and conferences that are needed to reach agreement depend upon the number of players? A very simple example of this dependence can be given. The final result of communication and discussion will be agreement upon some imputation (which will be an N-tuple of numbers, one number for each player). Suppose that this is a number between 1 and t. Then the probability that they will all agree is the same as the probability of being able to pick a point in this given N-dimensional space. This is $1/t^n$. In this very simple case the number of imputations that have to be discussed before agreement will be reached is seen to increase as a constant raised to the nth power.[11]

Four conditions on information and communication have been examined. All are seen to involve an increase of numbers with the Nth power, where N is the number of players. Any one of these conditions by itself is enough to warrant a modification of the von Neumann–Morgenstern type of game when attempting to set up an N-person economic model.

As soon as the number of players becomes fairly large, the amount of information at the disposal of a single firm becomes a minuscule segment of that required to play a von Neumann–Morgenstern type of game. By attaching even slight costs to the acts of storing, gathering, and processing information, any firm can compute that the cost of getting anything like complete information [148] will be astronomical. The firm must act under low-information conditions, which indicates that it would play a nonco-operative[12] game, since this really requires only a fairly good knowledge of one's own pay-off matrix. One can also look at this in the light of Wald[13] and say that information is so bad that each player plays as if he were in the position of playing a game against 'Nature'. Both these methods yield a solution analogous to that of pure competition. The assumption that is vital to make the game nonco-operative is one that has always been made implicitly in economic theory but has not been stressed very often explicitly, that is, that each player *must* be ignorant of the other players' individual supply curves or, even if he knows something about them, then the costs of computation, communication, and formation of coalition must be higher than the apparent gain from collusion. If one wished to consider a market with many players and complete information, pure competition might exist only if the costs of communicating and coalition formation were prohibitive or if coalition were prohibited by law (in which case implicit collusion might be expected).

It is because of lack of information and/or the expense of collusive organization in comparison to the estimated worth of the organization that monopoly can be treated by game theory as a 1-person rather than as an $(N + 1)$-person game. The N buyers can be replaced by the fiction of a demand curve because they are playing nonco-operatively.

The effect of the assumption of information limitations upon situations in which the number of competitors is small is difficult to analyze and has not been solved here. However, the problem can be formulated and discussed. With few competitors, the information state of each will most likely be such that he knows enough to realize that the coalition aspects of the situation he is in should be considered. But the state of information may be sufficiently bad for the knowledge by one player of the ignorance of another player to constitute an advantage. The game can be considered in two phases: first, the players seek more information, then they consider the aspects of coalition. One of the reasons given for the formation of cartels is that ignorance can be eliminated and with it the 'wasteful' aspects of competition under imperfect information.[14]

Consideration of game theory and information limitations suggest the following models:

Monopoly: A one-person nonzero sum game with a one-imputation solution.
Few competitors: Most of the multi-imputation solutions suggested by von Neumann and Morgenstern as long as the amount of information remains fairly

high. Then a zone which has not been analyzed as yet where the indeterminacy is due to the lack of adequately known methods of formulating action under various degrees of ignorance.

Many competitors: A nonco-operative game with a one-imputation solution.

What constitutes few or many competitors can be answered in terms of the quantity and costs of information. A market has sufficient firms for pure competition if the foreseeable pay-off due to coalition by any of them appears to be lower than coalition costs and the probable cost of obtaining more information.

Possibly one of the first steps for many firms in competition to take is to form a trade association among whose functions the dissemination of information is by no means minor. In this way, in a dynamic situation the amount of information available enables many to become few for the purpose of competition.

As methods of communication and processing [149] of information improve, the amount that any one firm can handle will grow; hence, the number of units that would constitute a situation of oligopolistic competition will grow.

This analysis has attempted to stress the importance of information as an economic parameter which merits explicit introduction into economic models. It has suggested that game theory, with communication conditions considered, serves to provide well-defined models of competition and offers an explicit statement of the unity implicit in theories of competition. [150]

Notes

1. This research was partially supported by the Office of Naval Research, Contract N60nr-27009.

2. To be more precise, the information referred to here is information concerning the rules of the game. In other words, complete knowledge of the pay-off values to every player of every possible outcome of the game is assumed. An example that demonstrates this is matching pennies. The players do not have information on how the pennies are going to land next time, but they know the eight numbers a_i, b_{ij}, where $i = 1, 2; j = 1, 2$, which represent the pay-offs to each player for all possible results.
3. Brems has worked out several interesting cases of duopoly based on period analysis (Hans Brems, *Product Equilibrium under Monopolistic Competition* [Harvard, 1951], chap. xiv).
4. F. H. Knight, *Risk, Uncertainty and Profit* (Boston, 1921), p. 197.
5. O. Morgenstern, 'Perfect Foresight and Economic Equilibrium,' *Zeitschrift für Nationalökonomie*, VI, Part III, 1935.
6. There is no need to keep to this simple assumption whenever using game theory. The complaints against the work of von Neumann and Morgenstern can be broken down into two varieties: (1) they can put money into their model only by assuming that it is linear with regard to utility, (2) they leave out a 'utility to gambling' when they consider probability combinations. The first complaint can be taken care of by letting each player have his own utility function for money and then redefining the solution to take care of this extra condition. The second complaint (which becomes much weaker when the first has been accounted for) has been taken care of by N. Dalkey in a paper on 'Probability and the Measurement of Utility', *Econometrica*, XIX, no. 1 (January, 1951), 53.
7. J. von Neumann and O. Morgenstern, *Theory of Games and Economic Behavior* (2nd edn; Princeton, 1947), p. 85.
8. Von Neumann and Morgenstern, op. cit., p. 47.
9. J. F. Nash, Jr, 'Equilibrium Points in *N*-Person Games', *Proceedings National Academy of Sciences*, 1950, p.36.
10. Von Neumann and Morgenstern, op. cit., pp. 102 and 264.
11. The aspects of communication and discussion are touched upon only very lightly here. Voting is an example of an *N*-person decision upon an imputation where a highly restrictive procedure has been erected in order to be able to cut down consideration of all possible imputations. If the fine structure

of game theory is not to be blunted, there should be no limitation on the number of imputations discussed within any solution set. A more complete discussion of the estimate of communication will be given elsewhere.
12. This term is used by Nash to describe situations where no collaboration or communication takes place (J. F. Nash, Jr, 'Theory of Non-cooperative Games,' Princeton University thesis, May, 1950).
13. Cf. L. J. Savage, 'The Theory of Statistical Decision', *Journal of the American Statistical Association*, XLVI (March, 1951), 58 and 59.
14. L. Marlio, *The Aluminum Cartel* (Washington, DC: Brookings Institution, 1947), p. 71.

INFORMATION, RISK, IGNORANCE AND INDETERMINACY*

By MARTIN SHUBIK

Introduction, 629. — I. The state of information and type of expectations, 630. — II. Ignorance, 634. — III. Risk, 635. — IV. Economic indeterminacy, 636. — V. Action under uncertainty, 637. — VI. Selective intuition, 638. — VII. Conclusion, 639.

This note discusses the role of the various types of uncertainties which can be characterized by risk, ignorance and indeterminacy, and their relation to the state of information in the formulation of economic decisions.

The state of information must play a key role in any plausible theory of the decision-making process — indeed, it is practically meaningless to talk of a businessman's expectations without explicit consideration of it. Although Knight,[1] Marschak,[2] Hart[3] and others have written on this topic, it has generally been neglected in the literature. This neglect is especially serious in view of the evident importance of the state of information as a cost factor. The nature of the expected rise in costs of information as the number of competitors in an economic situation increases has been discussed elsewhere.[4]

It is necessary first to mention some elementary aspects of information. A book on Greek drama may contain a great deal of information for a professor of classics and very little for a Chinese steelworker. Thus information pertains to use. Communication scientists[5] and engineers deal with information as a set of messages obtained by arranging a given set of symbols. "Frequently the messages have meaning: that is, they refer to or are correlated accord-

---

* This research was partially supported by the Office of Naval Research under contract No. N6onr-27009.

1. F. H. Knight, *Risk Uncertainty and Profit*.
2. J. Marschak, "Money and the Theory of Assets," *Econometrica*, VI (Oct. 1938), 311–25.
3. A. G. Hart, *Anticipations, Uncertainty and Dynamic Planning* (New York: Augustus M. Kelley, 1951).
4. A. G. Hart, *op. cit.*, p. 80; also M. Shubik, "Information, Theories of Competition and the Theory of Games," *Journal of Political Economy*, LX (April 1952), 145–50.
5. Colin Cherry, "A History of the Theory of Information," *Proceedings of the Institution of Electrical Engineers*, XCVIII, Part III, No. 55 (Sept. 1951), 383–93.

ing to some system with certain physical or conceptual entities."[6] However, "these semantic aspects of communication are irrelevant to the engineering problem."[7] It is precisely the meaning attached to and action taken upon receipt of a message that is of interest to the social scientist. The type of information discussed here is regarded as consisting of symbols or messages which are associated with some conceptual scheme such that their presence or absence would have a first order effect on the decision-making processes. Thus we may regard an entrepreneur who knows everything concerning costs, demands and prices to be completely informed even though he knows nothing about, say, oriental art. The very small chance that an advertising idea based upon oriental art may change the demand conditions in his trade can be regarded as an effect of the second order of magnitude.)

I. The State of Information and Type of Expectations

We regard an entrepreneur's state of expectations as being arrived at by drawing inferences from his available information. The table on the following page gives the different types of expectations and states of information on which they are based. The resons for and meaning of this classification are then discussed.

The dots in the diagrams abstractly represent all the economic information relevant to the situation. All the points contained within the outer rings represent the complete set of information. The inner rings contain the amount of information available to the entrepreneur. In the first diagram the inner ring contains all the points, in the second only some, and in the third none.

The following examples will help to give content to these states of information and types of expectations.

1.1. A firm knows its costs with certainty, and has a government contract which stipulates a price and guarantees to purchase the firm's total produce.

1.2. A firm knows its costs and has a government contract which stipulates a price and guarantees to purchase the firm's total produce. However, the firm's production depends upon some random variable such as the weather.

1.3. Two firms in a bilateral monopoly situation know each other's costs, and the demand for the final product of the purchasing firm is given by a government contract.

---

6. C. E. Shannon, *A Mathematical Theory of Communication* (Bell Telephone System Monograph B–1598, 1948), p. 1.
7. *Ibid.*

State of information	Type of expectation induced by presently understood logical methods of inference
1. Complete information 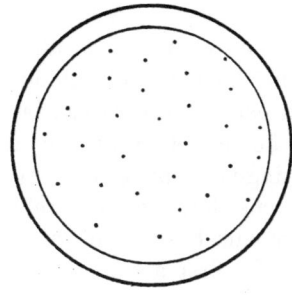	1. Certainty
	2. Risk
	3. Economic indeterminacy
	4. Risk and economic indeterminacy
2. Incomplete information 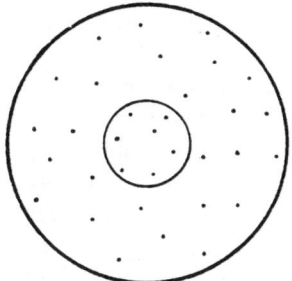	1. Partial ignorance
	2. Partial ignorance and risk
	3. Partial ignorance and economic indeterminacy
	4. Partial ignorance, risk and economic indeterminacy.
3. No information 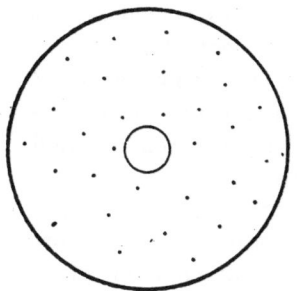	1. Total ignorance
	2. Total ignorance and risk
	3. Total ignorance and economic indeterminacy
	4. Total ignorance, risk and economic indeterminacy

1.4. Two firms in a bilateral monopoly situation know each other's costs, and the demand for the final product of the purchasing firm is given by a probability distribution of estimated demand.

2.1–2.4. The firms in the above cases are only partially informed about their cost curves.

3.1–3.4. The differences between these four cases can be seen only by the outside observer, say the "economist who sees all." Consider a native of Nepal who decides to go into business in England. He is offered four firms for purchase, one in each of the categories 1.1–1.4. Although he is totally ignorant of the economic aspects of any of the firms and thus is not able to distinguish between them, the differences are still there. It is fairly obvious that there is rarely an economic situation in which the entrepreneur has no information at all. If such a situation exists, then the first action of the entrepreneur must be to obtain information.

By complete information we mean the state that exists when an entrepreneur knows everything about the economic variables involved in his competitive situation. An example of this is an undifferentiated duopolist who knows his own costs, his competitor's costs and the market demand for the perfectly homogeneous product produced by both. A special case of complete information exists if he has only a probability distribution of expected demand. There is no reasonable economic theory which enables the duopolist to predict the move of his competitor. It is customary to leave out of economic theory the sociological and psychological "game" aspects of a duopoly (although statements on conjectural variation usually introduce extra-economic considerations). The duopolist may be regarded as being in possession of complete *economic* information even if he does *not* know how his opponent is going to act, providing he has the information mentioned above. It often pays the duopolist to assume that his opponent is an intelligent maximizing man; however, he may hold many other extra-economic theories of action.

Incomplete information is here taken to be the state that exists when the entrepreneur is not fully informed about the *economic* variables under consideration in the model. A lack of knowledge of a competitor's cost function is an example of a state of incomplete information. (In game theory the difference between complete and incomplete information can be described as complete or incomplete knowledge on the part of each player of all players' pay-off functions.) It is evident that most economic situations are characterized by more or less incomplete information. No information at all is merely a limiting case of incomplete information, and is illustrated in the last of the set of three diagrams.

The businessman must act subject to severe physical restrictions on his information caused by the cost (both in time and money) of gathering and deciphering it. A decision to invest in, say, a new plant can be regarded as the act of picking a strategy to use in the uncertain, but imminent battle ahead. This decision is obviously one of a complexity such that it would be foolish to expect that the chances of success could be gauged by a probability distribution of successful openings of firms of the same type. Yet to judge this problem as totally unamenable to statistical analysis, as does G. L. S. Shackle,[8] is equally foolish. Parts of the decision will be based on the analysis of statistical information gathered through marketing research studies: the number of people that pass by a certain street corner every day, for instance. Other parts will depend on such questions as: "Will Senator X succeed in getting an unfavorable law passed?" The businessman may be able to obtain a fairly clear picture of the chances of the law's passing, in which case correct hedging may be possible; or even better, he might be able to influence the result by bringing political or economic pressures to bear on the Senator and on others. We see then that an economic decision can be broken into components which are more or less measurable depending to a great extent upon the amount of information available and the ability to use it to frame a good method or law of action. (In the case of a machine, one already has all the physical laws; hence even though there may be only one machine of its kind in existence, its actions can be predicted fairly well. To do as well for a human being with free will would entail a complete understanding of motivation, then the setting up of a complex game or cross-purpose maximization problem. In certain cases one may be able to depend on large numbers in order to define an "average behavior"; otherwise the game theory aspects must be taken into consideration.)

Information in any economic situation is obtained by paying for it directly; by using free sources; or by having recourse to incompletely understood aspects of knowledge or memory, which constitute "general business experience." Given the state of information, the decision to follow one of a set of alternative courses of action may be made by some combination of processing statistical information, of acting on other information and of drawing on "business intui-

---

8. G. L. S. Shackle, *Expectation in Economics* (Cambridge: Cambridge University Press, 1949), pp. 109, 110.

tion."[9] Many businesses have, until quite recently operated "by guess and by God," using decision-making processes based on skimpy information, that defy present analysis. Nevertheless with the growth of large firms has come the realization that much that was formerly thought intangible is measurable. This has led to the growth of statistical and operations research departments. But even so, the core of art or intuition in decision-making remains undefined, and may remain so for a long time to come.

II. Ignorance

The amount of information regarding the future states of many factors influencing business drops off monotonically as the time period becomes more distant (although this is certainly not true for a large group of periodic events such as national elections). The effect of this will appear in any distribution in the adjustment of which these factors are being used. For example, the probability distribution of the demand for a good, derived from previous statistics becomes more and more tenuous with time until it ceases to have much meaning for the purposes of planning. Even if one were in possession of excellent statistics on the past operations of a business, economic life is so complex that the number of *ceteris paribus* conditions needed in order to make a forecast for more than a limited period is usually quite unreasonable. When a prediction is offered in the light of information, somewhere in the process a law must be assumed,[1] which enables us to use the past in order to advance into the future. The test of whether or not the law is a good one is, of course, pragamatic. The more and the better are one's data on the past, the more chance one has of picking a good law for predicting the future. Uncertainties involved in the making of predictions have been divided into three types in our table. It is evident from the above discussion that the primary uncertainty that will exist in most economic situations arises from some level of *ignorance* or lack of information. Even in a perfectly competitive economy, ignorance can cause uncertainty as to the outcome of a market situation, since lack of information weakens the individual's ability to draw successful inferences.

9. Cf. G. Burck, "The Jersey Company," *Fortune* (Oct. 1951), p. 102: "Good management, when all the books have been read and thrown away, consists essentially in having the knowledge and intuitive judgment to do the appropriate thing at the right time."

1. H. Cramer, *Mathematical Methods of Statistics* (Princeton: Princeton University Press, 1949), Chap. 13.

III. Risk

A second type of uncertainty arises from probability considerations. A forecast of the future usually does not entail the forecasting of a single event with probability 1. The original frequency tables will often have a fair sized dispersion. When dealing with random variables, we may have a probability distribution that gives an excellent frequency pattern in the long run, but we cannot say precisely what will happen in the next period. When we toss a coin the probability is 50–50 that it will come down heads, but we cannot answer the question: "If I now toss a coin once will it come down heads or tails?" Owing to the technological aspects of the economy, production plans cannot be changed instantaneously. Once a factory has been committed to a certain output, a certain amount of deviation may be possible, but it generally is not very great. The order to produce may be given on the basis of a most probable value for the demand (with some margin of leeway). However, it is always possible that the next period will turn out to be one where an event of very low probability occurs. Hence the amount of uncertainty is the amount of probability assigned to events outside the range to which one has become committed. This will depend on the dispersion and variance of the distribution. The type of uncertainty here is *risk*. If the probability distribution gave one event with probability 1 there would be no risk. Knight has claimed that one can insure against risk but not against uncertainty. And Marschak has commented on the distinction:[2]

This may be the rationale of Professor Knight's important distinction between "risk" and "uncertainty": the former is a known parameter of a frequency-distribution, the latter, the lack of knowledge of this (or any other) parameter.

Knight appears to have intermixed considerations of risk with considerations of ignorance. For instance, assume that a particular phenomenon can be predicted reasonably accurately by statistical methods if enough observations are known. If there are but a few observations, the sample may be too small for useful prediction purposes. One may not be able to buy insurance even though it is for a phenomenon that could be handled by statistical methods, because people do not possess the necessary information to hedge correctly. For this reason, although one can get insurance against the possibility of dying at an early age, or getting kicked in the head by a horse, or having one's factory burn down, it may be difficult to get insurance on having a certain demand for one's product next year.

Risk appears to be an uncertainty phenomenon that will appear

2. J. Marschak, *op. cit.*, p. 324.

even under a state of complete economic information. Knight's uncertainty concept appears to deal not only with ignorance, which is a phenomenon of incomplete information, but also with economic indeterminacy to which we now turn.

IV. Economic Indeterminacy

There remains a third type of uncertainty which arises out of game theory considerations. In dealing with the problem of forecasting we are often told that the forecaster himself has an influence upon what actually happens.[3] A stock market prediction published in the newspaper may influence many people to change their intended actions and thus help to make the forecast a reality. However, a salt company's forecast of the demand for salt in the next time period may have no effect whatever upon the demand. A theory of games interpretation can be given to the influence of forecasting. The people who read the stock market forecast may act as agents with free will. This situation may be formulated as an N-person non-zero sum game[4] in order to determine what can happen in the market. The firm making a market prediction may be in error thereby if it is dealing with a population whose actions cannot be treated statistically. In the case of a salt company, the assumption that the demand for salt can be fairly well approximated by looking at the data on the previous demand for salt and then postulating some law that will give the future demand may be justified by the observation that the purchase of salt is a semi-automatic act for most individuals.

Whenever any component of an economic decision depends upon the action of another individual or individuals acting with free will, then unless there are special mitigating circumstances, a game problem has to be considered. If we make the optimistic assumption that complete information exists, that every player knows the value of every possible alternative to everyone, then, as von Neumann and Morgenstern have shown,[5] in all but inessential games, even if a single solution exists,[6] it consists of more than one imputation.[7] The selection of a specific imputation from the solution requires taking into consideration extra factors such as the sociological and psychological

---

3. A. W. Marget, "O. Morgenstern on Economic Forecasting," *Journal of Political Economy*, 1929, p. 312.
4. J. von Neumann and O. Morgenstern, *Theory of Games and Economic Behavior*, p. 47.
5. Ibid., p. 231.
6. Ibid., p. 264.
7. Ibid., pp. 34, 37, 364.

aspects of the situation. Here we have a form of uncertainty that can best be called *economic indeterminacy*.[8] The economic model is open here, and cannot be used by itself to yield a determinate (in the sense of a single) answer.

V. Action Under Uncertainty

Action under risk presents no difficulties. The businessman can insure against it where its nature is sufficiently recognized so that insurance is available. Where its nature is not so recognized, the element of ignorance must be taken into account.

Action under ignorance typifies almost all economic acts. The immense quantity of data required for complete information even in the simplest of economic situations is such that there is not enough time to act on other than a sample of the information even if it were all available. The problem splits into two parts, the gathering of statistical information for some aspects of decision-making, and the gathering of information of a nonstatistical nature such as the knowledge of a competitor's aims or costs. The methods of gathering and processing statistical information present difficult problems of "statistical decision,"[9] as for instance: How big a sample should be gathered so that the marginal cost of information equals the marginal gain to be expected from the larger sample in estimating the distribution of the parent population? Modern statistical knowledge is more or less able to present a workable theory for solving this type of problem. Other information such as the state of credit, the expected state of the stock market, the expectation of panic buying if there is a political crisis, is needed in order to modify and delineate the boundary conditions of the statistical-decision and game-decision parts of the businessman's plans. Often it may be impossible to obtain much of the nonstatistical information needed; nevertheless businessmen do act, and it is here that we must fall back on studying such vague attributes as "business intuition," "experience," and the "ability to guess right."

Action under economic indeterminacy can only be studied in the light of the extra-economic aspects of an economic situation. An economist on Mars, upon being informed as to the ideas of economic

---

8. Cf. the discussion by E. H. Chamberlin on this point in "The Impact of Recent Monopoly Theory on the Schumpeterian System," *Review of Economics and Statistics*, May 1951, p. 135.

9. See L. J. Savage, "The Theory of Statistical Decision," *Journal of the American Statistical Association*, March 1951.

motivation on Earth, given complete information and computing facilities, would have to report that the model is too open for good prediction, but that it could be tightened up if he were given extra-economic information.

Economic indeterminacy exists almost everywhere in everyday life. It exists in the economists' general models of prediction prior to the adding of the special legal, sociological, and psychological restrictions necessary for the investigation of actual affairs.

VI. Selective Intuition

Action under low information has not been studied to any great extent as yet. In many ways the work belongs more to psychologists than to economists: nevertheless any operational concept of economic motivation must take this type of action into account. It appears to be worthwhile to try to formalize such attributes as "business intuition," "ability to guess right," etc. A tentative name suggested for the ability to make decisions under conditions of low information is "selective intuition." As an example of this property, consider a small entrepreneur with no money to spare on a statistics and research bureau, as against a large firm with large statistical and general information facilities. Suppose that each can draw up a frequency graph based on his previous knowledge of a series of events amenable to statistical analysis. The large firm has a far larger sample than the small one; hence if it were a matter of pure statistical treatment of a well defined random variable, the large firm would be able to get a far better estimate of the probability distribution than would the small entrepreneur. But many weightings must be given, owing to non-statistical considerations, before a probability distribution obtained from a frequency graph can be used. Suppose that in this case, the selective intuition of the directors of the large firm is poor, while that of the small entrepreneur is very good, and consider that an event x occurs in the next period. If both assigned a heavy probability weighting to x, this result might be interpreted as follows: that even though the large firm lacked directors with selective intuition and thus suffered from low ability to adjust for additional factors, the weight of actual statistics was so great that a relatively good prediction was made; and that even though the individual entrepreneur lacked good statistics, his intuitive evaluation of the scanty information on hand and his grasp of the significance of additional, hard-to-measure, factors, or his "sizing up of the situation" (possibly based on subconscious information obtained from "experience") enabled him to formulate a good prediction.

If both the large firm and the small entrepreneur were faced with trying to predict, say five years ahead, then so many of the *ceteris paribus* restrictions on the method of making a prediction from statistical data would have to be given up that neither the large firm nor the small entrepreneur would have much by way of understood methods of prediction. Here we would expect the small entrepreneur's prediction to be far better than that of the large firm. It can be seen that as soon as information becomes very scanty, unless the decision-maker has a very high selective intuition[1] he may do no better than if he made his guess at random. (Of course he will always have some information and "experience," and will probably act in some way other than using a straight randomization.) An example of this has been provided in the study by the Cowles Commission on stock market forecasting.[2]

VII. Conclusion

The informational aspects of any economic situation give rise to three types of uncertainty. They are *ignorance* owing to lack of information; *risk* owing to the spread of possible values for a random variable; and *economic indeterminacy* owing to the nature of most game solutions. Risk is the only one of these uncertainties that can be handled in a totally satisfactory manner by economic and statistical theory. Considerations of ignorance indicate a pressing need for the study of decision-making under low information conditions.

1. We might be able to get an ordering on the quality of selective intuition by setting up a questionnaire and asking a group of people for a forecast of some event that will take place once a week, then seeing whether or not any good correlation between their forecasts and the actual events are established which would indicate results better than those that would be obtained by random selection. The definition of selective intuition given here is, of course, vague, and its properties may easily vary with different situations as a function of individual background. For instance, there may be variations in ability to forecast technological, sociological or psychological events such as: What is the weather going to be next week? How will the American people feel about the Far Eastern situation next month? Will my business opponents feel that they should adopt a price-following position next month? etc. However, it is to be hoped that more specific definitions may be possible and that experiment may establish the existence of this attribute with its connections to experience and memory. Certainly a more precise definition of many such vague concepts as "bargaining ability," "business intuition," etc., would be of great use to economics.

2. A. Cowles, "Can Business Forecasters Forecast?" *Econometrica*, Jan. 1933.

Considerations of economic indeterminacy stress the openness of economic models with regard to sociological assumptions before they can be applied to specific economies, or economic situations.

MARTIN SHUBIK.

PRINCETON UNIVERSITY

[4]
Market Form, Intent of the Firm and Market Behavior
Von
Martin Shubik, New York, USA

I. Introduction

What is meant by competition? Does it refer to a type of behavior exhibited by individuals under all circumstances or only in special situations? Does the presence of certain features of a market guarantee that competition will exist?

The common practice among economists has been to *define* that a market form is competitive if there exist specific physical features of that market which tend to restrict the strategic scope of the individual to such an extent that he is forced to behave in a certain manner.

The usual assumptions are that in a market in which there are many small competitors producing a roughly homogeneous product, and in which new firms are not prevented from entering either by features involving scarcity of resources (such as managemental skills) or high costs of entry no firm will have control over the price that it can obtain for its product. Thus, although in theory, a firm is perfectly at liberty to charge anything it wants for its product if the unit price is above „market" it will not be able to sell its output. The profit maximization problem for a single firm in a competitive market, as defined above, becomes that of simple maximization given a market price over which the individual managements assume that they have no control.

Implicit in the assumption about the behavior of firms in a competitive market form is that for some reasons they do not band together and jointly influence price. Practically, this condition is fairly evidently fulfilled. It would be difficult for a group of wheat farmers to collude successfully unless an outside agency were to organize them, take care of the information difficulties, the costs of communicating and coordinating and the costs of persuading potential entrants not to come into the market.

All in all, the economist's definition of a competitive market is the definition of a *market form* the existence of which will cause the individual firm to follow one course of action. Profits are maximized by adjusting the output of goods. Each firm acts or behaves as though the market price were given. If the goods being sold are perfect substitutes for each other, then there is even no need for advertising. Stockholder A cannot gain by advertising that his 100 shares are better than those of stockholder B in the same firm. The demand for the identical grade of wheat from one farmer or the next will not be influenced by advertising.

Given the assumptions behind a competitive market form, the economist does not have to investigate the individual firm's motivations with respect to, or attitudes towards other firms. It does not matter if the management of a firm would be willing to collude with any competitor because the opportunity to do so profitably has been ruled out by the physical facts of the market form.

Few markets in the present economy fulfill the requirements of the competitive market form. Few large firms have this strategic straightjacket to limit their behavior in the market. Market forms for which the competitive market assumptions do not hold have been variously described as imperfect, monopolistic or oligopolistic. Behavior in such markets has been described by economists as imperfectly competitive, collusive, cooperative involving quasi-agreements, market-sharing, price control, price leadership and so forth. Institutional economists and lawyers have added to the list of descriptions of market behavior. Thus cutthroat competition, unfair competition and implicit collusion have made their way into the literature. In many games, the word competition is used with connotations of conflict.

On many occasions both economic and legal analysis would be furthered if it were possible to distinguish clearly between arguments based on assumptions concerning market forms, market behavior and/or motivation of the participants. It is the purpose of this paper to examine some of the implications that arise from assuming the existence of certain types of behavior given the existence of different market forms. In some cases the interrelationship between behavior and intent or motivation will be discussed.

II. Motivation or Intent

It has long been stressed that the assumption that the individual firm maximizes short run profits is a gross over-simplification of economic life. Many large corporations today state that there exists a complex of goals which motivate them. These include welfare considerations for employees, security and/or „fair returns" for small stockholders, concern for long-run economic progress and so forth. Many of these goals can hardly be defined without getting into a morass of socio-political and ethical considerations. The economist interested in „pure theory" can avoid most of the added difficulties involved in the consideration of many secondary goals by assuming that they only exist in order to further some plan for long-run profit maximization by taking into account some of the socio-political forces which could effect the long-run operations of any large firm. It is debatable as to whether this procedure is analyticly useful or is merely a method of pretending to solve a problem by giving it another name. For instance, one example where the long-run profit assumption will not suffice is provided where the managing group of a firm with some degree of marketing control regards any form of overt collusion or „gentlemen's agreement" as directly opposed to its business code, even though it might be maximal in the long-run from the viewpoint of extra income to the firm. The ethic of the „free trader" can become a goal in itself as well as a rule of the game.

In the examples given in section IV. certain simple assumptions will be made concerning the goals and motivations of the firms or departments involved (in many industries where large multi-plant and multi-product firms exist, a single component of a firm may be in a market with several independent competitors). There are situations in which the immediate goals of a department of a large firm may be reasonably approximated by a drastic simplification such as short-run profit maximization. If the department is small, has many competitors, is semi-autonomous, has incentive compensation, does not make its own major investment decisions and has an established product line which is not very vulnerable to technological change, the simple short-run profit maximization assumption may serve. It is also necessary to assume the absence of governmental regulations defining acceptable and unacceptable strategies of economic behavior.

III. Market Behavior

In order to simplify our discussion the examples will be limited to those which investigate the effect of two firms upon each other. The two firms may be representative of many others in the same market, hence the limitation, while convenient for purposes of illustration is not restrictive.

Suppose that the two competing firms or departments each has a well defined goal, aim or profit that is to be maximized. We denote the goal of the first firm by P_1 and that of the second by P_2. Furthermore let S_1 and S_2 stand for the set of strategies available to the first and second firm respectively. Depending upon the nature of the specific market and firm the set of strategies available to any individual will vary considerably. For instance, in some markets they may only consist of short run production decisions; in others pricing, production and advertising decisions; in still others a whole complex of pricing, investment, selling and product alternatives may present themselves as components which enter into the formulation of a strategy.

A specific strategy of the first firm, say the i th strategy of the first firm is denoted by s_i^1, and similarly the j th strategy of the second firm is denoted by s_j^2. The profits or payoffs of the two firms will, in general, be related to the strategies they, and other firms employ. If the actions of the other firms (when there are more than two) are given, the profits of the two firms depend upon or are functions of the strategies they each employ. Their payoffs can be denoted by $P_1(s_i^1, s_j^2)$ and $P_2(s_i^1, s_j^2)$ respectively.

Using the notation given above we are able to formulate very simply some different forms of behavior:

1. Completely cooperative behavior can be characterized by:

$$\underset{s_i^1}{\text{Maximum}} \ \underset{s_j^2}{\text{Maximum}} \ [P_1(s_i^1, s_j^2) + P_2(s_i^1, s_j^2)].$$

This states that the two firms select their strategies in a manner that is jointly maximal. They maximize the sum of their payoffs.

2. A non-cooperative behavior which may lead to an equilibrium in the market is illustrated by the simultaneous satisfaction of the two conditions:

$$\begin{cases} \text{Maximum } P_1(s_i^1, s_j^2) \\ \quad s_i^1 \\ \text{Maximum } P_2(s_i^1, s_j^2). \\ \quad s_j^2 \end{cases}$$

These conditions may be satisfied if each firm acts independently, assuming the move of the other to be given. Eventually their guesses about each other's strategy and their reactions can be consistent so that neither is motivated to depart from some state (unless a technological breakthrough or major market fluctuation makes the situation fluid once more). An equilibrium will exist if Firm 1, acting on the assumption that Firm 2 is using a specific strategy, is not motivated to change its strategy and vice-versa.

This type of equilibrium and the behavioristic assumptions made have been studied extensively in economic theory. The theories of Cournot, Bertrand, Edgeworth, Chamberlin and Stackelberg are all variations on this basic concept of equilibrium and the type of behavior leading up to it. The theories differ inasmuch as different emphasis is laid upon the various economic factors which go into determining the types of strategies available to the different firms.

3. Competitive behavior or „efficient production" can be described by conditions similar to those given in 2. if we limit the meaning of the payoffs P_1 and P_2 and the sets of strategies S_1 and S_2. Assume that the only choice available to each firm involves the selection of a production rate. Let the amount produced by the first firm be represented by q_1 and the amount produced by the second firm by q_2. Furthermore, suppose that the market price for their product is p and the total cost of production for the firms is $C_1(q_1)$ and $C_2(q_2)$ respectively. If P_1 stands for the short term profit made by the first firm, it can be expressed as:

$$P_1 = q_1 p - C_1(q_1),$$

and similarly for the second firm. If each firm attempts to maximize its short term profits on the assumption that it has no effect on price, but can only control its rate of production, then the following conditions must be satisfied:

$$\text{Maximum } [q_1 p - C_1(q_1)]$$
$$q_1$$
$$\text{Maximum } [q_2 p - C_2(q_2)].$$
$$q_2$$

There will, of course, be more equations if we wish to consider more than two firms, however the principle is the same. It can be shown that under certain conditions involving assumptions concerning the restriction of the types of strategies to be used, and the number of firms considered that the behavior rule given in 2. becomes equivalent to the one given here.

4 a. Two forms of „cutthroat" behavior are noted here as extremes in competitive malevolence which can be well defined. The firms may wish to „duel" with each other. Instead of wishing to maximize its own profit each firm may want to maximize the difference between its profits and those of its opponent. This amounts to trying to maximize a damage exchange rate. In other words, it involves an attempt to maximize the size of the profits lost by a competitor per dollar profit foregone by the firm itself. If both birms pursue this type of behavior, the following conditions will be simultaneously satisfied:

$$\underset{s_i^1}{\text{Maximum}} \ \underset{s_j^2}{\text{Minimum}} \ [P_1(s_i^1, s_j^2) - P_2(s_i^1, s_j^2)]$$

$$\underset{s_j^2}{\text{Maximum}} \ \underset{s_i^1}{\text{Minimum}} \ [P_1(s_i^1, s_j^2) - P_2(s_i^1, s_j^2)].$$

4 b. The end-all in malevolent behavior that is possible is obtained if the only goal of one firm is to inflict as much damage as possible on a competitor regardless of the cost to itself. Admittedly this is somewhat far-fetched, but it represents the final extreme of a form of behavior. If the first firm attempts to do this to the second, and the second firm realizes what is about to take place, the behavior can be described by:

$$\underset{s_j^2}{\text{Maximum}} \ \underset{s_i^1}{\text{Minimum}} \ P_2(s_i^1, s_j^2).$$

If the second firm attempts to inflict as much damage as possible upon the first, the condition will be:

$$\underset{s_i^1}{\text{Maximum}} \ \underset{s_j^2}{\text{Minimum}} \ P_1(s_i^1, s_j^2).$$

In general it is not possible for both of these aims to be satisfied simultaneously. A little reflection will show that the payoff or profit function for the firm wishing to inflict loss upon its opponent must be very strange indeed if it is willing to do so regardless of cost.

We have illustrated five different types of behavior each of which manifests itself in an attempt to maximize or minimize some combination of the payoffs to the firms involved. The list is by no means intended to be exhaustive. It is merely illustrative of certain behavior patterns which can be operationally defined. Given these different types of behavior we proceed to investigate the results of applying them to firms in various market forms.

IV. Market Forms

It is the contention here that, at least in the short run, market form is only slightly and indirectly determined or influenced by market behavior. The relation between form and behavior is analogous to that which exists between the rules of a game and the manner in which the game is played. The „market form" of a chess game is described by the resources of the players, the rules of chess, the description of the board, the manner in which men can move and the initial position of the men on the board. Similarly a market form of an industry is described by the number and resources of competitors and possible entrants; the physical description of the market, the legal structure of the country, the information state of the competitors and these describe the alternatives which are legally and technologically feasible for the players. The behavior of the players will be determined by the intents of the firms and the market form. In economic affairs, in the long run the behavior of the firms may actually change the market form. For instance, after a period of competition all competitors except one may be wiped out; or an informal coalition agreement which was a behavioristic feature of the manner of play may be changed into a law and hence become part of the „rules of the game". The cartelization laws in Germany provide one example of this: the guild codes and the codes of the early Roosevelt era provide other examples. These long-run effects must, no doubt, be taken into account when describing the long-run evolution of a market, but in most legal cases and in much of economics the inclination to examine the broad and majestic sweep of the long run must be tempered by carrying out an analysis of the short run. Questions such as: „are firms A and B colluding?" must be answered. Confining ourselves to the short run we turn to an examination of various market forms.

In the examples below the market forms are characterized by very simple matrices formed by considering the outcomes of two or three strategies available to each player. In practice, of course, there may be many thousands of strategies at the disposal of each firm (if it were able to calculate and recognize some of its possible lines of action). The simplification to two or three will not distort our argument because they will be picked in a manner that characterizes or typifies the type of market being examined.

a) **The „Competitive" Market Form.** The competitive market form is illustrated by the matrix in Figure 1. The numbers on the left-hand border and on the top represent the first and second strategies of the first

		Firm 2	
		1	2
Firm 1	1	(2,1)	(2,4)
	2	(6,1)	(6,4)

Figure 1

and second firms respectively. The pairs of numbers in the matrix are the payoffs to each firm as a result of having played some pair of strategies. For instance, if both of the firms use their first strategy, the first will obtain a payment of 2 and the second will obtain 1. If they both use their second strategy, then the first firm obtains 6 and the second obtains 4. The numbers in the matrices are the „profits" or net revenues for each firm. In order to be able to define them in all cases we must know the goal of each firm, i. e., its utility function or its objective. In many cases these profit numbers may be monetary sums, and should be interpreted as such in the competitive market model above. The use of specific numbers, monetary or otherwise, makes the examples easier to discuss, but in the following exposition the actual numbers are not important, the only properties that are used are those of ordering.

In Figure 1 the strategies can be interpreted as production decisions, as it is assumed that no firm has control over price and that advertising and other forms of product differentiation do not matter. We note that if the first player picks his first strategy he will obtain 2 no matter what strategy is picked by his competitor. Similarly if the second player picks a strategy his payoff is strategically separate from his competitor. If such a market actually exists, then all forms of behavior that were described in section 3 are *indistinguishable!* As their fortunes are not strategically interlinked it becomes meaningless to talk about joint maximization, cutthroat behavior and so forth. The economic structure forces each player to follow a certain course of action regardless of his feelings towards fighting or colluding. The five types of behavior are unidentifiable. In all cases both players select their second strategies and obtain 6 and 4 respectively.

The type of game portrayed by the matrix in Figure 1 is known as an inessential game[1]. Of course, in market such as wheat the individual actions of many firms do affect each other which brings us back to the point that a large enough organization of farmers might be able to influence the market, but as a good first approximation the interlinkage between two firms is as above.

b) The Rationalized Industry. Our attention will now be focussed upon various oligopolistic forms: Markets in which few firms compete. Figure 2 illustrates a market form that will be associated with a rationalized industry. Suppose that there are two firms whose total capacity is somewhat in excess of the total demand for their product. Furthermore,

		Firm 2	
		1	2
Firm 1	1	(5,5)	(0,2)
	2	(2,0)	(−3,−3)

Figure 2

suppose that the market is insensitive to advertising and that there are no major opportunities to innovate or improve product (an extractive

---

[1] J. von Neumann and O. Morgenstern: Theory of Games and Economic Behavior, Princeton: Princeton University Press, 2nd ed., 1947, p. 249.

industry such as coal might provide an example). The strategies of the two firms are respectively: first to accept implicitly or explicitly some share of a market in which both take care not to „flood", dump or undercut, but to react in some manner against any disturbance; second, the strategy may be to undercut, flood the market or otherwise engage in unrestrained competition.

The payoffs in the matrix indicate that if the firms „live in peace" they can each obtain a revenue of 5 from the market; if either takes another action they both fare less well, and if both indulge in strategies that have been traditionally called „competitive", they will both lose heavily.

Referring back to the different types of behavior described in section III. we note that behavior 1. joint maximization and 2. non-cooperative equilibrium both call for the players to use their first strategies and both give the same result, a yield of 5 to each. We cannot distinguish between these two behaviors in the market place if this is the market form. Behavior 3. is not defined unless we assume the strategies to be limited to production decisions; even if this were the case, the payoff matrix would show the disastrous effects of flooding the market, and the resultant behavior of the firms would be indistinguishable from the first two behaviors.

If the firms were intent upon damaging each other they would use their second strategies. Hence it is possible to distinguish the actions called for by behavior patterns 4 a and 4 b from the actions resulting from behavior patterns 1., 2. and 3.

c) **The „Dynamic" Oligopoly.** Figure 3 illustrates the competitive feature of a „dynamic" oligopolistic market in which features such as innovation and advertising play a role. The numbers in this matrix must be interpreted as expected values rather than as returns which will definitely result as an outcome of two strategies[2]. The first strategy may be

		Firm 2	
		1	2
Firm 1	1	(10,10)	(6,12)
	2	(12,6)	(7,7)

Figure 3

a plan to restrain outlays on innovation, research and selling effort combined with a recognition of a comfortable market share providing one's opponent does likewise; if the opponent does not conform to the desired action pattern, then some type of counter-measure will be put into force. The second strategy calls for a pattern of considerable spending in innovation and/or selling more or less regardless of the actions of the competitor.

The interpretation of the four possible outcomes is as follows: If both

---

[2] M. Shubik: Competition, Oligopoly and the Theory of Games (Forthcoming, Wiley, 1958), Ch. XI.

firms are „peaceful", they can each make a profit of 10; however, to the firm that is willing to lead in sales and/or innovation there is the possibility of an extra profit some of which may be obtained by getting a larger market share. The firm may not necessarily get this profit if its development or sales effort fails, but at least it increases its profit expectation. If both firms are committed to a program of innovation and sales effort from the start, their expected profit is 7 each.

The behavior characterized by joint maximization indicates that the firms should each employ their first strategies. All the other behavior conditions that can be defined will lead to the firms using their second strategies. In particular we note that in the first two market forms examined it was impossible to distinguish the results of a jointly maximal behavior from that leading to a non-cooperative equilibrium; here we are able to do so.

The simplification in Figure 3 is rather drastic. By making use of algebraic notation we could have given explicit consideration to different components of the competitors' strategies. Thus pricing, production and investment would appear together with innovation and advertising. Certain subsections (or sub-dimensions) of the competitors' strategies may actually result in payoffs which resemble those in Figure 2. For instance, this may be true of price competition. Both firms may be in a position to retaliate immediately to any price cut the net result of which is that price-cutting as a weapon to be used by itself becomes obviously unprofitable to both players. A slightly more complicated example in 4 e shows this.

d) The Strictly Competitive Market Form[3]. The feature which distinguishes many parlor games from most economic situations is that in the former the interests of the competitors are diametrically opposed, whereas in the latter this is usually not so. The amount of money won by one player in a two-man poker game equals the amount lost by his opponent. If one man wins a chess game his opponent loses it. This purity of opposition is seldom encountered in economic affairs. An example of where such a situation may be a good approximation to reality is in the selection of advertising programs. The payoffs in Figure 4 could arise if the firms were in a market in which they compete for each other's

		Firm 2	
		1	2
Firm 1	1	(2,1)	(3,0)
	2	(1,2)	(4,−1)

Figure 4

customers by using fixed budgets in different ways. We observe that Figure 4 portrays a *constant-sum game*. The sum of the payoffs resulting from the employment of any pair of strategies is constant.

---

[3] Ibid., Ch. VIII.

In this market situation the types of behavior described in 2., 4 a., and 4 b all result in the same action, both players use their first strategies. As all outcomes yield the same sum for the joint payoffs, jointly maximal behavior as described in 1. does not call for any particular course of action to be followed. All pairs of strategies are jointly maximal.

e) **A Heavy Industry Oligopolistic Form.** The market form described in Figure 5 is closely akin to that in 4 c, although we include some added market specifications here. Suppose that the competitors are in an indus-

		Firm 2		
		1	2	3
	1	(5,5)	(1,6)	(−5,2)
Firm 1	2	(6,1)	(2,2)	(−4,−2)
	3	(2,−5)	(−2,−4)	(−3,−3)

Figure 5

try in which they have considerable capital investment and plant, market share can possibly be expanded by innovation and sales effort, but retaliation to price-cutting is easy and can quickly result in the firms selling below cost. The three strategies of the players can be interpreted as follows: The first strategy calls for a peaceful market division with limited expenditures on innovation and sales and no price activity if it is observed that the competitor is doing likewise; if the competitor is observed to be active in innovation, then some type of counter-policy involving innovation is followed; if the competitor is active in price competition, then a counter-policy involving price is followed. The second strategy is a plan for innovation, possibly with contingent clauses stating what will happen if the competitor follows different policies; but its essential feature is that its central theme is to follow some innovation policy. The third strategy is based upon a price-cutting policy.

If the firms are 1. jointly maximizing, they will each pick their first strategies. If they follow the non-cooperative behavior as described in 2., they may reach one of two different equilibrium positions. These are obtained if both players use their second or if both use their third strategies. When the firms follow either of the „destructive" behaviors of 4 a or 4 b, they will each pick their third strategies. In this market it is always possible to distinguish between behavior of type 1. and the rest by observing the outcome. It is sometimes possible to distinguish between behavior of type 2. and the rest depending upon which strategies are employed.

V. Market Form or Market Behavior?

In the above examples, cases have arisen where, depending upon the market form the results of different forms of behavior are indistinguishable. Concepts such as implicit collusion, competition, the intensity of competition, collusion, cutthroat competition and so forth cannot necessarily be given meaning independent of market form. *Different market behaviors may give the same market results in some market forms. The same behavior may produce utterly different phenomena in different markets.* For instance, the charitable intent of one market gardener to provide low cost vegetables may lessen his short and long run profits considerably but have little effect on the profits of his competitors. The charitable intent of General Motors to provide low cost automobiles could result in the elimination of some competitors[4].

The problem of social control in an economy involves consideration of ethics and socio-political factors, but it also entails an understanding of logical problems which can arise in attempts to define, understand and control economic power.

---

[4] M. Shubik: A Game Theorist Looks at the Antitrust Laws and the Automobile Industry. Stanford Law Review, Vol. 8, No. 4, July 1956.

[5]

SIMULATION OF THE INDUSTRY AND THE FIRM

By Martin Shubik*

The high-speed digital computer[1] promises to provide the economist with means for constructing both the instruments for observation and the equipment for experimentation that have been the earmarks of the traditional sciences. Used in one way the computer supplies a viewing equipment to the economist in a manner analogous to the microscope for biologists (however, a great amount of work goes into setting up the "specimens" to be observed). Beyond its use as a viewing instrument, it provides a possibility for the construction and running of experiments. It has a use as laboratory apparatus. The various uses of the computer are not substitutes for economic analysis or observation. They are nevertheless supplements of considerable power.

Speaking broadly, there are five major areas of new development of interest to economists which depend heavily upon the advent of the computer. They are: (1) data-processing; (2) analytical methods; (3) simulation; (4) gaming; and (5) artificial intelligence. This paper deals primarily with the third of these, simulation, although there is a considerable interlinkage between the techniques in all categories; and the last two are sometimes treated as specialized types of simulation.

Data-Processing. In data-processing the important features of the new technology are the speed of operation and the ability to record, store and analyze quantities of raw data whose mere volume in most forms presents not only processing but even storage problems. The current generation of large computers performs an addition in under 5 millionths of a second.[2] New methods of storing information make it possible to keep a million "bits"[3] on a piece of plastic an inch square. It has become possible to examine mass census data in many different

---

* The author, on leave from Operations Research and Synthesis Consulting Service, General Electric Company, is visiting professor at Yale University in the current academic year.

[1] Possibly there will be a growing use for analogue computers; however, to date almost all the new applications of computer technology to economic problems have utilized digital computers.

[2] For instance, the IBM 7090 performs an addition in 4.8 micro-seconds.

[3] A bit is the unit in which information is measured. It is the amount of information required to distinguish between two possibilities. For example, yes or no, or equivalently 1 or 0, contain one bit of information.

aggregations and disaggregations. Furthermore, the investigation of economic models which previously called for many months of labor in carrying out statistical computations such as correlation can be carried out in a matter of days, even if all of the work in setting up the material for computation is included. Only a few hours may be needed for the actual computation. "Canned programs"[4] are already sufficiently numerous and available that technological unemployment is in sight for those wishing to write the type of master's thesis depending primarily upon tedious data-gathering and lengthy routine computations. More economic thought may be required as the data-processing times and difficulties decrease.

Analytical Methods. One of the main criticisms of economists using mathematical models has been that their models have tended to be "unrealistic." In many instances considerations of the economics involved have taxed the mathematics to such an extent that the problems have had either to be simplified in an undesirable manner or have not been mathematically tractable. The advent of the large computer is removing many of the limitations of mathematical analysis. Much of the work in the utilization of linear and dynamic programming would not be feasible without computing-machine assistance. Many problems in inventory, production or investment theory, for example, can be concisely stated but are not solvable by analytical methods. It is now possible to utilize numerical methods to obtain answers to many of these problems to any degree of accuracy.

Simulation. In the past few years the word "simulation" has begun to make its way into economic literature [24]. The word is old, especially in reference to other areas where analogue simulation has been used for some time. For example, the air frame industry has utilized physical models in wind-tunnel simulations; and for teaching purposes an hydraulic model at the London School of Economics has provided an analogue simulation of a macroeconomic system for many years. This particular model has been excellently illustrated by the English cartoonist Emett in *Punch* of 1953.

A simulation of a system or an organism is the operation of a model or simulator which is a representation of the system or organism. The model is amenable to manipulations which would be impossible, too expensive or impracticable to perform on the entity it portrays. The operation of the model can be studied and, from it, properties concerning the behavior of the actual system or its subsystems can be inferred.

The close interrelationship between simulation, gaming and artifi-

[4] A canned program is a program written to perfom a task which may be used many times in different problems; for example, a program for multiple correlation. Under such arrangements as SHARE whole libraries of programs are becoming available upon request.

cial intelligence will be considered in the next section. In the succeeding section, contemporary applications of simulation to the industry and the firm will be reviewed.

I. *Simulation and Allied Techniques*

Gaming and simulation are often classed together. It is important to make clear the distinction between these different methods designed for different purposes.

A. *Gaming.*

Gaming has been used as an experimental and training device by the military for many years [27]. Its use as an experimental tool for studying microeconomic behavior is recent. In a gaming experiment a model of an environment is provided. In other words, a simulated universe is constructed. In the military the simulated universe may be a sand table or a maneuver area; in a business game it may be a simulated oligopolistic market structure; in a small group behavior experiment it may be a specialized communications network. Players or decision-makers act within the simulated environment, and the experimenter, by observing them, may be able to test hypotheses concerning the behavior of the individuals and/or the decision system as a whole.

A distinction of interest to economists may be made between business games or "environment rich games" and more strictly controlled experimental games. For example, the recent work of Siegel and Fouraker [26] is of the latter variety. The many business games, almost all of them using a simulated structure of an oligopolistic market, provide examples of the former [14] [16].

B. *Simulation.*

In a simulation, either the behavior of a system or the behavior of individual components is taken as given. Information concerning the behavior of one or the other is inferred as a result of the simulation. Hence one distinction between gaming and simulation is that, although it is usually likely that a simulation will not include any human decision-makers (but merely the decision rules for their behavior), a gaming exercise or experiment will always include decision-makers whose behavior is to be influenced or studied.

The variety of problems explored by means of simulation has called for the development of specialized techniques. Most simulations employ only a high-speed digital or analogue computer. Those of interest to economists belong almost completely to the former type. There is, however, important work involving simulation utilizing both men and computing machines. Furthermore, problems in physics and engineering have utilized a statistical device known as the "Monte Carlo method." There is a highly specialized technical literature on this [20] [21]. As this is not of prime concern to the economist no further discussion of the Monte Carlo method is given here.

Man-Machine Simulations. This category consists of simulations which use both men and computing machines simultaneously, but are not necessarily concerned with the problems involved in gaming. The RAND Logistics Laboratory provides an example of this type of work [11]. Here a complete logistics system is being simulated in order to study the effects of different changes in policies and initial conditions on the over-all performance of the system. Many individuals participate in the actual simulation, not primarily to be trained or experimented with as individuals, but because the cheapest effective simulator of the individual who is in control of the ordering of spare parts for an air defense system may be the individual who is in control of ordering spare parts for an air defense system.

With respect to purpose, a distinction can be made between two kinds of simulation of interest to economists, although the two categories are not altogether well defined. Suggestive names are operations research or tactical simulation and exploratory or strategic simulation. In addition, a third technique which concerns the simulation of cognitive processes deals with areas which Marvin Minsky [22] and John McCarthy have called artificial intelligence. This topic is developed by Clarkson and Simon in a companion article. A brief further reference is, however, made below in this Section.

The distinction between the first two is between simulation used as a device to compare the results of alternative decision rules within relatively well-defined structures and simulation used primarily as a device to aid in the exploration of the behavioral properties and the validity of relatively ill-defined models. This is not a totally happy distinction; however, examples may serve to clarify.

Tactical Simulation. Under this category much of the work in the general area of applied microeconomics, known as operations research, has taken place. This has included traffic scheduling [17], waiting-line problems [4] and investigations of production scheduling [15]. In these instances it has been relatively easy to describe the physical model to be simulated. The validation problem, which is at the heart of the use of simulation as an aid to exploring the relatively ill-defined models of the economic system, does not present a critical barrier to the more "tactical" operations-research utilization of simulation.

Exploratory or Strategic Simulation. This category has to date included the work of Orcutt [23] and collaborators on some aspects of the whole economy, Hoggatt and Balderston [13], Kalman Cohen [5], Joseph Yance [28], Cyert, Feigenbaum and March [7], Jay Forrester [10], Benjamin Bryton and Martin Shubik [25], and others working at the levels of the industry or the firm and part of its immediate environment. There has also been military work, typified by that of RAND, which has already been noted, and there apparently exists a classified literature whose worth cannot be ascertained.

C. *Artificial Intelligence.*

The advent of the theory of games[5] helped to clarify much of the vagueness obscuring the analysis of decision processes. In particular it provided a unifying mechanism with which to view the "reaction curves" and conjectured behavior assumptions which had been used by economists as an aid in the study of oligopoly. One immediate result of the game theoretic formulation of a decision process in extensive form [18] was that it highlighted the paradox concerning rational behavior and complete information; thus theories of rational behavior often assume complete information (such as detailed knowledge of preferences portrayed by an indifference map; or knowledge of many production alternatives). Even if information concerning all alternatives were available, in general no "rational" decision-maker has a long enough life to look at all of them. A theory of rational decision-making must show how the decision-maker avoids having to deal with too much information.

There are too many possibilities in even the simple game of chess to permit a complete exploration of the alternatives by the most powerful computer of the present day working full time for many millions of years. Yet many people play a relatively good game of chess without this search through all possibilities. The growth of work in learning theory [3] combined with the realization of the impossibility of total search through all the alternatives implied by a game theory model has led to the investigation of "heuristic programming" and artificial intelligence models. These, in turn, bear relation to the study of industries and oligopolistic market structures if the manifestations of the behavior of the firm are considered in terms of "game learning" processes under conditions of uncertainty.

II. *Simulation of an Industry or Firm*

The actual and potential contributions of work on the simulation of the firm can be categorized as fourfold: (1) Simulation provides a new econometric device to produce models based on empirical investi-

---

[5] For a lucid exposition of the basic ideas of game theory see [18].

gation. (2) It serves as a computational aid and alternative to analysis in theory construction. (3) It may be used as a data-organizing device. (4) It may serve as a tool for anticipation and planning. The first two categories are of more interest to the theorist or academic economist, while the last two are more the concern of the economist or operations researcher in industry, government or other applied areas.

The mass of institutional detail which can be handled by a computer makes it possible to construct detailed micro-models of large systems, such as an industry. The time-series output of the simulation of these models provides alternative predictions to those derived by utilizing a set of structural equations estimated by econometric methods. Joint investigation of economic systems employing both methods offers a potential interlink between macro- and microeconomics.

Relatively casual empiricism (casual in a nonpejorative sense) has been the earmark of the work of many economic theorists. This is consistent with the use of intuition and the need for broad general searching and "scanning" prior to the construction of models whose testing may lead to new theory. Especially with conjectures and speculations concerning the behavior of the firm or a market, the work involved in tracing through the implications of a conjecture may be long and tedious or even not possible analytically. Quick small simulations enable the theorist to increase the number of casual conjectures that are worth further investigation. Simulation provides an aid to the intuition as a preliminary to doing the more extensive work of preparing for formal testing of new theory.

Possibly one of the most valuable contributions of simulation to date has been the discipline imposed by the necessity of precisely defining for the computer both the problems and questions to be answered. In spite of the current work on artificial intelligence the computer operates on most programs today in a docile and literal manner with very little exhibition of imagination. If it is fed nonsense it usually returns nonsense, or if the material is inconsistent or insufficiently defined it stops. These features force any utilizer of simulation techniques to have a well-defined problem. The discipline called for in making a program that will work is rigorous. The value of this is twofold, both in requiring the checking of models for consistency and completeness and in the designing of consistent data-organization processes.

As an applied tool for the firm the immediate value of simulation is probably more in providing a consistent data-organizing medium than in providing answers to policy questions for the next few years. Large economic organizations are, in general, not short of statistics, documents and paperwork. They are, however, invariably short of the statistics, documents or paperwork needed by the economist or other researcher in the form that will make them of use within his model.

It is premature at this time to discuss in very much more than hortatory terms the power of current simulators to decide policy questions. The first stages towards this possibility, however, are well under way. They are the transformations being wrought by new methods of data processing, in general, and simulation as a data-organizing device for economic models, in particular.

The advent of the computer has extended the possibilities for experimentation and for investigating the behavior of over-all systems by means of simulation. The application of empirical and experimental methods is slow and tedious. It may take several years before enough work can be done and validated for any new, large generalizations to be obtained. Even at this time, however, the current investigations indicate the value of the method as an exploratory tool, especially in its use in approaching situations, whose complexity had hitherto made analysis intractable, if not impossible.

Technological Considerations. The computer provides a tool of a degree of sophistication and complexity that makes it necessary for the economist to consider explicitly some of the technical problems entailed in its use. It is quite probable that well within ten years a basic knowledge of computers will have joined mathematics and statistics as "language requirements" for advanced degrees in many areas of economics. A worth-while discussion of many of the technical aspects of simulation is given by R. W. Conway, B. M. Johnson and W. L. Maxwell [6].

Although this is not the occasion to go deeply into technical problems, there are several of basic interest to those wishing to have even a casual acquaintanceship with the use of simulation in specific and of computers in general. To go from a verbal description of an economic process through to a computer simulation, in most instances, may entail the use of as many as five languages. The verbal description may be translated into a mathematical model which, in turn, will give rise to flow diagrams from which a computer program will be written in a special language which, in turn, is translated by the machine into the actual machine language. It is hypothetically feasible to proceed directly from a written description to the machine language, but in practice this is generally inadvisable.

Flow diagramming and special languages, such as FORTRAN [29] are relatively easy to learn and are in themselves of considerable value in defining models accurately. They are more precise than English and more flexible than mathematical notation. One or two days is sufficient to obtain a working acquaintanceship with flow diagramming and one or two weeks with a special language.

These techniques may be the most important contribution to the methodology of economics that the computer age has produced. They have enlarged the scope of modeling of complex systems by several orders of magnitude. Limitations of space do not permit an adequate exposition of this important point at this time.

Fundamentally, a simulation is a type of experimental investigation. A simulator can be regarded as laboratory equipment. As such, it can be assembled in many ways. Given unlimited time and money, it is no doubt true that just about anything can be simulated. However, when the economic restraints are considered attention must be paid to the costs of flexibility, for the latter is important to the economist interested in exploring different decision rules and changes in market structure. It is possible to design simulations in a modular manner; that is, so that without having to rewrite a complete program, the investigator will be able, for example, to substitute a new pricing or forecasting rule by merely rewriting a specialized subroutine. Otherwise, the change could easily cause lengthy, time-consuming and expensive revisions.

The past three or four years have seen dramatic changes in computer languages, speeds and flexibility of operations. For example, the existence of special languages has cut down the amount of time required to program any problem by a factor conservatively estimated at no less than five as compared with two or three years ago. Many of the difficulties of today probably will not exist within another three to four years.

Current Work. With respect to the firm as the central element of a simulation, current work is of three kinds: (1) simulations concerned primarily with the internal mechanisms of a firm; (2) simulations dealing with the firm and its relationship to its environment; and (3) simulations dealing with industries or aggregations of firms.

Most of the work has been done within the last two to three years and little of it has been published in finished form. Cyert and March [8] have begun to investigate the internal organization of a firm, specifically a department store, simulating the internal resource allocation and pricing behavior. The modeling problems for this type of simulation are great. Most of the other work at this level of micro-investigation has been gaming for research or teaching purposes on the small group-organization problems which underlie much of the internal structure of the firm.

Unpublished work by the writer deals primarily with a single firm and its relationship to its environment, taking into account some aspects of the forecasting, inventory and pricing behavior of the firm, its distributors and retailers. The major aims of the program are to serve as a data-organizing device by means of which information can be gathered and structured in a manner to provide models of sufficient worth to be of value in sensitivity analyses of changes in policy, decision rules and market structure. A further goal of this particular program is to carry out an investigation of different degrees of aggregation for the same economic models and to provide data for testing the simultaneous equation estimation procedure developed by R. L. Basmann [2]. The econometric estimation processes as applied to an industry and the simulation of an industry provide alternative approaches, the relative merits of which must still be investigated.

Of the work on the firm and its environment and on models of industry as a whole perhaps the "industrial dynamics" project of J. W. Forrester [10] is the best known. The published works of this project to date, however, appear to be primarily expository and hortatory. The prime contribution appears to have been the construction of a special computer language, DYNAMO, for the writing of simulation and difference-equation models. The value of this contribution can only be judged on technical modeling, programming and computing time considerations. Depending upon the previous training of the individual, the availability of courses and computer time, a case can be made in either direction. With the exception of the work of Yance [28] none of the simulations noted in this article have utilized DYNAMO. There does not appear to have been any specifically new contribution to economic thought or methodology in the industrial dynamics project's published works.

Both Kalman J. Cohen [5] and Joseph V. Yance [28] have produced simulations of the aspects of the shoe, leather and hide industries. The choice of these industries was due to the excellent information provided by the study of Ruth P. Mack [19]. In each of these studies an attempt has been made to compare the model with the actual performance of the industries over several years.

In the Cohen model a difference-equation system is set down to represent the aggregate consumer expenditure on shoes, retailers' behavior including sales anticipations, manufacturers' behavior, tanners' behavior and hide dealers' selling prices [5]. The Yance model has been described previously in this journal [28].

Cohen [5] stresses the difference between "one-period change models" and "process models." He claims that "most econometric models have been estimated and tested in such a way that they can be considered to be one-period change models." He contrasts this with the process model in which the lagged endogenous variables, after initialization, are produced by the system rather than brought in at each successive stage at their observed values. In other words, the process model lives with any errors made in previous periods. Cohen examines both of these types of model.

The work of Hoggatt [12] and Balderston and Hoggatt [13] is of a somewhat different nature. The interest appears to be more in the investigation of the sensitivity of market models as an aid to the construction of theory. In the Hoggatt paper the concern is with the birth and death processes of firms under various conditions of supply, demand, cost, price, exit and entry conditions. As has been noted by Hoggatt, "the most striking feature of these results is the richness in detail of the several dynamic models as related to the model of comparative statistics whose equilibrium point they share" [12]. The number of cases and conditions worked out by Hoggatt would have been unfeasible without a simulation.

The joint work of Hoggatt and Balderston has been directed towards studying the effect of changes in the structure of the information system and costs of information in a simulation of the West Coast lumber industry. This work grew out of the initial investigations by Balderston [1] of communication networks in intermediate markets. The results of interest at this stage are not so much the empirical accuracy of the model of a particular lumber industry, as the effects of changes in information and other conditions which otherwise could only have been obtained by copious and tedious computation, if they could have been obtained at all. In their model supplier firms sell to wholesale intermediaries who resell to customer firms. Taken into account are information, commodity shipments, cash payment flow, and birth and death processes for firms at all levels. Messages are paid for and are sent for search, ordering or confirmation of orders. The institutional rules for trading are specified. These include the decision rules for setting price, quantity and search patterns for trading partners. The study examines the relationship between the number of messages and costs and explicitly introduces a measure of market segmentation which arises from search costs. This serves as a valuable linking of concepts of economics and sociology as well as an important attempt to further the application of economic analysis to marketing.

III. Concluding Remarks

The advent of computers in general and the techniques of simulation in particular open up possibilities for the growth of a new scientific institutional economics. The next few years should see a considerable change in data organization and information processes within firms. The growth of operations research and management science, much of which is applied microeconomics, is bringing this about. Further simulation studies of the type discussed here promise to provide the way to add the richness (in terms of explicit consideration of information costs, marketing variables, organizational structure and so forth) needed to obtain adequate theories of the firm, pricing and market structure. The promise is twofold. The new methodology is beginning to offer the opportunity both to construct more complex theories and to validate them.

REFERENCES

1. F. E. BALDERSTON, "Communication Networks in Intermediate Markets," *Mgt. Sci.*, Jan. 1958, *4*, 154-71.
2. R. L. BASMANN AND C. L. CHILDRESS, "Interim Guide for the Practical Use of the GCL Method of Estimation of Simultaneous Equations and the ECOMP Computer Program," Feb. 15, 1960. Contract AT (45-1)1350 between Atomic Energy Commission and General Electric Company.
3. R. W. BUSH AND W. K. ESTES, eds., *Mathematical Learning Theory*. Stanford 1959.
4. A. COBHAM, "Priority Assignment in Waiting Line Problems," *Jour. Operations Research Soc. Am.*, Feb. 1954, *2*, 70-76.
5. K. COHEN, *Computer Models of the Shoe, Leather, Hide Sequence*. Englewood Cliffs 1960.
6. R. W. CONWAY, B. M. JOHNSON, AND W. L. MAXWELL, "Some Problems of Digital Systems Simulation," *Mgt. Sci.*, Oct. 1959, *6*, 92-110.
7. R. M. CYERT, E. A. FEIGENBAUM AND J. G. MARCH, "Models in a Behavioral Theory of the Firm," *Behavioral Sci.*, Apr. 1959, *4*, 81-95.
8. ——— AND J. G. MARCH, "Research on a Behavior Theory of the Firm," in *Contributions to Scientific Research in Management—Proceedings of the Scientific Program Following the Dedication of the Western Data Processing Center*, Los Angeles, Univ. California, 1959, pp. 59-68.
9. J. W. FORRESTER, "Advertising: A Problem in Industrial Dynamics," *Harvard Bus. Rev.*, Mar.-Apr. 1959, *37*, 100-10.
10. ———, "Industrial Dynamics—A Major Breakthrough for Decision-Makers," *Harvard Bus. Rev.*, July-Aug. 1958, *36*, 37-66.
11. M. A. GEISLER, "The Simulation of a Large-Scale Military Activity," *Mgt. Sci.*, July 1959, *5*, 359-68.
12. A. C. HOGGATT, "A Simulation Study of an Economic Model," in [8, pp. 127-42].
13. ——— AND F. E. BALDERSTON, "Models for Simulation of an Intermediate Market," presented to the Econometric Society, December 29, 1958, Chicago, Ill.
14. G. T. HUNTER, "Management Decision-Making Laboratory, Participants' Instructions, IBM Model 1," Internat. Bus. Machines Corp., Oct. 1, 1957.
15. J. R. JACKSON, "Simulation Research on Job Shop Production," *Naval Research Log. Quart.*, Dec. 1957, *4*, 287-95.
16. ———, "UCLA Executive Game No. 2: A Preliminary Report," Univ. California, Los Angeles Mgt. Sci. Research Project Discussion Paper No. 66, Apr. 2, 1958.
17. N. H. JENNINGS AND J. H. DICKENS, "Computer Simulation of Peak Hour Operations in a Bus Terminal," *Mgt. Sci.*, Oct. 1958, *5*, 106-20.

18. R. D. Luce and H. Raiffa, *Games and Decisions.* New York 1957.
19. R. P. Mack, *Consumption and Business Fluctuations: A Case Study of the Shoe, Leather, Hide Sequence.* New York 1956.
20. A. W. Marshall, *Experimentation by Simulation and Monte Carlo,* RAND P-1174. Santa Monica 1958.
21. H. A. Meyer, ed., *Symposium on Monte Carlo Methods.* New York 1954.
22. M. Minsky, "Artificial Intelligence," *Proceedings of the IRE,* forthcoming, Jan. 1961.
23. G. H. Orcutt, M. Greenberger, J. Korbel and A. M. Rivlin, *Microanalysis of Socioeconomic Systems: A Simulation Study.* New York: Harpers, 1961 (forthcoming).
24. M. Shubik, "Bibliography on Simulation, Gaming, Artificial Intelligence and Allied Topics," *Jour. Am. Stat. Assoc.,* (forthcoming) Dec. 1960.
25. ———, "Simulation of the Firm," *Jour. Indus. Engineering,* Sept.-Oct. 1958, *9,* 391-92.
26. S. Siegel and L. E. Fouraker, *Bargaining and Group Decision-Making: Experiments in Bilateral Monopoly.* New York 1960.
27. C. J. Thomas and W. L. Deemer, Jr., "The Role of Operational Gaming in Operations Research," *Jour. Operations Research Soc. Am.,* Feb. 1957, *5,* 1-27.
28. J. V. Yance, "A Model of Price Flexibility," *Am. Econ. Rev.,* June 1960, *50,* 401-18.
29. *Programmer's Primer for* fortran *Automatic Coding System for the IBM 704,* c. 1957 by Internat. Bus. Machines Corp.

APPROACHES TO THE STUDY OF DECISION-MAKING RELEVANT TO THE FIRM

MARTIN SHUBIK

IN THIS paper are reviewed some of the difficulties in utilizing much of economic theory as an aid to the study of decision-making at the level of the firm. New theories, techniques, and experimentation aimed at overcoming these difficulties are then discussed. The area covered is, of necessity, broad. No attempt is made to provide more than an indication of the type of work in progress and the nature of the problem to which it is addressed. References are supplied for those wishing a more detailed exposition of the many topics noted here.

I. THE PURPOSE OF AND DIFFICULTIES WITH MICROECONOMICS

In the past few years a series of new subjects and a host of new terminologies have sprung up in the border areas between economics, industrial management, industrial engineering, and psychology. Viewed from the university, the manifestations of change have come in the growth of interdisciplinary work under the broad category of the behavioral sciences. Viewed from industry, management science and operations research have been the precursors of change. They have served as the interlink between the firm, its concern for policy and specific problems, and a more academic approach to microeconomics, which is the study of the role of individual economic decision-makers (firms or consumers), including interactions between several individuals. It is natural to suspect that this proliferation of new work has taken place in order to fill a need not satisfied by the previously created body of knowledge. The limited success in attempts to apply microeconomic theory to the firm has called for a re-examination of the theory and its assumptions.

Until a few years ago the major applications of economic theory were aimed primarily at macroeconomics, or consideration of the economy as a whole studied in broad aggregates. Microeconomics has been, and, to some extent, remains, a jungle of special assumptions, special cases, unsatisfactory measurements, and tenuous theorizing. The successes of the application of macroeconomics have been considerable, and a growing body of both descriptive and normative macroeconomics exists today. The descriptive theory attempts to describe how the economy behaves, and, given the observations of behavior, predictions can be made about future behavior. The normative theory suggests how people "should" behave and, as such, can be linked up to political and social theories to provide an

101

economic basis for the planning of policy. The literature on macroeconomics is large and well known, and no further discussion of this topic will be given.

On the other hand, microeconomics has been a late starter in the revival of increased interplay between the discipline of economics and the functioning of the economy. The neoclassical theory of the firm, as expounded by Alfred Marshall, provided an elegant theoretical mechanism for describing the distribution of resources among firms in an industry, the long-run development of an industry, and the evolution of its price structure. This theory was based on the concept of a competitive market, that is, a market where all firms are sufficiently small to be able to disregard the specific interaction between any two of them, where adjustments, new investment, liquidation, hiring, and all the other myriad actions of the firm are accomplished smoothly, without restriction and in an open market governed by a smoothly functioning price mechanism known to all parties. This theory had the firm as its "atom" or decision-making unit. Little or no attention was paid to the internal structure of the firm or the many other microeconomic complications which will be discussed later.

This microeconomic theory of the firm within an industry has provided models for the consideration of long-run allocation of resources in parts of an economy where the forces of pure competition tend to dominate other considerations. Keynes's sardonic observation that "in the long run we are all dead" calls to our attention that the worth of a theory is determined by whether it is addressed to the relevant problems and provides insight and understanding. The Marshallian theory of the firm is only sufficient to offer very limited aid to individuals running firms or to those concerned with the control of industry. The reasons for this inadequacy are well known. The simplifications made to obtain the competitive model of the firm are too drastic to provide either a normative or descriptive microeconomic theory of more than limited use to the firm. Most industries contain large multiproduct firms whose actions have considerable influence on each others' destinies. They act in a haze of incomplete information or misinformation not accounted for by the competitive theory. There are many pressing internal problems and conflicting goals caused by corporate structure. These alone are sufficient to restrict the use of neoclassical theory of the firm in coping with problems of industrial organization or decision-making at the level of the firm or market.

The active participation of economists in advising firms has been increasing. The role of the economist and the use of his training in principles of micro- and macroeconomic theory for business planning has been ably and entertainingly exposited by Sidney Alexander in a discussion entitled "Economics and Business Planning."[1] However, the current trends in the development and application of microeconomic theory are being formed by individuals among whom the economists represent only one of many disciplines. The individuals engaged in operations research and management science include mathematicians, engineers, psychologists, physical scientists, and others, as well as economists. In spite of the varied backgrounds of those working in operations research or management science, their interests are primarily fo-

[1] Sidney Alexander, "Economics and Business Planning," *Economics and the Policy Maker* ("Brookings Lectures," Brookings Institution, 1958–59 [Menasha, Wis.: George Banta Co., November, 1959]).

cused upon the decision-making processes at the level of the firm. This may be regarded as an interest in the development of normative microeconomics. They are engaged in a search for methods to aid the decision-maker in a business in deciding how to act. Parallel developments are taking place influencing descriptive microeconomics. The new investigations by psychologists and others are yielding insights which lead to new and considerably different descriptions of economic man. The growth of new measurement devices and data-processing systems has changed considerably our ability to view and describe the environment. The latter development complements the former, for, as the student of or participant in the firm knows, the rule of decision may be excellent, but, if the description of the environment is incorrect, the application of the rule may be disastrous.

There have been many definitions of economics. Jacob Viner has suggested that "economics is what economists do." Herbert Simon has suggested that economics is "the science that describes and predicts the behavior of several kinds of economic man—notably the consumer and the entrepreneur."[2] Alexander notes that "economics is the study of the fundamentals governing business, and it stands in the same relationship to the practice of business as knowledge of physical laws and facts to the practice of engineering."[3] Another definition offered is that economics is the study of the optimal allocation of scarce resources (this might also fit psychiatry). The statements of Simon and Alexander are sufficiently operational that the need for more detail to fill out these condensations becomes apparent as soon as they are used by anyone wishing to understand economic problems. In spite of the desire of many busy individuals for condensations, complex concepts often require complex expositions. Thus, to most seekers of capsule knowledge, Viner's tautology serves as the appropriate reply.

The fourth definition has been more traditional to economics and can be made somewhat more specific to cover the standard economic theories of choice and decision for the consumer and the firm. Microeconomic theory involves the study of the optimization process for a "rational" individual decision-maker—an economic man—usually modeled as though he were confronted with a completely known set of certain or probabilistic outcomes. Economic man, as is implicit in the theory of consumer choice or of price in competitive markets, appears as a rational, omniscient, lightning-quick calculator resolutely bent on maximizing a known objective. For some descriptive purposes or over-all policy decisions this model of economic maximization contains sufficient truth to be worth considering. However, it is not a good model of the individual decision-maker with limited time at his disposal, limited perceptions and computing ability, and knowledge of only a few alternatives, all of which involve differing degrees of risk. Hence, when the questions we wish to ask involve decisions at the level of the firm, in many instances we will need different and/or richer models than are provided for by the concept of rational economic man.

The limitations of the economic view of man have been noted from many quarters in the behavioral sciences. For example, Freud has suggested that it is

[2] Herbert A. Simon, "Theories of Decision Making in Economics," *American Economic Review*, XLIX (June, 1959), 253.

[3] Alexander, *op. cit.*, p. 1.

only small decisions that are made rationally. The decision to select one out of several sizes of eggs at a supermarket may be adequately modeled by the theory of consumer choice under complete information. The decision to build a multimillion-dollar plant may be more adequately modeled by a theory of the monument-building propensities of retiring moguls than by the economic theory of investment. Kenneth Boulding, discussing the problem of economic behavior, notes the role of an "heroic man" whose ethic calls for action without calculation of cost. He suggests that this man is the motivator of saints and soldiers. "Without the heroic, man has no meaning. Without the economic, he has no sense."[4] Lasswell, writing about the decision process from the viewpoint of law and political science, maps out a dynamic system involving seven categories of functions which, although referring to an organization, could equally well refer to the individual.[5]

Starting with the model of economic man confronted with choices resulting in known certain outcomes, we note the new mathematical extensions of his abilities. Statistical uncertainty is introduced (i.e., although the individual is not certain about an outcome, he is given a probability of its occurrence, e.g., the occurrence of a number in roulette). In Section II of this paper the implications of the ability to compute and methods of dealing with this type of uncertainty are considered. The ability to maximize is hampered if more than one individual with goals that are not identical are involved. Section III considers rational man in conflict. Implicit in these two sections is the assumption that the individual or firm knows what he or it wishes to maximize. The value and plausibility of this as an assumption are discussed, and some problems in dealing with goals noted, in Section IV. Attempts have been made to construct a theory of organization which would help to account for many of the difficulties encountered in describing the goals of a firm, although little success has been achieved.

In Section V the lack of knowledge of the individual about his environment is considered. The limitations in his ability to perceive are noted, and consideration is given to problems of search, learning, and obtaining information from the environment.

Section VI is devoted to a discussion of new experimental and empirical methods which promise to be of importance in the study of decision processes.

It is easy to proliferate the distinctions and multiply the models of man in a manner as liberal as the supply of characters in Tolstoy's *War and Peace*. The problem is to do so in a manner that provides more than casual aid to the understanding of the decision-making process. The following areas of study are among those which are currently contributing actively to this understanding and which are referred to in subsequent sections:

Economic theory and statistical decision theory are among those which provide advice to economic man.[6] Game theory, games against nature, bargaining, and bidding help us to consider the actions of man in conflict or collusion.[7] Em-

---

[4] Kenneth Boulding, *Skills of the Economist* (Cleveland: Howard Allen, Inc., 1958), p. 183.

[5] Harold D. Laswell, *The Decision Process* (College Park: University of Maryland, 1956).

[6] R. G. D. Allen, *Mathematical Economics* (London: Macmillan Co., 1956); and A. Wald, *Statistical Decision Functions* (New York: John Wiley & Sons, 1950).

[7] R. D. Luce and H. Raiffa, *Games and Decisions* (New York: John Wiley & Sons, 1957); and L. J. Savage, *The Foundations of Statistics* (New York: John Wiley & Sons, 1954).

pirical investigation and theory of organizations, among others, cast light upon the nature of goals.[8] Work in experimental psychology, information theory, and artificial intelligence is expanding our understanding of man as an adaptive creature making his way in a partially known environment.[9] Experimental games, business gaming, and simulation are being employed as experimental and empirical tools to help to validate some of the theories noted above and to aid in the description of complex organizations.[10]

There is also a broad range of work encompassed under the general titles of operations research, management science, systems engineering, and applications of computers, as well as a growing literature on diplomatic, political, and economic negotiations which are not mentioned specifically.[11]

II. PROGRAMMING MAN: A DIRECT DESCENDANT OF ECONOMIC MAN

Before the atmosphere is clouded and the well-defined "dismal science" of the

[8] R. Eells, *The Meaning of Modern Business* (New York: Columbia University Press, 1960); and J. G. March and H. A. Simon, *Organizations* (New York: John Wiley & Sons, 1958).

[9] S. Siegel and L. Fouraker, *Bargaining and Group Decision-making: Experiments in Bilateral Monopoly* (New York: Macmillan Co., 1960); R. R. Bush and F. Mosteller, *Stochastic Models for Learning* (New York: John Wiley & Sons, 1955); Claude Shannon, "A Mathematical Theory of Communications," *Bell System Technical Journal*, XXVII (July, 1948), 379–423; and M. Minsky," Steps towards Artificial Intelligence," *Proceedings of the IRE*, January, 1961, pp. 8–31.

[10] Siegel and Fouraker, *op. cit.*; J. R. Jackson, "Learning from Experience in Business Decision Games," *California Management Review*, I (Winter, 1959), 92–107; and "Symposium on Simulation," *American Economic Review*, L (December, 1960), 893–932.

[11] R. E. Machol and H. H. Goode, *System Engineering: An Introduction to the Design of Large-Scale Systems* (New York: McGraw-Hill Book Co., 1957).

economists is replaced by a miasma encompassing the many facets of decision-making as viewed by the behavioral sciences, the more complex models of rational decision-making are considered.

ECONOMIC DECISION-MAKING UNDER CERTAINTY WITH MANY CHOICES

If a situation can be modeled so that it presents a well-defined but possibly highly complicated choice problem involving a great number of specified alternatives of known worth, linear programming and allied methods have been developed so that optimal solutions which were previously too difficult to calculate can now be found.[12] These have direct application to the selection of alternative production processes under limitations of capacity, to product-mix and transportation problems, and to other specific problems of the firm. With some modifications, linear programming can be made to deal with conditions of uncertainty in the form of a probabilistic knowledge of the environment.[13] With the aid of mathematical techniques and computers, the potential scope of economic man has been enlarged. This enlargement has already had considerable effect on the oil and chemical industries. Integer-programming and programming models of multiperiod processes have yielded results in production scheduling and the use and allocation of indivisible resources which have had an important effect on our ability to understand, measure, and enlarge the capacity of existing plants.

Even at the purely conceptual level the work on programming has helped to mold an operational investigation of capacity and illustrate the interlinkage be-

[12] S. Gass, *Linear Programming* (New York: McGraw-Hill Book Co., 1958).

[13] G. B. Dantzig, "Linear Programming under Uncertainty," *Management Science*, I (April–July, 1955), 197–206.

tween goals, physical resources, and rules of behavior in providing concepts and measures of capacity.

QUEUEING, INVENTORY, AND OTHER RELATIVES OF STATISTICAL DECISION THEORY

By introducing considerations of probability, an attempt can be made to extend the economic theory of choice to conditions involving uncertainty. The heroic assumption must be made that the situations to be modeled are such that it is valid and useful to utilize the theory of

"Nature"

		1	2
The Decision-maker	1	1	9
	2	2	7

FIG. 1

probability. This has given rise to discussions concerning subjective probability and a large literature on probabilistic preferences and the theory of utility.[14] Problems concerning gambling and risk preference have been examined, and several alternatives for optimal behavior have been suggested. For example, Savage has offered a man who wishes to minimize regret. In other words, after the event, when he looks back, he wishes to have acted in such a manner that he will be least sorry concerning the outcome.

Bayesian and "maximin" principles have also been suggested as manners in which the individual should cope with lack of knowledge. The simple examples given here illustrate the behaviors manifested in following these different principles. In Figure 1 a simple 2 × 2 payoff matrix is presented. The decision-maker

[14] Savage, *op. cit.*; and Ward Edwards, "The Theory of Decision Making," *Psychological Bulletin*, LI (September, 1954), 380–417.

must choose between one of two actions, knowing that "Nature," or the environment, may also make a choice which affects him. For example, if both select their second alternative, the payoff is 7 to the decision-maker. The principle he follows will depend upon his view of the forces and motivations present in his environment.

The Bayesian assumption says that all the actions of "Nature" are equiprobable, and thus the optimal behavior under this assumption will be to select the first alternative, with an expected payoff of:

$$\tfrac{1}{2}(1) + \tfrac{1}{2}(9) = 5 \ .$$

The maximin assumption has the decision-maker believe that the environment is "out to get him." In this case he will select his second alternative, assum-

"Nature"

		1	2
The Decision-maker	1	1	0
	2	0	2

FIG. 2

ing that, since the worst will happen, he can at least guarantee 2 for himself.

The "regret payoff" is illustrated in Figure 2 for the same situation. If he selected his second alternative and the environment did likewise, he would obtain 7 (Fig. 1) but could have obtained 9 (by selecting his first alternative); hence his regret is 2.

Shackle has constructed a "potential surprise function" which he feels dominates many major decisions which must be made in face of uncertainty which cannot be adequately portrayed by considerations of probability.[15] All the methods noted above depend upon assumptions as to how to deal with uncertainty. Which is the "best" assumption depends upon the application and knowledge of human behavior.

In spite of the limitations of models of probability, operations research and industrial statistics have, however, flourished by applying normative models of economic man acting under probabilistically portrayed uncertainty. Theoretical models of inventory, sequential sampling, and various queueing problems have been actively applied. These have already influenced the inventory levels of the whole economy and have had an effect on the understanding of reliability and risk in areas as diverse as individual credit risks, quality control in production, and stock-piling for emergency.

One area in which the applications have been fewer but the implications deeper is that of dynamic programming.[16] This methodology deals with situations where at each period the decision-maker chooses an action which influences a sequence of events stretching off into the possibly indefinite future. In subsequent periods he has the opportunity to modify the effects of previous decisions by current action. Although this theory still deals with statistical uncertainty and mathematical expectations, the rules of decision generated as dynamic program solutions have more of the flavor of over-all long-range strategic decisions. Dividend and investment policies can be studied as dynamic programs. However, the mathematics of functional equations used in dynamic programming is, unfortunately, difficult and still relatively underdeveloped.

III. RATIONAL MAN IN CONFLICT:
THEORIES OF GAMES

Possibly the most ambitious attempt to extend the concept of rational man has been with the various theories of games. The work of Von Neumann and Morgenstern helped to create many more problems than it solved.[17] But in doing so it provided a language for the study of decision-making (the description of the so called "extensive form" of a game).[18] This includes well-defined concepts of information, choice, and strategy.

The need for a precise language to describe situations of choice and conflict becomes clear when we consider even such relatively simple problems as chess. Furthermore, any attempt to compute a method of play in chess highlights some of the difficulties inherent in theories of rational decision-making. There are billions of alternative strategies to be considered. Methods must be found for scanning them in broad aggregates. It is to the finding of these methods that the new topic, sometimes termed "artificial intelligence," has been addressed; this is discussed further in Section V. It is not enough for a decision-maker to know that his problem is theoretically perfectly solvable if he is willing only to search through and examine some billions of alternatives each of which involves much computation. He must have the time and ability to do so, or he must employ an alternative approach. The end to the first

---

[15] G. L. Shackle, *Uncertainty in Economics and Other Reflections* (Cambridge: Cambridge University Press, 1955).

[16] R. Bellman, *Dynamic Programming* (Princeton, N.J.: Princeton University Press, 1957).

[17] John von Neumann and Oskar Morgenstern, *Theory of Games and Economic Behavior* (3d ed.; Princeton, N.J.: Princeton University Press, 1953).

[18] M. Shubik, *Strategy and Market Structure* (New York: John Wiley & Sons, 1959), chap. x.

procedure has been illustrated in a limerick:

> There was young fellow of Trinity
> Who solved the square root of Infinity,
> But it caused him such fidgets
> In counting the digits
> He chucked Maths and took up Divinity.

It was not only difficulties in the examination and calculation of strategies that were highlighted by Von Neumann and Morgenstern. Their analysis helped to isolate problems standing in the way of the development of satisfactory theories of decision in situations involving more than one individual. They offered two theories for the behavior of individuals: one for a special class of situations involving only two participants with diametrically opposed goals and the other for more general situations involving two or more decision-makers whose goals are not necessarily totally opposed.

The first theory has gained many adherents as a normative theory of behavior for such situations. If I were found in a situation which could be portrayed adequately as a strictly competitive two-person game, provided it were possible to do the computations I would be guided by this normative theory. There are many military problems which can be modeled as competitive games, and applications of the theory have been made. Unfortunately, important problems in the theory of the firm cannot be addressed in this manner.

The second theory of Von Neumann and Morgenstern has been far less satisfactory, not because of a lack of formal apparatus or mathematical elegance (on the contrary, the apparatus is formidable), but because it does not appear to provide models relevant to most of the problems of decision-making in the firm or in other situations. Many theories of behavior of individuals acting in an environment containing many others now exist. Today there are twenty-odd theories of behavior in n-person games, none of which is universally satisfactory. However, some of them have value in the analysis of special situations such as bargaining, competition among few firms, voting, and the study of the formation of power blocs.

A simple example of a problem involving voting and control is illustrated by a corporation with five shares held by four individuals. How much more "powerful" is the man who holds two shares than each of the others who hold one each? The Shapley value would impute power indexes of $\frac{1}{2}, \frac{1}{6}, \frac{1}{6}, \frac{1}{6}$ to the two-vote stockholder and the three single-vote stockholders, respectively.[19] This reflects their difference in being crucial to the formation of a controlling coalition.

The attempts to construct adequate theories of many-person games have highlighted difficulties elsewhere; for example, attention has been called to the meaning and the role of information in situations of conflict. Often the possession of special information is the key to a market. There is an old saying in bridge that a peek is worth two finesses. In many situations the major weapon of competitors may be special knowledge or information. In organizations, for example, one of the forms of control exerted by a subordinate is his ability to filter and bias the information he passes on to higher levels in the organization.

The very multiplicity of theories for the solution of games is an illustration that unsubstantiated mathematical conjectures are unlikely to serve as an ade-

[19] Lloyd S. Shapley, "A Value for n-Person Games," in H. W. Kuhn and A. W. Tucker (eds.), *Contributions to the Theory of Games*, Vol. II (Princeton, N.J.: Princeton University Press, 1953).

quate substitute for a systematic study of human behavior in its economic or other environments. The new mathematical methods have led to a new stress on the need for experimental and other empirical work on decision-making. The power of the methodology has made it easier to see the weaknesses and the gaps in the theories.

IV. GOALS AND ORGANIZATIONS
WHO MAXIMIZES WHAT?

The individual rational decision-maker of economic theory has been, on the one hand, a singularly simple individual and, on the other, an extremely complex one. His pristine simplicity comes about in his good fortune in knowing what he wants. As an entrepreneur-owner of a firm, his economic role is to maximize profits. The effects of life, death, cost accountants, partners or boards of directors and stockholders—all are of little significance to this economic man.

The picture of the individual profit-maximizer is not sufficient to provide either a descriptive or a normative model of the decision-maker in the firm. For example, the concept of profit is extremely difficult to define, and, once defined, it is generally hard to measure.

Some of the difficulties in defining the goals of the firm are illustrated in an unpublished study I made of the statements of corporate aims by twenty-five corporations, seventeen of which are among the five hundred largest manufacturing firms. They reveal a blend of economic, political, social, and ethical considerations which provide a self-image considerably different from that of economic man. The sources and detail of this material vary considerably, as is indicated below. Titles or captions of the statements of objectives are: Our Aim; Basic Objectives; Objectives; Creed; The Policy; Company Objectives; Company Policy; Principles; Our Credo; The Company's Philosophy; General Corporate Policies; Philosophy; Statement of Policy; and General Objectives of Management. The sources include statements from chairmen, both in the financial press and elsewhere; statements from management guides and manuals; approved releases by the board of executives; notes by company presidents; and statements from company reports. I classified the contents under topic headings, given below together with numbers indicating frequency of reference in the twenty-five statements.

Personnel.................	21
Duties and responsibility to society in general........	19
The consumer.............	19
Stockholders..............	16
Profit....................	13
Quality of product.........	11
Technological progress......	9
Supplier relations..........	9
Corporate growth..........	8
Managerial efficiency.......	7
Duties to government......	4
Distributor relations.......	4
Prestige..................	2
Religion as an explicit guide in business.............	1

Selznick has correctly observed that "once an organization becomes a going concern with many forces working to keep it alive the people who run it readily escape the task of defining its purposes."[20] Possibly the vagueness of some of the objectives noted above represents an attempt to escape defining corporate purposes. Yet, other work has indicated that the "cultural milieu" of a business provides considerable conditioning for management's goals. Recently, work on comparative management in different countries has highlighted the different

[20] Philip Selznik, *Leadership in Administration* (Evanston, Ill.: Row, Peterson & Co., 1957).

objectives of economic organizations. For example, the relation between the worker and the executive in the Japanese factory contrasts significantly with ours.[21]

The theorist may claim that the multiplicity of vaguely defined goals is merely a subterfuge for disguising the aim of long-run profit maximization. It is easy to construct mathematical models which treat all stated goals other than profit maximization as boundary conditions. It is not so easy to demonstrate that this provides the most satisfactory model of the behavior of the firm. It has been suggested that a useful model of behavior can be obtained by assuming that the object of the firm is to provide "the good life" for its executives. Though this may have an initial appeal, it is possibly even harder to define operationally than are most of the categories presented above.

Simon has suggested that economic man be replaced by a "satisficing man."[22] This individual, living in a complicated, vague and uncertain world, does not attempt to maximize but is content if his current levels of aspiration are met. Behind this model of the decision-maker are considerations of incomplete information, conflict, limited ability to process information, and limited perceptions and desires combined with an ability to learn and to modify the desires. Unfortunately, just like his *Doppelgänger*, *homo economicus*, "satisficing man" is hard to deal with, and not so well defined as he appears to be at first sight.

THEORY OF ORGANIZATION

The theory of organization is more a set of words than a reality. The work on many topics such as experimental gaming, game theory, bargaining, and business games, has direct bearing upon the development of a general theory of organization. In spite of the wealth of writings over the centuries from so many different disciplines as the military, political science, law, sociology, and anthropology, few general principles concerning organizations exist. Modern works as typified by Barnard, March and Simon, and Argyris provide insights and food for thought but do not supply a full or satisfactory theory.[23] The age of the organization chart is still with us, and, although the old chestnuts concerning span of control may no longer be believed (if they ever were), few new maxims have come to take their place. The observations of an ancient Chinese general writing in the fifth century B.C. or those of an Arabian administrator writing in the fourteenth century A.D. on problems of organization have in many parts a contemporary ring.[24] A modern writer on problems of organization would be hard put to exceed these contributions.

At a somewhat more restricted level, Marschak and Radner have confined themselves to the development of a theory of teams.[25] They consider only situations in which the several individuals involved have identical goals and no problems of interpersonal conflict. Their ma-

[21] J. C. Abegglen, *The Japanese Factory* (Glencoe, Ill.: Free Press, 1958).

[22] H. Simon, *Models of Man* (New York: John Wiley & Sons, 1957).

[23] C. I. Barnard, *The Functions of the Executive* (Cambridge, Mass.: Harvard University Press, 1956); March and Simon, *op. cit.*; and Chris Argyris, "The Individual and Organization: Some Problems of Mutual Adjustment," *Administrative Science Quarterly*, II (June, 1957), 1–24.

[24] Sun Tzu, "The Art of War," in *Roots of Strategy*, ed. Brig. Gen. Thomas R. Phillips (Harrisburg, Pa.: Military Service Publishing Co., 1944); Ibn Khaldun, *The Muquaddimah*, trans. Franz Rosenthal (New York: Pantheon Press, 1958).

[25] J. Marschak and R. Radner, *Economic Theory of Teams* ("Cowles Foundation Discussion Paper" No. 59 [New Haven, Conn.: Cowles Foundation for Research in Economics, 1958]).

jor investigation is centered upon problems of the design of optimal communication systems and information-handling procedures to optimize the performance of the team. As the costs of organization and communication are introduced explicitly, our understanding of many of the roles of distribution, marketing, and administration is increased and helps to reconcile the topic of economic decision-making with the more general problems of administration.

V. INFORMATION, LEARNING, AND INTELLIGENCE

LEARNING

As more recognition is given to the nature of the individual as a decision-maker with limited perceptions operating in an uncertain environment, the influence of and the need to understand the work of psychologists becomes more important. The developments in learning theory by Estes, Bush, Mosteller, Siegel, and others deal with many of the same problems met with in the design of forecasting systems with self-correcting feedback, that is, systems which are able to learn from their errors.[26] Probably one of the most valuable sources of data that many firms could tap, but few utilize, is a record of forecasts as compared with actual events. Memory is usually selective, and hence it becomes difficult to find out the degree of error actually present in a forecasting system, or the pattern of learning, without a careful compilation of these statistics.

Another topic in which the learning phenomenon presents itself in an important way is competitive behavior in a dynamic setting. For example, the conditions resulting in stable markets with price leadership are interlinked, implicitly or explicitly, with stimulus-and-response patterns of learning. Much of the economic literature describing lengthy chains of action and reaction between competitors in limited markets is of this variety.

The business-game exercises discussed below also involve learning. In general, it takes the participants several periods before they are able to develop a concept of the nature of the market within which they are operating and even longer to be able to anticipate the behavior of their competitors.

Much of the work in the development of a theory of learning has involved a simple binary-choice experiment, so called because the subject is required to choose between two items. For example, he may be confronted with two lights on a wall and be asked to predict which one will be illuminated next. The experimenter may use a random sequence to decide upon which of the lights will go on, or he may arrange for his choice to depend upon the actions of the subject, or many other conventions can be utilized.

Depending upon the convention and the manner of reward used, there now exist several theories to explain the behavior of the subjects. For example, under conditions of no or low monetary reward the Estes theory predicts that the subjects will track the relevant frequencies of the lights. Siegel has obtained results which indicate that, if there is a sufficient monetary or other utilitarian incentive, the game-theoretic predictions will be borne out. Furthermore, the speed of learning is apparently dependent upon the size of the incentive.

The binary-choice experiment can be

---

[26] R. R. Bush and W. Estes (eds.), *Studies in Mathematical Learning Theory* (Stanford, Calif.: Stanford University Press, 1959); Bush and Mosteller, *op. cit.*; and S. Siegel, "Level of Aspiration and Decision-making," *Psychological Review*, LXIV (1957), 253–62.

converted into a game situation by employing two subjects and letting the frequencies of the events depend upon their interactions. Even though this is a vast oversimplification of economic competition, many of the fundamental problems common to any learning situation are already present. In competitive situations messages may involve both learning and "teaching." The distinction between the two is not clear, although signaling in bridge and bluffing in poker serve as examples.

INFORMATION THEORY

The word "information" has been one of the most used and abused words in the writings on decision-making and planning in industry, the military, and elsewhere. There are many tales of the conscientious research worker who produces a five-hundred-page detailed report on a market study only to be told by his immediate superior to "boil it down" to ten or twenty pages for him, then to "boil it down" further to two or three pages for the next man in the hierarchy, and finally to summarize the whole report in a paragraph for the top decision-maker.

The moral is simple: What is "information" at one location in an organization is not necessarily "information" at another location. In order to seek a measure, it is necessary to consider the relationship between the input (of words, symbols, noises, or other stimuli) and the output. The detailed market study is of little or no use to the man who is not able to read the presentation in that form. For value to be obtained, either the transmission system must be modified, or the receiver or decoder must be changed, or possibly both. For optimal performance they must be compatible.

It is common for research workers and consultants to claim that their results and advice are emasculated by overcondensation and oversimplification. On the other hand, executives claim that many reports they receive are too long and full of irrelevant details. As more is learned about decision-making, and as the newer breeds of business-school graduates obtain positions in industry, both the nature of the information supplied and the ability of the executives to interpret it will change. For example, contrast the attitude toward statistics and probability today from that of twenty years ago. In some instances "one-number" forecasts have already been replaced or enhanced with considerations of confidence intervals. Although the attitude illustrated by "These graphs and charts are too confusing; call the treasurer and let's have a peek in the till" may be with us for a long time, the advent of the formal study of information has already made inroads into the understanding of the role of information in a decision system.

There are at least three areas in which work has been progressing. All of them are of relevance to the firm. They are, respectively, in computer technology, psychology, and economic or industrial decision systems.

Claude Shannon, in his concern with problems of communication, was the inventor of the fundamental unit of measure used in the evaluation of the capacity of systems to transmit and receive symbols.[27] His measure of information is a purely technical one, and he is not concerned with the semantics of the symbols being transmitted. In this theory there is no interest evinced as to what the symbols mean.

If a human being is regarded as an information-receiving device, several results in experimental psychology cast light upon man's ability to discern. In a

[27] Shannon, op. cit.

well-known and highly readable article, entitled "The Magical Number Seven, Plus or Minus Two," George Miller has discussed the ability of individuals to distinguish between stimuli.[28] He observes in experiments with loudness, taste, perception of visual position, and several other stimuli that the ability of an individual to make distinctions is limited to about seven items. The "channel capacity" of the individual is such that he does not appear to be capable of distinguishing among more than seven degrees of loudness, or pitch, or salinity, etc. If, however, the individual is confronted with objects which provide more than one clue, for example, the size and the color of a set of squares, his capacity increases. However, even there limitations soon set in, and, given many different clues for each item, there appears to be an upper limit at around one hundred and fifty items.

Miller notes that it is probably "only a pernicious Pythagorean coincidence" that the span of judgment is seven, which adds yet another item to the list of "the Seven Wonders of the World, the Seven Seas, the Seven Deadly Sins, the Seven Daughters of Atlas in the Pleiades, the Seven Ages of Man, the Seven Levels of Hell, the Seven Primary Colors, the Seven Notes of the Musical Scale, and the Seven Days of the Week."[29] (He omitted the "Seven-Man Span for Management Control.") In light of these considerations it is natural to ask: Even if we were able to generate any number of reports and alternative proposals at little or no cost, how many alternatives should be submitted to an executive? How many differences can a consumer perceive between the various products being offered? How many different wines can the amateur wine-taster distinguish?

Miller discusses several methods whereby individuals are able to increase their ability to discern. These involve considerations of memory and of the complex process of "coding." In industrial and military systems, especially, the design of optimum messages or signals, which is the problem of coding, is of prime concern.

Another set of closely related problems appear when information is looked at from the viewpoint of the economist. Marschak notes that "to an economist it seems natural to call value of information the average amount earned with the help of that information."[30] This immediately leads to considerations of demand-and-supply prices for information. For example, an as yet unsolved, but at least well-formulated, problem concerns the design of incentive systems to pay forecasters in such a manner that it will always be in their interest to supply the decision-maker with information of optimal value to him.[31] That this has not yet been achieved is easily observed by the plethora of newsletters, market forecasts, tipsters' sheets, and retrospective financial forecasting that can find a market.

COMPUTERS AND ARTIFICAL INTELLIGENCE

With increasing frequency the newspapers and business journals are filled with prophecies and warnings concerning the brave new world and the doom of middle-management personnel who will find themselves displaced by computers.

---

[28] George Miller, "The Magical Number Seven, Plus or Minus Two," *Psychological Review*, LXIII (1956).

[29] Miller, *op. cit.*, p. 96.

[30] J. Marschak, *Remarks on the Economics of Information* ("Cowles Foundation Discussion Paper" No. 146 [New Haven, Conn.: Cowles Foundation for Research in Economics, 1960]).

[31] J. McCarthy, "Measures of the Value of Information," *Proceedings of the National Academy of Sciences*, XLII (September, 1956), 654–55.

So long as an organization has at least three tiers of management, by definition middle management will not cease to exist. The odds are large, however, that within the next twenty years many functions which were regarded as within the domain of middle managers of the current variety will have been automated.

As machine programs become more intelligent, the automation will spread from more to less routinized jobs. Intelligent machines exist today, and the effort by man to make them more intelligent constitutes an important advance in our understanding of the nature of the creative process and of human intelligence itself.

The achievements of the intelligent machines are certainly not great when compared with those of human decision-makers. Nevertheless, they have already cast light on the decision processes. No machine has yet run a business, although most of them spend most of their time meeting payrolls. The few intelligent machines play a good game of checkers, a mediocre game of chess, write indifferent music, invent names for new substances, and prove theorems in geometry and in logic.

Even given a machine's capacity to do computations, it is impossible for it to make a complete search of the alternatives in a game such as checkers or chess. The programs used by machines to play these games recognize this limitation. No attempt is made to carry out an exhaustive search. Some aspects of openings and end games are memorized, but beyond that the machine plays according to general principles. The so-called techniques of heuristic programming provide the machine with patterns of exploration and general rules of learning and evaluation. From one point of view the complexity of the decision process in playing chess may appear to be considerably smaller than that in making a decision in the running of a firm. Nevertheless, as relatively little is known about the mechanism of either process, a gain in the understanding of even the simpler situation represents an advance. The use of machines to aid in production scheduling, inventory control, and linear programming is already well established. In some plants today machines are solving these problems routinely, while in other plants human beings are solving the same type of problem. The roles of many "creative" design engineers have now been taken over by computers in the design of heavy electrical equipment. In all these instances man has had to understand the nature of the process used to enable the machine to solve the problems.

Whether machines are capable of thinking, or of making better decisions than human beings can, or of "being creative" depends upon what we mean by thinking, decision-making, and creativity. In general, when we use these words they are ill-defined. Much of the work in teaching machines how to think has aided in teaching their mentors. An excellent non-technical summary of thought on machine thought is given by Daniel McCracken in a progress report on machine intelligence.[32]

EXPERIMENTAL GAMING

The use of games as a research device to study economic decision-making has been a new development of the past five years. The behavior of individuals in situations involving problem-solving, bargaining, learning, the formation of coalitions, and the performance of tasks under limited communication networks—all

[32] Daniel D. McCracken, "A Progress Report on Machine Intelligence," *Datamation*, VI (September-October, 1960), 10–13.

have recently been subjects for experimental investigation.

There have been many economic theories concerning the behavior of two rivals in a bargaining situation or bilateral monopoly. The study of Siegel and Fouraker has begun to produce verification of some of the basic conjectures concerning the behavior of bargainers.[33] In particular, they have studied the importance of knowledge of the other person's goals when bargaining. They observed that the lack of this information by either side had a profound effect on attempts to reach a "fair bargain."

Churchman and Ratoosh have experimented with a small group structured to simulate the organization of a firm.[34] In observing the behavior of the individuals engaged in jointly solving tasks set to the group, they have found that in some instances, even though one of the individuals is aware of the correct answer, this may not be sufficient to have the group take action. This type of behavior matches the reasons often offered for the employment of consultants. It is claimed that in many hierarchies, even though individuals within the firm may know the answers, it requires the presence of an outsider saying the same words to have the answers listened to and acted upon.

Several experiments involving two or more participants set in either an economic context or a more general bargaining situation have given results which indicate the value of the theories of behavior suggested by Nash and by Lloyd Shapley.[35] Nash has presented a theory of non-co-operative behavior which is closely related to the behavior postulated by economists when there is no collusion present. Shapley's theory suggests that an individual can expect to obtain a reward which represents an average of his incremental worth to any coalition in which he could become a member. This theory appears to have relevance to political situations and other areas of human affairs where coalition structures are of considerable importance.

BUSINESS GAMING

Although war games have been used for many years by the military for purposes of training and analysis, only recently has there been a development of their use as models of business situations. The American Management Association, International Business Machines Corporation, General Electric, Pillsbury Mills, and many other corporations as well as the University of California at Los Angeles, Carnegie Institute of Technology, and a host of other universities and business schools have constructed and are using these games.

They tend to be more complicated than the experimental games. The environment in which the play takes place is rich compared with the experimental games described above, although, of course, it is not as complex as is business itself.

The main use of these games has been for teaching and training purposes. They have been utilized for teaching at the functional level of a firm and for illustrating problems of organization, information, and competition. The proponents of these games have often claimed that they are very useful in teaching an appreciation for the interrelationship between different functions in an organization, and, in general, for giving individuals deeper

---

[33] Siegel and Fouraker, *op. cit.*

[34] C. West Churchman and Philburn Ratoosh, *Innovation in Group Behavior* (Working Paper No. 10 [Berkeley: Institute of Industrial Relations, University of California, January, 1960]).

[35] J. F. Nash, Jr., "Non-cooperative Games," *Annals of Mathematics*, LIV (September, 1951), 286–95; and Shapley, *op. cit.*

insight into complex problems of interaction.

There appears to be little doubt that these games are useful teaching and training devices, although they may not be so effective as many of their proponents claim. Nevertheless, there is a complete area which to date has barely been exploited and which may well prove to be a very fruitful outcome of the interest in gaming, that is, the use of business games for experimental purposes. They have a great potential as a device by which much can be learned about decision-making.

The very rigorously designed experimental games used primarily by psychologists take place in an environment that is so simplified that, although behavior can be observed with experimental precision, many individuals would balk at making the inference that the observed behavior would provide a good predicting device for the behavior of decision-makers in a far more complex environment. Business games provide an in-between stage in which to study behavior. Even though the situations tend to be far too complex for a thorough analysis, if care is taken beforehand, it is possible to design the game and have it played in such a manner that the data generated is of value in learning about the behavior of the players and in validating theories of behavior in relatively complex environments.

The business games in use at this time are restricted to numerical quantities. They are unable to cope with qualitative events, although in some of these games a news or business letter may be circulated during the play in order to simulate qualitative effects. In general, the larger games tend to utilize about twenty minutes to simulate the activity of a business quarter. The players usually play between two to three years of a market history. If the game is played with experienced personnel and is represented as a model of a business with which they are familiar, it may easily serve as a device to aid the players in verbalizing their perceptions of their industry or market. Experience in operations research and the science of management indicates that, in spite of hortatory writings, there remains a vast preponderance of experience and know-how which has been neither adequately recorded nor studied systematically. Business gaming now offers a new aid in doing this.

SIMULATION

The twin problems which beset those wishing to understand decision-making in the firm are, first, the observation and interpretation of behavior and, second, the portrayal and analysis of the environment within which decisions are made. Most of the behavioral sciences do not have the means to perform controlled experiments. The development of experimental gaming and business gaming, along with the advent of the high-speed digital computer, has helped to change this state in the study of areas of psychology, economics, and business.

The advent of new statistical methods and the great increase in human ability to handle very complex models brought about by the computer have provided for the observation of complex economic systems in a manner that was impossible ten years ago. For those who wish to study economic life, the advent of the computer has provided for the development of simulation and mass data-processing techniques which supply a viewing instrument that should have as profound an effect as did the microscope and telescope in other sciences.

A detailed discussion of the applications of simulation to the economy as a whole, to industries and the firm, and in the area of artificial intelligence is given in the December, 1960, issue of the *American Economic Review*.[36] Some idea of their scope may be provided by observing that a large-scale simulation of the demographic growth of the United States has been constructed. Simulations of the shoe and leather industries, the lumber industry, and several other industries, as well as detailed simulations of the workings of parts of a transportation system and sections of various firms, have been built.

The sheer amount of discipline required in producing the flow diagrams and in checking for consistency among different segments of these large complex models in itself is of considerable value to the understanding of the nature of a firm as well as of an industry or the whole economy. As a direct aid to decision-making in the firm, it is quite probable that within the next ten to fifteen years most large firms will be utilizing constantly updated simulators of themselves and their markets for planning and anticipatory purposes.

A simulator provides a means for testing out conjectures concerning the behavior of individuals en masse in a complex environment. As such, its use in the study of the over-all economy, individual industries, and marketing and distribution systems promises to be great. As the work in gaming continues, more and better theories of the individual will provide further inputs for simulations in an attempt to reach for the ultimate goal of understanding both behavior and environment together with their interaction.

VI. CONCLUSIONS

Necessity has forced the invention of a set of new words: decision theory, the behavioral sciences, information theory, the theory of organization, etc. The sub-

[36] "Symposium on Simulation," *op. cit.*

stance behind the names is nowhere near so great as we would wish it to be. However, it is growing fast and in response to the needs of the times.

The traditional areas of economics, psychology, mathematics, and statistics, to name a few, have not provided adequate theories to explain many problems of individual decision-making in a complex and uncertain environment. The model of rational man needs to be modified. The new work discussed here is beginning to replace him with a much less sure, more complex, and flexible individual whose problems and behavior are closer to those that we recognize in the world around us. The need to investigate the processes of decision and the means by which to understand them have both mushroomed. A very different and much deeper understanding of decision processes is in the making.

How does this work influence the firm, business schools, and universities now, and how will it influence them in the near future? Management science and operations research have already had their effect. More sophisticated uses of computers are constantly coming into being. Our knowledge of competitive behavior, learning, information processes, and artificial intelligence is still sufficiently limited that, by the time that it would be technologically feasible to replace current middle managers by computers, they will have been replaced by a very differently trained type of individual. The new theories have already raised many new and well-defined questions. The expanded possibilities of data-processing and the growth of a scientific and empirical attitude in economics, government, business schools, and management is resulting in a revolution in the gathering of statistics and the design of information processes.

Probably the greatest effect will come in the growth of experimentation and empirical investigation hand in hand with the growth of theory. Industry must regard itself in one sense as a vast experiment, and management itself as a subject of experimentation and observation. As the conjectures and hypotheses behind decision theory are confronted with demands for validation, it will be increasingly necessary for researchers to understand, work with, and design experiments with and on individuals and the institutions being observed. They, in turn, stand to benefit from uses derived from validated theories.

THE QUARTERLY JOURNAL OF ECONOMICS

Vol. LXXV   August, 1961   No. 3

OBJECTIVE FUNCTIONS AND MODELS OF CORPORATE OPTIMIZATION*

By Martin Shubik

I. Economic theory and the goals of the firm, 345. — II. The corporate economy, 356. — III. Problems in the specification of goals, 360. — IV. Some recent approaches to the study of economic intent, structure, and behavior, 368. — V. Power, intent and policy, 374.

I. Economic Theory and the Goals of the Firm

1. *Introduction*

The development of micro-economic theory has been entering a stage similar to that experienced by twentieth century physics in its break away from the simple comfortable closed systems of its precursor. *Homo Oeconomicus* operating as an entrepreneur or consumer in smoothly adjusting flexible markets still presents us with a logically and aesthetically satisfying body of theory. However, unfortunately there are many facts and facets of an economy which need to be explained, but which cannot be explained using this model. The behavior of individual economic units which still remains not sufficiently explained is of considerable importance to society in general and to the firms in particular. There is a need for an adequate economic theory to serve as a basis for analysis, prediction and as an aid in policy evaluation.

In recent years, linked with the growth of operations research and management science there has been an increasing self-consciousness on the part of many corporations which has led to a proliferation of investigations and self-analyses concerning goals. The development of certain parts of mathematical economics and industrial

---

* Research undertaken by the Cowles Commission for Research in Economics under Task 047-006 with the Office of Naval Research.

engineering have had the paradoxical effect of giving operational meaning to very complex aspects of individual maximization and at the same time illustrating in application, the limited use of models used in explaining behavior in many important areas of decision-making in the firm. The observation by Freud that only small decisions are made "rationally" appears to hold true in all adaptive organisms, be they individuals or economic organizations.

The problems of formulation of objectives have given rise to new terminology. For example, the phrase "objective function" has made its way into the literature of economics, operations research and management science by the route of linear programming. In a linear program the objective function is the expression to be maximized or minimized. The typical programming problem is one in which the prices of the final products are given, the capacity limitations on production are specified, technology is assumed to be known and the goal is to maximize a revenue or equivalently, minimize a cost subject to constraints.

Unfortunately in problems involving more than a few aspects of the running of a complex organization such as the modern firm, it becomes most difficult to describe the objective function. As the complexity of the organization grows the so-called objective function becomes more and more subjective and less quantitative as multiple goals, social and political goals and uncertainty and ill-perception of the environment are taken into account.

Does the firm maximize? What does the firm maximize? Should the over-all policy of the firm be described as maximization, optimization or something else? The first two questions are possibly of some interest to philosophers; however, the third is of direct interest to economists and others involved in shaping policy and predicting behavior of an economy. Preliminary to comment upon the third question, some of the models used in economic theory to describe the behavior of the firm are examined.

2. Intent, Behavior and Structure

Throughout this paper an attempt will be made to maintain three distinctions which are clear in the short run but become more difficult to differentiate in the long run. They are the distinctions among intent of the decision-maker, behavior and structure (or environment in which the individual acts). The manifestation of the individual's intent is expressed in his behavior within his environment. We observe his behavior and we observe the environment and from these observations we may be in a position to make statements such as "the individual behaves as though his intent were to maximize monetary profits." The intent per se is not observed. Furthermore many different intents may give rise to the same behavior within certain structures as is amply illustrated in the literature of attempts to control oligopolistic competition.

Another way of phrasing the above distinction is that intent is the operator upon structure which produces behavior.

In the long run, in human affairs, the dichotomy between behavior and environment becomes less distinct than in the short run. If enough individuals do not like the "rules of the game" they may be changed. However, as a first approximation, the distinction appears to be worthwhile.

3. *Pure Competition*

The simplest model of the firm is that which is referred to in economic theory as pure competition. The firm is considered to sell a single product to the market. The price for the product is known and fixed "by the play of the market forces" and is not influenced by the size of output, advertising or other acts of the firm. The aim of the firm is assumed to be the maximization of its short-term net revenue. In order to do this it selects a maximal rate of production. The policy can be expressed mathematically as follows: Let p be the price of the product in the market. Let $C(q)$ be the total cost of producing q units, and let $R(q)$ be the net revenue obtained from producing and selling q units.

$$\text{Maximize}_{q} R(q) = pq - C(q)$$

The q beneath the word maximize indicates that the strategic variable in control of the firm is production.

At best this model can be used to teach us something about the effect on price and output of changes in such parameters as the cost of inputs. It may be used to investigate problems of production, taxation and so forth. It serves as a precursor for the more elegant and complex techniques of programming under conditions of complete information.

It may be argued that this model is adequate to describe the behavior of the small individually owned enterprise, with low capitalization, no (or few and unimportant) market imperfections, no influence in the market and no problems caused by lack of perception, incomplete information or uncertainty.

If the structure of the economy had all of the properties noted above, then indeed the theory of behavior would have much to commend it as it requires almost no consideration of the intent of the entrepreneurs. Provided that we make the weak assumption that the individual prefers greater profits to lesser profits, *ceteris paribus*, then all additional assumptions concerning intent (such as the entrepreneurs' wish to do each other economic harm or that they are the best of friends and wish to display this in their behavior) are not germaine to the prediction of economic behavior. The structure is such that individuals are so weakly interlinked strategically that the interconnectness between any two is below the level of perception. Regardless of the intents of the individuals, the "invisible hand" of the market structure guides their behavior. Unfortunately there are few if any parts of an economy which have enough of the features to make the model of pure competition of more than limited use.

4. *Monopoly*

If a firm is in a monopolistic position its actions can influence the demand for its product (or products). For simplicity we consider only one product. Again economic theory considers that the firm seeks to maximize (short-term) net revenue.[1] Several complications such as pricing, production, advertising and product variation, may all be considered as strategic variables under the control of the firm and relevant to its maximizing process.

The mathematical expression for the behavior of the firm shows a distinct similarity to that of the firm in pure competition. Let $s = (s_1, s_2, s_3, \ldots s_n)$ stand for the strategic variables under the control of the firm, where s_1 is the production rate (the same as q in Section I.3 above), s_2 is the advertising budget, s_3 is the price, and so on. The demand for the final product may be a function of all these variables except the production rate and can be written as $D(s_2, s_3, \ldots, s_n)$. The cost of operating the firm now includes not only the costs of producing, but also advertising, product research and so forth. This can be expressed as $C(s_1, s_2, s_4, \ldots, s_n)$. The goal or objective of the firm is to:

$$\text{Maximize}_{s} R(s) = s_3 D(s_2, s_3, \ldots, s_n) - C(s_1, s_2, s_4, \ldots, s_n).$$

The revenue function is different from that under pure competition, and it is this difference which reflects the difference in the market

---

1. G. J. Stigler, *The Theory of Price* (New York: Macmillan, 1952).

structure. However, the "operator" on this function, or the intent of the individual is the same. The operator is maximization and, as before, this leads to an unambiguous behavior when applied to the market structure. Both the theories of pure competition and monopoly require that we know no more about the intents of the individuals than that they have a desire to maximize. Both the pure competitor and the monopolist act in isolation in which the only forces present are their own and the "impersonal market." In the first instance the individual is weak, in the second stronger. In both cases the behavior is that to be expected of a "rational maximizing utilitarian man" acting under conditions of complete information.

5. *Competition Among the Few: Oligopoly*

Even in the simplest models involving few players, each with a considerable influence in the market, but none with control, intent and behavior become hard to define and describe. Naively we might suggest that the firm still wishes to maximize its short-term net revenue, but this simple intent is no longer sufficient *by itself* to be used to predict behavior in the market.

In order to predict the behavior of the firm we must simultaneously consider the intents of all competitors. It is possible to define the short-term revenue function for a single firm in such a market, but it is a function whose value depends upon the strategic variables of the firm *and* its competitors. Any statement concerning maximization of revenue by a firm must contain some statement concerning the expected actions of the competitors. Thus instead of the straightforward maximization policies described in Sections I.3 and I.4, economists have discussed joint maximization;[2] "live-and-let-live" policies; quasi-maximization and various policies based upon conjectures as to the behavior of all other competitors.

In a market with few competitors, no matter how simple an economic model is constructed, the problems of defining goals and describing behavior become far more difficult and are far different from the simple maximization policies noted in Sections I.3 and I.4. Many models have been suggested but insufficient empirical work has been done to test the relevance of the models. A favorite model has been that of the market with a nonco-operative equilibrium;[3]

2. William Fellner, *Competition Among the Few* (New York: Knopf, 1949).
3. For a discussion of nonco-operative equilibrium see: M. Shubik, *Competition, Oligopoly and the Theory of Games* (New York: Wiley, 1959), Chap. 4.

this, in one form or another is at the basis of the models of Cournot,[4] Stackelberg,[5] Chamberlin,[6] and many others. Some of the market behavior patterns described allow for entry of new firms into competition and the departure of old firms;[7] however, costs of entry and difficulties of exit are not usually dealt with explicitly (although recently the work of Joe Bain has gone far to rectify this omission.[8] Two examples of intent and their resultant behaviors are given below.

Let $s_i = (s_{i1}, s_{i2}, s_{i3}, \ldots, s_{in})$ stand for the set of strategic variables under the control of the ith firm. $s = (s_1, s_2, s_3, \ldots, s_n)$ stands for the set of all strategic variables under the control of all m competitors. If each firm attempts to maximize its short-term net revenue under the assumption that its competitors' actions will remain fixed, then if $R_i(s)$ represents the net revenue function for the ith firm it acts to

$$\text{Maximize}_{s_i} R_i(s) = s_{i3} D(s_{11}, s_{12}, \ldots, s_{1n}; \ldots; s_{i2}, s_{i3}, \ldots, s_{in}; \ldots; s_{m1}, \ldots, s_{mn})$$
$$- C_i(s_{i1}, s_{i2}, s_{i4}, \ldots, s_{in}),$$

and similarly for the other firms. In general, as the products being sold may be different, the demand function D should be replaced by a D_i which is different for each player. Of course, empirically it would be fantastically difficult to deal with demand functions as complicated as these other than by very crude approximation.

In order to obtain a prediction of behavior in this model we have had to consider the simultaneous effect of m different intents interacting upon m revenue conditions or "payoff functions." Even in order to carry out purely theoretical calculations it is necessary to specify not only the individual's intentions concerning the impersonal environment, but precisely how he intends to utilize his information concerning the behavior of other individuals. Thus a full specification of the intent which leads to the equations above is: "I intend to maximize my short-term net revenue, as *defined by fixing the actions of all my competitors at their previous values.* I will ignore all information which appears to run contrary to my hypothesis concerning their behavior." If each entrepreneur followed this intent then under

4. Augustin A. Cournot, *Researches into the Mathematical Principles of the Theory of Wealth* (New York: Macmillan, 1897).
5. H. von Stackelberg, *Marktform und Gleichgewicht* (Berlin: Julius Springer, 1934).
6. Edward H. Chamberlin, *The Theory of Monopolistic Competition* (6th ed.; Harvard University Press, 1950).
7. *Ibid.*, pp. 196–97.
8. J. S. Bain, *Barriers to New Competition* (Harvard University Press, 1956).

most circumstances behavior leading to a nonco-operative equilibrium would be observed.

Without specifying assumptions concerning learning (i.e., the effect of information upon the actor) and implicitly or explicitly the socio-economic aspects of group interaction, it is logically not possible to describe completely a procedure to provide a prediction of behavior in a market with few competitors. The crux of any theory of oligopoly lies in these assumptions. The difficulties encountered and the inadequacy of theory to date has had two main sources. The first has been the logical inconsistency or incompleteness of the models of man in economic competition. This difficulty is being removed by improvements in methodology. The second is caused by lack of knowledge. Although it is easy to supply the necessary assumptions to produce logically consistent and complete models, it is extremely difficult to validate them. A certain amount of empirical work exists in the form of observations from institutional studies, and recently experimental work has commenced.

A second example of intent and the resultant behavior is given if we assume that each firm wishes to maximize jointly the net aggregate receipts of all. This is equivalent to:

$$\text{Max.}_{s_1}, \text{Max.}_{s_2}, \ldots, \text{Max.}_{s_m} (R_1 + R_2 + R_3 + \ldots + R_m).$$

Here there are m operators, operating simultaneously on one revenue condition. Although the intents in both instances illustrated here are quite different, as can be seen by the two expressions given above, under certain circumstances the observer will not be able to differentiate between them. A simple analogy serves to illustrate this fact. If two individuals are trapped in a cellar with water rising and a trap door providing the exit which requires two men to lift it, whether they are individual operators or joint maximizers the observed behavior will appear the same. The same "implicit collusion" can be the resultant behavior from many different intentions.

6. *Long-run Considerations*

The examples above all dealt with short-term or instantaneous problems of maximization. Attention has been paid to long-term maximization problems of the firm. The very simplest assumption is that the firm attempts to maximize the discounted value of its yearly income flow. In other words, the discounted value of the continuous stream of short-term net revenues. If R_t is the short-term net revenue obtained in period t, ρ is the discount rate, which is equivalent to $1/(1 + r)$ where r is the rate of interest and R is the discounted value of the yearly income flow, we can write:

$$R = \sum_{t=0}^{\infty} \rho^t R_t.$$

An immediate problem in time horizons is brought out by this formulation. As written above the income stream is summed (with the discount factor) from the present to infinity. Is the infinity merely a mathematical trick, or does it reflect the theoretically infinite life of a corporation? If the policy is for a firm owned by an individual with no heirs, would this be an adequate goal for him to pursue? There is no indication in this formulation of the possible value of terminating operations, selling out or liquidating. Admittedly, if all markets for assets were perfectly competitive there would be no need for this and the discounted income stream would reflect exactly the worth of the firm. However, few markets are perfect and in general this model is too simple to be of more use than as a teaching device.

7. Short-term Uncertain Demand

The short-term models in Sections I.3 and I.4 were based upon certain information concerning market conditions. The models of Section I.5 recognized the problem of interdependence, but did not treat it by assuming a probability distribution for the behavior of competitors (this can be done; however the reasons for doing so do not appear to be too convincing[9]). Fluctuations in demand, especially when there are many customers, can be treated by probability considerations. In recent years consideration of demand fluctuations has led to a considerable literature on inventory theory.[1] The simplest assumption is that when faced with uncertainty in the market the firm acts in a manner which maximizes its short-run expected revenue. We assume that p is the market price for the product (the same market form as in Section I.3). The value of a

---

9. It is unlikely that there will be a sufficiently good reason to ascribe a probability estimate to the actions of few competitors. If this is done, the literature on "Games Against Nature" offers several methods for behavior. See: R. D. Luce and H. Raiffa, *Games and Decisions* (New York: Wiley, 1957) and Shubik, *op. cit.*, Chap. 9.

1. See, for instance: J. Laderman, S. B. Littauer, and Lionel Weiss, "The Inventory Problem," *Journal of the American Statistical Association*, Vol. 48 (Dec. 1953).

unit of inventory is k. $\varphi(t)$ is the probability density that t units will be demanded. $\Phi(q) = \int_{t=0}^{q} \varphi(t)dt$ is the cumulative probability that q units or fewer will be demanded. $E[R(q)]$ is the expected revenue. We can express the policy of maximization of expected revenue as:

$$\operatorname*{Max}_{q} E\{R(q)\} = p \left\{ \int_{t=0}^{q} t d\Phi(t) + q[1 - \Phi(q)] \right\} - C(q) + k \int_{t=0}^{q} (q-t) d\Phi(t).$$

This consists of three parts. The first, which has two terms, is the expected revenue obtained from the market if the market demand is less than q (the production rate) and the expected revenue obtained if the demand is q or more. The second is the cost of production of q units and the third is the expected value of the inventory.

Should a businessman maximize his expected revenue? There have been many discussions as to whether or not expected income should be maximized as opposed to other quantities which take into account the variance and other risk considerations.[2] The problem amounts to what we are trying to portray. If the firm can be ruined by one bad season, this model of maximization of expected income is probably quite inadequate. An explicit consideration of the ruin conditions will be necessary. The model above is certainly a useful first approximation that enables us to carry out studies of inventory policies and goals which were previously beyond analysis. It enables us to answer problems of the type: "If a businessman wishes to maximize his expected revenue, what should he do?" A normative interpretation may be given to this objective function. We may state that individual owners, partnerships or corporation *should* maximize their expected short-term net revenue when faced with uncertainty of this sort. The third position which can be adopted in interpreting the objective function is that, in fact, businessmen of some class or category do act as though they were trying to maximize this objective function. It appears to be unlikely that this is the case, although with the appropriate specifications testable hypotheses can be formulated.

2. J. Marschak, "Rational Behavior, Uncertain Prospects and Measurable Utility," *Econometrica*, Vol. 18 (April 1950).

8. Interdependence, Uncertainty, Insensitivity and Misperception

"The firm should attempt to maximize its present value. The firm should attempt to maximize its present value. The firm should attempt to maximize its present value." The Bellman in The Hunting of the Snark remarked, "What I have said three times is true." This exhortation, however admirable and true, unfortunately provides limited aid either as a predictor to the policy-maker or as a maxim for the entrepreneur.

The firm does not consist of *homo oeconomicus in vitro* but in general consists of many men embedded in a market in a society. As such, an adequate theory of the behavior of the firm must at least implicitly reflect considerations of (1) the goals of the individual and the individual as a decision-maker, (2) the firm as an organization and economic unit, and (3) the interactions of firms in the socio-economic and politico-economic environment. This calls for a reconsideration of the properties of a model of economic man and the selection of the relevant variables and parameters in describing his environment.

Economic man is a creature who must act under severe limitations of incomplete information. Even if complete or perfect information were available to him, by choice he must select only a sample of it. His perceptions are poor and his channel capacity for processing information is low. His major asset is time. Total knowledge, even if available, must be abstracted and aggregated owing to costs and limitations of processing. The executive who tells the researcher to boil down his report (and thereby distort valuable information according to the researcher) is not always wrong. Economic man of necessity operates under conditions of *low information*. Leaving out the problems of interaction with others (which brings in consideration of organization and society) the properties of the individual as an adaptive organism with limited perceptions, limitations on data processing ability and goals dependent upon his learning appear to be of importance to the development of economic theory. The model of "satisficing man" of Herbert Simon is a construct which incorporates these features.[3] The new work in problem-solving methods, heuristic programming and artificial intelligence also involve these features.[4]

3. H. Simon, "Rational Choice and the Structure of the Environment," *Psychological Review*, Vol. 63 (Mar. 1956).

4. M. Minsky, "Artificial Intelligence" (forthcoming), *Proceedings of the IRE* (Institute of Radio Engineers), Jan. 1961.

For many years students of institutions have noted the role of the firm as a bureaucracy. The distinctions among manager, administrator, entrepreneur or leader are not always clear. Yet from the viewpoint of socio-economic policy which or what combinations of these roles are required by the heads of large enterprises may have an important economic effect. A broad economic theory of the firm must at least be cognizant of internal organization and the possibility that for many economic questions it cannot be assumed away as a minor imperfection. The works of Argyris, Barnard, Selznick, Simon and March and others have been addressed to the broad problems of organization of the firm.[5] Marschak and Radner, following more closely the developments of mathematical economics and statistics have developed a theory of "teams" which deals with problems of imperfect communication between individuals whose goals are identical, i.e., individuals in the same organization with a perfect harmony of interests.[6] In a later paper a model will be presented of an organization in which the individuals do not have identical goals and in which a basic strategic feature of the operation of the organization is the control of individuals at all levels over the processing, aggregating and disaggregating of information.

Having opened Pandora's box, let us quickly close it, consider that we are in a position to talk about the firm as a unit and consider the problems of the firm in a market. In my estimation, possibly the most important variable left out of economic analysis has been information and its cost. Once even rudimentary considerations are given to it, the wealth of institutional forms found in special markets, distribution systems and communication systems begin to conform to an over-all structure of an economic theory rather than remaining as a source for strange examples in a marketing textbook. As will be discussed further in Section IV there are good reasons to suspect that parts of economic theories of competition can be made to appear more plausible and can be at least partially validated experimentally if they are interpreted as actions under conditions of very low information.

5. C. Argyris, "The Individual and Organization; Some Problems of Mutual Adjustment," *Administrative Science Quarterly*, II (June 1957). C. I. Barnard, *The Functions of the Executive* (Harvard University Press, 1956). P. Selznick, *Leadership in Administration* (New York: Row Peterson, 1957). J. G. March and H. A. Simon, *Organizations* (New York: Wiley, 1958).

6. J. Marschak and R. Radner, *Economic Theory of Teams* (Cowles Foundation Discussion Paper No. 59, 1958).

The firm as a unit acts under a severe handicap of ignorance. Very often it does not know its own costs. It may easily make the assumption that its costs are linear up to a point because it is most likely cheaper to operate under that hypothesis than to pay the large sum required to find out otherwise. It has a birth and very often has a death. These are not continuous processes and hence cost of entry and of *exit* may be very important. The latter cost has not been given sufficient attention in economics. In the theory of war it has — thus Caesar destroyed the bridges when he crossed the Rubicon. An enemy when cornered may have no other choice but to fight. A dying firm with high investment in specialized equipment may be in the same position.

A long list of variables applying to different trades and markets is not hard to compile. Financial structure is usually important, distribution channels, corporate form and so forth all may play a vital role. These, however, are all factors which are needed to describe what a firm can do, not what it wishes to do, or what it does. Nevertheless structure may cause many intents to lead to the same behavior. As such it is an important part of economics to be able to observe the details of a market and to interpret them in terms of general properties of an organism such as viability, perception, flexibility, vulnerability and other terms used to describe the suitability of any organism or animal for surviving and possibly flourishing in its environment.

II. The Corporate Economy

1. *Economic Theory and Society*

Economic theory in general and twentieth century British-United States economic theory in particular have tended to bypass the sociological and political aspects of the economy which play a considerable role in determining the goals, nature and behavior of the firm and consumer demand. The micro-economic theory of price has probably suffered the most from this neglect.

By means of great oversimplifications an elegant theory of pure competition has been constructed which serves the purpose of demonstrating many important propositions concerning the allocation of resources and the determination of prices under special conditions. The models at the basis of this theory fit at most very few parts of any economy that exists today. Oligopoly theory leaves even more to be desired.

Even the most casual observations of the economy indicate that the goals, structure and behavior of the small individually owned unincorporated firm are sufficiently different from those of the large corporation that a theory which purports to be able to portray or predict behavior and price formation must explicitly take into account some of the differentiating institutional features. When firms employ tens or hundreds of thousands of individuals and have many thousands of stockholders the socio-political aspects of their existence can no longer be implicitly assumed away in an economic study.

In order to develop a theory of the firm in a modern economy, many conditions and roles in relation to the economic aspects of the firm must be specified. The aims and the effect of stockholders, managers and other employees may be predominant in determining the behavior of the present day firm. Financial structure, technological conditions and special institutional limitations largely influence market forms and hence the price structure. Recently there has been a recognition that the empirical content and the number of factors to be taken into account in the construction of a theory of oligopoly must rise.

The classification and differentiation of market forms has made its way into economic theory as can be seen by the works of Fellner, Machlup, Bain, and others.[7] Work such as that of Berle and Means and recently Eells are of aid in filling the gap in the theory of the firm due to a lack of analysis on the goals and organization of the firm.[8]

2. The Firm in the United States

As is well known, the corporate form is the most important type of business organization in the United States. The smaller corporations often are family owned affairs with a close connection between ownership and management. Almost all of the large corporations are run by managers for thousands of stockholders and employ many thousands of workers. The corporate form dominates manufacturing; partnerships are still to be found mainly in the professions and

7. Fellner, *op. cit.* F. Machlup, "The Characteristics and Classifications of Oligopoly," *Kyklos*, X (1952), 145. Bain, *op. cit.* U. S. Temporary National Economic Committee, *Competition and Monopoly in American Industry* (Washington: 1940), pp. 59–63, Monograph 21.

8. A. A. Berle and G. C. Means, *The Modern Corporation and Private Property* (New York: Macmillan, 1933). R. Eells, *The Meaning of Modern Business* (New York: Columbia University Press, 1960).

the individually owned unincorporated enterprise is predominant among shopkeepers and farmers.

In 1955 there were 4,180,000 businesses in the United States. Their form of business organization was as follows:[9]

Corporations	560,000
Partnerships Other Unincorporated Enterprises	3,620,000

If numbers are taken as the criterion, the small individual businesses still dominate the scene; when other criteria such as sales, assets or employees are examined this is not so. Thus, in 1948 less than 1 per cent of the firms classified as industrial enterprises controlled two-thirds of all the assets of firms classified as industrial enterprises.[1]

The relative importance of the large corporation to the economy can be seen from the compilation of the 500 largest manufacturing corporations in the United States, for which reports are available. In 1956 the Gross National Product was $415 billion. During this year the following figures helped to characterize the 500 firms:[2]

Sales	$174,300,000,000
Assets	$139,000,000,000
Net Profits	$11,500,000,000
Stockholders	11,100,000
Employees	8,800,000

With commitments of human and financial resources of such a size the large corporation becomes a social as well as an economic entity. Neither the economists studying the theory of the firm nor the managers running a large corporation can afford to ignore the multidimensional aspects of the corporation's goals and actions.

3. *Statements of Corporate Goals*

A study of the statements of corporate aims by twenty-five corporations, seventeen of which are among the five hundred largest manufacturing firms, reveals the blend of economic, political, social and ethical considerations which at least indicates the self-images of the corporations. The sources and detail of the material vary con-

---

9. Berry C. Churchill, "Business Population by Legal Form of Organization," *Survey of Current Business*, April 1955.

1. A. D. H. Kaplan, *Big Enterprise in a Competitive System* (Washington: Brookings Institution, 1954).

2. *The Fortune Directory*, Supplement to *Fortune*, July 1957.

siderably as is indicated below. Titles or captions of the statements of objectives are:

Our Aim; Basic Objectives; Objectives; Creed; The Policy; Company Objectives; Company Policy; Principles; Our Credo; The Company's Philosophy; General Corporate Policies; Philosophy; Statement of Policy; and General Objectives of Management.

The sources include statements from chairmen, both in the financial press and elsewhere; statements from management guides and manuals; approved releases by the board of executives; notes by company presidents; and statements from company reports.

The contents have been classified under the headings given below. These are accompanied by the numbers indicating frequency of reference in the twenty-five statements.

Personnel	21
Duties and Responsibility to Society in General	19
The Consumer	19
Stockholders	16
Profit	13
Quality of Product	11
Technological Progress	9
Supplier Relations	9
Corporate Growth	8
Managerial Efficiency	7
Duties to Government	4
Distributor Relations	4
Prestige	2
Religion as an Explicit Guide in Business	1

There are considerable difficulties involved in selecting categories to describe the statements of corporate objectives. The terms used here and the frequency figures are merely indicative of the scope of the corporate statements. Some examples serve to indicate the basis of selection.

Finally, we shall strive to serve our shareowners, employees, and the customers justly and honestly . . . in keeping with our American Tradition of democratic economic freedom.

We recognize that industry justifies its existence to the degree that it is of service to the community.

Our fourth responsibility is to the communities in which we live. We must be a good citizen — support good works and charity, and bear our fair share of taxes. We must maintain in good order the property we are privileged to use. We must participate in promotion of civic improvement, health, education and good government, and acquaint the community with our activities.

The three quotations above are classified under Duties and Responsibility to Society in General.

Therefore, our over-all objective is to conduct our business so that all groups in the community will regard us as the best company in our industry and want us to prosper and grow.

This statement is classified both under Corporate Growth and Prestige.

Throughout all the statements the words: "fair share, fair return, fair prices, equitable wages, fair and equal treatment and proper return to investment" appear with great frequency. Fair and proper may have meanings in law and ethics but unless specified in a welfare measure they do not have meaning in economic theory. The problems of a large corporation involve considerations both of economics and the social environment. In order to be able to assign operational economic meaning to corporate utterances, words such as "fair" must be interpreted in terms of the environment.

III. Problems in the Specification of Goals

1. *The Purposes of a Statement of Policy*

The intermixture of social man and economic man in a large enterprise can scarcely be unscrambled. Bearing this in mind we must examine some of the motivations for the statements of company policy before examining the statements. As public figures, presidents and other members of the upper hierarchy of corporations are often called upon to make public utterances. They talk to many social organizations and are called upon to make statements concerning their attitude towards labor, inflation, competition and so forth. They must prepare press releases and co-operate with the press or risk the consequences of failing to do so. Company policy must be explained to financial writers, to stockholders and creditors. Furthermore, a general policy must be mapped out for long-range operational purposes and more detailed policies must be described for the more specific problems of delineating the areas of action for departments or other subgroupings of a major corporation.

As is the case with an army the statement of objectives may serve both as a morale builder and a general "heuristic" or over-all rule for all members of the firm or corporation. There must be some sense of identification on the part of the members with the whole. Dewing has noted "that the association of human beings, in order to achieve a purpose is the fundamental and teleological basis for the coming into being and continuing existence of a corporation."³

How specific or how general must the statement of purpose be? Selznick notes that:

"to make a profit" is widely accepted as a statement of business purpose, but this is too general to permit responsible decision-making. Here again the more marginal the business, that is, the greater its reliance upon quick returns, easy liquidation, and highly flexible tactics, the less need there is for an institutionally responsible and more specific formulation of purpose. Indeed, the very generality of the purpose is congenial to the opportunism of these groups. But when institutional continuity and identity are at stake, a definition of mission is required that will take account of the organization's distinctive character, including present and prospective capabilities, as well as the requirements of playing a desired role in a particular industrial or commercial context.⁴

In light of this we can expect a hierarchy of statements in the specification of goals. For example consider a statement of goals by a large diversified manufacturing organization. (1) Any organization will usually have survival as a goal. (2) Any manufacturing organization may have as its intent the designing, manufacturing and selling of products at a profit. If, in a general sense, the concept of an "ethical organization" is introduced then (3) any ethical economic organization will consider as part of its intent, the maintenance of its reputation in the minds of employees, customers, investors, stockholders, vendors and the public at large. (4) The goals of a large diversified manufacturing organization, if it is to have a *raison d'être* must include the intent to contribute economic and social values best generated by this form of organization.

Some of the larger U. S. corporations have tended to assume the properties of institutions over and above their properties as economic organizations. It has been suggested that: "once an organization becomes a 'going' concern with many forces working to keep it alive, the people who run it can readily escape the task of defining its purposes."⁵ This conforms with the results of a study indicating that especially for large companies goals tend to reflect the company characteristics rather than the image of the management.⁶

3. A. S. Dewing, *Financial Policy of Corporations* (4th ed.; New York: Ronald Press, 1941), p. 5.
4. P. Selznick, *op. cit.*, pp. 147–48.
5. *Ibid.*, p. 25.
6. J. K. Dent, quoted in *Business Week*, Nov. 21, 1959, p. 114.

In cultures other than that of the United States, the institutional aspects of industry may be even more dominant over the simpler economic purpose of the economic organization than it is here. Abegglen, in writing about the Japanese Factory states:

> In a purely economic definition of employment terms, where the financial success of the factory is the overriding goal of management's policy, the national well-being and workers' welfare would be secondary considerations in much policy formation. Underlying the specific points made in support of the Japanese policy, and quite aside from the validity of these points, is the tacit recognition by management that the relationship between the company and the worker is not simply a function of the economic convenience of the two parties. The worker, whether laborer or manager, may not at his convenience leave the company for another position. He is bound, despite potential economic advantage, to remain in the company's employ. The company, for its part, must not dismiss the worker to serve its own financial ends. Loyalty to the group and an interchange of responsibilities — a system of share obligation — take the place of the economic basis of employment of worker by the firm.[7]

Given that economic organizations are complex and have the many facets we have discussed, what should the economist do about them? Should he base his predictions or recommendations upon a relatively austere economic model with traditional economic variables, should he modify the model or leave the field to the sociologist and anthropologist? In some instances the selection of the location for a monumental new plant or laboratory may have more to do with the mausoleum building propensities of a childless vice-president about to retire, than with the economics of transportation. In other instances actions may best be explained by looking at the firm as a club dedicated primarily to providing a satisfactory existence for its executives.

The actions of the firm have been of traditional interest to economists, and as an economist it is natural to suggest that the field should not be relinquished *in toto*. However, there is little doubt that economic model-building must take into account the many instances where other societal variables dominate the economic effects.

The fourteen categories noted in Section II.3 include many vague and ill-defined classifications which serve as a basis to set the general degree of co-operation internal to the organization, descriptive of the institution and the general milieu within which it operates. Within them, however, heavily disguised by slogans and catch words

7. J. C. Abegglen, *The Japanese Factory* (Glencoe, Illinois: The Free Press, 1958).

such as "trade ethics," the "American way of enterprise," "fair-dealing" and so forth, probably lie a few parameters which characterize the inherent degree of co-operation or limitations on the weapons of economic conflict which call for the American, or Japanese, or German large industrialist to conform to somewhat different rules.

There are great difficulties attendant upon attempting to formalize and obtain these measures; however the task does not seem to be impossible. This writer has attempted to devise a criterion for the degree of stability or co-operation within a market, although even from a theoretical standpoint it is not altogether satisfactory.[8] There are markets and customs which should give some insight into the formalization of parameters for co-operation which have received little if any treatment by economic theory. They are phenomena such as tipping, customs of bargaining and specialized features of trading, be they in used-car wholesaling, commodity markets, money-lending in underdeveloped areas or credit-rating in overdeveloped areas which reflect the relationship of economic man to the market and his society.

There are good reasons to suspect that in almost all oligopolistic markets, any reasonable economic theory will give rise to the possibility of *multiple equilibria*.[9] In a specific instance involving only two or three individuals in a relatively institution-free situation, psychological variables may be successfully introduced to remove the indeterminacy; however, in large markets shared primarily by large institutions a socio-economic measure of degree of co-operation is needed. The size and make-up of the "police-force" will tend to vary with the degree of communality or co-operation perceived of by the individuals.

2. *Problems of Measurement and Aggregation*

Returning to narrower methodological problems, the economist faced with reading and deciphering a general statement of corporate goals; or even advising a corporation in light of them; or helping to write them; may find it useful to apply five criteria to the examination of a statement of goals. They are: (1) Logical consistency, (2) Completeness, (3) Generality, (4) Measurability, and (5) Aggregation.

Statements which call for the simultaneous maximization of two or more objectives which are functionally interrelated are usually logically inconsistent. Thus it may be impossible to maximize profits

8. M. Shubik, *Strategy and Market Structure* (New York: Wiley, 1959), Chap. 12.
9. *Ibid.*, p. 226.

together with obtaining the largest share of the market. Often, however, these two goals may be fulfilled in concert, in which case the directive to maximize market share may be more specific for members of an organization to follow than the directive to maximize profits. An old favorite among statements of simultaneous optimization is the intent "to obtain the greatest output at least cost." This is nonsense inasmuch as it can be given no operational meaning. It is meaningful to ask for the greatest output for a fixed cost; or the least cost for a fixed output; but the greatest output at least cost is operationally undefined.

Corporate goals which call for: "fair returns to stockholders and employees, technological superiority over all competitors and a high rate of return to sales," are incomplete and general to the point of vagueness for the purposes of the economist. They must be regarded as general directives which serve to delegate responsibility. The individuals who have to act in accordance with the statements of general intent have the additional burden of interpreting these statements. Thus from the viewpoint of operations they are incomplete. Additional restrictions must be read in before action can be taken. For example, a general statement calling for "ethical conduct consistent with the competitive practices of U. S. enterprise" is sufficiently undefined that there are many paths to the criminal courts to be found by an individual in an oligopolistic market trying to give it operational meaning.

Completeness, generality and aggregation are somewhat different aspects of the hierarchy of goals within an adaptive organism such as a corporation. Armies and corporations tend to approach the problems of specification in the same way. There is an over-all statement of "mission" or corporate goals; then there may be more specialized statements of sub-missions or directives for special segments of the corporation, and so on down the chain to individual job descriptions.

The question of completeness is best illustrated in terms of a computer program example. The program or the statement of policy is complete if for every conceivable circumstance it will provide a course of action. In other words, in some sufficient sense it will instruct the machine how to handle any situation. A directive may be very general and yet complete if it contains a blanket statement concerning any situation which falls into a residual category such as "when in doubt as to what to do, ask your manager." The goals in this instance will provide a rule for behavior under all circumstances if they are consistent and if the president or chief officer to whom all unresolved problems will come has a method for dealing with them that is implicitly covered in the statement of the goals.

The test for completeness is fundamental to all models in the behavioral sciences. Many of the difficulties and much of the confusion in writings on the theory of oligopoly have been caused by inconsistent or incomplete models; in other words, by models which purport to cover a class of events, but which either produce logically inconsistent outcomes for some of them, or which are unable to provide a specification of the outcome for certain events.

Generality and aggregation are different aspects of the same phenomenon. A corporation, its goals and its environment are all multivariate systems. Statements concerning goals are, of necessity, general or aggregative statements concerning multivariate systems. It has been fashionable to talk about "the aggregation problem" in certain branches of economics. There is no "the aggregation problem." It is the general problem of model building and abstraction which underlies all scientific endeavor.

There is a possible distinction between a general statement and an aggregative statement. The latter usually refers to a simplification of a complex, but nevertheless known or completely knowable situation. The former may include instructions as to how to handle the unknown; it has learning and adaptation built into it. The vague and general statements of companies contain elements of aggregation to avoid a detailed specification of known objectives, but much of the generality is an implicit recognition that they are fundamentally *searching* organisms who neither fully know their environment nor totally know what they are after.

3. *Some Techniques for the Economic Formulation of Goals*

In this section a few of the formal problems involved in an attempt to specify goals are sketched. These are primarily aimed at the "second level" of organization, in other words at the managers, engineers or economists faced with making the broad utterances operative in a narrow economic sense.

(a) *Multiple Objectives*

Let there be n items which enter into the goals of the firm. Leaving out the difficulties of measurement we assume, for the moment, that the intensity of fulfillment of the ith goal is x_i where

$i = 1,2,3, \ldots, n$. Let there also be m means by which the goals are achieved. The intensity of application of the jth mean is denoted by y_j, $j = 1,2,3, \ldots, m$.

We may write:
$$x_i = \Phi_i(y_1, y_2, \ldots, y_m) \qquad i = 1, 2, \ldots, n,$$
where not all the y_j necessarily appear in every Φ_i. We also must specify the boundary conditions which apply to the y_j. At its simplest the form may be:
$$A_j \leqslant y_j \leqslant B_j.$$
The bounds may be limits on production rates, labor supply and so forth.

If we wish to account for all n items directly in the goals of the firm we can write, generally as the aim of the firm:
$$\max_{y_1} \max_{y_2} \ldots, \max_{y_m} U(\Phi_1(y_1, y_2, \ldots, y_m), \Phi_2, \Phi_3, \ldots, \Phi_n)$$
subject to:
$$A_j \leqslant y_j \leqslant B_j,$$
where U may be some complicated function of the x_i. It is the general utility function for the firm.

(b) *Functionally Related Incompatible Goals*

Often a statement may be made that "we wish to maximize profits and market share." These items may be negatively correlated through the effect of the independent variables. In this case the production rate and the cost of the sales effort necessary to increase market share may decrease profits. This is so basic and evident that a formal demonstration will be omitted.

(c) *Measurable Goals*

The multiple goals of a firm may be specified only by an ordering relationship, equivalent to the indifference relations encountered in the theory of consumer behavior. Thus if profits and market share are the multiple goals, we can say that one state is preferred to another but we cannot specify by how much. This gives us a well-defined method for determining policy providing *both* the "transformation function," which delineates the relationship between market share and profits for a given level of resources of the firm, and the utility or value function are known. Figure I illustrates this situation.

CORPORATE OPTIMIZATION

T_1T_2 is the transformation function which indicates the maximum profit that can be made for a specific market share. The point M is the solution point which specifies the most preferable profit, market share combination which is attainable. The dotted line on which T_1

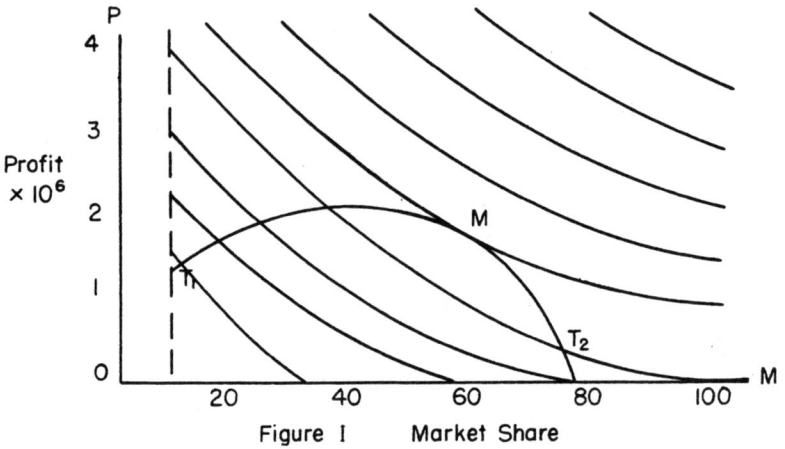

FIGURE I

lies is drawn to indicate that it is not meaningful to define the preference relations — when market share is 0, and it is not particularly useful or relevant to define the relationship for very small market shares.

(d) *Aggregated Goals*

In some situations the correlation between the factors which control the values of two of the stated goals of the firm may be sufficiently close to 1 that maximization of the value of one goal maximizes the other to a good enough approximation. An example of this would be provided if, as has already been noted, growth and profitability or (in a market of fixed size, with a fixed price and constant average costs of production and marketing) profitability and market share were both stated as goals.

(e) *Goals and Boundary Conditions*

A firm may state that it wishes to "maximize profits, maintain growth and treat employees and stockholders fairly." This statement contains no evaluation of the worth of fulfillment of the different aims and does not indicate the interrelationship that may exist between them. If, as is invariably the case, an over-all valuation or utility function for the many features of corporate aims does not exist we have to devise methods to give operational meaning to them. In the example above we can select one feature and assume that the firm wishes to maximize it, subject to *boundary conditions* which require that the other corporate aims meet certain specifications. We can represent the aim of the firm as:

(1) Maximizing profits subject to maintaining a specific growth program, dividend rate and employment policy which will satisfy stockholders and employees sufficiently that they do not act to change our environment.

Alternatively, if the dominant interest of the firm is to take care of its employees, its goal may be stated as:

(2) Maximizing disbursement to employees subject to maintaining a specified growth pattern and dividend and profit policy which will satisfy stockholders.

(f) Goals of the Adaptive Organism

The recent work in psychology,[1] artificial intelligence and problem solving[2] has dealt with models of organisms as feed-back systems with levels of aspiration varying according to their success. A formal expression of such a system is far more easily given by means of a flow diagram or a programming language such as Fortran[3] than it is in either English or mathematics. These methods are more precise than English and more flexible and "richer" in expression than mathematics. It is my belief that they are sufficiently important to be mentioned here, but an adequate development of the above would require a separate paper.

IV. Some Recent Approaches to the Study of Economic Intent, Structure, and Behavior

In the recent past developments have taken place in several diverse yet important aspects of the study of the firm. They have been in the methods and techniques of observation and data processing; experimentation and methodology. Two topics which might be termed "experimental economics" and "mathematical institutional

1. S. Siegel, "Level of Aspiration and Decision-Making," *Psychological Review*, Vol. 64 (1957).
2. Minsky, *op. cit.*
3. *Fortran*, IBM Reference Manual.

economics" utilizing respectively the techniques of gaming and of simulation, in my opinion give promise of becoming extremely important in the investigation of the behavior of the firm, its structure and the structure of its environment. These combined with advances in methodology should serve to provide a basis for further developments in theory.

1. *Experimental Economics: Gaming*

Until recently it has been all but impossible to validate conjectures concerning oligopolistic intent and behavior *in vivo*. Formal theories concerning oligopolistic behavior have been in existence for well over a hundred years, yet it has only been in the last three or four years that strict formal experimental attempts have been made to see whether at least the conjectures concerning intent and behavior hold up in the highly simplified and overly-antiseptic atmosphere of a laboratory.

Two major types of gaming are currently being employed. The first is commonly referred to as business gaming. A complex model of several firms interlinked in a synthetic or simulated oligopolistic market is constructed. Usually three to five or six teams consisting of anywhere from two or three to seven or eight persons per team participate in the game. Each team simulates the activities of the management of one of the firms. In the course of anywhere between twenty minutes to several hours it is possible to simulate the activity of a fiscal quarter. In the course of a day several years of simulated history can be experienced. The prime purpose of these games to date has been training and teaching.[4] Their extensive use in business schools and industry has already indicated that they are very useful as aids to the understanding of organizational and market structures. Furthermore they provide illustrations of oligopolistic problems of intent and behavior.

Although these games are still considerable simplifications of actual industrial situations, they are very complex and "environment rich" as compared with strict laboratory games. Most of them require a high speed digital computer to perform the calculations required after every move. In at least one instance it has been possible to examine the structure of a relatively complex four team business game and to calculate predicted behavior given certain

---

4. For a listing of much of the literature on gaming see: M. Shubik, "Bibliography on Simulation, Gaming and Artificial Intelligence and Allied Topics," *Journal of American Statistical Society*, Vol. 152 (Dec. 1960).

hypotheses concerning the intent of the competitors. An investigation of the structure of the business game constructed by George Feeney[5] and the behavior patterns exhibited by many teams playing this game indicates that the average payoff received by the players if they exhibit a nonco-operative behavior (in the sense of Cournot or Chamberlin or more generally in the sense of Nash[6]) serves as a good predictor of the upper bound of the average payoffs achieved by teams playing the game.

The second type of gaming is strictly controlled experimental economics. M. Flood and others[7] instigated and carried out experimental investigations of general n-person games. Sociologists,[8] social-psychologists and psychologists have all used experimental gaming techniques. In 1948 E. H. Chamberlin reported on an oligopoly experiment.[9] Since then the work of Jeremy Stone, Austin Hoggatt, David Stern, Seigal and Fouraker and the current experiments of Seigal, Fouraker and Shubik have been aimed at formal controlled experiments to test hypotheses concerning economic behavior.[1] Stone and Seigal and Fouraker were primarily concerned with bargaining. The remaining investigations have been aimed at duopolistic or oligopolistic situations.

In the experiments to date, actions and payoffs associated with a short term nonco-operative behavior appear to be fairly good predictors, at least in symmetric games, although Hoggatt's results are for nonsymmetric games. This also appears to be the case in the complex business game constructed by Feeney. There is one important difference between the roles of the players in the "environment

5. G. J. Feeney, "Simulating Marketing Strategy Problems" (circulated) (New York: Jan. 1959).
6. J. F. Nash, Jr., "Non-Cooperative Games," *Annals of Mathematics*, LIV (Sept. 1951), 286–95.
7. M. Flood, "Game Learning Theory and Some Decision-Making Experiments," *Decision Processes*, R. M. Thrall, C. H. Coombs and R. L. Davis (eds.) (New York: Wiley, 1954).
8. O. K. Moore and M. L. Berkowitz, "Game Theory and Social Interaction," ONR Technical Report, Contract No. SAR/NONR-609. (New Haven, Nov. 1956.)
9. E. H. Chamberlin, "An Experimental Imperfect Market," *Journal of Political Economy*, LVI (Apr. 1948), 95–108.
1. J. S. Stone, "An Experiment in Bargaining Games," *Econometrica*, Vol. 26 (April 1958). A. C. Hoggatt, "An Experimental Business Game," *Behavioral Science*, Vol. 4 (July 1959). D. H. Stern, "An Experimental Test of the Dynamic Cournot Hypothesis" (Processed) (Princeton: Sept. 1958). S. Siegel, and L. Fouraker, *Bargaining and Group Decision-Making: Experiments in Bilateral Monopoly* (New York: Macmillan, 1960).

poor" or strictly experimental games and in the business game. In the experimental games competition was observed under conditions of *incomplete information* (or in game theoretic terminology, lack of complete knowledge of the rules of the game). Player 1 was not aware of the payoff to Player 2 and vice-versa. In the business game the players knew that all firms had the same payoff structure. In this instance, however, the amount of information processing required to take advantage of this knowledge was such as to swamp the channel capacities of the teams. It appears that this overburden served to make the information conditions in this game equivalent to the others. Supplying one team with a computer to do the data processing for it, changes the characteristics of this game.

In the simple experimental games it appears that complete information by both players changes the behavior and the outcome considerably. The tendency (which should be of surprise to no one) is to move above the distribution obtainable at the nonco-operative equilibrium state towards (but not necessarily at) a point on the Pareto Optimal Surface.

As is well known, under reasonable conditions it can be shown that as the number of participants in an oligopolistic market becomes large and the relative size of the individual small, the nonco-operative equilibrium point (or points) tend to the equilibrium of pure competition. The experimental work at this time appears to indicate that competitive or nonco-operative theories are probably reasonably good *provided that* they are interpreted as theories postulating *low levels of information* of the players. J. M. Clark apparently appreciated this fact,[2] although sight was lost of it in the philosophical discourses which ensued on the topic of "perfect information."

2. *Mathematical Institutional Economics: Simulation*

Even if intent were known, without a sufficient understanding of environment it would not be possible to predict behavior. Microeconomics has been severely limited by difficulties in the description of environment. In general, mathematical economic models have been too "simple" and restricted in their descriptions (the aggregations have been too drastic owing to problems of analysis or to the inherent lack of richness of mathematics as a language). On the other hand, descriptive works and institutional studies have tended

2. J. M. Clark, *Studies in the Economics of Overhead Costs* (University of Chicago Press, 1923), p. 417.

to be too discursive for analytical manipulation. The new techniques of high speed digital computer simulation provide a means for large-scale data-organization and detailed model building which preserve at least some of an institutional "richness" and at the same time permit manipulation and experimentation. It is too early at this time to do very much more than to call attention to the work done so far and to express the belief that this approach will be of increasing importance. The symposium on simulation which has recently appeared in the *American Economic Review* provides a sketch of the applications of simulation relevant to economics.[3]

3. Models of the Firm: A Suggestion

A few of the techniques for handling multiple goals dependent upon interrelated means have been noted. At the level of the economy, Tinbergen's masterful essay, *On the Theory of Economic Policy*, covers most of them more generally.[4] However more specifically at the level of the firm, the desire for survival (and the decision to come into existence) are heavily economic decisions and merit explicit introduction into models of the firm.

In an economy with frictions, indivisibilities, uncertainty, costs of entry, losses upon exit and various institutional forms for business firms the decision to come into being or to continue to exist is often a vitally important decision. The objective function of the group in *control* of the firm must be defined and costs of entry and exit calculated if specific consideration is to be made of birth and death processes. One simple model of the firm with these additional features is suggested below.

The fortunes of the firm are divided into two distinct accounts. The first is called the *Corporate Account*, and the second the *Withdrawal* or *Stockholder Account*. The distinction between the two accounts is clearest in a corporation where the stockholders have limited liability. A forced liquidation may kill the firm and wipe out the stockholder's equity in the corporate account; however the previous income stream into the withdrawal account still remains. The firm may die eventually, but during its lifetime it may yield enough to merit owning.

---

3. "Simulation: A Symposium," *American Economic Review*, L (Dec. 1960), 893–932.
4. J. Tinbergen, *On the Theory of Economic Policy* (Amsterdam: North Holland Press, 1952).

Let the corporate account at the start of period t be C_t. The net revenue of the firm during period t is denoted by R_t. The amount paid into the withdrawal account is w_t. This gives the condition:

$$C_{t+1} = C_t + R_t - w_t.$$

Bankruptcy and liquidation conditions can be specified immediately. If $C_t \leqslant A$ (where $C_{t-n} > A$ for $n = 1,2,3,\ldots$) then the firm is bankrupt and is forced to liquidate. The liquidation value may be some number $B = B(C_t) \leqslant A$. The difference between C_t and B is a measure of the liquidation loss caused by market imperfections.

Money paid into the withdrawal account represents immediate income to ownership. In order to value the income stream a discount factor is introduced. $p = \dfrac{1}{1+q}$ where q is the rate of interest.

Three different corporate goals are now defined in which the attitudes of risk towards the possible death of the firm and the possibility of paying out income are illustrated.

(a) *The Individual Entrepreneur*

The objective function of an individual entrepreneur whose goal is to maximize the value of the discounted income stream from his incorporated firm will be:

$$\text{Maximize} \left\{ \sum_{0=t}^{r} p^t w_t + p^r B \right\}$$

where r is the period at which the firm is finally bankrupted.

The examples in III.3(e) on p. 367 can now be formalized.

(b) *The Unincorporated Enterprise and the Cautious Corporation*

In some cases a small businessman may place a high value on his "independence." He will be willing to run his own enterprise as long as he is able to pay himself a sum sufficient to compensate for the wages he could earn in an alternative employment. As there will be a large disutility to ruin (probably greater than the mere monetary loss) the goal of the businessman can be described as:

$$\text{Maximize} \prod_{t=1}^{r} \left\{ 1 - P_t(c_t \leqslant A) \right\}$$

subject to $w_t \geqslant a_t$ for $t = 1,2,\ldots,r$.

$P_t(c_t \leqslant A)$ is the probability that the firm will be bankrupted in the tth period. a_t is the minimum withdrawal required by the businessman in the tth period. The above expression indicates that the probability of survival for r periods is maximized subject to certain withdrawals being made each period.

This policy could also be adopted by a corporation whose managers wish to keep their survival chances as high as possible subject to having to declare dividends sufficient to satisfy the stockholders.

(c) *Dividends and Risk*

A management-run corporation may express its policy by attempting to maximize the value of the discounted income stream paid to stockholders over some time period subject to the boundary condition that the risk of bankruptcy is kept below a given level.

$$\text{Maximize} \sum_{t=0}^{r} p^t w_t$$

$$\text{subject to:} \prod_{t=1}^{r} \left\{ 1 - P_t(c_t \leqslant A) \right\} \geqslant k.$$

The boundary condition indicates that the probability of survival for r periods will be at least k.

A simple modification of the above model suffices to distinguish between firms in existence and "firms-in-being."[5] It is evident that the models of maximization by the firm suggested here are related in some ways to the Gambler's Ruin problems of statistics. This relation is explored elsewhere in an investigation of "games of economic survival."[6] A necessary but only preliminary attempt is made to consider oligopolistic markets in terms of many person games of economic survival.[7]

V. Power, Intent and Policy

The need for an understanding of intent of the firm beyond weak conditions of maximization varies with the *power* of the firm in its market places and society. The central problem of political science and of administration becomes the central problem of the economics of the oligopolistic firm and the industries within which it operates. That is the meaning and measurement of power. The less the firm is able to influence its environment, the less needs to be known about the motivation of the management of the firm for most purposes of policy.

5. Shubik, *Strategy and Market Structure, op. cit.*, p. 259.
6. M. Shubik and G. L. Thompson, "Games of Economic Survival," *Naval Research Logistics Quarterly*, Vol. 6 (April 1959).
7. Shubik, *Strategy and Market Structure, op. cit.*, Chaps. 10 and 11.

Current industrial organization is such, and future industrial organization will probably be such that under any reasonable criterion of power, powerful firms will exist. The multiple goals of the powerful firm are of importance both to it and to society. Its very possession of power prevents it from ignoring them. Although few economists or political scientists would deny that General Motors is a powerful firm and Russia is a powerful country, the measurement of power (or even its definition) in general is hardly adequate. This is easily seen in the difficulties encountered in attempts to measure the influence of multi-market firms or financial entities.

It is possible that the needs of society are best served by inherently powerful firms provided that their behavior falls within certain bounds. If this is so then the control of behavior may precede the control of power. Concomitant with this is the burden to be borne by the firms that impeccable intentions acted upon by the powerful may result in undesirable behavior.

YALE UNIVERSITY

INCENTIVES, DECENTRALIZED CONTROL, THE ASSIGNMENT OF JOINT COSTS AND INTERNAL PRICING*

MARTIN SHUBIK

Cowles Foundation for Research in Economics, Yale University

1. The Problems of Control and Cost Accounting

Cost accounting in the modern corporate economy is recognized as a tool with many purposes. It must serve, to a greater or lesser extent, financial and legal requirements, the technical needs of branch managers of industrial plants, and at the same time is used by top management as a basis for policy decisions. The growing recognition of the importance of cost accounting control[1,2] has highlighted several problems which belong jointly to the fields of concern of managers, engineers, cost accountants, and economists.

Information which is portrayed and set out with one purpose in mind may be worse than useless when used for another purpose. The detailed reports and statistics which are of the utmost necessity to a branch manager concerned with technical problems may confuse and mislead a controller interested in investment policy and other areas where finance and tax structure come to the fore and technical, physical details are of secondary importance. The boiled down aggregate information used by a board of directors may seem to be a travesty of the facts to an operating engineer, but in a world where information is expensive, time dear and decisions cannot be postponed, abbreviations and condensations must be made.

A goal of good management should be to design a reward system for those who take risks in making decisions in such a manner that the rewards to the individual correlate positively with the worth of the decision to the organization (taking into account the attitude of the top management to variance as well as to expected gain). In many organizations cost accounting supplies much of the information used for control at several levels. In this paper we examine some of the control problems that arise if joint costs are assigned by various cost accounting and some internal pricing conventions.

A method for assignment of costs which has desirable incentive and organization properties is then discussed. This method is based upon a result in the theory of games obtained by L. S. Shapley.[3] A self-contained exposition of the features of game theory required for this paper is given in section 2 following.[4]

---

* Received May 1961.

[1] Lang, T., W. B. McFarland, and M. Schiff, *Cost Accounting*, NewYork: Ronald Press, 1953, Chapters 3, 27.

[2] Brummet,R. L., *Overhead Costing*, Ann Arbor, Michigan: University of Michigan Press, 1957.

[3] Shapley, L. S., "The Value of an N-Person Game," in Kuhn, H. W. and Tucker, A. W..

2. Basic Concepts Relevant to the Study of the Assignment of Joint Costs

The theory of cooperative games, as developed by von Neumann and Morgenstern depends upon a measure of the interrelatedness and increase in joint rewards obtained by a group of individuals who are willing to act together, as compared to acting individually. The profitability of a corporation may be viewed as depending upon the sum of the joint rewards which can be obtained by the optimum coordination of all branches. This analogy will be specified even more closely in section 3. The *players* in a von Neumann and Morgenstern game may be regarded as the branches or departments of a corporation or the sections in a factory.

The measure of the complementarity in a game (i.e., the worth of joint coordinated action) is given by its *characteristic function*.[5] This function is a *super-additive set function*, and although its technical name may at first frighten the non-mathematician, the meaning of it is relatively easy to explain. It is called a set function as it is defined, i.e., it takes on values for a set of entities. In this case the entities consist of every possible combination of departments in a corporation. For example suppose that a corporation consisted of a central office and two departments; if each of these were regarded as an independent entity denoted by 1, 2, and 3 then the characteristic function would be defined for seven values. These are values for 1, 2, and 3 individually; the pairs (1, 2) (1, 3) and (2, 3) and the firm as a whole (1, 2, 3). For completeness we assign a value of zero to a coalition consisting of no one. This gives eight values to the characteristic function of a firm considered as three entities.

The characteristic function is called *super-additive* because the value of the amount obtained by any grouping of participants is always as much or more than they can obtain by individual action. For instance, a coat may be worth more than two halves of the same coat. The characteristic function provides a handy way in which complementarity can be described between different objects or groups.

Consider a firm with several branches, say different plants. They share the common overheads of the firm, and the actions of one branch may affect the direct profits of another (vertical integration might cause this, or there may even be competition in the market between differentiated products of the same corporation; for instance the different automobiles produced by General Motors). One way in which an index of the importance of any branch can be measured is by calculating the effect upon profits if it closed down, and the optimum alternative use were made of the resources it relinquished. In a similar way we can evaluate the importance of any set of branches to the corporation as a whole.

Eds., *Contributions to the Theory of Games*, Vol. II, Princeton: Princeton University Press, 1953, pp. 307–317.

[4] von Neumann, J., and O. Morgenstern, *Theory of Games and Economic Behavior*, Princeton: Princeton University Press, 3rd ed., 1953.

[5] Ibid., p. 238 ff.

Let $v([i, j])$ stand for the profit that branches i and j of a firm can make on the assumption that the remaining branches have closed down.[6] In general, $v(S)$, the characteristic function, describes the profit made by the set S, of departments or other separate components of the firm which are to be considered as acting in unison.

As a simple example consider a factory consisting of two departments, 1 and 2. The only cost that they share in common is a joint overhead for the factory. Furthermore suppose that if either department closes down, there is no alternative use that can be made of the excess plant facilities. Assume that the net receipts for Department 1, leaving out the joint cost assessment, are x, and for Department 2, are y. Let the joint cost be C. The set consisting of the two departments is denoted by $\{1, 2\}$. The value of the profits they can obtain together is:

$$v(\{1, 2\}) = x + y - C.$$

The amount that the firm obtains if the second department is closed is:
$v(\{1\}) = x - C$, with the first department closed it can obtain
$v(\{2\}) = y - C$.
We note that:

$$v(\{1\}) + v(\{2\}) = x + y - 2C,$$

hence:

$$v(\{1, 2\}) > v(\{1\}) + v(\{2\}).$$

Although formally the inequality above can be defined, in this example care must be taken in interpreting the meaning of $v(\{1\}) + v(\{2\})$. Both departments cannot simultaneously realize their own survivor's value. If instead of two departments, the example had been that of a husband and wife deciding to file a joint tax return or to both "go it alone" then the above sum would have a direct physical counterpart.

In order to avoid difficulties such as the one above, it is necessary to divide the firm into separate decision-making entities and to specify the powers of the various decision makers to close down a plant, to go out of business or to "secede from the union." This is discussed in more detail in section 3.

By utilizing the characteristic function, the von Neumann and Morgenstern theory of games leads to a concept of solution in which all players act in a manner to jointly maximize profits and then use their bargaining power as represented in the possible coalitions to arrive at an imputation of the proceeds. The method suggested for splitting the profits[7] is somewhat complicated and does not concern us here. No unique imputation is given, although certain bounds are placed upon the shares received by each individual.

---

[6] We are implicitly assuming that the strategy space of a top manager is limited to such a manner that he has the choice of closing down or producing optimally with those branches which do not close down. This assumption is discussed further in the text.

[7] von Neumann, J., and O. Morgenstern, *op. cit.*, p. 264.

The method for assigning a portion of joint proceeds to each player which has been advanced by Lloyd Shapley does provide for a unique division. Furthermore it will be shown that this method satisfies a certain set of properties which an accounting system should have if decentralized decisions are to be based upon the internal imputation of profits to semi-autonomous sections of an organization.

3. Incentives, Control and Cost Accounting

Broadly speaking it is often deemed desirable to be able to delegate as many decisions as possible to the branches of a firm. In many organizations of large size the exponential growth of messages and red tape cause diseconomies in centralized decision-making for those decisions which depend heavily upon on-the-spot knowledge. If decision-making power is to be delegated it is preferable to have an organization which is designed to encourage initiative. One way of doing this is to have the reward structure designed so that the selection of choices which are best for the individual decision-maker will always coincide with those which are best for the organization. For instance, a branch manager may be aware of a change which may have the effect of increasing corporate profits, but decreases the size of his own department and may even reduce the profits assigned to it by the accounting system. If his success and income are measured and determined by the accounting profits assigned to his department then it may not be in his interest to select the decision optimal for the firm.

Of course there are many sociological and psychological aspects to an incentive structure in a corporation, church, university or commissariat. Thus gold medals, memberships in golf clubs, prestige, pride in workmanship and so forth all play an important role.[8] Furthermore in even the most impersonal and mechanized systems single number measures of the performance of an individual are rarely used. For purposes of this paper, however, the sociological, psychological and psychiatric aspects of the individuals are taken as given. As bonuses and "incentive compensation" to executives in many corporations are based upon the profits imputed to their operations the economic and accounting problem may be of interest by itself.

There are many technical and conceptual difficulties to be faced in the accounting treatments of fixed costs, variable costs and joint costs. There is a wide variety of practice in accounting methods. Lee Brummet notes five for example:

(1) Complete absorption costing
(2) Expected or average activity standard costing
(3) Practical capacity standard costing
(4) Direct standard costing
(5) Prime standard costing.[9]

---

[8] Argyris, C., "The Individual and Organization; Some Problems of Mutual Adjustment," *Administrative Science Quarterly*, June 1957.

[9] Brummet, R. L., *op. cit.*, p. 32.

Joint costs have been assigned as a percentage related to the direct labor costs of each operation; charged as a rate per direct labor hour; a rate per unit of product; a percentage of direct material cost or a percentage of prime cost (direct labor + direct material cost).[10] No exegesis of accounting methods is to be presented here. Many important and vexatious accounting problems are ignored. However viewing one of the roles of accounting as helping: "to provide management with cost information necessary to business decisions and related policy",[11] it is observed that under several of the methods above it is possible that a department be assigned costs which make its "paper profits" negative even though it may be a vital and efficient part of the firm. It is also possible that an improvement in the efficiency of a department may damage its individual profit statement even though it increases the over-all profitability of a firm.

There should always be an incentive for a manager to implement an efficiency or report a new idea if it benefits the firm as a whole no matter what changes it may cause to take place in his own operation. Under some methods of cost assignment, for example, if the decision to discontinue a product line rests with individual department heads it is possible that individual rational action based upon the cost assignment may add up to corporate idiocy. A simple example of this is given in section 5.

Ideally the assignment of joint costs to individual products or departments is not necessary from a purely economic point of view if the decision to maximize for the company as a whole is made in a single office. This is usually impossible in practice, hence a cost accounting and internal pricing scheme can serve as an administrative device in the design of a viable and economic decision-making system.

4. Decentralization, Decisions and Information

The concept of decentralization deals with the possibility of delegating decision making to more than one location in an organization. An optimally decentralized system will have the property that the net effect of all individual actions will be more favorable to the firm than the actions selected by any other array of decision centers. This must take into account costs of messages and organization and the possibilities of committing errors when decisions which appear to be locally optimal are not of benefit to the organization as a whole.

The limiting case for the possibilities for decentralization comes when all decision centers or units are independent. This is merely another way of saying that an action by anyone or any group has no effect on any other unit or combination of units. This is true for small numbers in a purely competitive market which may be viewed as a decentralized organization. It is not completely true as can be seen by problems in agriculture and other "chronically competitive markets." If the characteristic function of an organization is *flat*, i.e., if

---

[10] Lang, T., W. B. McFarland, and M. Schiff, *op. cit.*, Chapter 13.

[11] "Report of the Committee on Cost Concepts and Standards," *The Accounting Review*, April 1952, p. 175.

the sum of the amounts which can be obtained by any two coalitions acting together is precisely the same as the amounts that they can obtain by acting independently, then obviously there is no need whatsoever to coordinate their actions as each unit is an autarky and neither gains from nor adds to any joint venture sufficiently to merit other than individual action.

Interesting and important cases for decentralization arise when the joint welfare is influenced by individual action, or the action of coalitions. The degree of influence is reflected in the characteristic function; which, if its values are appropriately defined display both the technological and decision structure of the firm.

In a game, a player is characterized as an individual decision-maker with some degree of free choice. By analogy we may consider a general manager in a corporation as a player in a position to choose among a set of actions pertaining to his department or part of the corporation. He is a "dummy player" if, in fact, his actions are irrelevant to the functioning of the organization. This happens when some other individual is in a position to over-rule and change any of his decisions. This is so in a completely centralized organization; or is apparently so until we consider the information conditions. In theory as well as in practice the selection of what type of message to send up to the decision-maker is a decision in itself and gives the individual a degree of power which varies as the difference between his knowledge and the knowledge of his superior, and the importance of this to the decision.

Effective complete centralization requires either that the central office is completely informed and merely uses the remainder of the organization as an instrument for execution and not for information gathering; or that all individuals are assumed to be unbiased gatherers of data. In other words the central office, if it is not totally informed, must assume that individuals within the organization will not be motivated to distort the information they send or to take actions based on goals which do not correlate with those of the central office. This calls for a concept of an organization as a *team*[12] rather than a series of arrangements between individuals with possibly differing goals.[13] The former can be regarded as a limiting case of the latter. Our interests here are concerned with simple problems arising from the latter concept.

The specification of a characteristic function as a model of the potentials of sectors of a firm contains within it both considerations of the decision structure of the firm and the potential worth of the resources. This can be seen when an attempt is made to assign a worth to what can be achieved by a subset of departments. In order to do this several questions must be posed concerning the

---

[12] Marschak, J., R. Radner, "Economic Theory of Teams," Cowles Foundation Discussion Paper No. 59, 1958.

[13] An organization may be regarded as a game with restraints on the players and the sending of messages as one of the major weapons of control. A concentration camp fits this model better than does a team.

location of responsibility for key decisions. A partial list of relevant decisions is given below:

1. The decision on major investment
2. The liquidation of a department
3. The abolition of a product line
4. The introduction of a new product
5. The introduction of other innovations (such as a change in distribution)
6. The merger of several departments
7. The splitting of a department into several independent entities
8. Pricing, purchase of raw materials and sales of final product

If, for example, the managers of each department had decision responsibility for all of the above (which might be the case if the organization being described were a weak cartel rather than a corporation) then the meaning of the value attached to any subset would be the value that a subgroup of participants in the cartel agreement could obtain by acting together by themselves outside of the cartel agreement.

If only some of the decisions are to be delegated while others remain under the control of an executive or central office, then it may be desirable to introduce the office as a player. Returning to and reworking the example of a factory with two departments given in section II; it can be regarded as an organization with three participants. Suppose that there is a president and executive office which has delegated decisions 2, 3, 4, 5 and 8 above to the two managers of the departments, but maintains its decision-making power on the others. Furthermore suppose that the managers are instructed to maximize the profit assigned to their departments under the accounting system used by the firm. We assume that the central office has dictated a method of accounting which calls for all costs and revenues to be imputed. As the managers are in a position to liquidate their departments unilaterally and to discontinue product lines, they can guarantee themselves individually a profit of not less than zero. Let the central office be player 1 and the departments be 2 and 3. The characteristic function for this firm with structure shown in Figure 1 is:

$$v(\{\Theta\}) = 0$$

$$v(\{1\}) = 0 \qquad v(\{2\}) = 0 \qquad v(\{3\}) = 0$$

$$v(\{1, 2\}) = x - c \qquad v(\{1, 3\}) = y - c \qquad v(\{2, 3\}) = 0$$

$$v(\{1, 2, 3\}) = x + y - c$$

We assume that the central office can obtain a value of zero by liquidating and employing the proceeds elsewhere, hence $v(\{1\}) = 0$.

A good decentralized system should have the property that each decision center will make a decision which is optimal for the whole with a minimum of cost for coordination and information and message costs. In the example above,

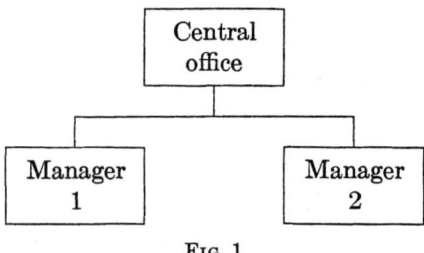

Fig. 1

the role of the executive or central office is to assign joint costs. It must do so in a manner that will guarantee that if a department should exist for the good of the firm as a whole, then it will not get an assessment that makes its net revenue negative. For example, suppose $c = 10$, $x = 4$ and $y = 7$, then an assignment of costs of 5 each to the two operating departments will motivate player 2 to shut down even though his operation is of value to the business, hence this is not a good assignment.

If $c = 10$ and $x = y = 4$, then the assignment of costs should be such that the firm should be motivated to liquidate (or otherwise change drastically).

Looking at this firm as an administrative system, the only information needed by the central office from the departments is their individual net revenues, and the only information that it will send them is the size of their assessments (it is presumed that the executive office has some other economic, financial or service function which it renders to the firm as a whole).

For another example we consider a firm without overheads or other joint costs, but with two departments producing the same item at costs $C_1(x)$ and $C_2(y)$ which is then sold by the central office which acts as a marketing agency for the firm as a whole. Here the problem is to assign shadow prices to be paid by the central office to the departments and to impute the remaining profit. We assume that the only decisions which are decentralized are production levels and individual technology. The characteristic function will be:

$$v(\{\Theta\}) = 0$$
$$v(\{1\}) = 0, \quad v(\{2\}) = 0, \quad v(\{3\}) = 0$$
$$v(\{1, 2\}) = \operatorname*{Max}_{x} [x\varphi(x) - C_1(x)]$$
$$v(\{1, 3\}) = \operatorname*{Max}_{y} [y\varphi(y) - C_2(y)]$$
$$v(\{2, 3\}) = 0$$
$$v(\{1, 2, 3\}) = \operatorname*{Max}_{x} \operatorname*{Max}_{y} [(x + y)\varphi(x + y) - C_1(x) - C_2(y)]$$

where $p = \varphi(q)$ is the final demand schedule for the product. It has been shown by Dantzig and Wolfe that for the appropriate limitations on technology, a firm decentralized in the manner above need only send messages concerning shadow prices and outputs in order to reach a joint optimum. This is the "decomposition" principle for solving a linear program, and although it is primarily a computational device it may be viewed as an administrative arrangement.[14]

The two examples given above are treated in detail in section 6. We turn in section 5 to the development of the general method for imputing costs and assigning prices to satisfy incentive criteria.

Although the topic is not pursued further at this time, it should be noted that the problem to which this paper is addressed is closely related to the more general problems of organization and stability in an economy.[15] This in turn is related to the role and effectiveness of group action in administrative systems.[16]

5. An Incentive System for Decentralized Control

A corporation is characterized as a set of n decision centers. The characteristic function of a corporation reflects not only the technological features of complementarities between products, common overheads, joint costs and other technological interrelationships, but also the decision structure of the various centers.

We limit ourselves to considering only firms which should not completely liquidate. A firm should not liquidate if there is at least one subset of decision centers S which can earn as much or more than the income obtained from investing the proceeds of liquidation.

Let the set of decisions of the ith center be denoted by D_i. An individual decision is $d_i \in D_i$. In general the characteristic function is calculated for all values as follows. We define $\psi_{S_j}(d_i, d_j, \cdots, d_m)$ a function of s variables which represents the payoff to the firm as a whole on the assumption that a particular set of s centers denoted by S_j are active and the remainder have been dissolved.

$$v(S_j) = \max_{d_i \in D_i}, \max_{d_j \in D_j}, \cdots, \max_{d_m \in D_m} \psi_{S_j}(d_i, d_j, \cdots, d_m)$$

In particular, $v(I) = \max_{d_1 \in D_1}, \max_{d_2 \in D_2}, \cdots, \max_{d_n \in D_n} \psi_1(d_1, d_2, \cdots, d_n)$.

For many large corporations with diversified businesses some of the structure of the functions ψ_{S_j} can be specified simply. For instance, if a firm sells two products which share no joint variable costs, incur no joint economies in marketing and have negligible influence on each others' markets, an example is diesel locomotives and Christmas tree lights, the function $v(S_j)$ where S_j consists of department 1 making locomotives and department 2 making lights, can be written as:

$$v(S_j) = \max_{d_1 \in D_1} \max_{d_2 \in D_2} [\psi_1(d_1) + \psi_2(d_2) - C].$$

The only connection between the departments is a joint fixed cost.

---

[14] Dantzig, G. B., and P. Wolfe, "Decomposition Principle for Linear Programs," *Operations Research*, Vol. 8, No. 1, January 1960, pp. 101–111.

[15] Arrow, Kenneth J. and Gerard Debreu, "Existence of an Equilibrium for a Competitive Economy," *Econometrica*, Vol. 22, July 1954, pp. 265–290; Arrow, Kenneth J. and L. Hurwicz, "On the Stability of Competitive Equilibrium," *Econometrica*, Vol. 26, October 1958, pp. 522–552.

[16] Shubik, M., "Extended Edgeworth Bargaining Games and Competitive Equilibrium," Cowles Foundation Discussion Paper No. 107, 1961.

The more obvious forms of interconnection also serve to enable us to specify the calculation for the characteristic functions without great difficulty. These include vertical integration, aspects of horizontal integration, joint variable costs, such as transportation or the use of a commonly owned computing machine. Interconnectivity in the market is also reflected in the characteristic function. For example, many consumer durables may compete with each other in the market.

The ascribing of a value to the one decision unit acting by itself, $v(\{i\})$ depends upon whether the decision system allows the manager to close his plant or production.

We assume that it is desirable not to assign negative profits to any decision center whose existence is of value to the firm as a whole. This can be achieved by using a characteristic function where for any i

$$v(\{i\}) = \max[0, \max \psi_{1_i}(d_i)].$$

This is tantamount to allowing a manager to close production or dissolve his unit if he is assigned a negative profit. If the system only assigns him a negative profit when in fact, the liquidation of his activity is for the good of the firm, this has a desirable property for a well decentralized system.

We present the five properties or *axioms* for a good assignment of the proceeds of a joint profit (and hence, implicitly the imputation of joint costs, internal prices and revenues) to different decision centers of a firm. A verbal statement of each axiom is given first, this is followed with the precise mathematical formulation.

Axiom 1: The profit assigned to a given center depends at most upon the various revenues which can be earned by all alternative uses of all centers or combinations of centers.

Symbolically, if we use the notation ϕ_i to stand for the profit assigned to the ith center, we can write:

$$\phi_i = F(v(\Theta), \cdots, v(S), \cdots, v(I))$$

where $v(S)$ is the characteristic function which portrays all complementarities inherent in the optimal use of any combination of the facilities of the firm.

Axiom 2: The profit assigned to a center depends symmetrically upon all centers in a firm. In other words, if two firms are identical except that their departments or decision centers are called by different names, then the accounting system will assign the same profit to the centers which are physically the same despite the difference in names.

Symbolically if we let Γ stand for the game characterized by $v(S)$ and Γ' the game such that

$$v'(S) = v(S^*),$$

where S^* is like S but with i replacing j and j replacing i, then

$$\phi'_i = \phi_j \qquad \phi'_j = \phi_i$$

Axiom 3: The accounting system imputes all the profits earned by the firm.

$$\sum_{i \in I} \phi_i = v(I)$$ where I is the set of all decision centers.

Axiom 4: A homogeneous expansion of fixed costs, variable costs and profits will result in a homogeneous rise in the accounting profits imputed to all processes.

$$\text{If} \quad \Gamma' = \beta\Gamma, \quad \beta > 0, \quad \text{then} \quad \phi' = \beta\phi$$

For example, if the currency unit were changed so that one new franc is worth one hundred old, the new profit assignment ϕ' if measured in francs is such that $\phi' = \frac{1}{100}\phi$.

The fifth axiom envisions a strange situation which might arise if two independent firms jointly share a facility. For instance suppose that each rents a certain plant and each have managers to run it, one for the day shift and the other for a night shift! Furthermore, let us imagine that neither firm has any use for more than one shift from the facility they both rent. If we were confronted with the strange arrangement then:

Axiom 5: If two independent firms are considered as a unit, the profit imputed to the operations utilizing this facility will be the sum of the profits that each firm imputes to its own operation which utilizes the facility separately. The profits imputed to any department or decision center which is not jointly used by each of the firms will not be changed by the consideration of both firms as a unit.

If Γ consists of the game obtained by considering the games Γ' and Γ'' together, then:

$$\phi_i = \begin{cases} \phi'_i + \phi''_i & \text{for } i \in I' \cap I'' \\ \phi'_i & \text{for } i \in I - I'' \\ \phi''_i & \text{for } i \in I - I' \end{cases}$$

The proof that these five axioms lead to a unique formula based on the characteristic function is given by L. S. Shapley.[17] We will not be concerned with this mathematical problem here, but rather with the interpretation of the result. The formula is:

(1) $$\phi_i = \frac{1}{n!} \sum_i (s-1)!\,(n-s)!\,[v(S) - v(S - \{i\})].$$

It assigns a share of the joint profits to each center (and hence automatically imputes joint costs). The rationale behind the formula can be seen in terms of addition to productivity. The addition to profits caused by a center acting jointly under all possible conditions with the other centers (i.e., every possible arrange-

---

[17] Shapley, L. S., *op. cit.* The axioms used in this paper are an earlier formulation by Shapley in a RAND paper RM-670 which are equivalent to those used in his latter publication.

ment with some shut down and others operating) is evaluated and an average is taken.

The economist will recognize that this amounts to assigning a profit to each center according to its expected marginal or incremental value productivity. This can be seen immediately by examining the terms in (1). First,

$$[v(S) - v(S - \{i\})]$$

is simply the contribution which department i makes to a coalition S if it is a member of S. Second, the term

$$(s - 1)! \quad (n - s)!$$

is the number of orderings of the remaining departments of S and $I - S$, where the latter is the set of all departments of the firm excluding S. $n!$ is the total number of permutations of all members of I.

We now show that the method of imputation obtained by using the Shapley Value defined in (1) upon the characteristic function of the firm has desirable incentive properties.

Theorem 1: The profit assigned to a Department which should be in operation if resources are efficiently allocated by the firm will never be less than $v(\{i\})$ for the ith Department.

This is trivially proved. In the formula given in (1) the sign of ϕ_i depends upon the terms $[v(S) - v(S - \{i\})]$, but as the characteristic function is super-additive all terms are at least as large as $[v(\{i\}) - v(\Theta)] = v(\{i\})$. This completes the proof.

Theorem 2: An increase in efficiency or flexibility (see example 3, section 6 for a definition of flexibility) or any action taken by a center which is of value to the firm as a whole will never cause the profits assigned to that center to fall.

This is easily demonstrated. If the game Γ is defined by $v(S)$ and the new game is defined by $v'(S)$ where $v(S) = v'(S)$ for all S not containing i, $v(S) \leq v'(S)$ for all S containing i, then for all S:

$$[v'(S) - v'(S - \{i\})] = [v'(S) - v(S - \{i\})] \geq [v(S) - v(S - \{i\})].$$

This completes the proof.

6. Some Examples Calculated and Interpreted

Example 1. Common Fixed Overhead

As a first example we take the first case presented in section 4, consider a factory that produces two products which use all the same facilities with the same intensity. Each product is under an independent manager. Each process takes up one half of the factory floor space, railyard, etc. The same number of man hours are used on each production method. An apparently natural way to assign joint fixed and variable costs between the two decision centers is to charge one half of the costs to each as they all utilize one half of the resources of the factory. If we assume that the costs of the raw materials are the same for the products, then all the cost accounting methods noted would assign overhead equally. The characteristic function for this example is given below:

$$v(\{\Theta\}) = 0$$
$$v(\{1\}) = v(\{2\}) = v(\{3\}) = 0$$
$$v(\{1,2\}) = \text{Max}((x-c, 0)), \quad v(\{1,3\}) = \text{Max}((y-c, 0)),$$
$$v(\{2,3\}) = 0, \quad v(\{1,2,3\}) = x+y-c$$

If $x + y > c$ then the firm runs at a profit. Suppose, however, that $x < c/2$. Standard accounting in this instance would compute the overhead evenly, giving $\phi_2 = x - c/2 < 0$, $\phi_3 = y - c/2$.

The first manager would be motivated to close down. To be fanciful let us suppose that this firm were highly decentralized in communication, that $c/2 < y < c$ and that there is no alternative use for the closed plant. On the next assignment, all the costs will be put on the second manager (who, after all, is using the plant himself). This gives $\phi_3 = y - c < 0$, hence he is motivated to close down even though the plant as a whole with the two products could make a profit.

Applying the Shapley value we obtain:

$$\phi_i = \frac{1}{3!}[(2!)(1!)(v(\{i\}) - v(\Theta)) + (v(\{i,j\}) - v(\{j\}))$$
$$+ (v(\{i,k\}) - v(\{k\})) + (1!)(2!)(v(\{i,j,k\}) - v(\{j,k\}))]$$

which gives:

$$\phi_1 = \tfrac{1}{6}[2(0) + 2(x+y-c)]$$
$$= \tfrac{1}{3}(x+y-c)$$
$$\phi_2 = \tfrac{1}{6}[2(0) + 2(x+y-c)]$$
$$= \tfrac{1}{3}(x+y-c)$$
$$\phi_3 = \tfrac{1}{3}(x+y-c)$$

Suppose $c = 10$, $x = 4$, $y = 7$. This gives $\phi_1 = \phi_2 = \phi_3 = \tfrac{1}{3}$ thus the assessments are $3\tfrac{2}{3}$ and $6\tfrac{2}{3}$ respectively.

If the values had been $y > x > c$ then we would have had

$$\phi_1 = \frac{3x + 3y - 4c}{6}, \quad \phi_2 = \frac{3x - c}{6} \quad \text{and} \quad \phi_3 = \frac{3y - c}{6}$$

For $c = 10$, $x = 16$, $y = 17$ this gives $\phi_1 = \tfrac{59}{6}$, $\phi_2 = \tfrac{38}{6}$, $\phi_3 = \tfrac{41}{6}$ giving assessments of $\tfrac{58}{6}$ and $\tfrac{61}{6}$. In both instances the two operating departments are assessed more than the total overhead. They pay a levy to the central office, but their net revenues are always positive. The central office requires only one number from each of them, their net profit before assessment.

Example 2. Common Marketing and Technological Improvement

In the second example in section 4 we considered a centralized sales operation with two decentralized factories. We will modify the example to a trivially simple linear program which will nevertheless be useful in demonstrating the appropriate decentralization properties.

Suppose the sales operation handles two products, 1 and 2, and the market will buy up to 10 units of each at prices π_1 and π_2. Both factories produce both items. Factory 1 has technology coefficients α_1 and α_2 (both $< \pi_1$ and π_2 respectively). Factory 2 has technology coefficients β_1, $\pi_1 > \beta_1 > \alpha_1$ and $\beta_2 < \alpha_2$. There is some limit larger than 10 on their productions.

$$v(\{\Theta\}) = 0$$
$$v(\{1\}) = v(\{2\}) = v(\{3\}) = 0$$
$$v(\{1, 2\}) = 10\pi_1 + 10\pi_2 - 10\alpha_1 - 10\alpha_2$$
$$v(\{1, 3\}) = 10(\pi_1 + \pi_2 - \beta_1 - \beta_2)$$
$$v(\{2, 3\}) = 0$$
$$v(\{1, 2, 3\}) = 10(\pi_1 + \pi_2 - \alpha_1 - \beta_2).$$

Suppose that the marketing board sends out shadow prices p_1 and p_2 and gets back information on production possibilities. By merely solving three local linear programs production will be optimally allocated. In particular the prices $p_1 = \alpha_1$ and $p_2 = \beta_2$ will satisfy. They cause the correct specialization and give the market operation a profit of $10(\pi_1 + \pi_2 - \alpha_1 - \beta_2)$ and the others obtain profits of zero.

Suppose there is a potential shift in technology which can be installed by the manager of the first plant. It replaces α_1 by $\alpha'_1 \ll \alpha_1$. If he puts this in, then in the optimum production search via shadow-pricing, the prices $p_1 = \alpha'_1$ and $p_2 = \beta_2$ will serve to allocate production. The accounting profits of the first plant are still zero. Is there a measure which will more or less automatically reflect the worth of the action of the first manager? Calculating the set of values we obtain:

$$\phi_1 = \tfrac{10}{6}[(\pi_1 + \pi_2 - \alpha_1 - \alpha_2) + (\pi_1 + \pi_2 - \beta_1 - \beta_2)$$
$$+ 2(\pi_1 + \pi_2 - \alpha_1 - \beta_2)]$$
$$= \tfrac{10}{6}[4\pi_1 + 4\pi_2 - 3\alpha_1 - 3\beta_2 - \alpha_2 - \beta_1]$$
$$\phi_2 = \tfrac{10}{6}[(\pi_1 + \pi_2 - \alpha_1 - \alpha_2) + 2(\beta_1 + \beta_2 - \alpha_1 - \beta_2)]$$
$$= \tfrac{10}{6}[\pi_1 + \pi_2 - 3\alpha_1 + 2\beta_1 - \alpha_2]$$
$$\phi_3 = \tfrac{10}{6}[(\pi_1 + \pi_2 - \beta_1 - \beta_2) + 2(\alpha_1 + \alpha_2 - \alpha_1 - \beta_2)]$$
$$= \tfrac{10}{6}[\pi_1 + \pi_2 - 3\beta_2 + 2\alpha_2 - \beta_1]$$

We observe that if α_1 is replaced by $\alpha'_1 \ll \alpha_1$, both ϕ_1 and ϕ_2 rise in value. There is an extra information cost implicit in this method however, inasmuch as extra computations were needed to obtain the value of subsets such as $v(\{1, 2\})$.

The ϕ_i can be used to calculate shadow prices or awards which are both consistent with the optimal production under current technology and provide an incentive for improvement. It should be noted that throughout this paper the discussion switches from costs to prices and profit allocations. If information and computation were free and all men had the same goal there would be no need to allocate many joint costs or revenues. It is suggested here that allocation, whether involving costs or revenues, is part of the same problem which is the utilization of these imputations for the appropriate incentives in a decentralized decision system.

Example 3. Incentives for Flexibility

Suppose a firm has two identical departments. One say, produces pink refrigerators, the other white ones. Let us furthermore suppose that they each have the same costs and face identical inelastic markets and each can more than cover total overheads. Thus:

$$v(\{\Theta\}) = v(\{1\}) = v(\{2\}) = v(\{3\}) = 0$$
$$v(\{1, 2\}) = v(\{1, 3\}) = x - c, \quad v(\{2, 3\}) = 0$$
$$v(\{1, 2, 3\}) = 2x - c.$$

As everything is symmetric for the two departments we expect and find that the imputation to both centers is the same. Suppose that there were a probability p that demand for both products would decrease, leaving both with excess capacity. Thus expected revenues are down symmetrically. Suppose, however, that new product entry has been decentralized; if one of the departments has a business plan ready to utilize the expected excess capacity while the other does not, the general managers know that the imputation scheme will acknowledge this, immediately any change in state occurs.

Example 4. Cost-Plus Internal Pricing

Under some methods of dividing joint profit an improvement instigated by one operation may not only not improve its own profit imputation but can actually have an adverse effect. An example is provided by "cost plus" pricing in a vertically integrated organization. Suppose there is a sales office and a factory. The factory produces a product not produced elsewhere, hence there is no "lowest priced alternative supply" method for establishing a price. A common prac-

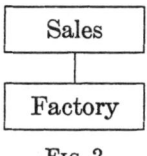

Fig. 2

tice is to use a cost plus formula. Suppose the sales office faces an inelastic demand for its product, hence at any (sufficiently low price) it will buy the same number from the factory. Say its selling price is π and that it has a fabricating, packaging or selling cost of k per unit. The cost at the factory is c per unit. The markup is $(1 + \Theta)$. Then if q units are sold

$$P_1 = (\pi - k - c(1 + \Theta))q$$
$$P_2 = c\Theta q.$$

Now suppose that the factory has a technological breakthrough which halves costs $c' = c/2$, the new imputation is:

$$P_1 = \left(\pi - k - \frac{c}{2}(1 + \Theta)\right)q$$
$$P_2 = \frac{c\Theta q}{2}$$

The innovator is penalized for his action. A manager whose bonus depends on the "profits" of his department might think twice before acting here.

$$v(\{\Theta\}) = v(\{1\}) = v(\{2\}) = 0$$
$$v(\{1, 2\}) = (\pi - k - c)q$$
$$\phi_1 = \phi_2 = \frac{(\pi - k - c)q}{2}$$

Any improvement by either is shared in this scheme.

Example 5. Inferior Goods

The next two examples envisage relatively complex relations between the components of the firm. If such relations exist they have to be known and their effects on profits coordinated for optimum behavior of the firm.

Consider a firm with three centers which produce and market, and with a headquarters whose expenses vary directly as the volume of business. Suppose that the first two sell products to the third, say potatoes, rice and meat. Suppose that a drop in the price of either of the first two will be more than compensated by the rise in revenues from the third. The initial characteristic function could be as follows:

$$v(\{\Theta\}) = v(\{i\}) = 0 \quad i = 1, \cdots 4$$
$$v(\{1, 2\}) = R_2 - C(q_2),$$
$$v(\{1, 3\}) = R_3 - C(q_3) \quad v(\{1, 4\}) = R_4 - C(q_4)$$
$$v(\{1, 2, 3\}) = R_2 + R_3 - C(q_2 + q_3)$$
$$v(\{1, 2, 4\}) = R_2 + R_4 - C(q_2 + q_4)$$

$$v(\{1, 3, 4\}) = R_3 + R_4 - C(q_3 + q_4)$$
$$v(\{1, 2, 3, 4\}) = R_2 + R_3 + R_4 - C(q_2 + q_3 + q_4).$$

All other coalitions not noted have a value of zero.[16]

Suppose that there is an important improvement in the technology for producing potatoes. If the manager of the potato board is in control of both technology and pricing he has a choice. He can introduce the efficiency and maintain his price. This replaces R_2 by $\tilde{R}_2$ where $\tilde{R}_2 > R_2$ and all other costs, levels of production and revenues remain the same. We can see from the calculation of the ϕ_i below that this is of benefit to the manager. He also can reduce the price of his product. This will reduce his indidviual net revenue vis-a-vis the market, however, *if* the executive office is able to gauge the overall effect of his action his assessment wil be such that this will constitute his most profitable course. In a decentralized system we can imagine that, at least as a first approximation he can send a message stating that unless his minimal estimates of his effect on the values of the characteristic function are regarded as reasonable, he will merely maintain his price.

If there are strange complementarities or complex relationships between departments which are present, then it is reasonable to suspect that at least those most concerned will attempt to evaluate them. In general, such an attempt is not going to call for the re-evaluation of $2^n - 1$ values for coalitions, but for observations on a very limited number.

$$\phi_i = \frac{1}{4!}[(0)!(3)!(v(4) - v(4 - \{2\})) + (1)!(2)!(v(\{i,j,k\}) - v(\{j,k\}))$$
$$+ (1)!(2)!(v(\{i,j,l\}) - v(\{j,l\})) + (1)!(2)!(v(\{i,k,l\}) - v(\{k,l\}))$$
$$+ (2)!(1)!(v(\{i,j\}) - v(\{j\})) + (2)!(1)!(v(\{i,k\}) - v(\{k\}))$$
$$+ (2)!(1)!(v(\{i,l\}) - v(\{l\})) + (0)!(3)!(v(\{i\}) - v(\{\Theta\}))]$$

$$\phi_1 = \tfrac{1}{24}[6(R_2 + R_3 + R_4 - C(q_2 + q_3 + q_4)) + 2(R_2 + R_3 - C(q_2 + q_3))$$
$$+ 2(R_2 + R_4 - C(q_2 + q_4)) + 2(R_3 + R_4 - C(q_3 + q_4))$$
$$+ 2(R_2 - C(q_2)) + 2(R_3 - C(q_3)) + 2(R_4 - C(q_4))]$$

In this example we will assume that
$$C(\sum_i q_i) = \sum_i q_i, \quad \text{hence:}$$

$$\phi_1 = \tfrac{1}{2}(R_2 + R_3 + R_4 - (q_2 + q_3 + q_4)),$$
$$\phi_2 = \tfrac{1}{2}(R_2 - q_2), \quad \phi_3 = \tfrac{1}{2}(R_3 - q_3), \quad \phi_4 = \tfrac{1}{2}(R_4 - q_4)$$

Suppose the manager does not change price, then the values become:

$$\phi_1 = \tfrac{1}{2}(\tilde{R}_2 + R_3 + R_4 - (q_2 + q_3 + q_4)),$$
$$\phi_2 = \tfrac{1}{2}(\tilde{R}_2 - q_2), \quad \phi_3 = \tfrac{1}{2}(R_3 - q_3), \quad \phi_4 = \tfrac{1}{2}(R_4 - q_4)$$

Now we consider the case where he cuts price. This changes his revenue by ΔR_2. Suppose it has no effect on player 3 but sends up the revenue of 4 by ΔR_4 (where $|\Delta R_4| > |\Delta R_2|$), furthermore we assume that the output of player 2 is reduced by Δq_2 and the output of player 4 is raised by Δq_4, these affect costs. The new value for ϕ_2 is given by:

$$\phi_2 = \tfrac{1}{2}(R_2 - q_2) + \tfrac{1}{24}[8(\Delta R_4 - \Delta R_2) + 8(\Delta q_2 - \Delta q_4) + 4(\tilde{R}_2 - R_2)]$$

The first term in the square bracket represents the effect of the overall changes in revenue upon the imputation to the second manager. The second term (which is negative here) measures the change in the structure of joint costs; and the third term takes account of the value of improvement even under conditions of absence of the fourth player (in which case price should not be cut and the second player would take in a revenue of $\tilde{R}_2$).

An example for which a price cut is marginally better is given. Suppose:

$$R_2 = 10, \quad q_2 = 4, \quad \tilde{R}_2 = 15, \quad \Delta R_2 = 5$$
$$\Delta q_2 = 1, \quad \Delta q_4 = 2 \text{ and } \Delta R_4 = 12.$$

Initially $\phi_2 = \tfrac{1}{2}(10 - 4) = 3$.

If he puts in his improvement but does not cut price

$$\phi_2 = \tfrac{1}{2}(15 - 4) = 5\tfrac{1}{2}$$

If he cuts price:

$$\phi_2 = 3 + \tfrac{1}{24}[8(7) + 8(-1) + 4(5)] = 5\tfrac{5}{6}.$$

In the three instances the overall profits to the firm are respectively $v(\{1, 2, 3, 4\})$, $v(\{1, 2, 3, 4\}) + 5$ and $v(\{1, 2, 3, 4\}) + 6$.

The ϕ_i represent the final allotments, hence the actual assessments are obtained by subtracting the net revenues collected by each decision center from the ϕ_i.

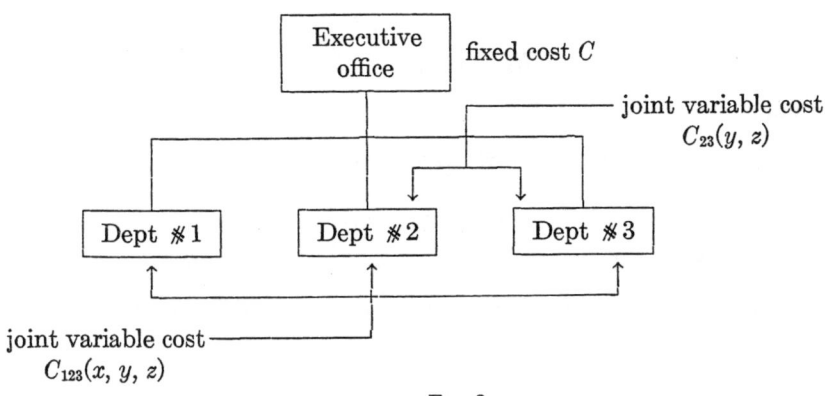

Fig. 3

Example 6. Joint Cost Upon Joint Costs

A further example where, if an attempt to impute joint costs might easily lead to an undesirable incentive system is indicated in Figure 3.

This example is not developed further here. It involves a straightforward application of the Shapley value to the characteristic function that can be easily written down.

It must be emphasized that in this case because of the complexity of the interrelationship more joint knowledge of the characteristic function is needed. The firm is basically less decentralizable than others.

7. Conclusions

This paper has attempted to emphasize a decision-making point of view to any scheme designed to impute joint costs or interrelated revenues. These problems are not separable without a loss in terms of the use of a system for internal imputation as a means for control.

Although it does not appear that the computations required to calculate the necessary information concerning the characteristic function needed at various levels present a major problem; it is desirable that methods be designed to do so and that the costs and information flows involved in doing this be included in the model.

[9]

OWNERSHIP AND THE PRODUCTION FUNCTION *

L. S. Shapley and Martin Shubik

I. Introduction, 88. — II. Mathematical-institutional economics and agrarian reform, 89. — II. A simple production function, 91. — IV. The feudal system, 93. — V. The capitalist and landless peasants, 94. — VI. A small landowner capitalist society, 97. — VII. The village commune, 100. — VIII. Corporate ownership, 102. — IX. Threats and point product, 104. — X. Summary, 104. — Appendix, 105.

I. Introduction

The simple concept of property implicit in many classical models of a competitive economy is — we suggest — an insufficiently basic representation of the phenomenon of ownership. More fundamental is the concept of an individual's operational or strategic control over certain goods or processes, as subject to laws (natural or man-made) defining his rights and powers. So long as there are no possibilities of public interaction caused by private use, the simple "chattel" view of ownership may suffice; but in more complex economic situations, such as when the rate of production of A influences (e.g., through a waste product) the costs of B, an adequate solution of the competitive model may be impossible unless constraints are imposed on the individual's strategies, over and above the physical limitations of technology. The nature of the solution will depend crucially on the nature of these constraints.

We shall elaborate this thesis by considering a series of simple, highly symmetric models, based on the same technological facts but incorporating different institutional constraints. Several solution concepts will be invoked, to illuminate different aspects of the situation; the game-theoretic "characteristic function" will play the central role. Our object is not an exhaustive analysis of any of the models, but only an exploration in sufficient depth to highlight the salient features that differentiate each one from the others.

* This paper was written under USAF contract AF 49 (638)-1700 and issued in preliminary form as RAND Corporation Memorandum RM-4053-1. An earlier version had the partial support of ONR contract NR 047-066 and was issued as Cowles Foundation Discussion Paper 167. The views expressed should not be interpreted as representing official opinion or policy of the supporting agencies.

II. MATHEMATICAL-INSTITUTIONAL ECONOMICS AND AGRARIAN REFORM

Although the scope of this paper does not permit us to go into the detailed application needed to explain the considerable differences that exist in land policies and practices in various parts of the world, we can at least sketch an outline of the connection between our mathematical approach and institutional investigations.

The problems of the distribution and tenure of land tend to be an intermix of economic, political and social factors. Our approach stresses the strategic role of the individual. This role is determined not only by strictly economic considerations but by political and social factors as well. In the sections that follow we have selected a few, suggestive forms of ownership to analyze, but we have left out, for instance, slave and forced labor systems such as were used to develop the large plantations and the mines of Latin America.[1] However, in the various land tenure systems, the principle is basically the same and has been summed up by Barraclough and Domike:

> ... Income from land, however, cannot be realized without labor. Rights to land have therefore been accompanied by law and custom which assure the landowners of a continuing and compliant labor supply. These land tenure institutions are a product of the power structure. Plainly speaking, ownership or control of land is power in the sense of real or potential ability to make another person do one's will.[2]

In our succeeding discussion we have abstracted out of our consideration several important economic features, such as the varying quality of land, which go a long way in explaining the low intensity of use, for example, of the lands of Patagonia or parts of the Altiplano. Relative productivity of land and differing transportation costs, however, constitute only partial data for the explanation of land prices, agricultural wage rates and the tenure systems. What other reasons might be relevant to structural phenomena such as the million-acre size of sheep-farms or *estancias*?[3] Factors such as a monopoly position in the water supply or in transportation, combined with the probability that the large landowner's family may also supply the judges and the politicians of the region, help to

---

1. See for example Boleslao Lewin, *La Insurrección de Túpac Amaru* (Buenos Aires: Editorial Universitaria de Buenos Aires, 1963).
2. Solon L. Barraclough and Arthur L. Domike, *Evolution and Reform of Agrarian Structure in Latin America* (Santiago: Instituto de Capacitación e Investigación en Reforma Agraria, 1965).
3. The current holdings of the Compania, Explotadora de Tierra del Fuego are approximately two and a half million acres.

delineate the power and threat structure which our theory attempts to utilize.

The extremes to which the social aspects of the economic power structure may extend are indicated in the observation that:

> ... In some cases, the administration's consent is required even to receive visitors from outside or to make visits off the property. Even corporal punishment is still occasionally encountered on some of the most traditional plantations and "haciendas". Tenants and workers depend on the "patron" for credit, for marketing their products and even for medical aid in emergencies ... With the abolition of compulsory servitude during the last century, "peones" and tenants now have the right to leave, but with few alternative job opportunities, little education, and because they are usually in debt, this possibility often appears to be as much of a threat as an opportunity for improving his lot.[4]

The evaluation of the price of land by traditional economic theory is based upon at least the existence of a fair-sized market. When a preponderance of the supply of land is locked up in the hands of a very few owners, who are able to maintain this position through a combination of political, social and economic features, the economic concept of the price of land becomes hard to define without including institutional costs as well as economic alternative costs. An interesting illustration of the role of the institutional factor is to be seen in the different costs of beach property of similar quality in various islands of the Carribean, depending upon which government controls the island.

When for instance, as in the case of Hawaii, one ranch owns 262,000 acres out of a total farmland area of about 2,500,000 acres, questions of land prices and taxation levels, not to mention land reform, become clashes involving changes in ways of life rather than mere matters of price and economic use within a fixed institutional framework.

In Figure I the Lorentz curves for Peru, Chile, the United States, and Argentina are drawn. The Gini indexes of concentration (the area between the curve and the diagonal, as compared with the total area under the diagonal) are approximately .71 for Argentina, somewhat higher for the United States, and around .87 for Chile and .95 for Peru.

Although this article is not aimed at policy discussion, we must stress that the existence, for example in Latin America, of a more or less stable system of large *latifundia* controlling most of the land and great numbers of *minifundias* (holdings too small to employ a single family at a socially acceptable minimum income) points to

---

4. Barraclough and Domike, *op. cit.*, p. 11.

OWNERSHIP AND THE PRODUCTION FUNCTION

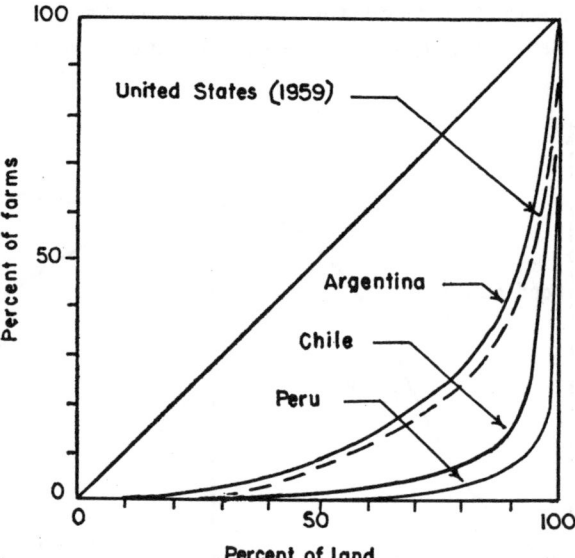

Sources: Latin American statistics are based on Interamerican Committee for Agricultural Development studies as reported by Barraclough and Domike op. cit., p. 5. U.S. statistics are from: *Statistical Abstract of the United States, 1963*, p. 613.

FIGURE I — Typical Land Distribution Curves

problems of policy that are far more involved with institutional change and the specification of welfare goals, concerning education, political power and so forth, than with the classical problem of resource optimization within a given system.

III. A Simple Production Function

A very simple example of a production process, which has been used to illustrate increasing, constant, and then decreasing marginal returns from an input, is that of several farm laborers working a single field.[5] As we increase the number of laborers there is at first a rise in the amount of food they produce per man, until, at some point they begin to run short of land to work. Then, even with the best of organization, the added product due to added labor begins to drop off. Conceivably, the added product might even become negative, as when "too many cooks spoil the broth."

Figure II illustrates such a situation, where s is the number of laborers and $f(s)$ is the output of food from a fixed area of land L.

---

5. For example David Ricardo, *Principles of Political Economy and Taxation* (Everyman ed., London, 1923), Chap. II, esp. p. 45.

When we want to consider the amount of land as variable, we shall instead write $\phi(s, l)$, where l is the amount of land, and $\phi(s, L) \equiv f(s)$.

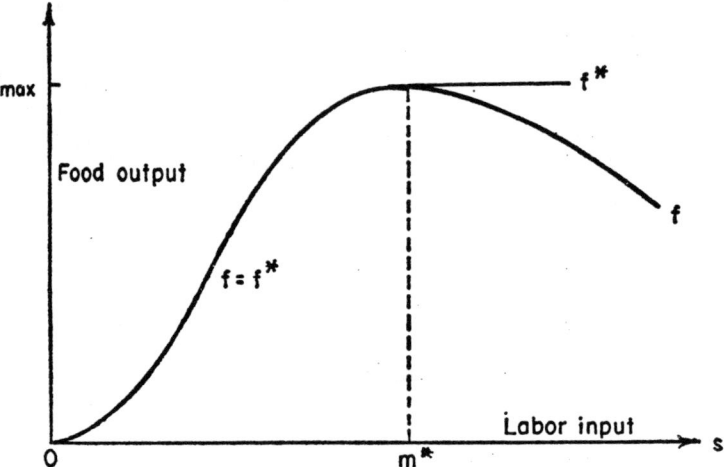

FIGURE II — The Production Function for One Unit of Land

Thus, the technological possibilities, though not the strategic possibilities, are completely determined by the functions f or ϕ.[6]

Taking this simple example, with the limited resource *land*, the variable resource *labor*, and the output *food*, we shall construct several different game-theoretic models reflecting different social conditions for the ownership and working of land. Land, food, and labor (but not people!) are assumed to be homogeneous and infinitely divisible. We also assume for simplicity that all the laborers have the same linear utility for food. Whether they actually consume the food that they acquire, or sell it in an outside market, is immaterial for our purposes. These simplifying assumptions make little or no conceptual difference to the examples we shall discuss.

In order to develop our examples, we make primary use of the so-called *characteristic function*, which specifies the best outcome that each subset of the participants ("players") can achieve un-

---

6. For this paper, we shall usually assume that f is "S-shaped," that is, convex over an initial interval and concave thereafter. It may be bounded or unbounded. Its maximum value, if any, will be denoted by f_{max}, and the first s at which $f(s) = f_{max}$ by m^*, as in Figure II. If there is no maximum, we formally set $m^* = \infty$. It will be convenient to introduce the auxiliary function f^*, defined by $f^*(s) = \max f(t)$, $t \leq s$, which is, of course, necessarily monotonic. The existence of first and even second derivatives for f and f^* will sometimes also be needed.

For the assumptions regarding ϕ, see Section VI below.

OWNERSHIP AND THE PRODUCTION FUNCTION

aided.[7] In our present application, this function, denoted by $v(S)$, will specify the amount of food that any coalition S of individuals is able to obtain for its members under the technical, social, and legal conditions of the model, regardless of the actions of those outside the coalition. This amount will therefore depend upon the ownership and use conditions for the land, and the degree of freedom of action allowed the individual participants. In contrast with the characteristic function, the production function only specifies a technical optimum, and has no ownership or strategic implications whatsoever.

Before proceeding we must consider the meaning of a *negative marginal value* of labor. This possibility cannot be dismissed unless we assume that there is a costless method of disposing of any unwanted surplus of labor, i.e., that the master engineer in charge of production can keep unwanted labor off the field at no cost. In a closed economy with fixed technology and fixed nonhuman resources, this assumption would permit an extreme case of the Malthusian view, where the productivity of added labor in the end would not only drop below subsistence wage but actually diminish total product. In our models we shall permit negative marginal productivity, but not insist upon it.

IV. The Feudal System

Consider an economy consisting of $n + 1$ individuals, of whom n are landless peasants with nothing to contribute but their labor, and one is the lord owning the land. We must distinguish several cases.

In a strict feudal relationship we do not have a true characteristic function, as there are actually no coalition possibilities available to either the lord or his serfs. Neither do they have strategic choices; rather, they have duties toward each other, which will define the division of the total product. It is possible that production might be higher if some of the serfs could be removed from the domain. This, however, may not be feasible. Hence, if the marginal product of the serfs fails to cover their subsistence needs — as might happen under crowded conditions — then an increase in

---

7. See for example, R.D. Luce and H. Raiffa, *Games and Decisions* (New York: Wiley, 1957), Chap. 8, or the authors' "Concepts and Theories of Pure Competition" to appear in M. Shubik (ed.), *Studies in Mathematical Economics: Essays in Honor of Oskar Morgenstern* (Princeton: Princeton University Press, 1966), or their *Pure Competition, Coalitional Power and Fair Division*, The RAND Corporation, RM–4917, June 1966.

numbers would result in a loss in the amount of product accruing to the feudal lord.

This strictly feudal model is trivial from the game-theory viewpoint, since everyone's behavior is prescribed. In subsequent models, the *rules* of the game will always leave some scope for individual or collective initiative, and we shall have to turn to the *solutions* of the game, which embody various principles of individual or societal rationality, to obtain predictions of behavior. Indeed, the basic purpose of the theory of games is to provide systematic techniques for resolving the indeterminacies inherent in multilateral decision processes.

V. The Capitalist and Landless Peasants

The relationship that exists between the landlord and landless peasants of a capitalist society gives rise to a true characteristic function with the superadditivity property.[8] Let us regard the landlord as player 1 and the peasants as players $2, 3, \ldots, n+1$. Then, for a coalition S, we have

$$v(S) \begin{cases} = 0 & \text{if 1 is not in } S, \\ = f(s-1) & \text{if 1 is in } S \text{ and } s-1 < m^*, \\ = f_{\max} & \text{if 1 is in } S \text{ and } s-1 \geq m^*, \end{cases}$$

where s denotes the number of individuals in S. (We assume that the landlord does not work.) If the number of laborers is greater than m^*, it is assumed that the landlord has no responsibility for them and can keep them off the land.

In the description of the characteristic function given here, we have not specified the subsistence-level requirements of the individuals as an input or cost to be met. Thus we ascribe a value of $v(S) = 0$ to every coalition of peasants alone. This implies that (at least in the short run) they are not able to obtain an alternative employment to cover subsistence. Otherwise, if alternative employment were available that just covered the subsistence requirement, k, we might instead assume $v(S) = ks$ if 1 is not in S. In a chronically overpopulated area, the original assumption might be reasonable, and in a model that introduced dynamic aspects into the relationship between the workers and landowners, the zero value (below subsistence) for some coalitions would more accurately

---

8. Superadditivity: If the coalitions S and T have no members in common, and if U is their union, then $v(U) \geq v(S) + v(T)$.

depict the "threat" potentials in the game. Such models, however, are outside the scope of this paper.[9]

Though the characteristic function ignores many details of process, it embodies enough of the essential "rules of the game" to permit meaningful application of several "solution" concepts. First, viewing the model as an open market, we find that as the number of laborers becomes large, the marginal productivity of labor declines, as does its price. The rewards to the landlord therefore rise, if we impute returns to satisfy the conditions of competitive equilibrium.[1] If $f(s)$ is a bounded function, then not only the wage rate, but the total wages paid, approaches zero, and in the limit the landlord reaps all of the gain from the economic activity. In the case where m^* is a finite number ($f(s)$ has a maximum), then he attains this position as soon as n exceeds m^*.

Two game-theoretic solutions supplement this economic solution. Consider first the *core*. It may be defined as the set of all imputations $(x_1, \ldots, x_{n+1})$ that "satisfy" every coalition S, in the sense that $\Sigma_s x_i \geq v(S)$. First we note that the imputation that gives all to the landlord is always in the core, since he is absolutely essential to any production. Next, we note that if the coalition $\{1, 3, 4, \ldots, n+1\}$ is to be satisfied, then x_2 cannot exceed $f^*(n) - f^*(n-1)$. The same applies to $x_3, \ldots, x_{n+1}$. In particular, if $f(s)$ is bounded, each peasant's maximum share in the core tends to zero as n increases; indeed, the total share to the peasants tends to zero (see Figure IIIa). Thus, the core of the game in this case shrinks down upon the competitive solution, a phenomenon that has been noted in other contexts.[2]

If the production function is *not* bounded, then the laborers may retain a positive fraction of the total output, even in the limit. For example, if we take the function $f(s) = 2\sqrt{s}$, then it can be shown that the imputation $(\sqrt{n}, 1/\sqrt{n}, \ldots, 1/\sqrt{n})$, which gives exactly half to the landlord and half to the peasants, satisfies all coalitions and is therefore in the core. (This imputation is also the competitive equilibrium solution.) Of course, the core also includes imputa-

---

9. In a hacienda-dominated agricultural economy the lack of alternative employment for the laborers has been suggested as an important factor in the stability of the system. See S. W. Mintz and E. R. Wolfe, "Haciendas and Plantations in Middle America and the Antilles," *Social and Economic Studies*, Vol. 6 (Sept. 1957), pp. 380–412.

1. See for example, K. J. Arrow and G. Debreu, "Existence of an Equilibrium for a Competitive Economy," *Econometrica*, Vol. 22 (July 1954), pp. 265–90.

2. For references see the authors', "Concepts and Theories of Pure Competition," op. cit., or *Pure Competition, Coalitional Power and Fair Division*, op. cit.

tions more favorable to the landlord, e.g., $(2\sqrt{n}, 0, \ldots, 0)$ as previously noted, but it includes nothing appreciably more favorable to the peasants if n is large.

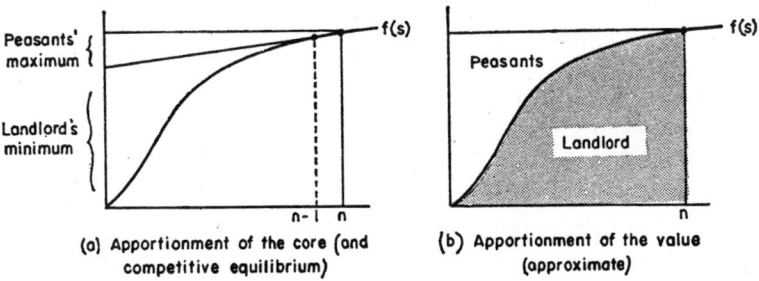

(a) Apportionment of the core (and competitive equilibrium)

(b) Apportionment of the value (approximate)

FIGURE III — The Landless Peasants

The other game-theoretic solution, the *value*, is a single imputation that measures a sort of marginal productivity of individual players, but is based upon the characteristic function rather than the production function. It may be defined, for a given player, as his expected marginal worth in a coalition that is formed at random.[3]

In the present case, it is easily determined that the value to the landlord is given by

$$\Phi_1 = \frac{1}{n+1} \sum_{s=0}^{n} f^*(s), \text{ or approximately } \frac{1}{n} \int_0^n f^*(s) \, ds,$$

since when he enters the randomly forming coalition, it may already contain anywhere from 0 to n members, with equal probability. Since the sum of all values must be $f^*(n)$, we see that the total value to the peasants is $f^*(n) - \Phi_1$. This is depicted in Figure IIIb; the values are in the same ratio as the two areas.

When $f(s)$ is bounded above, the value apportionment converges to the other solutions: the landlord gets everything in the limit. But if m^* is finite, the peasants' share does not drop to zero as soon as $n > m^*$, as in the other solutions; they still receive some small credit for their ability to thwart the owner of the land by forming a large enough coalition.

In our unbounded example $f(s) = 2\sqrt{s}$, it is interesting to note that the value solution awards ⅔ of the total produce to the landlord in the limit, as compared with ½ in the competitive solution and anywhere between ½ and 1 in the core.

3. See Shapley and Shubik, *Pure Competition, Coalitional Power and Fair Division, op. cit.*

VI. A Small Landowner Capitalist Society

Suppose we conceive a symmetric, equalitarian society, in which each individual owns $\frac{1}{n}$ of all the land as his share, as well as his own labor. Before the characteristic function can be constructed some further technological assumptions are needed concerning cultivation of smaller plots of land. That is, a production function $\phi(s, l)$ must be specified for $0 \leq l \leq L$, where L is the total amount of land and $\phi(s, L) \equiv f(s)$.

One way to do this is to make ϕ homogeneous in the first degree: $\phi(ks, kl) = k\phi(s, l)$; this implies

$$\phi(s, l) = \frac{l}{L} f\left(\frac{Ls}{l}\right).$$

However, such a function makes sense only if f is concave throughout, rather than S-shaped, since otherwise some labor forces would be able to produce more on a small plot of land than a large one. Accordingly, for this example only, we shall have to assume that f is concave, as in Figure IV.

The characteristic function is given by [4]

$$v(S) = \phi^*(s, Ls/n) = \frac{s}{n} f^*(n),$$

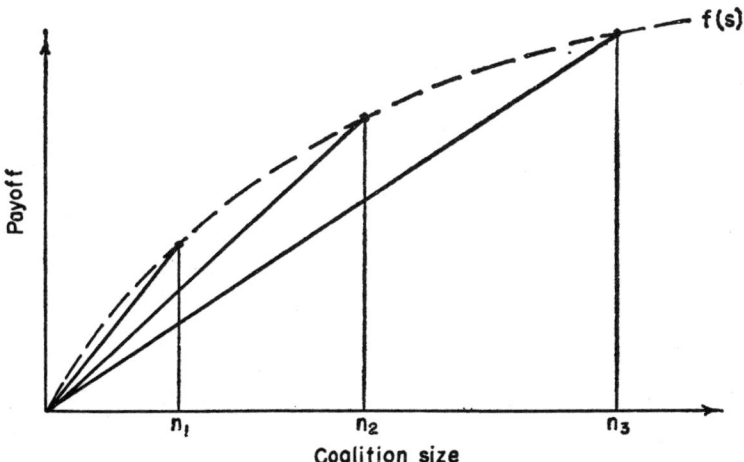

Figure IV — The Small Landowner's Model with Homogenous Production: Typical Characteristic Functions

4. In a symmetric game like this and those that follow, the function on sets S is replaced by a function on integers s, since all coalitions of the same size are equivalent.

where $\phi^*(s, l)$ is defined in analogy with f^* as the maximum of $\phi(t, l)$ for $0 \leq t \leq s$. Figure IV illustrates some typical characteristic functions for games of this kind. The important thing to observe is that each function is additive; hence each game is what the game-theorists call "inessential," and all significant solution concepts coincide. The payoff to each player is just $1/n$ of the total output, $f^*(n)$, and he can always achieve this payoff, unaided if need be, by cultivating his own plot of land. There is no gain in collusion.

We note that with f concave, the per capita income is always a decreasing function of the population, n.

A more interesting variant is obtained if we retain the S-shape and drop homogeneity. Instead of the production process being infinitely divisible, we assume that there is a decrease in efficiency, for some reason, when plots of land smaller than L are cultivated. To keep things simple, we shall use the form

$$\phi(s, l) = \frac{l}{L} f(s), \quad \text{for } 0 \leq l \leq L.$$

It is easily verified that this function satisfies $\sum \phi(s_i, l_i) \leq \phi(s, l)$ for $\sum l_i = l \leq L$ and $\sum s_i = s \leq m^*$; that is, it is not possible for subplots of land to be worked at different intensities with a gain in total output. Thus, there is no longer any inconsistency in an S-shaped production curve.

The characteristic function is given by

$$v(S) = \phi^*(s, \frac{Ls}{n}) = \frac{s}{n} f^*(s),$$

as illustrated in Figure V. This is not additive, but superadditive, and the n-person game is therefore "essential." A core exists, since no subcoalition of s members can produce more *per capita* than the society as a whole.[5] Indeed, if $n \leq m^*$, then all subcoalitions will be distinctly less efficient, and the core will contain many asymmetric imputations in addition to the "equal split" payoff. But if $n > m^*$, as at n_2 in the figure, then coalitions of size $s \geq m^*$ will be able to block all imputations other than the equal split, and a one-point core results. This may be translated: if the marginal productivity of labor is greater than zero, there will exist asymmetric distributions of product that cannot be successfully challenged by any dis-

---

5. The existence of a core, in symmetric games of this kind, depends on whether the payoff point for the all-player coalition (heavy black dot) is "visible" from the origin. If the core exists, it always contains the symmetric imputation $(f^*(n)/n, \ldots, f^*(n)/n)$, which is also the value of the game. If the all-player point is "clearly visible," as at n_1 in Figure V, then the core will contain other imputations as well.

affected coalition. But if the marginal productivity is zero (or negative), then only the unique symmetric distribution is coalitionally stable.

The value solution also gives the equal split, as might be expected from the symmetry of the model. If we suppose that the landholdings are unequal in size, however, the result is more interesting. By a computation that we do not detail, we find that the fraction of the total product that an individual receives in the value imputation (which purports to represent a "fair division"), lies *between* his fraction of the total land and his fraction of the total labor. The influence of the landholdings in the solution depends on the shape of the production curve — specifically, on the ratio of the area beneath it to the area to the left. (Compare Figure

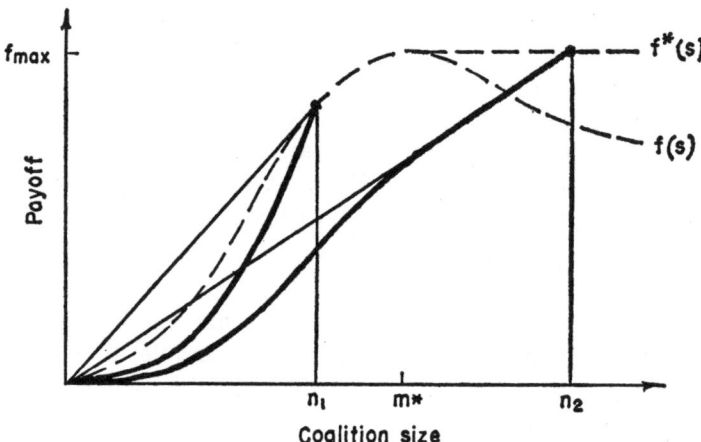

FIGURE V — The Small Landowner's Model with Increasing Returns to Scale

IIIb.) A small area to the left, for example, means that labor is unlikely to be in short supply, so that an individual's value to a coalition depends chiefly on how much land he brings in. Conversely, a small area under the curve means that labor is generally the critical input, and the value solution gives little weight to any differences in landholdings.[6]

Finally, we remark that the present, nonhomogeneous model

---

6. A good approximation to the value, when all of the plots are small, is given by
$$\Phi_i = \left\{A \frac{l_i}{L} + (1-A)\frac{1}{n}\right\} f^*(n),$$
where l_i is the size of the i-th plot and $A = \int_0^n f^*(s)ds/nf^*(n)$, the relative area under the production curve.

has no competitive solution unless $n \geq m^*$, whereupon labor gets a zero wage and land rents for $f^*(n)/L$. Indeed, if $n < m^*$, then the marginal return on labor is positive and we have increasing returns to scale, thus:

$$\phi(\lambda n, \lambda L) = \lambda f(\lambda n) > \lambda f(n) = \lambda \phi(n, L),$$

for λ slightly greater than 1 (assuming $\partial \phi / \partial l$ continuous at $l = L$). Hence no prices for land and labor can ever lead to the required production level.

VII. THE VILLAGE COMMUNE

Rather than being held individually, the land may be held jointly as in a primitive village or a kibbutz, with the use of the land controlled by majority vote. In order to define such a situation fully, we must specify the obligations of the majority and the rights of the minority.

For a first, extreme case, we assume that the majority exercises absolute control, and that once it has decided, the minority is in no position to obstruct, either by abstaining from ordered work or by carrying out other threats against the society. The characteristic function for this game is then given by:

$$v(S) = \begin{cases} 0 & \text{if } s \leq n/2, \\ f^*(n) & \text{if } s > n/2. \end{cases}$$

Figure VI illustrates some characteristic functions for different sizes of n. Games of this type are called simple majority games.[7] Any winning coalition takes all, where "all" is defined by the adjusted production function f^*.

In a situation of this kind, the problem of imputation of wealth becomes more socio-political than economic. The political mechanism of the *vote* is used to decide upon the disposition of jointly produced goods, in contrast to the economic mechanism of the *market* determining the disposition of individually owned resources. This prompts the observation that political processes in general may be interpreted economically as choice mechanisms for jointly deciding the individual allocation of products obtained from a jointly owned resource.

When we apply our three solution concepts to this game, we find, first, that the *competitive equilibrium* is not defined. Even if we permit the sale of votes, no equilibrating price schedule would exist. Secondly, the *core* of this game is empty. There is no imputa-

---

7. See Shapley, "Simple Games: An Outline of the Descriptive Theory," *Behavioral Science*, Vol. 7 (Jan. 1962), pp. 59–66.

tion which some group of more than half of the players would not be able to improve upon, by exploiting the rest. Finally, the *value* of the game, by symmetry, must award exactly $f^*(n)/n$ to each player. This is not a particularly informative result. However, if we imagine a nonsymmetric version of this game, as when the players are family units with unequal numbers of votes or unequal supplies of labor, then the value profile gives an interesting measure of the relative "political power" of the players, and provides a plausible, though hardly enforceable, formula for translating "voting strength" into economic terms.[8]

The absence of a core implies that there is no distribution of the output that is free from social pressure:

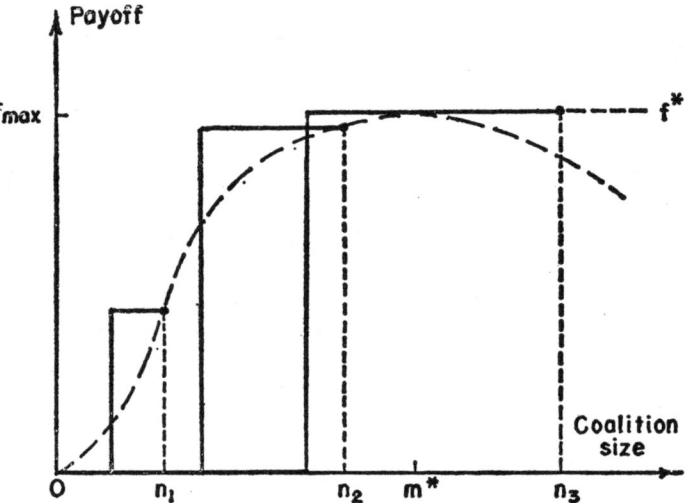

FIGURE VI — The Village Commune

some coalition will always be able to obstruct any proposal. The imputation of proceeds actually observed, in a situation like this, can therefore best be explained in sociological terms. The von Neumann and Morgenstern "stable-set" solutions, may be regarded as sociologically oriented, and might well be appropriate here.[9]

8. See Shapley, *Values of Large Games*, III: *A Corporation with Two Large Stockholders*, The RAND Corporation RM-2560PR, Dec. 1961 and Shapley and Shubik, "A Method for Evaluating the Distribution of Power in a Committee System," *American Political Science Review*, XLVIII (Sept. 1954), 787–92.
9. See, e.g., Luce and Raiffa, *op. cit.*, Chap. 9.

VIII. Corporate Ownership

We next assume that the land is jointly owned under majority rule, but that the power of the majority is not absolute. For example, it may be agreed that a majority vote controls how the land is to be utilized and that the minority must abide by the decision and not hamper the work in any manner; yet, each individual is guaranteed by law an equal share in the total proceeds. (In the nonsymmetric case, both votes and dividends would be proportional to the individual's share of ownership.)

The characteristic function (symmetric) is then:

$$v(S) = \begin{cases} 0 & \text{if } s \leq n/2 \\ \dfrac{s}{n} f^*(n) & \text{if } s > n/2. \end{cases}$$

Figure VII shows typical characteristic functions for different sizes of n. As before, the adjusted production function f^* provides us with the locus of the $v(\{1, 2, \ldots, n\})$. These characteristic functions represent a sort of compromise between those of Figures IV and VI.

When we apply our different concepts of solution, we find that the *core* consists of a single imputation (unless $n = 2$). This is because a coalition of $n - 1$ players can always prevent the other player from getting any more than $f^*(n)/n$, his symmetric portion of the total product. By symmetry, the *value* coincides with the

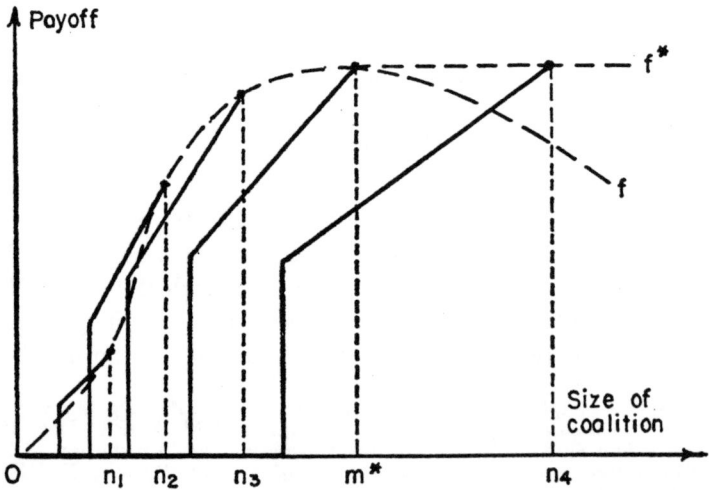

Figure VII — Characteristic Functions for First Corporate Model

core, although this is not necessarily the case with one-point cores in nonsymmetric games.

In order to consider the *competitive equilibrium*, which is not a purely game-theoretic concept, we must introduce the possibility of selling votes. If we permit this, then a competitive price for votes can be established, which will in this case yield the same symmetric imputation as the value and the core.

While these solutions resemble those obtained for the "inessential" model of Section VI (Figure IV), there are nevertheless some significant underlying differences. (1) Technically, we note that the present characteristic function does not depend on the homogeneity of the extended production function $\phi(s, l)$, whereas in the small landowners model homogeneity was crucial. (2) Strategically, it is apparent that individuals and small groups are powerless under the "corporate ownership" assumptions, and must rely for their share of the output on the rationality of the controlling majority's behavior; the opposite is true for the small landowners. (3) Sociologically, there is an incentive toward formation of large coalitions (e.g., political parties) to govern the corporation; this tendency is not present in the small landowner society. Other game-theoretic solution concepts have been defined that would shed more light on these differences, but it would lead us too far afield to discuss them here.

Thus far we have chosen extremely simple forms of strategic control, in order to provide a group of contrasting, easily-analyzed models. We must not, however, lose sight of the fact that much more complicated patterns of control are generally found in practice, when many "owners" have a hand in the same productive process. Even when the rules are unambiguous, and can be stated in a few words, the logical ramifications of the rules are likely to be so intricate that an extended mathematical analysis may be required to calculate the characteristic function.

To illustrate this point, several variants of the corporate model are presented in the Appendix. They are derived from the present case by modifying in various ways the rights and responsibilities of the participants. Questions of wage and hiring policy, and the managerial function, can then be introduced naturally, without expanding the technological side of the model. Mathematical details aside, it will be clear from a glance at Figures VIII, IX, X, and XII that the characteristic function responds quite sensitively to the nuances in the control and decision structure.

IX. Threats and Joint Product

In the previous sections, plus Appendix, we have provided many examples having the same technical economic background, but different legal and socio-political structures of ownership. One important feature is the "threat" potential of the individual or of the minority group, as it is explicitly or implicitly included in the legal structure. It is easy to see that we could construct many other variants by allowing minority groups to obstruct in different ways the production plans of the rest, e.g., by refusing to work, or voting against all production decisions. However, as the characteristic function does not take account of the cost *to the threateners* of carrying out their threats, a game-theory analysis of such variants would not be fruitful without additional apparatus. This is particularly true in the so-called "variable threat" case, where there is a choice among several threat strategies of different costs and damage potentials.[1]

Tied in closely with the problem of threats is that of joint product (often of a social nature) and external economies and diseconomies. These all relate to situations in which the activities of individuals who are not members of a coalition can strategically affect the payoff to those who are. For example, in the case of a surplus of labor, the presence of the extra people, if they have any strategic choice whatever, affects the payoffs to any operating coalition. In the case of external economies or diseconomies — for instance, where the production of A influences the costs of B — the payoff to a coalition containing B may vary continuously with the threats of A. This is not true in the classical price-system models of trade and production, nor in most of the examples given here. There appear to be at least two properties necessary for the establishment of a price system: the absence of variable threats, and the presence of a core. The design of price-system economies must involve introducing rules or laws which limit threat possibilities and ensure the presence of a core; in many societies, this would entail changes in basic institutions.

X. Summary

The foregoing discussion suggests that the concept of property implicit in many of the standard theoretical models of a competitive economy is, in general, too simple and insufficiently basic. The

---

1. See, e.g., J. Harsanyi, "A Simplified Bargaining Model for the n-Person Cooperative Game," *International Economic Review*, Vol. 4 (May 1963), pp. 194–220.

more elemental aspect of ownership is the strategic scope of the individual "owner." This will be influenced by the laws and customs defining his rights and powers over his property. Given the usual assumptions made when discussing a modern society with competitive markets, the simple concept of a chattel — such as a book or an orange — may suffice. However, as soon as there are possibilities of public interaction caused by private use, this simple view of ownership is insufficient. In these more complex and often socio-economic situations, if we wish to have certain conditions satisfied, this may require imposing rules upon the game to restrict appropriately the strategy sets of the players. In a society these amount to the imposition of legal constraints and sometimes the changing of institutions.

Appendix

For the more venturesome reader we present four additional models. They are related more to corporate management than to land tenure as such, but they serve to show how more intricate considerations of ownership and strategic control can be handled in game-theoretic analysis. The mathematical derivations omitted here may be found in the authors' RAND memorandum with the same title.[1]

Preliminary Discussion

A possible elaboration of the corporate ownership model treated in Section VIII would be to separate the control of the land from the control over labor. We may suppose that a majority has control over the use of the land, but by law is constrained to pay a uniform wage to all labor employed. Furthermore, individuals are free to accept or reject the offer for work at the wages named. After wages have been paid, the group in control must prorate any remaining "profits" equally to all owners, whether or not they belong to the controlling group, and whether or not they worked.

To complete the model, we must specify how the production decision is reached. If all applicants at the proferred wage are employed, then there is the possibility that the corporation will not be able to meet its payroll.[2] Accordingly, we shall constrain the corporation not to operate at a loss and give it the option of hiring

1. *Ownership and the Production Function*, The RAND Corporation RM–4053–1, Sept. 1965.
2. We are assuming that both wages and dividends are paid in kind; there is no money or credit, and no stored inventory to draw upon.

only some of the applicants, or even none at all. We can still distinguish two cases: (1) the controlling majority is free to hire its own members first, and (2) the controlling majority chooses only the *number* to be hired — i.e., the production level.

The logical structure underlying the two characteristic functions may be expressed as follows, always assuming that $s > n/2$:

(1) $\quad v(S) = \max_{w} \; \max_{s'} \text{-} \min_{s''} \; \max_{h}$

$$\left\{ \frac{s}{n} [f(h) - wh] + w \min(s', h) \right\},$$

(2) $\quad v(S) = \max_{w} \; \max_{s'} \text{-} \min_{s''} \; \max_{h}$

$$\left\{ \frac{s}{n} [f(h) - wh] + \frac{ws'h}{s' + s''} \right\}.$$

Here w denotes the wage rate, s' the number of members of S who apply for work, s'' the number of nonmembers of S who apply for work, and h the number actually hired, these variables being constrained by $w \geq 0, 0 \leq s' \leq s, 0 \leq s'' \leq n\text{-}s, 0 \leq h \leq s' + s''$. The hyphenated "max-min" indicates that s' and s'' are to be chosen simultaneously, although, as it will turn out, the order of choice makes no difference. In both (1) and (2) the first term represents the dividends received by the members of S, the second term the wages.

Model 1: Direct Control; Discriminatory Hiring

The indicated computations in (1) are easily carried out. On the one hand, it is clear that a majority coalition can always arrange to hire just its own members, and collect the entire output in the form of wages, with no corporate "profit" to distribute. On the other hand, since they cannot force the others to apply for work, it is clear that they can do no better. Hence (1) reduces to

(1a) $\quad v(S) = \begin{cases} f^*(s) & \text{if } s > n/2, \\ 0 & \text{if } s \leq n/2. \end{cases}$

(See Figure VIII).

This game has a core for small values of n; it is generally not unique.[3] As n increases, the core disappears when the marginal productivity of the n-th laborer drops below the average productivity (i.e., when $n > m_1$ in the figure). For large values of n, if $f(s)$ is bounded, the game grows to resemble the "village-commune" model with its "winners-take-all" feature (Figure VI). This is because the protection accorded the minority stockholders has been circumvented by discriminatory hiring, and their last remaining recourse — refusal to work — has less and less effect as the production function levels off and there is a surplus of labor.

3. See footnote 5, p. 98.

OWNERSHIP AND THE PRODUCTION FUNCTION

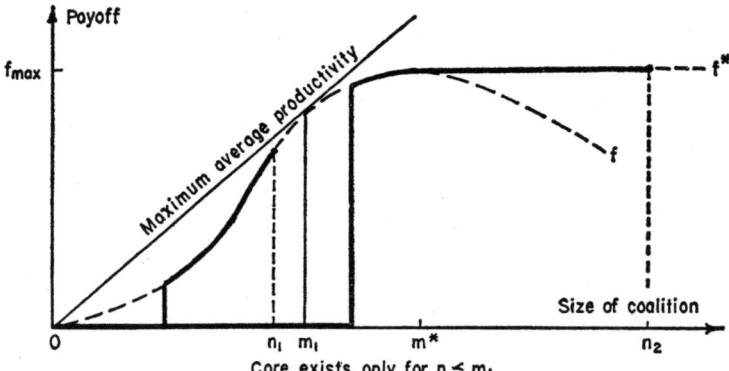

FIGURE VIII — Corporate Ownership with Discriminatory Hiring

MODEL 2: DIRECT CONTROL; NONDISCRIMINATORY HIRING

The simplification of (2) is slightly less straightforward. The most effective strategy for a majority coalition S turns out to be: (a) set the wage at $w = \min (f^*(s)/s, f^*(n)/n)$, (b) apply for work *en masse*: $s' = s$, and (c) maximize total output: $h = \min (m^*, s + s'')$; and the characteristic function turns out to be

$$(2a) \quad v(S) = \begin{cases} \min \{ f^*(s), \dfrac{s}{n} f^*(n) \} & \text{if } s > n/2, \\ 0 & \text{if } s \leq n/2. \end{cases}$$

We omit the details.

As Figure IX shows, this characteristic function reveals a sort of hybrid of the two previous corporate-ownership models (Figures VII and VIII). The antidiscrimination rule restores the core for large n — and with it the possibility of a socially stable outcome, while the option of refusing employment destroys the uniqueness of the core for small n.

MODEL 3: INDIRECT CONTROL; DISCRIMINATORY HIRING

As a further variation on the theme of corporate ownership, we introduce a *production manager*, whose fixed objective is to maximize the profits of the corporation. The manager is merely a mechanism, not a player or potential coalition member. He chooses the production level, but the owners still set the wage rate by majority vote.

We consider first the discriminatory case, in which the members of a majority coalition can demand priority in being hired. The logical structure of the model is given by

$$(3) \quad v(S) = \max_{w} \max_{s'} \min_{s''} \{ \dfrac{s}{n} \max_{h} [f(h) - wh] + w \min (s', h(w)) \},$$

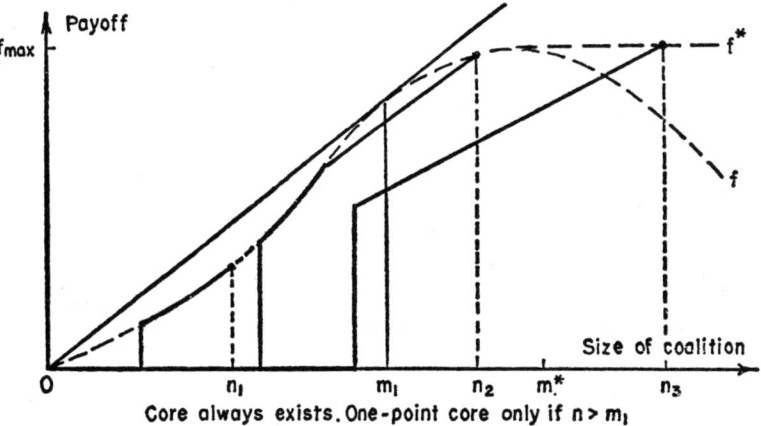

FIGURE IX — Corporate Ownership with Nondiscriminatory Hiring

for coalitions S with $s > n/2$. (Compare (1) and (2) above.) Here $h(w)$ maximizes $[f(h) - wh]$ subject to $0 \le h \le s' + s''$,[4] and the other variables are as before. This expression differs from (1) only in the scope of the innermost "max."

The reduction of (3) begins with the observation that the max-min is attained by $s' = s$, $s'' = 0$; and proceeds through a rather lengthy argument to the following characteristic function:

(3a)
$$v(S) = \begin{cases} \max_{m_1 \le h \le s} \{\frac{s}{n} f(h) + \frac{n-s}{n} hf'(h)\}, & (s > m_1, s > n/2), \\ f(s) & (s \le m_1, s > n/2), \\ 0 & (s \le n/2), \end{cases}$$

where m_1, as before, is the point of maximum average productivity of labor. In the top line (largest coalitions), the strategic wage rate is given by the marginal productivity relation $w = f'(h(w))$; in the second line the wage is simply $f(s)/s$, permitting the ruling coalition S to take the entire output in wages. The transition between the two cases (at $s = m_1$) is continuous.

This characteristic function is illustrated in Figure X for $n = 30, 40, 50$, under the assumption that f is given by the following cubic:
$$f(s) = 1.2s + .045s^2 - .001s^3.$$
This has $m_1 = 22.5$, $m^* = 40$, and there is a point of inflection at $s = 15$.

For $n \le m_1$, the case of ever-increasing average productivity, Model 3 has the same characteristic function as Model 2, and a core

---

4. We take the largest maximizing h if there is a choice. Thus, we assume that the manager will choose to operate at zero profit rather than shut down.

always exists, in general not unique. For larger n, there is almost never a core; in fact there is no chance for one unless $f(s)$ is linear

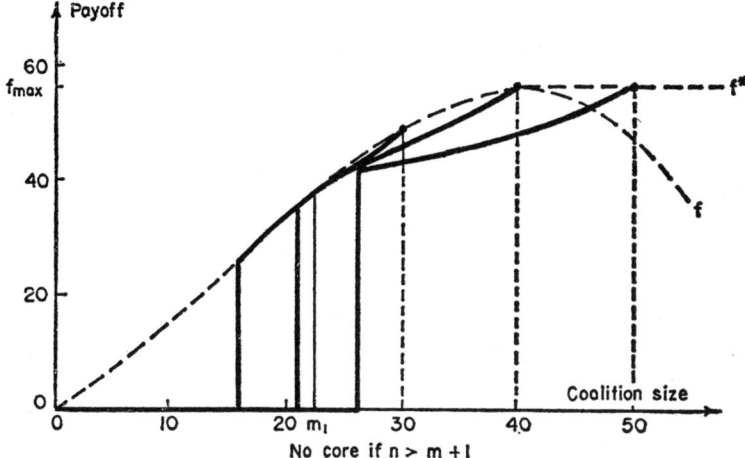

FIGURE X — Corporation with Manager (Discriminatory): Characteristic Functions for $n = 30, 40, 50$

in the interval max $(m_1, n/2) \le s \le n$, and even this very special condition is only necessary, not sufficient, for a core.

MODEL 4: INDIRECT CONTROL; NONDISCRIMINATORY HIRING

Finally, we consider the nondiscriminatory-hiring version of the managerial game. We have (compare (2), (3) above)

(4) $\quad v(S) = \max_{w} \max_{s'} \min_{s''}$
$$\{\frac{s}{n} \max_{h} [f(h) - wh] + \frac{ws'h(w)}{s' + s''}\}, \quad (s < n/2),$$

with $h(w)$ as in (3). Again this can be simplified considerably; the result is

(4a) $\quad v(S) = \begin{cases} \dfrac{s}{n} f(h) = \dfrac{s}{n} f^*(s) + \dfrac{n-s}{n} sf'(h), & (s > s_0, s > n/2, \\ f(s), & (s \le s_0, s > n/2), \\ 0 & (s \le n/2). \end{cases}$

Here h is defined by $(n - s)f'(h) = f(h) - f^*(s)$; and s_0 by $f(s_0)/s_0 = f(h_0)/n = f'(h_0)$. Figure XI may be helpful in visualizing these relationships. The strategic wage rate is indicated by the slope of the lines labeled "w."

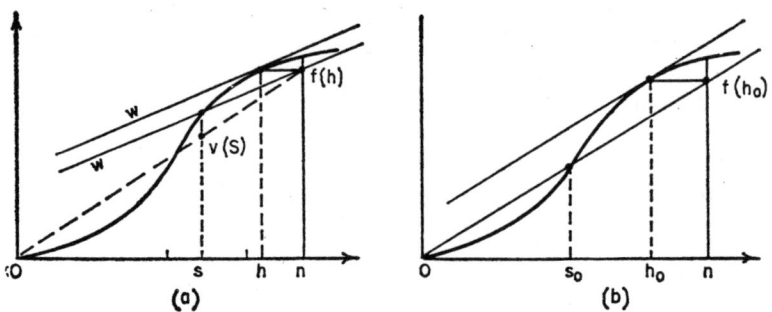

FIGURE XI — Graphical Derivation of the Characteristic Function

Figure XII is based on the same cubic production function as Figure X. For $n = 24$, s_0 is about 16; for larger n, s_0 is less than $n/2$ and hence inoperative as a case discriminator. The curve for $n = 40$ is first barely convex, then barely concave, with an inflection at about $s = 30$; the curve for $n = 50$ is slightly convex up to $s = 40$, then linear.

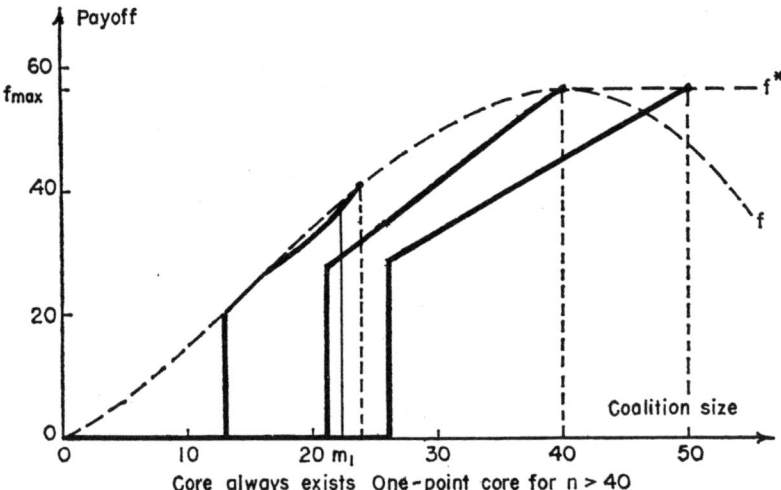

FIGURE XII — Corporation with Manager (Nondiscriminatory): Characteristic Functions for $n = 24, 40, 50$

To verify that the game always has a core, it suffices to note, from (4a), that $v(S) \leq \frac{s}{n} f^*(n)$. In words: no coalition can ever get on its own more than its proportionate share of the total optimal product.

Conclusion

Both with and without a manager, an open market for labor leads to a stable economy, in which there is no particular advantage to a coalition of less than all the participants which seizes control (Figures IX and XII). The possibility of preferential hiring practices destroys this stability (Figures VIII and X) if the economy is in the region of decreasing average productivity ($n > m_1$), even though the production decisions themselves are made for the good of society as a whole.

THE RAND CORPORATION

YALE UNIVERSITY.

A Curmudgeon's Guide to Microeconomics

By MARTIN SHUBIK
Yale University

Note to the compleat scholar:

In an article such as this it is scarcely possible to prove or completely substantiate every point. As much as possible I want to avoid "ordeal by footnote," thus, I will not go into detail on every assertion made. But in those instances where I feel that a lengthy discourse would have been called for to establish my assertion, I will indicate this by a "* * *" at the end of the sentence; I am prepared in these cases to back up my claims at length.

THIS IS A curmudgeonly article. It is frankly partisan and gives a biased view of what microeconomics is and what is good or bad about microeconomics in its current state. Suggestions made for its improvement are probably equally biased. Furthermore, this article does not pretend to be complete.

Microeconomics is many things to many people. Perhaps one of the reasons why many perceive it to be in an unsatisfactory state is that many of these people with utterly different views and goals think that there should be one microeconomics for all.

In the fight for academic territory, status, organization, tenure slots, and prestige, defining the scope of a discipline becomes important. Given a definition one can raise funds, issue brochures, organize a union, exclude the unwashed and the unqualified, and settle down to enjoy the benefits of the organized, civilized pursuit of an academic discipline.

At this moment, within the confines of the United States, there exist departments of economics which use the same name: but psychologically and academically they dwell galaxies apart. These different departments have managed to mark the cards their way and each is successfully playing its game. There are: economics departments in business schools; business schools in economics departments; institutionally and historically oriented economics departments; mathematically oriented departments; sensible adherents to Marshall, diagrams, and common sense; departments plainly blighted by operations research; and some for whom measurement is an article of faith. At one extreme there is a handful of missionaries of the computer, some of whom are interested in microeconomics as part of the development of modern behavioral science. Scattered in many areas are the methodologists, who, not unlike the Society of Jesus, occasionally suffer martyrdom, or are banned from certain dominions, but always survive and carry the faith forward. Last but not least there are those who still (or who once more) regard microeconomics as part of political economy.

Styles and tastes change. Academic pursuits are often more vulnerable to the effects of style than is the garment industry [14, Black, 1963]. In the dress industry they know that style plays an important role. In the teaching-research-entertainment industry known as academia, many of us pride ourselves on an objectivity in the pursuit of truth which rules out style.

A change in style may come about through a change in climate ("Max, mark every item in the new spring line as 'socially relevant'"). Crime, grime, unemployment, inflation, deflation, the dollar crisis, cost effectiveness, and the balance of payments all have their seasons.[1]

Powerful individuals with new ideas and changes in technology obviously play central roles so that these changes are not merely fads. The pressures of the problems of society are real; but it takes both moral and intellectual stature to be able to distinguish between today's fashions and the basic problems that an economist might help to solve.

Earlier in this century in the United States the institutionalists were driven from the temple. The era of the economic theorist had begun. There was a considerable tendency in the textbooks to throw away the footnotes and the asides which made many of the 19th Century writings reasonably commonsensical, even if long-winded and tedious. Now, in place of footnotes, diagrams have proliferated. In many departments calculus is here to stay. Modern textbooks in subjects such as price theory or theory of the firm have virtually no institutional content whatsoever. Furthermore, they scarcely contain the type of discussion that one might find in A. Marshall [64, 1966] or his precursors which embedded the subject in the society and the polity so that even the naive would be somewhat warned against spurious generality.

[1] For the ambitious young microeconomist who likes to travel, I recommend comparative pollution studies: New York, London, Paris, and Rome. This will be in high style for several years. The study of poverty is currently a fashionable way of avoiding the state of poverty.

The consumer with his indifference curves confronts the price system which functions smoothly for the individually operated profit-maximizing firm. The owner combines factors by setting marginal everything equal to marginal everything else. Supply and demand curves are obtained and for three quarters of a semester the God-given, institution-free, frictionless markets of upper-middle class Western, utilitarian man function in neo-Newtonian fashion.

A few weeks are devoted to frictions in the system. Oligopoly theory may be mentioned, and as virtually no economics textbook is honest enough to admit that this is more of a name than a coherent body of theory, a week or two is wasted. A more modern and adventuresome course will throw in a quick dollop of the general equilibrium system (failing, however, to observe that it is logically inconsistent with the partial equilibrium analysis that the students have just been taught).

As a final measure of the victory of theory, the advanced student of microeconomics may be treated to mysteries of welfare economics complete with nonexistent social welfare functions [3, Arrow, 1951] and ill-defined or logically impossible compensation criteria.

Throughout this process, if the student is fortunate enough to be exposed to a first-class economics department, he will know that everything he has learned is not only institution-free, but value-free as well. The scientific objective study of eternal universal forces of the market has no room for value-loaded statements comparing the welfare of one individual with another.

Since the defeat of the institutionalists, there have been many new developments in economics that I believe are going to result in the joining together of detailed institutional studies, advanced mathematical-economic theory and political economy. I expect that a new microeconomics is about to emerge. It can be described (in a rather ponderous manner) as mathematical-institutional-political economy.

Ten to twenty years from now textbooks in microeconomics will be enormously different. Advanced theory and new methods of computation and simulation are going to permit the reintroduction of institutional complications into general microeconomics. With a reasonable amount of luck, an important lesson from the false starts in general economic theorizing can be learned. Instead of looking for a general theory of welfare economics, or a theory of oligopoly, etc., there may be a greater stress upon special theories which will be of limited scope, but with considerable application.

An exploration of a dead end can be extremely useful if we realize that it is a dead end, and why it is so. Examples are provided by the development of indifference curve analysis, the theory of consumer choice, and the theory of choice. The problems of graphical or calculus analysis of consumer choice made a reformulation, in terms of a more general theory of choice, imperative [30, Debreu, 1959; 3, Arrow, 1951]. The lack of an adequate manner in which to handle risky alternatives leads to a modern theory of preference for risky alternatives [72, von Neumann and Morgenstern, 1947]. The simplicity of the model of consumer choice may be adequate for some purposes; but in many markets is not adequate for the construction of models with which one might predict demand. Both the growth of the behavioral theory of the firm [28, Cyert and March, 1963] and operations research in marketing [9, Bass, 1961] owe part of their impetus to a realization of the inadequacies of the accepted theory of consumer choice.

Once the footnotes and qualifications are removed from a verbal description of consumer behavior, the mathematics and diagrammatics look purer and more "scientific." At this point the hardy soul who secretly wanted to be a 19th century physicist can indulge in the luxury of spurious generalization and pursue the unified pure theory of microeconomic behavior.

Having attained the degree of abstraction of indifference curve analysis by implicitly accepting a host of psychological assumptions of dubious worth, the ambitious pure theorist can then carry his theorizing several steps further. If he is unhampered by any common sense or feeling for economics, political science, psychology, or sociology, he may be satisfied with the level of generality of the indifference curve analysis.

It is a short step to abstract items such as goods or even services from individual choice-making and to change one's pursuits to the description of the general theory of choice under all conditions. At this point the microeconomist has modestly subsumed a goodly part of general behavioral science and philosophy as a subset of economics.

It is foolish and demeaning to argue whether or not so-and-so's work is really economics. I would be the last to argue against those who choose to work in general choice theory. However, at the same time, I count myself among the first to argue that general choice theory is by no means the only theory of choice that economists need look at.

It may be desirable to have a theory of choice which covers with the same neat axioms Mrs. Jones' decision to buy an extra pound of bacon or Mr. Jones' decision to murder Mrs. Jones or to climb an impossible mountain peak. It would be nice: *but* for almost all purposes of economic theorizing it is not necessary. Furthermore, there is every indication that the price paid to include these phenomena under one unified theory is too high for an economist to pay.

Microeconomics is not merely a branch of history. For example, a detailed institutional study of the wool trade emphasizing why it is utterly different from the cotton trade may be interesting to some, but it is of limited worth. Nor on the other hand is microeconomics merely an exercise in logic or a branch of mathematics. It is the sum of many subjects and has different uses to different individuals. We must now turn to specifying some of these uses and special subjects.

Microeconomics: What It Is and Who Might Care

If I belonged to the Alice-in-Wonderland-school-of-economics, I might try the old ploy that "microeconomics is what microeconomists do." This formulation has the right amount of "inmanship" and might be worth a free sherry at someone's Society of Fellows or at High Table. In the gamesmanship department I prefer the definition that microeconomics is "economics done by those of us economists equipped with small minds." More properly, though, microeconomics is concerned with the study of economic decision-making at the level of the individual or family and the firm. It is part of political economy inasmuch as there are deep economic problems connected with jointly held assets as well as individually held assets; and many conscious decisions concerning the production and distribution of goods and services are politicoeconomic in nature.

There is no neat taxonomy as to what falls into the domain of microeconomics and what falls elsewhere. The problems of dynamics and aggregation often blur distinctions. At what level does a microeconomic model become a macroeconomic model, or vice versa?

It used to be that if you ignored the price system you had a macroeconomic model. Now, one can have a price system in a model, and it is still a macroeconomic model. Does the distinction start if one aggregates above the individual level to the family, or from the family to groups? It really does not matter very much. What matters is the question to be answered. In a rough approximation, I am happy to call an economic study microeconomic if its basic decision-making units are treated individually rather than in aggregate.

I admit that the above suggestion does not provide an airtight or elegant taxonomy. Elegance in this instance, I will leave to hairdressers and logical perfection to logicians. Sometimes one wishes to treat the corporation as the elemental decision-making unit; on other occasions it may be desirable to examine it as a hive of individuals, an organization of extreme complexity. I do not believe we should exclude studies of either type as pertaining to microeconomics.

How should we treat decision-making by banks or by the Treasury? I leave the jurisdictional dispute to the reader. It seems to me that unless he has an extremely important operational reason to answer the question, it does not need answering. (An example of an "extremely important reason" might involve an extra academic slot in a department. The definition might determine who gets the slot).

Microeconomics is, or overlaps with, many subjects. In this article I wish to variously consider the following: the theory of the firm; the theory of consumer choice; the theory of demand; the theory of price; oligopoly theory; general equilibrium economics; the theory of money; and an old fashioned subject called political economy.

Attention will also be given to several "unclean topics," fit only to be taught in the business school or in departments beyond the fringe—such as operations research departments. These topics include, besides operations research, inventory theory, production scheduling, and rude mechanical small-think topics such as integer programming, dynamic programming, and the construction of algorithms devoted to actually trying to run economic systems.

Finally, to add a touch of science fiction, the economics of information and organization, gaming, simulation, and artificial intelligence as they impinge upon microeconomic understanding shall be mentioned.

We must also pay some attention to the development of tools and methodology that are more or less free from substantive content. Here we consider mathematical and logical methods, statistics and econometrics, and the role of computers, communications, data banks, and the revolution in data processing technology.

Who should be studying what in microeconomics and why? In a modern complex society such as ours virtually every college graduate (and possibly high school graduate) *qua* citizen could benefit considerably from a basic course in micro- and macroeconomics. An understanding of what is an economic question is of great value and it is a source of deep wonderment to me that it is possible for many individuals to obtain a Ph.D in nuclear physics without obtaining this understanding. A book such as Samuelson's basic text [82, 1948] does a good job of covering this need. In my opinion, though, the trouble with microeconomics lies at a different level. Those who intend to be active in business, public life, or law need to know more microeconomics than they obtain from an elementary course. They need not only to be able to recognize an economic problem; they need to be able to abstract the nature of an economic problem from the institutional reality in which they operate.

It appears to me that some economics departments take an almost perverse pride in being able to turn out Ph.D's who never learned how to read a balance sheet; similarly, business schools are proud of their track record in turning out graduates who can read a balance sheet but who would not dream of relating their reading to a formal economic model.

Corporate lawyers, at a more specialized level, might benefit somewhat from an understanding of the economics of the firm. Production managers might benefit from a mastery of economic theories of production. Marketing managers might add to their understanding from basic courses in the theory of the firm and in marketing (if there were adequate courses to teach them). Private or governmental lawyers and economists concerned with control of industry might benefit from the study of the structure of industry, the theory of price, and oligopoly theory.

Bankers, money managers, and microeconomists might all benefit from a study of the microeconomic theory of money. Here, however, we are spared the added work as it is recognized that there is no such theory worthy of the name.

Economic motivation is a legitimate area of study for psychologists, psychiatrists, sociologists, anthropologists, and political scientists. In short, there are probably more good reasons to study many of the aspects of microeconomics than there is good material to examine.

In summary, I suggest that an intermediate or advanced course in microeconomics must achieve two goals. It needs to teach the concepts and the methodology associated with the study of problems of individual economic choice. But it needs to stress the problems of realism and relevance. It needs to provide a map for those whose interests are specialized. Why and how microeconomic theorizing is relevant to the future lawyer, legislator, economist, manager, operations researcher, banker, bureaucrat, or citizen, are important questions. They can be answered in a nontrivial way if stress is laid upon model building, formulation of questions, and derivation of criteria to decide when you have obtained an answer.

There is a central core to microeconomic theory. It contains both the arts and sciences of how to build models of economic phenomena and how to analyze them. When this is fully appreciated, we will no longer have the ego needs to talk about *"the* theory of consumer behavior" or *"the* theory of this or that." We will be able to live with many special purpose theories designed by and for individuals who are interested in one economic problem but are concerned with different facets of it.

Consumer Choice and Demand Theory

With the arrogance that characterizes our profession it is customary to refer to a set of moderately dull exercises on some constructs arising from mediocre, casual utilitarian psychological theorizing as "the theory of consumer choice." Several generations of graduate students have been taught to jump through these hoops.

Samuelson, Houthakker, and several others have worried about the operational significance of indifference curve analysis [83, Samuelson, 1947]. Revealed preference analysis has at least addressed itself to the problem of how one might deduce the indifference curve structure of a consumer's preferences from his market behavior. Binary choice experiments [67, Mosteller and Nogee, 1951] have been carried out with a small modicum of success. Through the work of Debreu [30, 1959] and others the mathematical basis for a theory of consumer choice has been changed from indifference curve analysis to a set theoretical basis. This represents a step forward inasmuch as several highly undesirable implicit assumptions associated with continuity, divisibility, and perfect discrimination are no longer present.

My objection to all of the above material is not that it does not represent a contribution to economic knowledge, but that it is presented to students of microeconomics as though it were *the relevant* way to study consumer behavior, rather than one partially explored path in almost virgin territory.

How much use is indifference curve analysis in buying a hi-fi set, or perfume, or a house? Most of my friends (and I suspect most of Herb Simon's friends and Dick Cyert's friends) are rather poor models of utilitarian man. Perhaps someone has a few friends who have been correctly trained. But for the most part my friends are not technical experts on hi-fi sets or automobiles. To a great extent they are not always even sure of what they want; they are not particularly expert on judging quality; their decisions are made under uncertainty, and they regard the amount of time they spend making their decisions as often being a considerable cost to themselves.

If I were a marketing manager with a bent for economic theorizing and was trying to sell most of my acquaintances (as well as myself) a complex consumer durable, I would certainly be building a different model than that which we peddle as *the theory of consumer choice.*

Beyond the pale, those infected by operations research worry about Markov models of consumer purchases for brands [103, Telser, 1962]. Many of these special models are probably quite poor, but at least individuals who are seriously concerned with estimating the sales of this or that consumer durable cannot afford the lordly overview that will suffice for the instruction of graduate students in microeconomics.

What is the relationship between the theory of consumer choice that we teach and the existence of demand curves? Empirically, the question boils down to how much variance can be explained by price and income information. Evidence is hard to come by. The nature of much of the competition in an economy such as ours is via product variations and innovation [61, Mansfield, 1963], so the problem of measurement becomes horrendous.

As a theorist, it is nice to be able to show how a demand curve might be generated from information on individual preferences, but this is scarcely a start on the problems of consumer choice or demand. Lancaster [56, 1966] and others have discussed consumer demand for commodities in terms of satisfaction of an array of desired services rather than in terms of the specific physical aspects of the item. For example, one might want comfortable transportation from point A to point B without really caring whether the transportation is by sedan, automobile, or helicopter. But when we adopt this type of abstract approach to consumer demand, the problems of measurement become extremely difficult. When we deal with problems of quality and service, not only do we not have good measures, but very often we do not even have a good dimensional analysis to describe the units of measurement. For example, what are the dimensions in which we should measure a good transportation service?

Numerical taxonomy [100, Sokal, 1966], the theory of measurement, and special models are going to loom large in the development of adequate theories of consumer choice and demand in the next decade.

One of the crowning successes of the great logical breakthrough in the theory of consumer choice in the 1930s was the discovery that it did not require the postulate of a cardinal measure of utility or worth. It is a remarkable fact that in a world without uncertainty and with the appropriate technical conditions specified, the existence of a price system can be established using only the ordinal properties of consumer preferences. If, however, one wishes to consider the possibility that at least on occasion an individual may be required to choose among uncertain prospects, von Neumann demonstrated that by adding some highly plausible axioms to those accepted as true for a consumer choosing among certain outcomes, then a cardinal scale for preferences could be constructed.

It is worth noting that if one is committed to the proposition that consumer preferences cannot be determined beyond an ordinal measure, then nothing whatsoever can ever be said about welfare schemes involving any form of fair division or equitable settlement.° ° °

The Theory of the Firm

The partial equilibrium analysis of Marshall, cleaned up and somewhat more formalized in the last forty years, still provides part of the standard fare of the student of microeconomics. The firm is the primitive concept of the theory. It is assumed implicitly or, on occasion explicitly, that the firm is run by an individual owner who is a profit maximizer. Conditions analogous to those of consumer choice (without an income effect) are developed to show how the firm will choose among inputs as their price is varied.

Probably one of the most important technical considerations which made the economics profession adopt the concept of continuous substitutability among all input factors is that continuous isoquants are easier to draw than discontinuous ones. Furthermore, if you intend to present the theory using calculus, it is useful to have curves with a couple of derivatives defined at every point.

The advent of linear programming has enabled a little reality and better modeling to be pirated into the economics of production via operations research groups and applied mathematics courses. In some of the more venturesome microeconomic texts linear models of production are now making their appearance. But it is of interest to note that the change was initiated by Dantzig and was made from outside the profession [29, Dantzig, 1963]. Dorfman [34,

1951; 36, 1958] was one of the first dues-paying members to take it seriously.

What questions are to be answered by the theory of the firm and the partial equilibrium analysis of the industry? For whom are these questions to be answered? The grand dynamics of the industry adjusting, when treated verbally and diagrammatically, provides an interesting overview and would certainly help a student structure his thinking when he takes his seat in the House of Commons or goes to write for *Fortune*. But does it help others?

Some students are interested in details, in questions such as: How many extra-marginal firms are there? What really is meant by an industry? How long does it take a new firm to enter? How many firms are single product firms? Is production scheduling an important variable in the understanding of the firm or should it be left to engineers? Is product variation important and how do you describe it? For these students there exists virtually nothing. The microeconomic texts, in their haste to present the general picture concerning models we know how to handle, give virtually no guidance as to the relevance and importance of factors left out or simplified. The student might even turn to books labelled "managerial economics," and these are almost always cookbooks with few insights and a low degree of abstraction.

If an intermediate or advanced course in microeconomics is meant to give an abstract overview of the general role of the firm and industry in production and the allocation of resources, it also should provide a guide to the student to indicate the importance of details that have been suppressed, and to serve as a bridge between the general and the particular.

Even three or four pages on the relationship between the theory of the firm and accounting, and the theory of the firm and finance, would provide at least a glimmer of insight to the student. The recent work of Marris [63, 1964], Cyert and March [28, 1963], and a few others has started to add detail—this at the cost of abandoning the firm as a primitive concept and describing it as an organization.

Logical consistency between one theory in microeconomics and another is a luxury and not a necessity. The theories are or should be constructed in order to answer a limited set of questions. The aggregations and selection of variables for one theory will be different from those of another. Thus the theory of the firm and industry in partial equilibrium in a freely functioning price system is not necessarily going to match a theory of oligopolistic competition, and that theory may easily fail to match a general equilibrium treatment of the economy.

There is no particular paradox in the failure of the different theories to dovetail. Frequently a great amount of insight can be gained by asking why they fail to be consistent with each other. The examination of the inconsistency can have a high payoff in isolating fundamental difficulties. It is important to be able to convey to others where the weaknesses in modeling are and why consistency between theories is not necessary. For example, in the theory of the firm we draw average cost curves which decrease and then increase after some point; yet in proving the existence of a price system in general equilibrium analysis, production functions are homogeneous of order one; or in a more modern manner we assume that production sets are convex.

Even if we do not choose to consider the internal organization of the firm, is the model of the individually owned profit maximizing entity an optimal model? Simon [98, 1957], Baumol [10, 1959], Shubik [95, 1961], and many others have questioned this.

It is a rare textbook that bothers to point out to a student that there are several institutionally different forms under which a firm might do business. Furthermore, for many very good financial, legal, and organizational reasons the goals of the managers of one may be different from the goals of the owner-managers of another [90, Shapley and Shubik, 1967]. The more elementary the textbook is, the more likely there will be information on different organizational forms. However, as soon as our study becomes "advanced", we do not bother to differentiate between General Motors and the local candy store. Different institutional forms exist in Samuelson's basic text, but not in his *Foundations* [83, 1947] or in Henderson and Quandt [49, 1958]. Vickrey, in his usual careful manner, at least warns the student of his text that "the gradual development of the modern corporation has made the entrepreneur of classical economic theory a somewhat unreal figure in a large part of the typical modern industrial economy" [106, 1964].

General Equilibrium Theory

In a review of Samuelson's *Foundations of Economic Analysis* Boulding suggested that "coping stone" would have been a better word than "foundations." Samuelson's book was elegant and written when he was young: he pays his respects to Hicks' *Value and Capital* [51, 1946] which in many ways might be regarded as elegant and literary, to boot. Both of these books were undoubtedly important intellectual contributions to the development of microeconomic theory. Yet at the same time, in my opinion, there is a pervading sense of sterility in them. An overpowering aura of specious generality is notably present in Hicks. He gives the game away when he says, "This is a work on Theoretical Economics, considered as the logical analysis of an economic system of private enterprise, without any inclusion of reference to institutional controls. . . . For I consider the pure logical analysis of capitalism to be a task in itself, while the survey of economic institutions is best carried on by other methods, such as those of the economic historian."

If Hicks had in fact produced the pure logical analysis of capitalism, it would indeed have been a signal breakthrough. I, however, regard a study, even at its most abstract level, of a system consisting of an indefinite number of utilitarian men, completely informed, trading only in individually owned commodities in a world with no indivisibilities, no externalities, no government, no taxation, and no money in frictionless instantaneous markets as something less than a pure logical analysis of capitalism.

At the risk of belaboring the point, I even feel a little queasy when I observe Schumpeter [87, 1934] make a rather persuasive case for considering innovation and the introduction of new products as part of the capitalist process. Yet in Hicks' work the number of commodities is fixed and known at the start.[2]

Economics is, at its best, an applied science (if it can be considered a science at all). The quality of the model or abstraction must be questioned as well as the quality and elegance of the analysis. The very power and elegance of Hicks' analysis may have set the subject back as far as it set it forward. I am not sure that it did happen, but it may have happened because the power of the analysis—combined with the pontifical style—made it appear that Hicks' abstraction was somehow central, universal, and of broad application. Men of lesser intellect, but more common sense, were served warning in his introduction that they could muck about with other models if they

[2] For the reader who is adept at the game this observation should cause no difficulty. Any microeconomist worth his salt will merely enlarge the utility function of his economic man to include commodities that will not be invented for another couple of hundred years. We give our utilitarian man known preferences for the commodities-to-be; then enlarge the commodity space by attaching a date, place, and quantity to all goods and services-to-be—and *le voilà* we have taken care of innovation!

wanted to be economic historians, but if it was *reine Wissenschaft* that they were after, this was it.

To some extent one might argue that a sense of sterility is also present in Debreu's *Theory of Value* [30, 1959]. But this line would be unfair as it is scarcely cricket to criticize a man for doing no more or less than he states that he intends to do. *Theory of Value* is one of the few truly elegant thin books in economics. Debreu's work reminds one of a diamond polisher. The rules are clear, formal, and well defined. The argument is as sharp as the set of pencils that can be found geometrically centered on his perfectly clean working desk. His presentation of the existence of a price system in a closed economy is a work of beauty. Yet the theorist interested in the development of political economy or even just economics must regretfully remember that even the work of Debreu is of limited generality.

Radner [79, 1968] has recently started to enlarge the analysis by including uncertainty in a general equilibrium model, and Hahn [47, 1969] has tried to introduce money into a general equilibrium model. It is too early, though, to judge the degree of success of these attempts.

Until very recently the theorizing on general equilibrium was in the form of nonconstructive proofs of a price system. In other words it was possible to show that it existed, but it was not possible to show how to calculate it. H. E. Scarf [85, 1969] has recently devised a method for doing so. It means that if we had part of the economy with technical and taste conditions so that as a reasonable first approximation the conditions needed for the existence of a price system could be assumed to prevail, we might actually be able to take a first step towards predicting prices or producing prices to control it in a more or less decentralized manner.

The work of Arrow and Hurwicz [4, 1962], Gale [44, 1960], McKenzie [66, 1959], Nikaido [73, 1964], and many others gives evidence of the increasing sophistication of the use of mathematics in the study of general equilibrium. The basic ground rules for defining the models have not changed very much; the new methodology being developed is capable of dealing with models of greater generality and is far more concerned with computational methods.

The National Association of Manufacturers as part of its tribal rituals feels it necessary every now and then to sing the praises of the price system. It is fair, just, equitable, democratic, etc., etc. . . . A question that occurs to the theorist is, "Is it unique?" If it is not unique, then which one of these fine price systems is fair, equitable, democratic, and so forth? It is quite difficult to even phrase the question: "What are the odds that an economy obeying the conditions needed for the existence of a price system may have more than one price system?" On the surface this may look like a meaningful question, but it is not well defined.*** Debreu has recently been working on the problem of making this into a meaningful question.

One of the ways in which general equilibrium economics strives for its apparent generality is by pretending that firms do not exist; or if they do, they exist as a nebulous mass of profit maximizing automata [for instance, 30, Debreu, 1959, Chapter 3]. All firms are in-being in every industry simultaneously. However, nowhere in any of the discussions or proofs of the existence of a price system do numbers matter (beyond there being more than one economic decision-making unit).*** Partial equilibrium economics, the theory of the firm, and oligopoly theory, on the other hand, appear to be loaded with concepts such as the firm as an institution, the number of firms in an industry, and entry into and exit from an industry.

What has happened to the firm in the

shuffle from partial equilibrium analysis to general equilibrium analysis? The size and number of firms in existence disappear in the general equilibrium analysis and one type of institutional assumption that appears in partial equilibrium analysis and oligopoly theory is replaced by another trickier institutional assumption that is carefully hidden from view by displaying it prominently in the foreground.

The basic institutional assumption in general equilibrium economics which can fool the unsuspecting into believing that it is somehow institution-free *is that a price system exists.*°°° The price system is not deduced as a consequence of the interaction of the forces of competition. It is assumed to exist. We first constrain all of the economic agents to act as automated price-takers and then show that a set of prices that clears the market at a Pareto optimal point exists. When we study oligopoly theory, or, for that matter, life, we do not constrain our economic actors to behave like puppets. If we try to deduce the existence of a price system, then, as I discuss later, it is necessary to introduce numbers of firms and consumers into the general equilibrium model.

Very little work has been done which views a closed economy as composed of noncooperating, eligopolistic firms and customers. The attempts of Mrs. Robinson [81, 1969] and Triffin [104, 1940] were not satisfactory. Negishi [70, 1961] and Arrow [2, 1969] recently have begun to consider this problem.

I believe that the failure of attempts to get very far with the reconciliation of oligopoly theory and general equilibrium theory have a great deal to do with modeling the assumptions concerning the management and the strategic roles of the firms as well as assumptions concerning money and the circulation of income [89, Shapley, 19–].°°°

General equilibrium economics is undoubtedly a splendid intellectual achievement. But it is not by any means on the level of Newtonian mechanics. In a world with large complicated corporations, selling thousands of goods and services (and often selling whole systems), the way we stick to our simple models (which at best cover one simple limiting case) is ludicrous. I am reminded of the story of the drunk who had lost his keys at night and spent his time searching for them under a streetlamp fifty yards from where he had lost them because that was the only place where he could see anything.

Oligopoly Theory

Oligopoly theory provides one of the clearest examples for the malaise in microeconomics. It is here that the contrast between institution-free and detail-rich approaches is the most striking. Furthermore the problems of dynamics appear in their starkest form.

There is no oligopoly theory.°°° There are bits and pieces of models: some reasonably well analyzed, some scarcely investigated. Our so-called theories are based upon a mixture of common sense, uncommon sense, a few observations, a great amount of casual empiricism, and a certain amount of mathematics and logic.

A curious aspect of that which we call oligopoly theory has been the role played by mathematics. A. A. Cournot [25, 1838], some hundred and thirty years ago, was able to formulate a careful and mathematically precise model for the study of oligopolistic competition. His model was of limited value because its economic features were not sufficiently rich or realistic to capture many of the important qualities that characterize oligopolistic markets. E. H. Chamberlin [18, 1948], whose book was a major contribution to the field, made his contribution because his economic model was considerably better than that of Cournot. (His mathematical analysis and precision of formulation were actually weaker than Cournot's.)°°°

Writing at approximately the same time as Chamberlin, Mrs. Robinson [81, 1933–69] managed to write a book on imperfect competition in which the work of Cournot was not even mentioned. The mathematical apparatus she assembled to do this job was such that von Neumann once remarked that if the archaeologists of some future civilization were to dig up the remains of ours and find a cache of books, the *Economics of Imperfect Competition* would probably be dated as an early precursor of Newton. It is of interest to note that the concept of equilibrium in Mrs. Robinson's book is the same as that in Cournot's work.

What is the status of oligopoly theory or monopolistic competition in our teaching? The more elementary texts devote very few pages to the subject and I am grateful that these pages are mostly descriptive with virtually no calculations. The intermediate and advanced texts devote a chapter or two of neoscholastic obeisance to the liturgy. With action and reaction curves and marginal cost and revenue curves of a dozen varieties, diagram drawing has its finest hour when a new crop of seniors or fresh graduate students are given the one or two week special on oligopoly and monopolistic competition.

It would be better for both, the students and their mentors, to drop the teaching of the couple of weeks once-over lightly of oligopoly and, instead, spend some time showing the students where the difficulties lie.

A classic example is provided by the usual diagram used to demonstrate how, by a series of actions and reactions, duopolists can reach the Cournot equilibrium. The drawing of reaction curves and period-by-period moves is fundamentally an illustration of a dynamic process. We merely require the students to accept the following: no matter how many periods the process takes and no matter how stupidly wrong each oligopolist happens to be, he will continue to be steadily stupid until the end. The lengths of the periods of action and reaction do not matter. They are merely a technical detail hardly worth the while of a serious pure theorist. Such a detail is left to the operations researcher or other rude mechanic concerned with fine points. The financial implications of the actions and reactions need not be taken into account. The possibility is ignored that if the firm had blundered about in the manner suggested for any length of time, it might go bankrupt before attaining an equilibrium.

I submit that we do our students a great injustice by teaching oligopoly theory in a manner that is simultaneously neither sufficiently mathematical, nor sufficiently rich in institutional detail. The understanding of oligopolistic markets is tantamount to the understanding of the economic power of the firm. The power of a firm in one market may depend delicately upon its price control; in another, upon product variation; in yet another, upon its retailing and distribution setup. The important strategic variables of a firm may be advertising, control of resources, financial strength, advantages in production processes, or dozens of others, depending upon economic circumstances. There is no Royal Road. There is no *a priori* way for us to wave our hands and reduce the study of oligopoly to this Stackelberg [101, 1952] diagram or that.

If the goal of microeconomic theory is to turn out microeconomists to perpetuate the religion, then undoubtedly the ability to play mental ping-pong with second, third, or higher degree difference equation models of ghostly, incompletely defined actors called firms is useful training. When an economic theorist is called in to work on a serious problem in the economics of competition in the United States, he leaves the ping-pong game and, like Adelman [1, 1959], Dirham and Kahn [32, 1954], Kaysen [53, 1956], Mason [65, 1957], and many others, settles down to do a serious specific investigation to identify what in fact are the major control variables in the market to be studied.

Schumpeter [87, 1934] stressed innovation. Bain [7, 1956] has analyzed barriers to entry. Others have discussed model change or stressed the financial aspects of the firm [17, Brems, 1951; 63, Marris, 1964]. Hence, even at the level of theory, matters are beginning to change. Yet little of this has trickled down through even the more advanced courses in microeconomics.

The single product atomistic firm in a static environment using only production or price as its strategic variable gives us an attractive model to start the analysis. Its appeal is great. Undoubtedly price and production are often important economic variables. Furthermore, we know how to build the models and how to study them. This competence is a feature that is of considerable importance. For example, a linear approximation to some aspect of economic reality may not be very fine, but often it is worth making because the loss in accuracy is more than compensated for by the gain in analytical power. Once more it must be stressed that this point cannot be judged in advance. There is no *a priori* method for judging the value of tradeoff.

A great deal can be learned by studying models with several single product firms interacting via price or production in an antiseptic environment with no entry or free entry. Personally I enjoy duopoly theory. I like it better than double acrostics. *De gustibus non disputandum est.* However, if I forgot the distance between the highly simplified models I study and the actual firms and markets of our society, I would be doing myself and students a great disservice.

Having already charged through the chinashop, I might as well knock the few remaining teacups off the shelves. It is easy to criticize—but where to from here? Anyone can inveigh against what is wrong, but what do you suggest to improve matters?

I suggest that in the next ten to twenty years students of oligopoly and other areas of microeconomics will first be trained in the methodology of model building.*** Dimensional analysis, the theory of measurement, and numerical taxonomy will be serious topics for fledgling economists. The theory of games will provide a methodology for examining the basic concepts of economic power, maneuverability, and viability.***

Digital computer simulation of specific market structures will add new dimensions to the detailed empirical study of market forms.*** Specific market simulations will facilitate the use of much more technical and engineering information, as well as marketing and organizational information. The important time lags and delays that are often critical in characterizing the viability of organizations will be adequately represented. Organizational studies comprising a mixture of economics, operations research, social psychology, and artificial intelligence will start to provide measures to characterize the internal flexibility and decision-making capabilities of economic organizations. Gaming with models of specific market structures will be used for experimental, operational, and teaching purposes. A student of economic theory will, more or less automatically, be able to translate a balance sheet into an abstract economic model. By playing in games with relatively good simulations, students will be able to obtain an appreciation of how the critical weaponry varies from market to market.

The underlying theory and methodology will be more abstract and general than the theory we teach now. However, the concern for detail, different variables, and the explicit treatment of institutional details will be far greater than it is now. This is not going to come about easily. Attempts to blend qualitative and quantitative material usually lead to the extremes of institutionalist story-telling on the one hand or a faith in cost effectiveness on the other.

In spite of the difficulties, I believe that computational methods for both arithmetical and logical operations, combined with a revolution in our ideas of the meaning of measurement, will make the coming together of detail and formal models a reality.

Simulation, Gaming, the Behavioral Theory of the Firm, Artificial Intelligence

New techniques and methods, especially those involving the use of expensive toys such as high speed digital computers, are often looked at by the old guard with a great deal of skepticism. There are several good reasons for this. New "gimmicky" areas are often the hunting ground of the charlatan. The man who knows nothing old and is incapable of learning any of the accepted wisdom may try to get ahead by taking on the role of the prophet of the new.

Another explanation comes from capital theory. Investment considerations make the problem of whether you can or "you can't teach an old dog new tricks," irrelevant. It is too expensive to teach an old dog new tricks. The payout time may not be long enough when compared with the cost. If you are over fifty you know what a senior debenture is [55, Graham, 46], but probably do not know what a high speed digital computer simulation is. At the club there is an aura of suspicion as the senior debenture holders eye the bright (or merely brash?) young men who want to do things radically differently.

A computer simulation of an economic institution or entity is a model of the entity written in such a manner that, given the initial conditions which describe the starting state of the system, the computer can produce a time series of future states. Furthermore, by varying the appropriate inputs, time series of contingent future states can be obtained. In other words, plans for different contingencies can be generated [97, Shubik, 1960].

The construction of a useful simulation of an economic organization calls for a combination of talents. The first is the ability to view systems in terms of abstractions. It is extremely important to be a good model builder. Harvey Wagner [107, 1969] entitles a chapter on simulation in his book "When all else fails." I am not quite so skeptical; however, the point is well taken. Before you build a simulation, you make a model of the model (this is best done on the back of an old envelope or on a cheap pad of yellow paper). It may turn out that the model of the model is adequate for the purpose of answering your questions. In this case you do the appropriate economic theorizing with the preliminary verbal diagrammatic or mathematical model. If detail appears to be important, an analytical model may not suffice and a simulation should be built. At this point an ability to model and coordinate detail is called for. An understanding of the role of accuracy and the costs of obtaining information must be manifested.

The mere discipline of drawing flow diagrams and in checking the logical consistency and completeness of a simulation serves as a powerful organizing device in studying a firm or an industry.°°° A computer simulation is far more flexible than a mathematical model and at the same time far more organized and precise than most verbal descriptions. It is a device which, if used with care, enables one to build a model combining logic, mathematics, and a richness of detail.

The literature on simulation is growing exponentially. Among some of those whose work is of interest to microeconomists are Orcutt [75, 1961], K. J. Cohen [21, 1960], F. E. Balderston and A. C. Hoggatt [8, 1960], and J. P. Crecine [26, 1969].

Closely related to the development of simulation has been the growth of work on the behavioral theory of the firm. (I would prefer to describe it as some behavioral theorizing about the firm.) Cyert and March, C. P. Bonini [15, 1963], and several others have concerned themselves with constructing detailed models of the organization of firms or parts of firms. They have introduced many new variables which to most economists (including myself) belong to other disciplines. But then again, if propensities to consume or save can make their way into macroeconomic theory, we can probably make room for some usage of "satisficing" and aspiration levels.

What is artificial intelligence, and how can it possibly be of relevance to economics? What our subjects need is more real intelligence, not science fiction fads!!

Artificial intelligence is directed towards producing machines and computer programs which perform tasks that are regarded as requiring human intelligence. Proving theorems or playing chess using computer programs are examples of artificial intelligence [40, Feigenbaum and Feldman, 1963].

Clarkson [19, 1962], a few years ago, wrote a computer program to simulate the behavior of an investment trust officer. He was able to produce a program that picked such marvelously uninspired and safe portfolios for its customers that by now it may have become a member of the Union League Club.

At Yale we have been experimenting with an artificial player for an oligopoly business game [60, Levitan and Shubik, 1967]. Is it possible to build a good player? We think it is. What is meant by a good player? A player who plays in a manner that his human competitors cannot detect that they are in fact playing a program would be a good criterion. Eventually, it would be of interest to see if it is possible to construct a program that performs better than most human players.

For those who like to play with dynamic oligopoly models, the construction of artificial players for business games poses a serious scholarly challenge. The mythology of the profession is filled with discussions of price leaders, leader-follower relationships in the market, aggressive price-cutting, implicit collusion, conscious parallelism, etc. ... If these concepts are really well defined, it should be possible to build an artificial player who will manifest the appropriate behavior. If it is not possible to do so, the concept was not well defined or the problem lies elsewhere. In either case a greater understanding of the process can be obtained.

In many of the business schools of the United States running the business game is now a more or less accepted pastime. This is not necessarily an unqualified recommendation for the business game. But it does indicate that at least a great number of individuals and institutions perceive some use for the business game as a teaching device.

The business game at Carnegie-Mellon [22, Cohen and other, 1964] is probably about the most complex and has been extensively used in teaching. The University of California at Los Angeles game [Cf. 50, Hensaw and Jackson, 1966] has been used considerably, as has the FAME game of IBM [59, Levitan and Shubik, 1959].

There also is a growing interest in using business or oligopoly games for research purposes. (This is part of a larger field of experimental gaming encompassing many disciplines.) Specifically pertaining to microeconomics, there is work by E. H. Chamberlin [18, 1950], Siegel and Fouraker [42, 1963], Hoggatt [52, 1959], James Friedman [43], Shubik [93, 1969] Vern Smith [99, 1962], David Stern [102, 1967, pp. 255–82], and many others.

Experimental studies have attempted to investigate how people behave in games based upon oligopolistic market structures under varying conditions of information and communication. A useful exercise whereby students can be taught an appreciation for the structure of industry is to have them build a game based upon a specific market. If they are limited to using, say, revenue functions which must be quadratic, the model they will construct will be sufficiently manageable that it becomes possible for them to examine the predictions of various theories of oligopoly in the context of their simulated market.*** For example, given five firms of different sizes selling somewhat differentiated products in a market represented by a quadratic payoff structure, what will be the Chamberlinian equilibrium prices and outputs? In the context of an experimental game that several of us are currently using it is possible to make such calculations [60, Levitan and Shubik, 1967].

The discipline of microeconometrics has scarcely begun. I suspect that in the course of the next two decades there will be many market models built, maintained and improved. As these models are built they will serve not only individual firms, but they will also begin to provide a new and organized empirical basis for far more complex and meaningful theories of oligopoly. It may well be that a library of game models and some time spent in war gaming with oligopolistic structure would offer those interested in the problems of the social control of industry a new approach.***

Welfare Economics and Public Finance

The state of welfare economics at this moment is in extreme ferment. The new welfare economics, complete with only ordinal preferences and the weak welfare principle, managed to do damage to the subject under the guise of objectivity and generality. But the pendulum appears to be swinging back. Welfare economics, a subject of key importance to our understanding of any economic system, may be escaping the dreadful dead-end trap of the utterly sterile attempt to construct a value-free science with spurious generality.

Welfare economics provides another example of microeconomic investigation splitting into two camps at extreme cost to both sides. The critically important subject of public finance and taxation was all but banished to business schools. The practitioners in public finance were not regarded as having the same intellectual standing as the pure theorists. (This was not the condition at the time of the Italian school of public finance or even in the era of Pigou or Edgeworth.) Fortunately, a few stalwarts, such as Musgrave [68, 1959], managed to preserve the study of public finance during its evil days.

The *kudos* in the profession was such that a really bright young man was best rewarded by writing down general social welfare functions. Otherwise he could discuss compensation criteria which were vacuous and counterfactual, logically wrong, or required an axiom such as the existence of a transferable utility*** which was ruled out by the rules of the big league, "New Welfare Economics Association."

It is interesting that the disease of the new welfare economics produced the very antibodies which finally killed it. The deathblow was delivered by the publication of Kenneth Arrow's *Social Choice and Individual Values* [3, 1951]. He produced a proof that if one accepted certain axioms usually accepted in the new welfare economics, it was not possible, in general, to construct a community welfare function. Arrow's book is another of the very few elegant thin books in political economy.

Does Arrow's work represent a setback to the development of welfare economics? On the contrary, it helps to get us back on the tracks by showing that pure logic combined with pure welfare theory does not get us very far. There are many different ways of setting up other axioms from which *it is possible* to construct a community welfare function [89, Shapley and Shubik]. The argument is not so much a mathematical one, but to decide if one wishes to make different assumptions about society.***

Economic behavior takes place in a society complete with political institutions. The concept of a community welfare function is probably more a sociological and political concept that it is economic. There is no logical or operational need for a community welfare function to be logically consistent with all individual economic or political choices. If you define exactly what you include in the welfare function, it may still be a useful construct even though not necessarily consistent with individual behavior.***

The English Utilitarian economists were not afraid to suggest such welfare functions as the sum of all individual utilities. If a man is willing to specify what he means by a statement, at least there is something to argue about. One might even combine it with diminishing marginal utility to argue that optimum world population should be as large as possible above the starvation level! I do not favor this view with wild enthusiam, but at least I know what the assumptions are and could discuss it.

Welfare economics deals with the production and distribution of heroin, hospital care, and hydroelectric dams. It is concerned with policy in advertising, the distribution of information, the effects of externalities, and the handling of public goods and services, some of which we do not even know how to describe. (For example, in what units are we meant to measure foreign policy or justice?) The recent flurry of interest in the construction of social indicators [105, U.S. Dept. H.E.W., 1969], I believe, represents the beginning of a serious attempt to become far more specific and operational in the development of welfare economics.

Owing to political necessity, the advent of new methods of data processing and gathering, operations research methods, and new techniques for computation, there has been a considerable growth in detailed studies of public goods problems; for example, there are investigations of water and air pollution, transportation, education, health, and urban redevelopment. As these studies increase in number, the steps towards the construction of new theories in the broad area of welfare economics might comprehend generalizing from specifically understood problems rather than trying to construct a general theory out of the whole cloth.

There are some topics of considerable importance which are described in aggregate by two or three words but in fact comprise a group of extremely different problems hidden behind a single title. Welfare economics is one of these (cancer and time-sharing are examples of two others). Furthermore, there are some topics which have the property that as soon as we try to become specific, problem after problem can be isolated, and many of them are sufficiently important and different to merit separate study.

I suspect that eventual pure theory of public goods (which is but a small part of welfare economics) is going to need to develop its theorizing around at least twenty or thirty different classes of public goods (possibly many more) [94, Shubik, 1966]. There are fundamental differences such as: Can you force an unwilling individual to consume the good? Can you prevent an individual from using the good? Is the item appropriable? Does it have a natural unit of measurement? Is it indivisible? Could it be individually owned? ... and so forth. Samuelson's [84, 1959] pure public good specifies only one dimension, among many.

A theory which purports to cover the size and quality of the law courts, public libraries, drug and liquor control, the economics of pollution, patents and the economics of information sources, public transportation, and myriads of other topics had better be rich in subtheories and studies of special cases unless its only operational value is to serve as a hurdle in the promotion race at academic institutions. For this purpose the old "new welfare economics" was admirably suited.

Marketing and Finance

In the caste system of American academia finance has historically been rather low. It has been on the fringes of respectable economics departments—regarded as belonging to business schools. Marketing, as a subject, has not merely been regarded as of the lower castes, but as virtually untouchable. Both of these subjects have not, until recently, been noted for their intellectual content. They are, however, critically important to the understanding of two aspects of the firm in an economy, such as ours. Furthermore, marketing in the sense that it deals with detailed studies of consumer behavior is important to our understanding of the consumer.

It was once observed that war is too important to be left to the generals. In the same sense, finance and marketing are far too important to be left to institutional studies alone and to be only located in the outer reaches of the profession.

The work of Markowitz [62, 1959] and others, the coming into being of the *Journal of Quantitative Finance* and, in a different vein, the work of Marris, show the signs of change. It is my belief that this work should and will proliferate. In particular, I believe that the financial aspects of oligopolistic competition have been considerably underestimated except in popular left-wing sociological writing. There has been little in the form of careful analysis and theory construction. In Bain's book, *Barriers to Competition* [7, 1956], there is, at least, a glimmer of the size of financial indivisibilities. How important is the interest rate as compared with capital rationing to the firm or individual?

An indication of the size of financial and institutional indivisibilities in our society is given by the rash of take-over bids in the stock market during the last few years where assets may be revalued by forty or fifty percent almost overnight. This is a far cry from a minor readjustment in a smoothly functioning capital market!

To many economists advertising is a dirty word. We draw a demand curve on the blackboard, look wisely at the class, and say that advertising shifts this demand curve. We go to the board and draw another line showing that the demand has been shifted. How it is shifted, why it is shifted, what are the underlying mechanisms—these questions we leave implicitly as exercises for the students. This treatment of advertising reminds me of the story about the owl who was the wisest creature in the forest. A centipede with ninety-nine sore feet came to consult with the owl. The advice he was given was to walk one inch above the ground for the next two weeks in order to give his sore feet enough time to heal. The centipede observed that it was a splended suggestion, was precise, insightful and logical. He then asked the owl, "How am I going to walk one inch above the ground?" The owl replied, "I have solved your conceptual problem. Do not bother me with technical details."

The firm spends money on advertising. *Ergo* the demand curve may shift. We have solved your conceptual problem! Pure theory has triumphed again! Do not bother us with technical problems such as the details of how and why it shifts. We need ten or twenty first-class Ph.D. theses in this area and there appear to be virtually none. O. J. Firestone's small book [41, 1967] contains some useful lists of the pros and cons of advertising. These lists could serve as a starting point for several investigations and for the construction of many different models.

In the last few years several books on quantitative models in marketing have appeared. These together with attempts to simulate specific market structures are opening up paths that the serious microeconomist concerned with the behavior of the firm and the consumer will have to follow. *The* theory of consumer behavior is going to be replaced by several theories of consumer behavior, incorporating information costs, search, habit, and a host of other factors introduced in an explicit manner into our studies.

Operations Research, Linear, Convex, Integer, and Dynamic Programming

In the business schools, in departments of applied mathematics, in the Siberian salt mines, and in every mine and mill ex-applied mathematicians, statisticians, and economists have appeared as operations researchers. The second generation is already here; unlike their intellectual parents who began life elsewhere, they are being graduated from operations research departments. It used to be said, but no longer holds, that a management scientist or operations researcher is an ex-economist, statistician, mathematician, or engineer applying to the Ford Foundation for a grant to spend in a business school.

Along with this growth of personnel has come a great amount of detailed work on topics such as inventory theory, production scheduling, waiting lines, transportation studies, optimal assignment problems, product-mix problems, and so forth. I do not urge every microeconomist to rush out and join in. But it is absolutely false and ostrich-like to believe that this work is of only minor concern to the microeconomist. Even at the level of pure microeconomic theory topics such as inventory theory have already made a considerable difference. Incidentally, it is interesting to note that in the case of oligopolistic competition, if we make the assumption that production takes time and that inventory carrying charges are positive, then the Chamberlinian large group oligopolistic equilibrium does not exist.*** Operations researchers are uncultured and anxious men; nonexistence might bother them. However, microeconomists skilled in dealing with community welfare functions should not be disturbed.

The advent of the computer and advances in mathematics have made it possible to carry out computations on economic models with hundreds and even thousands of variables. I have already noted Scarf's algorithm for calculating the price system. Lemke's algorithm [57] for bimatrix games preceded it. The work of Gomory on integer programming [45, 1968] is undoubtedly of critical importance in increasing our understanding of the role of indivisibilities.

The work of Bellman [12, 1957] and Pontryagin [77, 1964] has begun to provide a highly promising mathematical basis for the investigation if capital theory. This work has recently been discussed in an article by Dorfman [35, 1969] on control theory and dynamic programming.

Advances in linear programming, as Gale, Kuhn, Dorfman, Samuelson, Solow, and others have shown, have made some economic theorists far more conscious of boundary solutions. How many basic textbooks draw their indifference maps so that they cut the axes?

The work in linear systems has also been manifested in von Neumann's growth model [71, 1945] and in the input-output analysis of Leontief [58, 1949]. In the latter case it is interesting to note that the methodology and organization it provides have stimulated data gathering and economic introspection the world over on a scale that can only be compared to the work on national accounts sparked by the writings of Keynes. Is input-output analysis microeconomics or macroeconomics? I leave this existential question to the reader.

Wagner's book on operations research [107, 1969] provides an excellent source for the thoughtful microeconomist who wishes to obtain an overview of what these topics cover. Baumol's book on *Economic Theory and Operational Analysis* [11, 1961] provides more explicit bridging at a more elementary level.

Lest I am misinterpreted, I want to stress that operations research is not microeconomics and microeconomics is not operations research. The areas of interaction, however, are large, and it behooves us to be aware of and understand them.

Game Theory

For this writer to give an appraisal of the role of game theory in microeconomics is not unlike asking a bear to guard the honey supply. Having expressed this *caveat emptor*, I can now proceed with the utmost of objectivity (as befits a pure theorist of microeconomics).

The theory of games must be considered on at least three levels; they are, as mathematical discipline, as it is applied to microeconomic theory, and as it is applied to other behavioral sciences or to dynamic processes such as bargaining and negotiation in economics and other behavioral sciences.

The theory of games provides a language for the study of multi-person decision-making in detail. The technique of the game tree complete with its information sets offers a manner to describe the anatomy of interlocking decision-making at a level that was not available previously.

In many ways one of the major weaknesses of the theory (and at the same time one of its major strengths in its application to microeconomics) is that it is a direct descendant of the model of economic man. The player in game theory models is meant to have known preferences and is basically *homo oeconomicus* involved in a complex cross-purposes maximization problem. Von Neumann and Morgenstern [72, 1947] stressed these limitations of their models in their book.

The demands of model construction when applying game theory are very exacting. Perhaps it is the level of explicitness required in model building that turns one of game theory's weaknesses into a strength. Completeness and consistency are required to such a degree that ridiculous models of human affairs are immediately revealed as ridiculous by the model. To return to the example of adjustment to a Cournot equilibrium: in order to translate Cournot's contribution into a formal game theoretic model, one would be forced to choose between a static or dynamic formulation. If the dynamic formulation were chosen, the imperatives of the methodology would force the theorist to specify the period-by-period time lags, the possibility of bankruptcy, and other aspects that could easily be missed in a verbal or diagrammatic treatment.* * *

A new branch of investigation—that can best be described as conflict study—has to a great extent developed out of the recognition of the inadequacy, for some purposes, of formal game theoretic models. The works of Boulding [16], Rapaport [80, 1960], Schelling [86, 1960], and many others fall in this category. They are concerned with gamesmanship, the societal and psychological aspects of bluffing and threats. The roles of fear, suspicion, trust, or manifestly irrational behavior are questioned in a manner that involves a mixture of virtually all of the behavioral sciences laid upon social-psychology and political science. Restricting ourselves more closely to microeconomics: it is when we try to build formal models of the bargaining process or conflict and cooperation within the firm that the needs for a behavioral theory that takes into account learning, teaching, perception, and modification of goals, become evident. The recent book by John Cross [27, 1969] provides an excellent example.

This article is not the place to discuss game theory *qua* mathematics. The growth of the mathematical literature has been large; our interest is in its effect on microeconomics. Nevertheless, even at the level of mathematics, Debreu [30, 1959] has noted that game theory was outstanding among the influences "which freed mathematical economics from its traditions of differential calculus and its compromises with logic."

What is a solution to an economic problem? What do we mean by solution? In the methodology of the theory of games there are around thirty or forty solution concepts, several of which are of key interest to the economist. Some of these solutions reflect meanings of efficiency, decentralization, individual strategic power, the countervailing power of groups, fair division or "just distribution" and social stability [92, 1968].

There are many solutions to an economic problem. You may wish to regard your problem as solved if the outcome is Pareto optimal—a weak criterion of efficiency. Others may be more interested in meeting criteria concerning the equity of distribution. Still others may have a notion of social stability and be willing to regard their problem as solved when the stability has been achieved.

Can certain outcomes satisfy more than one criterion? For example, is it possible that an outcome which is the solution to efficient allocation and decentralization is also the outcome called for by satisfying the demands of groups in the structure of countervailing power? Questions such as this can be specified and answered by game theoretic methods.*** This specific question, in the course of the last eighty years, has been answered at various levels of generality by Edgeworth [38, 1881], Shubik [91, 1959], Scarf and Debreu [31, 1963], and Aumann [5, 1961]. The *core* of a market game is that set of distributions of goods which cannot be challenged by the economic power of any subgroup of the society. It is a remarkable fact that if the appropriate conditions hold on technology and tastes that a price system will produce a distribution of goods and services that lies within the core.

The various attempts at the construction of compensation principles in welfare economics were nothing more or less than poorly defined pre-game theoretic attempts to define *blocking coalitions* or (in nontechnical words) the distributions in a society that could be ruled out by deals between subgroups. A cheap and easy way to wade through this rococo literature is now provided by the methodology of game theory. You need to ask four questions: 1) What are the sidepayment conditions? 2) Is the Pareto optimal surface well defined? 3) Can I describe *the* set of distributions obtainable by any subgroup of the society? and 4) Is the core nonempty?

It is quite possible that there is no distribution satisfying the conditions required for the existence of a nonempty core for an economic model. When this condition prevails there will be neither a price system nor a taxation scheme unchallengeable by at least one group in the society.*** When the core is empty, it will always be possible for some subgroup of society to have an attractive bribe for some other subgroups, no matter what distribution is suggested. In other words, there will be no way to satisfy all the conflicting claims of all subgroups.

Indivisibilities, externalities, and public goods provide many economic situations in which a price system will not be satisfactory to guarantee the efficient or decentralized allocation of resources. When these cases are present, other criteria must be used. Nash [69, 1953], Harsanyi [48, 1956], Shapley [88, 1953], and others have been concerned with conditions for fair division.

The value of this approach has been that it is explicit about the assumptions concerning possible normative criteria for distribution.

In my opinion, one of the most important features of game theory in its application to economics has been its concern with numbers of decision-makers. For example, it has been shown that if we consider an economy in which we increase the numbers of butchers, bakers, and candlestick makers so that there are many individuals in every trade, then, viewing the distribution of resources within this economy as a struggle among groups, the core will shrink as the number of participants increases until it becomes a single point and that point can be interpreted as giving rise to a price system.

This limit theorem result deduces the existence of a price system as the result of the interactions of groups. It does not make the institutional assumption that markets exist. This point may appear picayune, but it is central. If you really believe that somehow "pure economic theory" should be as institution-free as possible, then you must be extremely careful in accepting innocent looking assumptions concerning individual behavior, the nature of property, and the existence of markets.

Competitive equilibrium theory, which is meant to be the pride and joy of microeconomic theory, suffers from three sets of sloppy assumptions (leaving out the bad psychology concerning individual preferences). The definition of individually owned goods and services is extremely limited. Only in the hands of a theorist, with the precision and honesty of Debreu, is the major danger avoided and then because he takes great care in making precise how general or lacking in generality his assumptions are. In general teaching, however, I believe most students are led astray into believing that the theory is far more general than is the case. A couple of hours devoted to the difficulties associated with the concept of property would help us see microeconomics in better perspective.

The second set of sloppy assumptions concerns the existence of markets. Often a totally false picture is given to the students which amounts to what I call a fake limit theorem where assumptions are changed in the middle of the exposition. The general equilibrium theory, as it is taught, does not depend upon the number of firms or consumers. Frequently we begin by discussing the Edgeworth analysis of bilateral trade. This involves only two individuals and we often state that we cannot say much more than that the solution should be on the contract curve. Our argument is that with only two traders costs of information, communication, haggling, etc. . . . will be such that a price system will be adhered to. In the passage of the analysis from two to many, a host of *unspecified assumptions* concerning communication, information and trade are made. Because we do not bother to specify them, this effective vacuum somehow qualifies the theory as being institution-free. Going one step further, we find that the only use made of the number of traders in our theorizing is to justify the unspecified assumptions. The actual mathematics used holds for two or two billion; and there is no limit theorem proved since the result does not depend on the numbers.

The third sloppy assumption is closely tied in with the second. We *assume* that the individual traders will act as price-takers. We do not deduce that they will act as price-takers. They are, in fact, no longer economic agents with any freedom of choice or with strategies. They are *automata* constrained to act in a particular manner. A game theoretic approach to this same problem leaves the individuals as economic men with freedom of choice and then *deduces* that, as a result of numbers, the economic power of the individual is sufficiently weakened that he might as well behave as a price-taker.

This last point might not appear to be of particular importance to economic theorizing, except for a few of its consequences. Because of the constraints placed upon the individual we do not even bother to calculate the consequences of his committing an error.*** In particular, this means that when we model monetary systems we leave out bankruptcy. Furthermore, one of the prime reasons why oligopoly theory is inconsistent with general equilibrium economics is because almost all of oligopoly theory in one form or another is modeled as a non-cooperative game with the individuals possessing freedom of choice.*** This freedom includes the freedom to make wrong moves; to end up with inventories or be caught out-of-stock. Yet when we pass to general equilibrium models the individuals are price-taking *automata*, forced to act so that supply equals demand—forced to do the right thing and forbidden to go bankrupt.

The methodology of game theory makes one describe the whole of the payoff set. In other words, you must be able to specify what happens given every feasible set of moves. You may then deduce that many of the outcomes will be avoided. This situation, however, is different from assuming that they are avoided. The cost price of going the other way is that you may easily fake yourself into believing that you have a well defined model of an economic activity when you have nothing of the sort.

Another specific example, in which a game theoretic formulation wipes out a paradox caused by sloppy formulation of the assumptions concerning a competitive system, is given when one tries to explain pecuniary externalities to a group of students being indoctrinated to the mysteries. This requires a few pages of demonstration and has been done elsewhere [96, Shubik, 1970].

The methodology of game theory can be applied to political problems involving voting and power as well as to economic problems. It is my belief that many of these topics fall directly into the domain of political economy, especially when we observe how much of the decision-making concerning public goods, school bond issues, zoning, etc., depend upon voting. The book of Farquharson on the *Theory of Voting* [39, 1969] is the third elegant thin book I can legitimately mention in this article. Arrow's approach to voting is analogous to Debreu's approach to general equilibrium. The individual decision unit is not regarded as a player in a game of strategy, but is an isolated maximizing unit. Farquharson, in contrast to Arrow, regards the voters as players who may easily vote for their second choice when they think that their first choice is going to lose. The literature in this area is proliferating. The earliest workers on voting included Condorcet [24, 1785] and Lewis Carroll, [108, Williams, 1931; 23, Collingwood, 1945; 33, Dodgson, 1873]; among the more recent have been Duncan Black [13, 1958], Downs [37, 1957], Kramer [55], Plott [76, 1957], Riker [78, 1962], Wilson [109, 1968], and many others.

Note that up to this point not one word has been said about maximin solutions, saddle points, or zero-sum games. There is some interesting mathematics concerning games involving two players in which the winnings of one equal the losses of the other. The study of such games has some military applications but for the behavioral sciences in general, and microeconomics in particular, they are not terribly relevant. Yet with the notable exception of Vickrey, if game theory is mentioned in a microeconomic text, most of the analysis presented is of two person zero-sum games which are almost irrelevant.

Some of the early predictions of the effect of game theory on microeconomic theory were premature. However, I believe that the next ten to twenty years will see a considerable growth in its influence.

The Theory of Money

To a preponderant extent the theory of money has been in the bailiwicks of the macroeconomists and those concerned with policy. Every now and then an attempt to integrate money into a general equilibrium model is made. To date none has been completely successful. In an article such as this I will take poetic license to comment on some of the problems and possibilities in the integration of monetary theory in microeconomics.

I believe that money is an institutional phenomenon. A pure microeconomic theorist in his rush to stay pure has great natural tendencies to throw the baby away with the bath water, or at least the cash down the drain.

A central theme in this article is that there is no such thing as institution-free economics. Explicitly or implicitly we slip in assumptions concerning the nature and role of property, political, legal, and social organization. This act does not mean that it is not possible to theorize at levels of great generality. It does mean however that between any two economies there may be subtle differences caused by law or custom which may influence our theorizing. This point is particularly true when dealing with the phenomenon of money.

Money cannot even be defined in isolation. It must be considered part and parcel with the laws for financial operations. Bankruptcy laws are as much a part of the monetary system as are dollar bills. Economies with different financial laws differ from each other much in the same way as geometries based upon different axiom systems differ from each other. Money and laws for financial operations form part of the rules of the economic game. When the rules are different, the game is different.

In W. Bagehot's classic discussion [6, 1962) of Lombard Street, there is an important clue to why so many economists interested in a pure theory of money have been thrown off the track. A tool or device or an institution, coming into being for one purpose, may turn out to have a completely different, yet far more important purpose. Undoubtedly such is the case with money. It is easy to list some of money's simpler features: a measure of value, a means of exchange, a store of wealth. We then may spend our time trying to put money into the economic system by adding a transactions cost into a general equilibrium model. There are undoubtedly many *ad hoc* ways of doing so, none of which, so far, seeming to have been aesthetically attractive to the theorist (although Hahn is currently working on a model that may be successful). But what I have thus far noted is only part of the story.

Money is an invention of man. It was introduced in many different *ad hoc* ways into different economies. Once it was there, it took on a life of its own. In particular, in a general equilibrium model with no government and no financial intermediaries, with no bankruptcies, no uncertainty, and no game theoretic maneuvering, it appears to me to be reasonable to expect that even if someone sneaks in a transactions cost money into a competitive equilibrium system he will get out a very mouselike money. It will be neutral and serve as a means of exchange, a measure of value, and a minor store of wealth, but that is all. Some further exercises will be generated in contemplating the effect of velocity of transactions on the worth of this type of money; and if our traders trade at ever increasing velocities, if the theory is any good, with a shriek of triumph the money will vanish at the point of infinity where it no longer will be needed as all exchanges will have become instantaneous.

The serious theory of money started

when some ruler appointed himself the issuing authority. It took on new dimensions when the first goldsmith decided that he could lend gold belonging to someone else, because he could replace it in time to satisfy the claims on him. These were strategies by players in a nonsymmetric game.

A complete theory of money will need at least three distinguished players. They are: some abstract form of governmental body whose preferences and powers must be stated; a distinguished player representing an abstraction of a financial intermediary. The financial intermediary, as a good first approximation in some cases (provided it is not organized as a mutual), can be regarded as maximizing its profits. The third set of players consists of everyone else. Were we restrict our model to them alone we would have only a general equilibrium barter model.

By being slightly venturesome but still only using players of the third type, I have suggested that we may be able to develop a pallid theory of money comprising features such as transactions costs. Once uncertainty is introduced, I suspect we must include at least one more type of player, the financial intermediary. Furthermore, we must specify new rules of the game. What are the limits on his behavior and what are the bankruptcy laws of the society?

The final step in the development of a theory of money calls for the introduction of a governmental agency with further rules of manipulation and the use of the monetary mechanism as a means of policy. This inclusion calls for a game theoretic or strategic approach to money. For example, a hyper-inflation is a game theoretic concept. It can only take place when individuals have strategic power and are consciously playing in a money game.

Envoi

Microeconomics is fun for many in its band of followers, many of whom have an "irrational passion for rationality." Amongst those of us who tend towards enjoying pure theory (whatever that is), there is also some extra psychic kick to be obtained from elegance.

Unfortunately, microeconomics is probably not an elegant subject when really well done. In order to keep us going, every now and then it is a good idea to set almost all parameters equal to one or zero, keep everything convex, and make a few further simplifications here and there, and for this price we can be rewarded with some nice convergence theorem. The Invisible Hand appears, pats us on the head, and we feel elegant. Beyond that it is necessary to go back to the miasmal swamp of reality.

Our theorizing and our results are probably better than any other social science. Yet we still have an enormous distance to go. New mathematical methods, additional data-gathering, and added computational capability combined yield greater support than ever before for the development of an understanding of microeconomic phenomena. This step taken, our greater wisdom will provide us with the opportunity to be able to put detail and institutions back into microeconomic theorizing.

REFERENCES

1. ADELMAN, M. *A and P: A Study in Price-Cost Behavior and Public Policy.* Harvard Economic Studies, Vol. 113. Cambridge: Harvard University Press, 1959.
2. ARROW, K. J. "Firm and General Equilibrium," Technical Report No. 3, Office of Naval Research 047–004. Harvard University (mimeographed), May 1969.
3. ———, *Social Choice and Individual Values.* Cowles Commission Monograph, No. 12. John Wiley & Sons: New York, 1951.
4. ARROW, K. J. and HURWICZ, L. "Competitive Stability under Weak Gross

Substitutability: Nonlinear Price Adjustment and Adaptive Expectations," Int. Econ. Rev., May 1962, 3, 233–55.
5. AUMANN, R. J. "The Core of a Co-operative Game Without Side Payments," *Transactions of American Mathematical Society.* March 1961, 98, 539–52.
6. BAGEHOT, W. *Lombard Street.* Edited by HARTLEY WALTERS. Homewood, Ill.: Irwin, 1962.
7. BAIN, J. S. *Barriers to New Competition: Their Character and Consequences in Manufacturing Industries.* Cambridge: Harvard University Press, 1956.
8. BALDERSTON, F. E., and HOGGATT, A. C. "The Simulation of Market Processes." Working Paper No. 22, Management Science Research Group, University of California at Berkeley, Oct. 1960.
9. BASS, F. M., et al. (eds.). *Mathematical Models and Methods in Marketing.* Homewood, Ill.: Irwin, 1961.
10. BAUMOL, W. J. *Business Behavior, Value, and Growth.* New York: Macmillan, 1959.
11. ———, *Economic Theory and Operations Analysis.* Englewood Cliffs, N.J.: Prentice-Hall, 1961.
12. BELLMAN, R. *Dynamic Programming.* Princeton, N.J.: Princeton University Press, 1957.
13. BLACK, D. *Theory of Committees and Elections.* Cambridge: Cambridge University Press, 1958.
14. BLACK, R. D. C. "Economic Fashions," Inaugural Lecture Delivered before the Queens University of Belfast on Dec. 4, 1963. Printed by Marjory Boyd, Printer to Queen's University of Belfast, 1963.
15. BONINI, C. P. *Simulation of Information and Decision Systems in the Firm.* Englewood Cliffs, N.J.: Prentice-Hall, 1963.
16. BOULDING, K. *Conflict and Defense: A General Theory.* New York: Harper & Row, 1962.
17. BREMS, H. *Product Equilibrium under Monopolistic Competition.* Cambridge: Harvard University Press, 1951.
18. CHAMBERLIN, E. H. *The Theory of Monopolistic Competition: A Re-Orientation of the Theory of Value.* 6th edition. Cambridge: Harvard University Press, 1948.
19. CLARKSON, G. P. E. *Portfolio Selection: A Simulation of Trust Investment.* Englewood Cliffs, N.J. Prentice-Hall, 1962.
20. CLARKSON, G. P. E. and SIMON, H. A. "Simulation of Group Behavior," *Amer. Econ. Rev.,* Dec. 1960, 50.
21. COHEN, K. J. *Computer Models of the Shoe, Leather, Hide Sequence.* Englewood Cliffs, N.J.: Prentice-Hall, 1960.
22. COHEN, K. J., et al. *The Carnegie Tech Management Game: An Experiment in Business Education.* Homewood, Ill.: Irwin, 1964.
23. COLLINGWOOD, S. D. *The Life and Letters of Lewis Carroll* (contains excerpts of letters of May 1881 & July 1884). Dates are given by ROGER LANCELYN GREEN in "Lewis Carroll and the St. James's Gazette," *Notes and Queries,* April 7, 1945.
24. CONDORCET, MARQUIS DE. *Essai sur l' Application de l' Analyse à la Probabilité des Décisions Rendues à la Pluralité des Voix.* Paris, 1785.
25. COURNOT, A. A. *Recherches sur les Principes Mathematiques de la Theorie des Richesses.* Translated by N. T. BACON. New York: Macmillan [1838] 1897.
26. CRECINE, J. P. *Governmental Problem Solving: A Computer Simulation of Municipal Budgeting.* Chicago: Rand McNally, 1969.
27. CROSS, J. G. *The Economics of Bargaining.* New York: Basic Books, 1969.
28. CYERT, R. M. and MARCH, J. G. *A Behavioral Theory of the Firm.* Englewood Cliffs, N.J.: Prentice-Hall, 1963.
29. DANTZIG, G. B. *Linear Programming and Extensions.* Princeton, N.J.: Prince-

ton University Press, 1963.
30. DEBREU, G. *Theory of Value: An Axiomatic Analysis of Economic Equilibrium.* Cowles Commission Monograph, No. 17. New York: John Wiley & Sons, 1959.
31. DEBREU, G. and SCARF, H. "A Limit Theorem on the Core of an Economy," *Int. Econ. Rev.*, Sept. 1963, 4, 235–46.
32. DIRHAM, J. B., and KAHN, A. E. *Fair Competition: The Law and the Economics of Anti-trust Policy.* Ithaca: Cornell University Press, 1954.
33. DODGSON, C. L. A Discussion of the Various Methods of Procedure in Conducting Elections. Preface dated Dec. 18, 1873. Only copy in U.S. at Princeton University.
34. DORFMAN, R. *Application of Linear Programming to the Theory of the Firm: Including an Analysis of Monopolistic Firms by Non-Linear Programming.* Berkeley: University of California Press, 1951.
35. ———, "An Economic Interpretation of Optimal Control Theory," *Amer. Econ. Rev.*, Dec. 1969, 59, 817–31.
36. DORFMAN, R.; SAMUELSON, P. A.; and SOLOW, R. *Linear Programming and Economic Analysis.* New York: McGraw-Hill, 1958.
37. DOWNS, A. *An Economic Theory of Democracy.* New York: Harper, 1957.
38. EDGEWORTH, F. Y. *Mathematical Psychics.* London: Kegan Paul, 1881.
39. FARQUHARSON, R. *Theory of Voting.* New Haven: Yale University Press, 1969.
40. FEIGENBAUM, E. and FELDMAN, J. eds., *Computers and Thought.* A collection of articles. New York: McGraw-Hill, 1963.
41. FIRESTONE, O. J. *The Economic Implications of Advertising.* London: Methuen, 1967.
42. FOURAKER, L. E., and SEIGEL, S. *Bargaining Behavior.* New York: McGraw-Hill, 1963.
43. FRIEDMAN, J. "On Experimental Research in Oligopoly." Cowles Foundation Discussion Paper No. 246. Yale University: Cowles Foundation for Research in Economics.
44. Gale, D. *The Theory of Linear Economic Models.* New York: McGraw-Hill, 1960.
45. GOMORY, R. E. "Some Polyhedra Related to Combinatorial Problems." R C 2145. Yorktown Heights: IBM Research, July 25, 1968.
46. GRAHAM, B., et al. *Security Analysis.* 4th edition. New York: McGraw-Hill, 1962.
47. HAHN, F. H. "Money in General Equilibrium Models." Paper presented at 1969 meeting of Allied Social Sciences Association, New York, Dec. 29, 1969.
48. HARSANYI, J. C. "Approaches to the Bargaining Problem Before and After the Theory of Games: A Critical Discussion of Zeuthen's, Hicks', and Nash's Theories," *Econometrica*, April 1956, 24, 144–57.
49. HENDERSON, I. M. and QUANDT, R. E. *Microeconomic Theory: A Mathematical Approach.* New York: McGraw-Hill, 1958.
50. HENSAW, R. C. and JACKSON, J. *The Executive Game.* Homewood, Ill.: Irwin, 1966.
51. HICKS, J. R. *Value and Capital: An Inquiry into Some Fundamental Principles of Economic Theory.* 2nd edition. London: Oxford University Press, 1946.
52. HOGGATT, A. C. "An Experimental Business Game," *Behavior Science*, July 1959, 4, 192–203.
53. KAYSEN, C. *United States vs. United Shoe Machinery Corporation: An Economic Analysis of an Anti-Trust Case.* Harvard Economic Studies, Vol. 99, Cambridge: Harvard University Press, 1956.
54. KOOPMANS, T. C. *Three Essays on the State of Economic Science.* New York: McGraw-Hill, 1957.

55. KRAMER, G. H. "An Impossibility Result concerning the Theory of Decision Making," in J. L. BERND, ed., *Mathematical Applications in Political Science*, Vol. 3, Charlottesville, Va.: University Press of Va., 1967, 39–51.
56. LANCASTER, K. J. "Change and Innovation in the Technology of Consumption," *Amer. Econ. Rev. Pap. and Proc.*, May 1966, 56, 14–23.
57. LEMKE, C. E. "Bimatrix Equilibrium Points and Mathematical Programming," *Management Sci.*, Aug. 1965, 681–89.
58. LEONTIEF, W. W. *Input-Output Economics*. New York: Oxford University Press, 1949.
59. LEVITAN, R. E. and SHUBIK, M. "FAME, the Financial Allocation and Marketing Executive Game." Prepared by IBM at the Thomas J. Watson Research Center, Yorktown Heights, New York, 1959 (mimeographed).
60. ———, "A Business Game for Teaching and Research Purposes." Part IV—Mathematical Structure and Analysis of the Non-Symmetric Game, Cowles Foundation Discussion Paper No. 219, March 2, 1967. Part V—The Non-Symmetric Game: Joint Maximum Efficient Solution and Measures of Collusion and Welfare, CFDP No. 225, May 31, 1967. Part VI—The Cournot Equilibrium in a Nonsymmetric Oligopolistic Market, CFDP No. 224, May 16, 1967. Part VII—The Nonsymmetric Game: The Beat-the-Average Solution, CFDP No. 227, June 26, 1967.
61. MANSFIELD, E. "Size of Firm, Market Structure, and Innovation," *J. Polit. Econ.*, Dec. 1963, 71, 556–76.
62. MARKOWITZ, H. M. *Portfolio Selection: Efficient Diversification of Investments*. New York: John Wiley & Sons, 1959.
63. MARRIS, R. *Economic Theory of Managerial Capitalism*. New York: Free Press of Glencoe, 1964.
64. MARSHALL, A. *Principles of Economics*. 8th edition. London: Macmillan Papermacs, 1966.
65. MASON, E. S. *Economic Concentration and the Monopoly Problem*. Harvard Economic Studies, Vol. 100. Cambridge: Harvard University Press, 1957.
66. McKENZIE, L. W. "On the Existence of General Equilibrium for a Competitive Market," *Econometrica*, Jan. 1959, 27, 54–71.
67. MOSTELLER, F. and NOGEE, P. "An Experimental Measurement of Utility," *J. Polit. Econ.*, Oct. 1951, 59, 371–404.
68. MUSGRAVE, R. A. *The Theory of Public Finance: A Study in Public Economy*. New York: McGraw-Hill, 1959.
69. NASH, J. F. "Two-Person Cooperative Games," *Econometrica*, Jan. 1953, 21, 128–40.
70. NEGISHI, T. "Monopolistic Competition and General Equilibrium," *Rev. Econ. Stud.*, June 1962, 28, 196–201.
71. VON NEUMANN, J. "A Model of General Economic Equilibrium," (translated by G. MORGANSTERN). *Rev. Econ. Stud.*, 1945, 13, 1–9.
72. VON NEUMANN, J. and MORGENSTERN, O. *Theory of Games and Economic Behavior*. 2nd edition. Princeton: Princeton University Press, 1947.
73. NIKAIDO, H. "Generalized Gross Substitutability and Extremization," In M. DRESHER, L. S. SHAPLEY, and A. W. TUCKER, eds., *Advances in Game Theory*. Princeton: Princeton University Press, 1964.
74. ORCUTT, G. H. "Simulation of Economic Systems," *Amer. Econ. Rev.*, Dec. 1960, 50.
75. ORCUTT, G. H.; GREENBERGER, M.; KORBEL, M.; and RIVLIN, A. M. *Microanalysis of Socioeconomic Systems*: A simulation study. New York: Harper, 1961.
76. PLOTT, C. R. "A Notion of Equilibrium and Its Possibility under Majority Rule," *Amer. Econ. Rev.*, Sept. 1967, 57, 787–806.
77. PONTRYAGIN, L. S., et al. *Mathematical*

Theory of Optimal Processes. New York: Pergamon Press, 1969.
78. RIKER, W. H. *Theory of Political Coalitions.* New Haven: Yale University Press, 1962.
79. RADNER, R. "Competitive Equilibrium under Uncertainty," *Econometrica,* Jan. 1968, 36, 31–58.
80. RAPOPORT, A. *Fights, Games and Debates.* Ann Arbor: University of Michigan Press, 1960.
81. ROBINSON, J. *The Economics of Imperfect Competition.* 2nd edition. London: Macmillan, (1933) 1969.
82. SAMUELSON, P. A. *Economics: An Introductory Analysis.* 1st edition. New York: McGraw-Hill, 1948.
83. ———, *Foundations of Economic Analysis.* Cambridge: Harvard University Press, 1947.
84. ———, "The Pure Theory of Public Expenditure," *Rev. Econ. Statist.,* Nov. 1954, 36, 387–89.
85. SCARF, H. E. "An Example of an Algorithm for Calculating General Equilibrium Prices," *Amer. Econ. Rev.,* Sept. 1969, 59, 669–77.
86. SCHELLING, T. C. *The Strategy of Conflict.* Cambridge: Harvard University Press, 1960.
87. SCHUMPETER, J. A. *Theory of Economic Development: An Inquiry into Profits, Capital Interests and the Business Cycle.* Cambridge: Harvard University Press, 1934.
88. SHAPLEY, L. S. "The Value of an N-Person Game," *Annals of Mathematics,* Study 28, 307–17. Princeton: Princeton University Press, 1953.
89. SHAPLEY, L. S. and SHUBIK, M. *Competition, Welfare, and the Theory of Games.* (Manuscript in process), Part III.
90. ———, "Ownership and the Production Function," *Quart. J. Econ.,* Feb. 1967, 81, 88–111.
91. SHUBIK, M. "Edgeworth Market Games," pp. 267–78 in R. D. LUCE and A. W. TUCKER, *Contributions to the Theory of Games.* Vol. 4. Princeton: Princeton University Press, 1959.
92. ———, "Game Theory II—Economic Applications," pp. 69–73 in *International Encyclopedia of the Social Sciences,* New York: Macmillan and The Free Press, 1968.
93. ———, "A Note on a Simulated Stock Market," *Amer. Inst. Decision Sciences,* Nov. 1969.
94. ———, "Notes on the Taxonomy of Problems concerning Public Goods," Cowles Foundation Discussion Paper No. 208. Yale: Cowles Foundation for Research in Economics, April 19, 1966.
95. ———, "Objective Functions and Models of Corporation Optimization," *Quart. J. Econ.,* Aug. 1961, 75, 345–75.
96. ———, "Pecuniary Externalities: A Game Theoretic Analysis," Cowles Foundation Discussion Paper No. 288. Yale: Cowles Foundation for Research in Economics, Jan. 27, 1970.
97. ———, "Simulation of Industry and Firm," *Amer. Econ. Rev.,* Dec. 1960, 50.
98. SIMON, H. A. *Models of Man, Social and Rational: Mathematical Essays on Rational and Human Behavior in a Social Setting.* New York: John Wiley & Sons, 1957.
99. SMITH, V. L. "An Experimental Study of Competitive Market Behavior," *J. Polit. Econ.,* April 1962, 70, 111–37.
100. SOKAL, R. R. "Numerical Taxonomy," *Scientific American,* Dec. 1966, 215, 106–16.
101. STACKELBERG, H. V. *The Theory of the Market Economy.* Translation and introduction by A. T. PEACOCK. London: William Hodge, 1952.
102. STERN, D. "Some Notes on Oligopoly Theory and Experiments," in M. SHUBIK (ed.). *Essays in Mathematical Economics in Honor of Oskar Morgenstern.* Princeton: Princeton University Press, 1967.

103. TELSER, L. G. "The Demand for Branded Goods as Estimated from Consumer Panel Data," *Rev. Econ. Statist.*, Aug. 1962, 44, 300–24.
104. TRIFFIN, R. *Monopolistic Competition and General Equilibrium Theory.* Cambridge: Harvard University Press, 1940.
105. UNITED STATES DEPARTMENT OF HEALTH, EDUCATION AND WELFARE. *Toward a Social Report.* 1969.
106. VICKREY, W. S. *Microstatics.* New York: Harcourt, Brace, & World, 1964.
107. WAGNER, H. M. *Principles of Operations Research with Applications to Managerial Decisions.* Englewood Cliffs: Prentice-Hall, 1969.
108. WILLIAMS, S. H. and MADAN, F. *Handbook of the Literature of the Rev. C. L. Dodgson (Lewis Carroll).* Oxford, 1931.
109. WILSON, R. "An Axiomatic Model of Logrolling," Working Paper No. 143, Graduate School of Business, Stanford University, August 1968. (mimeographed).

[11]
ON DIFFERENT METHODS FOR ALLOCATING RESOURCES

The subject of economics has often been described as dealing with the allocation of individually owned scarce resources. Price theory, which forms a considerable part of micro-economic theory, deals with the use of a price mechanism in solving the problems of production and allocation of resources among a population.

An important feature of the political process is the production and distribution of the product of jointly owned scarce resources within a society. Thus, for example, although one might wish to use an economic mechanism to solve the problems of production and distribution of automobiles, we may regard it as desirable to use a political mechanism to solve the production and distribution of heavy military weaponry.

In a complex, modern industrial society it is not easy to make clean differentiations among strictly economic, political or social problems. In some cases, one might be able to solve the problems of supply using economic institutions. On the other hand, the creation of effective demand may not be feasible via an economic mechanism; and vice versa.

What are the alternatives to the market place? Where are they used and how important are they to our economy? Are the alternative methods for the allocation and production of resources used in all societies? Do they appear in the same mixes in the Soviet Union as in China or as in the United States? Do they appear in the same mixes in underdeveloped countries as they do in developed countries? These questions are critical in helping us to understand the development of modern politico-economic systems.

How deeply does the democratic form of government depend upon a market structure primarily characterized by small enterprises and a price system? Is there any relationship between the price system and democratic institutions? It might well be that the viability of the various Eastern European governmental structures will depend upon the introduction of price systems in certain parts of their economies.

In the list below, eight different mechanisms are suggested for the allocation of resources. They are by no means totally independent of each other and furthermore each of them stands for a gross aggregation of more refined mechanisms. Nevertheless the list in this crude form is suggestive of the different approaches used in our society and in every other society for resource allocation. We do not have a clear characterization of the importance of these different methods for distribution within our own society; yet they are undoubtedly critical to an understanding of politico-economic and socio-economic allocation.

The following eight processes appear to be the major means for the allocation of resources:

1. The economic market with a price system.
2. Voting procedures.
3. Bidding.

4. Bargaining.
5. Allocation by higher authority, fiat or dictatorship.
6. Allocation by force, fraud and deceit.
7. Allocation by custom, including gifts and inheritance.
8. Allocation by chance.

Influencing most of these processes are the effects of persuasion and other methods which concentrate upon changing individuals' preferences and hence how they act when allocating resources.

The market mechanism of the impersonal price system, where each individual buys and sells at prices set by a faceless market, appears to be well suited for solving the technical and administrative problems for the decentralized efficient allocation of resources. This is especially true for satisfying the individual wants of consumers for goods and services of a personal nature which also can be produced by relatively small individual firms.

The smooth functioning of a price system calls for many participants on both sides of the market. Thus, for example, the supply and demand conditions for small restaurants or tailors meet the requirements. The demand for automobiles is generated by many individuals, however there are only a few firms selling automobiles. When there are only a few suppliers of a good the market is said to be *oligopolistic*, and there are many good economic reasons to suspect that the price mechanism will not function freely [1, 2].

Even if there are potentially many suppliers and customers, special considerations may cause a society to prevent an open economic market with a price system from developing. Thus the opium and other drug trades potentially could function as a price-guided market but are restricted from doing so.

Voting is usually related to the control of a jointly owned resource where the actual operation of the resource is governed by a group of trustees acting as fiduciaries for the original owners. Thus, for example, corporations sell shares and the stockholders are the owners, yet the top executives and trustees are responsible for the running of the corporation even though the stockholders are called upon to vote on occasion. In a similar manner, one may argue that the citizens of the United States use their vote in order to appoint trustees or fiduciaries for the running of their political system.

Votes on bond issues, taxation schemes, subsidies, and grants are all examples of the political mechanism being used as a method for defining effective demand for joint goods or services.

The political voting mechanism is far more closely related to the formation of societal values than is the market system. Speaking loosely, we might characterize the market system as dealing with the effective demand for individual utilities whereas voting systems deal with the aggregation and the creation of effective demand to satisfy societal values.

No country really uses its price system alone to determine attitudes towards justice, public health, foreign aid, or education. The price of beans, baby bonnets, and basketballs may be determined by the market without needing to become involved in deep political problems. This is not the case for moon shots, urban

MARTIN SHUBIK

development or decisions on national communications systems. Undoubtedly individual values are related to societal values; however, it is important to note that they are not the same.

There is a growing literature on the theory of voting and on political parties operating in societies with voting systems [3–6]. There are wide varieties in the voting methods employed and the implications of the differences among them have still been only partially explored.

Another important device for the allocation of resources is bidding. In a society in which few large customers, such as the federal government or state governments, the army or other large institutions buy whole complex systems from the few firms capable of producing them, bidding is often used to decide the contract award. Usually the mechanism is the sealed bid in one of its many variants.

In some economic activities the open auction is used instead of the sealed bid. These include art and antique markets as well as furs, used cars, and tobacco. The Dutch auction is employed for selling tulip bulbs. Bidding and auctioneering appear to have originated in the West. There are Roman and Babylonian references; but the earliest oriental reference known to this author is from seventh century China [7]. An interesting survey of auctions has been given by R. CASSADY [8]. In recent years a considerable amount of mathematical and operations research work has been devoted to the study of bidding [9–12]. This has been motivated to a great extent by the economic importance of the construction of large-scale systems. There is every indication that their importance will grow.

It is important to note that one of the major differences in emphasis between the price mechanism and the bidding process is that implicit in the first is the assumption that all parties are capable of evaluating the worth of the transaction. Bidding, on the other hand, lays stress upon the importance of evaluation in the economic process.

In some instances bidding procedures may become so complex and the items to be bid on may be so large that a two-stage bidding process is called for where the first bid is for a contract to prepare a study of the evaluation of the program for which the second bid will be submitted.

Bargaining is used as a means of allocation in both the political and economic arenas. The most prevalent use of bargaining as an allocation process in the economy is in the domain of labor management disputes. On the international level tariff negotiations, trade treaties, disarmament conferences, and so forth, provide many examples.

Large-scale bargaining is often carried on with intermediaries and much of the process is devoted to reformulating and clarifying issues as well as specifying the positions of the participants. The process itself is both costly and lengthy. The expense of the bargaining process is easily indicated by the importance of professional arbitration. A valuable theoretical analysis of the time and cost aspects of the bargaining process has been done by JOHN CROSS [13].

There is a considerable difference between two-sided bargaining and multi-person bargaining. It is fairly well established that two-person bargaining in spite of its costliness appears to be a useful and viable form for the settlement of a broad range of problems. It is not clearly known how the degree of effectiveness of bargaining as a process drops off as the number of parties is increased.

An important feature in bargaining is that in many areas formal institutions and procedures exist to take care of the process. The work of IKLÉ on how nations negotiate [14] is a valuable companion to the work of CROSS inasmuch as the latter stresses the economic aspects of bargaining whereas the former lays emphasis on its international diplomatic aspects.

In the discussion of bargaining above, the type of bargaining could be described as formal bargaining inasmuch as, in general, there are relatively well-defined 'rules of the game'. Professionals and intermediaries are often employed and there is a more or less formal structure to the process. There is also informal bargaining. This more properly belongs with haggling and is discussed below.

Informal bargaining and haggling have more of a historical and anthropological interest than they have an influence on a large modern economy. In many South American, African and Asian countries there are still relatively simple markets where one can haggle over how many chickens a pig is worth. The exchange may be effected by barter.

Undoubtedly there is a large group of individuals who derive considerable joy from bargaining, 'buying wholesale' and getting discounts. There are those who devote much time and psychic energy to the uneconomic pursuit of discounts and 'bargains'. The process is only uneconomic if we fail to realize that bargain-hunting is to some as much of a sport as baseball is to others.

The haggling in an oriental bazaar or in a horsetrading market [15] might be regarded as the preliminary steps towards the formation of a price system. In the mythology of economic theory the 'tâtonnement' in the market as horsetraders rush from post to post trying to get a better deal results eventually in a narrowing of the range of prices until finally with enough buyers and enough sellers to a good first approximation only a single price exists for a specific commodity. This view of the market brings out the relationship among haggling, bargaining and the price system. Haggling is bargaining in a more cultural than formal institutional setting; and the price system may be regarded as a limiting case of multilateral bargaining when the participants are very many, the commodities standardized and plentiful and communications are good.

Just because one form of trading merges into another under certain circumstances this does not tell us that they are similar. Quantitative differences often imply qualitative differences. This appears to be more or less the rule when examining trade, exchange and distribution.

Dictatorship, or the decision of a recognized higher authority or elite group is used in many instances to determine allocation. This type of allocation mechanism is linked both to custom and force. For example, the awarding of prizes and scholarships and in many instances the promotion of individuals within a hierarchy may be a fiat decision. Here, once more, it must be stressed that the decision may be closely related to the evaluation process. The awarding of a prize or promotion in a hierarchy usually calls for a type of evaluation far more complex and many-faceted than does the buying of an automobile.

During wartime or in other periods of emergency or scarcity, rationing systems are designed which present a complex mixture of allocation by fiat and by market price (or in some cases Black Market price). At the bureaucratic level, rationing and quota systems also generate an undercover bargaining system. Thus, the Soviet factory manager may be forced into intricate bargains with his suppliers or the central bureaucrats.

There is a perennial debate concerning what are the areas of activity in any society which can be handled best by fiat decision rather than by voting, bargaining or some other process.

Although it is a far cry from dictatorship, arbitration is a means for calling upon a third party in a manner that the original group recognize it as having higher authority. Given that the parties have agreed to submit their dispute to arbitration, within bounds the outcome depends upon the whim or wisdom of the 'higher authority'.

Force, fraud and deceit are often overlooked as methods for the allocation of resources. Yet in a country such as the United States undoubtedly several billions are allocation in this manner. There are many definitional problems in the classification of fraud and crime. This is especially true in the area of 'white collar crime', where the type of deceit and fraud is more gentlemanly than common theft.

A distinction must be made between the force of a dictator or hierarchy and that of a robber or invader. The former exercise power that is legitimized within the society, the latter do not have that legitimacy.

Although the distribution of goods by force is usually not legitimized by society and references to it are for the most part pejorative, there is a minor exception: That is prizes from physical, intellectual or other competitions. Boxing matches and chess tournaments both come under this category.

Custom accounts for a great amount of allocation [16], especially in more traditional societies. In some places the church is entitled to its tithe. In England and in other monarchies the king is entitled to certain resources. The size of tipping in different societies is basically a custom-determined method for allocating resources. Inheritance customs are extremely influential and gifts play an important role in distribution.

At this point it is also important to note that it is not only goods and services that are distributed in society. Some of these mechanisms also work well for honor and prestige. Others are not as efficient. Thus fiat, custom and even voting or political means are better for obtaining an appropriately legitimized title than is the price system. The title devaluation in the Austro-Hungarian Empire and in Imperial Brazil provide examples.

Gifts, honor and prestige all pose problems far beyond those of 'normal' economic goods. It has long been stressed by anthropologists [17] and others that although the exchange relationship and the gift and honors relationship may have much in common they also are considerably different. Even though the economist or psychiatrist would immediately say, 'Something must be given in return for the gift', nevertheless both will recognize that that which may be given in return is paid in a totally different currency. The currency may be enormously difficult to characterize.

Last but not least plain dumb luck should not be overlooked. Some people actually win lotteries. Others encounter other types of windfall. Still others specialize in having everything go wrong in their vicinity. Chance plays its role in human affairs. The perceptive and prepared individual will undoubtedly exploit good fortune better than others, however, he still needs the good fortune. A mediocre Poker player with good cards stands a good chance of winning more than a good Poker player with rotten cards.

In some instances where an impasse is reached with respect to the division of an indivisible article the allocation may be decided by 'tossing for it'. However, possibly one of the most important aspects of random allocation is in the distribution of external economies or diseconomies. The grower of roses or the man burning rubber tires very often does not care what happens to the aroma that does not reach him [18]. In cases involving torts due to negligence the interpretation of the responsibility for random occurrences is often critical [19].

The price system is not God-given. A simple-minded cost effectiveness analysis which fails to account for the nature and costs of the actual distribution process may easily cause large-scale errors. Given the nature of a modern economy it has not been demonstrated that the trend is necessarily towards a larger functioning of the price system. The problems of large-system implementation and of external economies and diseconomies certainly point in the opposite direction.

Yale University MARTIN SHUBIK

REFERENCES

[1] M. SHUBIK, *Strategy and Market Structure*, John Wiley, New York 1959.
[2] L. S. SHAPLEY and M. SHUBIK, *Competition, Welfare and the Theory of Games*, in process.
[3] K. J. ARROW, *Social Choice and Individual Values*, John Wiley, New York 1951, 2nd ed., 1963.
[4] R. FARQUHARSON, *Theory of Voting*, Yale University Press, New Haven and London 1969.
[5] A. DOWNS, *An Economic Theory of Democracy*, Harper and Row, New York 1957.
[6] M. SHUBIK, 'A Two Party System, General Equilibrium and the Voter's Paradox', in: *Zeitschrift für Nationalökonomie*, Vol. 28, 1968, pp. 341–54; 'Voting or a Price System in a Competitive Market Structure', in: *American Political Science Review*, forthcoming, 1970.
[7] LIEN-SHENG YANG, 'Buddhist Monasteries and Four Money-Raising Institutions in Chinese History', in: *Harvard Journal of Asiatic Studies*, Vol. 13, June 1950, pp. 174–91.
[8] R. CASSADY, *Auctions and Auctioneering*, University of California Press, Berkeley and Los Angeles, California, 1967.
[9] W. VICKREY, 'Counterspeculation, Auctions and Competitive Sealed Tenders', in: *Journal of Finance*, Vol. 19, 1961, pp. 8–37.
[10] J. H. GRIESMER and M. SHUBIK, 'Towards a Study of Bidding Processes', Parts 1, 2, and 3, in: *Naval Research Logistics Quarterly*, Vol. 10, Nos. 1, 2, and 3, pp. 11–21, 151–73, 193–247.
[11] L. H. GRIESMER, R. LEVITAN and M. SHUBIK, 'Towards a Study of Bidding Processes', Part 4, in: *Naval Research Logistics Quarterly*, Vol. 14, No. 4, pp. 415–34.
[12] V. L. SMITH, 'Experimental Auction Markets and the Waloosion Hypothesis', in: *Journal of Political Economy*, Vol. 73, August 1965, pp. 387–93.
[13] J. CROSS, *The Economics of Bargaining*, Basic Books, New York 1969.
[14] F. C. IKLÉ, *How Nations Negotiate*, Harper and Row, New York 1964.
[15] E. BOEHM-BAWERK, *Positive Theory of Capital*, Stechert-Hafner, 1891, p. 203.
[16] K. POLANYI, C. M. ARENSBERG and H. W. PEARSON (Eds.), *Trade and Market in the Early Empires*, The Free Press, Glencoe, Illinois, 1957.
[17] MARCEL MAUSS, *Le Don*, The Free Press, Glencoe, Illinois, 1954.
[18] L. S. SHAPLEY and M. SHUBIK, 'On the Core of an Economic System With Economies', in: *American Economic Review*, forthcoming, 1970.
[19] F. H. BOHLEN, *Cases on the Law of Torts*, 4th ed., Bobs-Merrill, Indianapolis, Indiana, Chapt. 3, 1941.

[12]

The "Bridge Game" Economy: An Example of Indivisibilities

Martin Shubik

Yale University

Most of the mathematics associated with the more general cases of production with indivisibilities than are dealt with here has been presented elsewhere in an article dealing primarily with nonconvex preference sets (Shapley and Shubik 1966). That analysis, however, could have been applied to production.

We assume that there are n people in an economy with one unit of one producer good, "time," and each with a utility function or a value for the one possible consumer good, "the Bridge game."

For simplicity we assume that a Bridge game requires the participation of four players, each for one unit of time. The production function for Bridge games can be described as:

$$x = \left[\frac{t}{4}\right],$$

that is, the output is the largest integer in the number $t/4$.

For further simplicity we assume that the utility function for each individual is $U_i(x_i) = x_i$ where x_i is the amount of Bridge playing he obtains. In this trivial case x_i will be 0 or 1, as each individual has the time for only one game.

We do not need to assume that the individuals' values for a Bridge game are comparable or that there exists any monetary mechanism. If we did, then we could state the characteristic function of the economy as below and illustrate it with a simple drawing, as is shown in figure 1. In this function,

$$v(s) = 4\left[\frac{s}{4}\right] \text{ for all } s \text{ where } s = |S| \text{ and } S \subset N,$$

Research undertaken by the Cowles Commission for Research in Economics under contract nonr-3055(01) with the Office of Naval Research.

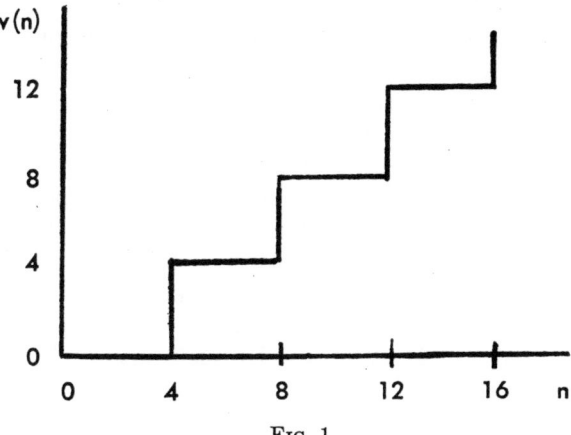

Fig. 1

$v(s)$ may be read as "the value that can be obtained by a set of players S." As the game is symmetric we may save ourselves a small amount of notation and use only "s," the number of players in the set "S," as all sets of the same size have the same total value even though they may have different players.

If we do not wish to compare utilities we must use a "characterizing function" whose values must be described by a vector with a component for each member of the coalition. For clarity the characterizing function for the five-person game is given. Here we use the following notation: $V(\overline{123})$ stands for the outcome achievable by the set consisting of players 1, 2, and 3.

$$V(i) = (0) \quad \text{for } i = 1, \ldots, 5$$

$$V(\overline{ij}) = (0, 0) \quad \text{for all pairs } i, j.$$

$$V(\overline{ijk}) = (0, 0, 0) \quad \text{for all triads } i, j, k.$$

$$V(\overline{ijkl}) = (1, 1, 1, 1) \quad \text{for all tetrads } i, j, k, l.$$

$$V(\overline{12345}) = \begin{cases} (1, 1, 1, 1, 0) & \text{or} \quad (1, 1, 1, 0, 1) \quad \text{or} \quad (1, 1, 0, 1, 1) \\ \text{or} \quad (1, 0, 1, 1, 1) & \text{or} \quad (0, 1, 1, 1, 1) \end{cases}.$$

It is easy to observe that if we treat the Bridge-game problem as an economy, a price system only exists when the number of players is divisible by four (or trivially when the number of players is less than four).

In the terms of game theory, the core (Shubik 1959; Debreu and Scarf 1963) exists only for $n = 1, 2, 3, 4, 8, 12, \ldots, 4k$. For $n = 1, 2, 3$ the core is trivial; no group can obtain anything. For $n = 4$ it is every imputation[1] in the game with side payments and the single point $(1, 1, 1, 1)$ for the no side-payment game. For $n = 4k$ where $k > 1$, the core is a single point for all of these games.

In this model our economy suffers from the effects of the indivisibility or the integral aspects of the Bridge game. As the numbers increase, the price system and the core appear and disappear periodically.[2]

Does it seem reasonable to have the price system disappear when the economy has 1,000,001 people instead of 1,000,000? A way to avoid this undesirable situation is to introduce a lottery ticket for the Bridge game. We assume that for any size n, each individual sells his time in exchange for a lottery ticket which carries the probability of $4/n[n/4]$ that he plays in a Bridge game and $1 - 4/n[n/4]$ that he is left out. As the number of

---

[1] In the side-payment game an imputation $a = (a_1, a_2, \ldots, a_n)$ will add to

$$\sum_{i=1}^{n} a_i = 4\left[\frac{n}{4}\right].$$

For a core to exist the above condition must be satisfied together with the inequalities

$$\sum_{i \in S} a_i \geq v(s) \text{ for all } S \subset N.$$

These conditions are illustrated for $n = 4$, 5, and 8.

$n = 4$	$n = 5$	$n = 8$
$v(1) = 0$	$v(1) = 0$	$v(1) = 0$
$v(2) = 0$	$v(2) = 0$	$v(2) = 0$
$v(3) = 0$	$v(3) = 0$	$v(3) = 0$
$v(4) = 4$	$v(4) = 4$	$v(4) = 4$
	$v(5) = 4$	$v(5) = 4$
		$v(6) = 4$
		$v(7) = 4$
		$v(8) = 8$

$a_i \geq 0$	$a_i \geq 0$	
$a_i + a_j \geq 0$	$a_i + a_j \geq 0$	
$a_i + a_j + a_k \geq 0$	$a_i + a_j + a_k \geq 0$	only the imputation
$\sum_{i=1}^{4} a_i = 4$	$a_i + a_j + a_k + a_m \geq 4$	$a_i = 1$ for $i = 1, \ldots, 8$
	$\sum_{i=1}^{5} a_i = 4$	
any imputation	no imputation	

[2] It is of interest to note that this unsatisfactory state of affairs is not encountered with the value (Shapley and Shubik 1969) of a game to a player. For the n-person game the value of the ith player Φ_i is given by

$$\Phi_i = \frac{4}{n}\left[\frac{n}{4}\right],$$

which fluctuates between 1 and $1 - 3/n$ for $n = 4k + 3$ and $k \geq 1$.

players increases, the probability of getting a Bridge game becomes arbitrarily close to 1.

The interpretation of the "lottery ticket" in this case is quite natural. If we leave out special social structure, then everyone has the same chance to find a game. In spite of the fact that the indivisibility may still cause an annoyance and one, two, or three people may fail to play, the odds become insignificantly small as the numbers increase. The price of the lottery ticket becomes approximately the same as the guaranteed game.

As long as the number of types of indivisible factors of production is finite and their capacities are finite the core will still appear and disappear in a periodic manner as the population mix of the different owners of resources is or fails to be in the correct ratio. As the size of the economy grows sufficiently there will still remain the possibility of a mismatch of resources, but relative to the whole economy, the amount of the mismatch approaches zero.

The implications of this example for the economy as it is are that, when indivisibilities are small relative to the economy as a whole, they do not matter very much and lottery tickets could be formally introduced and sold to preserve the price system; otherwise the lottery aspect will come about by a relatively minor amount of queuing, or social convention to correct for the minor aberration from the price system. In many instances the indivisibilities in society are large relative to the economy as a whole. Hence the approximation and limit argument will not apply.

The approach here appears to be related to the type of work on integer programming by Gomory (1968) and is also related to the comments on nonconvexity of Farrell (1959, 1961a, 1961b).

References

Debreu, G., and Scarf, H. "A Limit Theorem on the Core of an Economy." *Internat. Econ. Rev.* 4 (1963): 235–46.
Farrell, M. J. "The Convexity Assumption in the Theory of Competitive Markets." *J.P.E.* 67 (1959): 377–91.
———. "A Reply." Ibid. 69 (1961): 484–89. (a)
———. "Rejoinder." Ibid. 69 (1961): 493. (b)
Gomory, R. E. "Some Polyhedra Related to Combinatorial Problems." RC 2145. Yorktown Heights, N.Y.: IBM Research, July 25, 1968.
Shapley, L. S.; and Shubik, M. "Quasi-Cores in a Monetary Economy with Nonconvex Preferences." *Econometrica* 34, no. 4 (1966): 805–27.
———. "Pure Competition, Coalitional Power and Fair Division." *Internat. Econ. Rev.* 10, no. 13 (1969): 337–62.
Shubik, M. "Edgeworth Market Games." In *Contributions to the Theory of Games*, edited by A. W. Tucker and R. D. Luce. Vol. 4. Princeton, N.J.: Princeton Univ. Press, 1959.

A Note on the Shape of the Pareto Optimal Surface*

1. THE PROBLEM

Given n individuals each of which has a complete preference ordering over k outcomes, we may represent the preference ordering of any individual on a utility scale. Thus any one of the k outcomes can be represented as a point in an n-dimensional Euclidean space.

As we only postulate an ordering over the outcomes many utility scales will reflect the preference structure. Any scale that can be derived from any other by an order preserving transformation will serve. Thus for example if: $a > b > c > d$ then a scale which assigns values 4, 3, 2, 1 or another scale which assigns values 4, .003, .0003, .0001 will serve equally well.

Given the freedom in selecting scales is it possible to select them in such a manner that the k points lie on a hyperplane?

2. MOTIVATION

In the various theories of social choice such as voting, the economic theory of market exchange, bargaining, fair division procedures and various game theoretic solution theories, the solution suggested usually depends upon one of three sets of assumptions. They are (1) assumptions concerning the structure of the outcomes and the relationship between the outcome structure and the preference structure of the individual; (2) assumptions concerning a preference structure given *in abstracto* with no particular connection to or assumptions made about the structure of the outcomes; and (3) assumptions concerning the utility of the outcomes to each individual with a minimum of consideration given to the structure of the outcomes and the preference structure.

Examples of solutions pertaining to each of the three types of assumptions are given. The fair division procedure to decide how to divide a homogeneous cake between two individuals exemplifies (1). One individual cuts and the other chooses. The presumption is that the cake is divisible and homogeneous and (usually) that less cake is not preferred to more.

---

* The research was supported by the Office of Naval Research. The research was also partially supported by a grant from the Ford Foundation. The research was also partially supported by a National Science Foundation grant GP-32158X.

The economic model of exchange in a market with prices provides another example of (1). The existence of a price system depends upon the shape of the preference contours over different bundles of commodities.

The Arrow [1] discussion of voting and preferences provides an example of (2). No particular properties of the outcomes are specified.

The Harsanyi [2] and other value solutions, [3] the bargaining set [4] and the core [5] are usually defined for situations in which utility functions for individuals can be specified up to a linear transformation. Thus, the solutions considered are composed of an imputation or set of imputations in the n-dimensional utility space. The details and physical aspects of the outcomes giving rise to the imputations in the solution need not be considered, beyond making the assumption that there are sufficient conditions to enable us to derive utility functions from preferences over outcomes.

How much structure is imposed on the shape of the Pareto optimal surface in the utility space as we make assumptions about the structure of the outcomes and about the structure of individual preferences over these outcomes?

In particular the interest in the question above comes when we search for intrinsic ways to compare utilities or to achieve transfers or sidepayments. For example, if it were in general easy to obtain a flat Pareto optimal surface we might claim that a natural way of transferring and possibly even comparing welfare exists. If it is rarely possible to "flatten" the surface this appeal to an intrinsic measure loses its force.

This note examines part of this question for the case where no structure or description of the outcomes is given beyond observing that there are k outcomes.

3. A Planar Pareto Optimal Surface

Although our concern is with any number n of people and k outcomes we commence with the case for $n = 3$.

For 3 people and k outcomes with a complete preference ordering for each person, what is the smallest value of k such that there exists an ordering for which there is no way to assign nonnegative utilities that sum to one for each outcome that is consistent with the preference orderings.

Notation. Let u_j^i be the utility of the ith person for outcome j $i = 1, 2, 3$ and $j = 1, 2,..., k$. Assume that the ordering for person 1 is $(1, 2,..., k)$ where 1 is most preferred and k is least preferred. The orderings for persons 2 and 3 are $(m(1), m(2),..., m(k))$ and $(n(1), n(2),..., n(k))$ respectively. For a given ordering, utilities are sought such that:

(1)
$$u_1^1 > u_2^1 > \cdots > u_k^1$$
$$u_{m(1)}^2 > u_{m(2)}^2 > \cdots > u_{m(k)}^2$$
$$u_{n(1)}^3 > u_{n(2)}^3 > \cdots > u_{n(k)}^3$$
$$u_1^1 + u_1^2 + u_1^3 = 1$$
$$\vdots$$
$$u_k^1 + u_k^2 + u_k^3 = 1$$
$$u_j^i \geq 0, \quad i = 1, 2, 3 \quad \text{and} \quad j = 1, 2,..., k.$$

The problem is to find the minimum k for which there exists an ordering such that (1) has no solution.

Assumption 1. If each person prefers outcome i to outcome j, then it is not possible to find utilities that satisfy (1). Therefore, we will assume that no outcome is preferred to another outcome by all persons. This condition implies that for every pair of outcomes i, j—at least one person prefers i to j and at least one person prefers j to i.

The question of whether utilities exist that satisfy condition (1) for a given ordering can be resolved by solving a linear programming problem.

(2)
$$\max t$$
$$s/t \quad u_1^1 \geq u_2^1 + t$$
$$u_2^1 \geq u_3^1 + t$$
$$\vdots$$
$$u_{k-1}^1 \geq u_k^1 + t$$
$$u_{m(1)}^2 \geq u_{m(2)}^2 + t$$
$$\vdots$$
$$u_{m(k-1)}^2 \geq u_{m(k)}^2 + t$$
$$u_{n(1)}^3 \geq u_{n(2)}^3 + t$$
$$\vdots$$
$$u_{n(k-1)}^3 \geq u_{n(k)}^3 + t$$
$$u_1^1 + u_1^2 + u_1^3 = 1$$
$$\vdots$$
$$u_k^1 + u_k^2 + u_k^3 = 1$$
$$u_j^i \geq 0, \quad i = 1, 2, 3 \quad \text{and} \quad j = 1, 2,..., k.$$

PARETO OPTIMAL SURFACE

The optimal solution to this linear program is a solution to (1) if $t > 0$ and there is no solution to (1) if the optimal solution has $t = 0$. The linear program has a feasible solution ($u_i{}^j = 1/3$ for all i, j and $t = 0$), the optimal solution is bounded above by 1, thus (2) always has an optimal solution. Since the coefficients of (2) are all integer, a well-known result of linear programming gives:

LEMMA 1. *If there is a solution to (1), then there is a rational solution.*

LEMMA 2. *Assume that for every possible ordering with $k - 1$ outcomes satisfying assumption 1 there is a solution to (1). For any ordering with k outcomes such that one outcome is most preferred by at least one person and is least preferred by at least one person, there is a solution to (1).*

Proof. Without loss of generality, assume outcome 1 is most preferred by person 1 and least preferred by person 2. Remove outcome 1 from the orderings, then by the hypothesis there exists a solution to (1) for outcomes $2, 3, \ldots, k$ denoted by $\bar{u}_j{}^i$. Set $\hat{u}_1{}^1 = 2$ and $\hat{u}_j{}^1 = \bar{u}_j{}^1, j = 2, 3, \ldots, k$. Set $\hat{u}_1{}^2 = 0$ and $\hat{u}_j{}^2 = \bar{u}_j{}^2 + 1, j = 2, 3, \ldots, k$. For the third person, one of these possible cases occurs:

(A) If outcome 1 is the most preferred set $\hat{u}_1{}^3 = 2$ and $\hat{u}_j{}^3 = \bar{u}_j{}^3$, $j = 2, 3, \ldots, k$.

(B) If outcome 1 is least preferred set $\hat{u}_1{}^3 = 0$ and $\hat{u}_j{}^3 = \bar{u}_j{}^3 + 1$, $j = 2, 3, \ldots, k$,

(C) If $1 = n(p)$, set $\hat{u}_1{}^3 = [\bar{u}^3_{n(p+1)} + \bar{u}^3_{n(p-1)}]/2$ and set $\hat{u}_j{}^3 = \bar{u}_j{}^3$, $j = 2, 3, \ldots, k$.

The sum of utility for outcomes $2, 3, \ldots, k$ are equal, denote the sum by t. Let s denote the sum for outcome 1. If $s = t$, dividing each $\hat{u}_j{}^i$ by t gives a rational solution to (1). If $t > s$, add $t - s$ to $\hat{u}_1{}^1$, dividing each $\hat{u}_j{}^i$ by t gives a rational solution to (1). If $t < s$ add $s - t$ to each $\hat{u}_j{}^2, j = 2, 3, \ldots, k$, dividing by each $\hat{u}_j{}^i$ by s gives a rational solution to (1). Q.E.D.

LEMMA 3. *For any ordering with 2 outcomes that satisfies Assumption 1, there is a solution to (1).*

Proof. Allowing a renumbering of the outcomes and persons, there is only one case to consider.

$$u_1{}^1 > u_2{}^1$$
$$u_1{}^2 > u_2{}^2$$
$$u_2{}^3 > u_1{}^3$$

Then $u_1{}^1 = u_1{}^2 = 1/2$, $u_2{}^3 = 1$, $u_2{}^1 = u_2{}^2 = u_1{}^3 = 0$ satisfies (1). Q.E.D.

LEMMA 4. *For any ordering with 5 or fewer outcomes that satisfies Assumption 1, there is a solution to* (1).

Proof. Each person has a most preferred and least preferred outcome. Since there are 5 or fewer outcomes, among these 6 numbers must be at least one outcome appearing twice. Assume this is Outcome 1. Outcome 1 must be most preferred by some person and least preferred by another person in order to satisfy Assumption 1 (if Outcome 1 was most preferred by two persons and the third person preferred Outcome 1 to Outcome j then Assumption 1 would be violated; similarly if Outcome 1 was least preferred by two people). Now for $k = 3$, Lemmas 2 and 3 imply the result for $k = 3$; this result implies the result for $k = 4$ and this latter result implies the result for $k = 5$. Q.E.D.

LEMMA 5. *For 6 outcomes, there is an ordering satisfying Assumption 1 for which there is no solution to* (1).

Proof. Assume the following is a solution to (1)

$$u_1^1 > u_2^1 > u_3^1 > u_4^1 > u_5^1 > u_6^1$$
$$u_3^2 > u_6^2 > u_5^2 > u_2^2 > u_1^2 > u_4^2$$
$$u_5^3 > u_4^3 > u_1^3 > u_6^3 > u_3^3 > u_2^3$$

Sum the 9 strict inequalities that involve an outcome with an odd number being preferred to an outcome with an even number, this yields:

$$u_1^1 + u_3^2 + u_5^3 + u_3^1 + u_5^2 + u_1^3 + u_5^1 + u_1^2 + u_3^3 >$$
$$u_2^1 + u_6^2 + u_4^3 + u_4^1 + u_2^2 + u_6^3 + u_6^1 + u_4^2 + u_2^3.$$

This can be simplified to $3 > 3$ which is a contradiction. Thus, there is no solution to (1) for this ordering. Q.E.D.

Lemmas 4 and 5 completely resolve the case $n = 3$. It is also possible to answer the question for problems with a different number of persons where problem (1) is extended in an obvious manner.

LEMMA 6. *For two persons, any positive number of outcomes and any ordering that satisfies Assumption 1, there is a solution to* (1).

Proof. The only ordering satisfying assumption 1 is:

$$1 > 2 \quad > \cdots > k$$
$$k > k - 1 > \cdots > 1.$$

A solution satisfying (1) is $u_j^1 = (k - j + 1)/(k + 1)$ and $u_j^2 = j/(k + 1)$, $j = 1, 2, ..., k$. Q.E.D.

PARETO OPTIMAL SURFACE

LEMMA 7. *For $n \geq 2$ persons, 2 outcomes and any ordering satisfying Assumption 1, there is a solution to (1).*

Proof. The first k persons $1 \leq k < n$ have ordering $1 > 2$ and the remainder have $2 > 1$. A solution to (1) is $u_1^j = 1/k$ and $u_2^j = 0$, $j = 1, 2, ..., k$, $u_2^j = 1/n - k$ and $u_1^j = 0$, $j = k+1, ..., n$. Q.E.D.

LEMMA 8. *For $n \geq 2$ persons, 3 outcomes and any ordering satisfying Assumption 1, there is a solution to (1).*

Proof. Person 1 has ordering $1 > 2 > 3$. If there exists a person with outcome 1 the least preferred outcome, then an obvious generalization of the proof of Lemma 1 together with Lemma 7 gives the result. If no person has outcome 1 as the least preferred outcome, then in order for assumption 1 to be met there must be a person with $2 > 1 > 3$ and another person with $3 > 1 > 2$. The above argument is repeated with outcome 3 replacing outcome 1. Q.E.D.

LEMMA 9. *For $n \geq 4$ persons and 4 outcomes, there is an ordering satisfying Assumption 1 for which there is no solution to (1).*

Proof. The first 4 persons have orderings

$$1 > 2 > 3 > 4$$
$$1 > 4 > 3 > 2$$
$$3 > 2 > 1 > 4$$
$$3 > 4 > 1 > 2$$

which satisfy Assumption 1 and each of the remaining $n - 4$ persons has one of the above orderings. Summing the $2n$ inequalities with an odd numbered outcome preferred to an even numbered good contradicts the existence of a solution to (1). Q.E.D.

Summary. For $n \geq 2$ persons, the smallest number of outcomes k for which there exists an ordering satisfying Assumption 1 for which there is no solution to (1) is given by:

n	2	3	4	5	6	...
k	∞	6	4	4	4	...

We have assumed strict preferences; however, our results extend to the case where there is at least one good strictly preferred by at least one person. It is easy to see that the counterexamples in the proofs of Lemmas 5 and 9 remain counterexamples.

Further Problems

Given n individuals and k outcomes there are in total $(k!)^n$ preference states for the society of n individuals. What fraction of these could have arisen from trading economies with indivisible goods?

A trading economy with n individuals trading in m goods where for any good i there are b_i units, will have a preference structure for each individual that can be described by a modified lattice. An example to illustrate this is given by $n = 2$, $m = 2$, $b_1 = 2$, $b_2 = 2$.

Assumption 2. The preferences of an individual depend only upon the commodities he obtains, not on the holdings of others.

Assumption 3. An individual does not prefer less to more. Figure 1 shows the preference structure among the 9 outcomes that exist in this 2 person 2 goods 2 units of each good trading economy.

Assumption 4. Diminishing rate of substitution between goods.*

This is the assumption of strict convexity in "level sets" or indifference curves. In the example in Fig. 1 it would require that (1, 1) be preferred to (2, 0) or (0, 2).

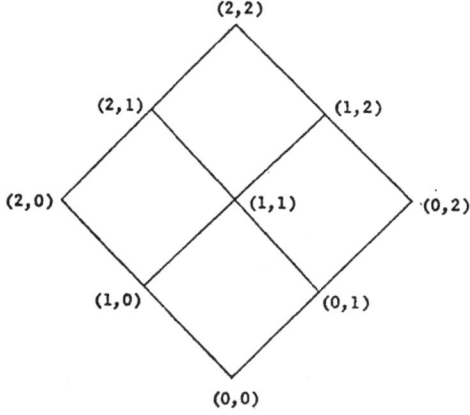

FIGURE 1

It is easy to see that only an extremely small fraction of preference states can arise from trading economies. For a two good economy with b units of each good and only integral amounts of goods possible, there are $(b + 1)^2$ possible states for each person, $(0, 0), (1, 0),..., (b, b)$. Of the $[(b + 1)^2]!$ preference states, fewer than $2^{(b+1)(b+1)}$ satisfy Assumption 3.

---

* We could modify this to a nonincreasing rate of substitution.

For preference states that arise from trading economies (that is, satisfy Assumptions 2, 3, and 4) with n persons what is the smallest number of goods k so it is possible to obtain a Pareto Optimal set that does not satisfy (1)? By Lemma 6, for 2 people there is no k. For one good every preference ordering yields a Pareto Optimal set that satisfies (1). Two examples show that for n persons ($n \geqslant 3$) k is 2.

EXAMPLE 1. For 3 persons in a trading economy with 24 units each two goods, consider the following 6 outcomes where x, y is the amount of Good 1 and Good 2.

	Person 1	Person 2	Person 3
Outcome 1	18, 0	6, 0	0, 24
2	0, 15	4, 9	20, 0
3	9, 5	15, 4	0, 15
4	0, 11	0, 4	24, 9
5	8, 0	14, 0	2, 24
6	0, 6	0, 18	24, 0

If persons 1, 2, 3 determine their preferences for the outcomes using functions $x + y$, $x + y$, $2x + 3y$ respectively, then the resulting preferences are the same as in the proof of Lemma 5. Thus there is no solution to (1).

EXAMPLE 2. For 4 persons in a trading economy with 10 units each of two goods, consider the following 4 outcomes:

	Person 1	Person 2	Person 3	Person 4
Outcome 1	6, 0	0, 6	4, 0	0, 4
2	0, 5	4, 0	2, 5	4, 0
3	4, 0	0, 4	6, 0	0, 6
4	0, 4	5, 0	0, 4	5, 2

If persons 1, 2, 3, 4 determine their preferences using functions $6x + 5y$, $5x + 6y$, $2x + y$, $x + 2y$ respectively, then the resulting preferences are the same as in the proof of Lemma 9. For $n > 4$, use the same outcomes for persons 1, 2, 3 and give each of the others $1/(n - 3)$ times the outcomes for person 4.

The utility functions ($x + y$, $2x + 3y$, etc.) used to define preferences in Examples 1 and 2 do not have strictly convex level sets. However, slight modifications of the functions do yield utility functions with the same preference orderings and strictly convex level sets.

References

1. K. J. ARROW, *Social Choice and Individual Values*, 2nd ed. Wiley, New York, 1963.
2. J. C. HARSANYI, "A Bargaining Model for the Cooperative n-Person Game," *Ann. Math. Stud.* **40** (1950), 325–355.
3. L. S. SHAPLEY AND M. SHUBIK, "Competition, Welfare and the Theory of Games" (unpublished manuscript), Ch. 7.
4. R. J. AUMANN AND M. MASCHLER, "The Bargaining set for Cooperative Games," in *Advances in Game Theory*, (M. Dresher, L. S. Shapley, and A. W. Tucker, Eds.), Princeton, 1964.
5. D. B. GILLIES, "Solutions to General Non-Zero-Sum Games," *Ann. Math. Stud.* **40** (1959), 47–85.

RECEIVED: March 19, 1973

GORDON H. BRADLEY[†]
*Naval Postgraduate School,
Monterey, California 93940*

MARTIN SHUBIK
Yale University, New Haven, Connecticut 06520

[†] This work was done while this author was at Yale University.

[14]
ON THE ROLE OF NUMBERS AND INFORMATION IN COMPETITION *

1. On the Gap Between the Verbal Description of the Competitive Market and Mathematical Economics

In the verbal descriptions of price in a competitive market the role of many competitors in bringing about the existence of a competitive price system is critical. Yet in the purported mathematization of these descriptions the role of the presence of many competitors disappears. The results of Arrow and Debreu [1] are utterly independent of the number of competitors assumed to be in the market.

The mathematical economist who may reserve his rigour for theorem proving rather than for assumption testing may argue « well when there are many small competitors in a market, then effectively each is so insignificant that he can be treated as a passive pricetaker ». It is precisely this piece of legerdemain that has resulted in a general equilibrium price theory that is unable to distinguish between a competitive and a centralized economy using a price system. Furthermore by getting rid of numbers the way they did, the propounders of the general equilibrium theory managed to maintain if not enlarge the split between oligopoly theory and general equilibrium theory. The general equilibrium model cannot be easily enlarged to include mixed markets with oligopolistic and competitive sectors existing in the same economy.

2. How to Take Numbers into Account : Replication

My first interest in trying to formulate a clearly well defined model of the role of numbers in a market came about through reading the basic work of Cournot [6] on competition among firms in a market and *Mathematical Psychics* by Edgeworth on the recontract of parties

---

* The research described in this paper was undertaken by a grant from the Office of Naval Research.

to a bargain [8]. In both instances it was shown that as numbers increase a competitive price would emerge. However what do we mean by « as numbers increase » ? Do we mean more, but different people in the market or do we mean more of the same type of people-doppelgangers or replicas ?

Edgeworth elected to consider closely related but not identical individuals : « It may appear that the quantity of final settlements is diminished as the number of competitors in increased. To facilitate conception let us assume that the field consists of two Xs, not equally, but nearly equally, natured ; and of two Ys, similarly related » [18].

Boehm Bawerk in his discussion of a horsetrading market had price emerge as the number of traders increase, but he too introduced different traders [4]. Shapley and Shubik have considered the emergence of price in the Boehm Bawerk market in terms of a game theoretic analysis [21]. Boehm Bawerk's analysis is closer to that of Edgeworth than Cournot in the sense that the former two were dealing with closed economic systems, i.e. with systems in which both sides of any market were considered as containing active economic agents. This contrasts with the open model of Cournot where the only agents of interest are the firms. The consumers are given by an aggregate demand function.

Cournot, in contrast with the other work noted was clear in his description of a replication process. « To make the abstract idea of monopoly comprehensible, we imagined one spring and one proprietor. Let us now imagine two proprietors with two springs of which the qualities are identical, and which, on account of their similar positions, supply the same market in competition » [6].

The work that is related to that of Cournot in the sense that large group competition was modeled as a noncooperative game, is that of of Chamberlin [5] [1]. When Chamberlin discusses monopolistic competition however each firm is differentiated from the other, albeit symmetrically. Chamberlin, like Cournot, studied an open system. When he talked about the entry of new competitors he did not make it clear how the differentiation among the large set of products related to the smaller [25].

In all of the above discussions the intuitive idea of a limiting process for the study of competition was present. However even in Edgeworth, attention was not paid to the details of definition of related economic models with different numbers of economic agents.

---

1. Cournot used quantity of an identical product as the independant variable. Chamberlin used the price of differentiated products and also included entry, but the solution concepts were mathematically the same.

Around 1952 in the context of the two different models of open oligopoly and bilateral monopoly Shapley and Shubik discussed and formulated the method of replication player types [17], [24], [25].

From the very start a choice had to be made between duplicating economic agents or fractionating agents. One method would lead naturally towards a framework where would always be with large finite sets of agents and where although between successive stages the power of the individual might be attenuating it nevertheless would at any stage be there. The other method would lead naturally towards considering a measure theory approach where a powerless trader could be considered as having a measure of zero. In the context of political models of voting Shapley suggested this for « oceanic games » [18] where the individually powerless mass of small voters could be considered as an « ocean » or continuum of voters. Considerably later Aumann adapted this approach to economic models [2].

The reasons for opting for the replication method when dealing with the oligopoly models of Cournot, Bertrand [3] and Edgeworth [9] had to do with the modeling of the firm as an institution. A simple example will help to illustrate this point.

Consider a single firm, a monopolist with a classical « U-shaped » average cost curve as shown in Figure 1. The mere drawing of the average cost curve of this variety is a method of modeling a host of subtle institutional assumptions concerning the firm. It is an economist's

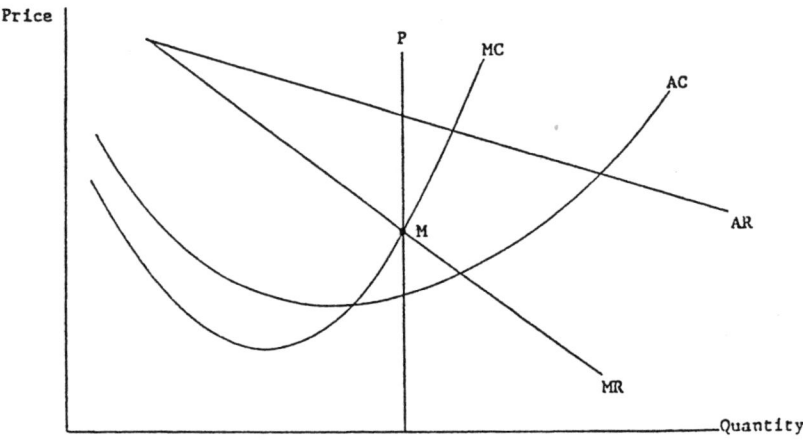

FIGURE 1

dream and an accountant's and operations researcher's nightmare. In the story that is usually told to justify this shape, real world considerations such as capacity constraints, limits on management, organization, internal information costs and controls, minor indivisibilities, overheads and all the host of factors which define a firm as an institution are included. This contrasts starkly with the neat production function that is homogeneous of order 1. For the single product firm with given input prices and no overheads such as plant, management or other forms of organization this would yield a constant average cost function.

The difference between the model of the firm as an institution and as a noninstitutional operator of a constant returns input-output formula is precisely that between a restaurant and many simple recipes. As a good approximation a steak dinner for two requires twice as many ingredients as for one. Usually however the requirements on the kitchen and chef differ in a nonlinear manner.

Returning to Figure 1, we can calculate the monopoly price that the firm should charge to maximize net revenue. It is indicated by the point P, where the line MR is the marginal revenue curve and the line AR is the demand. Suppose we now wish to answer the question « what is the effect of more competition on this price ? » Our problem is how to make this question sufficiently precise that when we have answered it we know what we have answered.

Let us be even more specific, suppose we wish to contrast the power exercised by a duopoly with that of a monopoly. Intuitively it would be nice to cut the monopoly in half, but how ? What do the two new average cost curves look like when we try to saw AC in half ?

We might try to introduce another monopolistic firm that is identical with the original, but then we will have so much overcapacity in the market that the question we can asnwer is not really the one we have in mind. Let us try to operationalize the question in a different way.

Suppose a single firm operates in a market. Now suppose that the market were doubled in size, i.e. at any price twice as much could be sold as could be sold previously. Now suppose that in the new market there were two firms each identical with the single monopoly firm that existed in the smaller market. It is easy to see that if the two acted as one [2] then the monopoly price will be as before and the

---

2. We also need some conditions forbidding monopolistic discrimination against market segments. This point has been discussed elsewhere [24].

individual revenues and sales of the two firms will be each the same as the monopolist's in the smaller market.

Now if we apply a solution concept other than monopoly, for example the Cournot noncooperative behavior, or a variant such as those of Bertrand, Edgeworth or Chamberlin then we may obtain some insight into the question : « How is monopoly power weakened when we contrast the gains of the identical firm operating in a market in which the size of its absolute share is constant, but its percentage share decreased as $1/n$? »

Using this method of replication it is possible to show that for a host of somewhat different models involving price or quantity as the strategic variable and including product differentiation, then the non-cooperative solution (which is Nash's mathematical generalization [11] of Cournot's original solution) shows an approach to the competitive level as numbers increase. This is sketched in Figure 2.

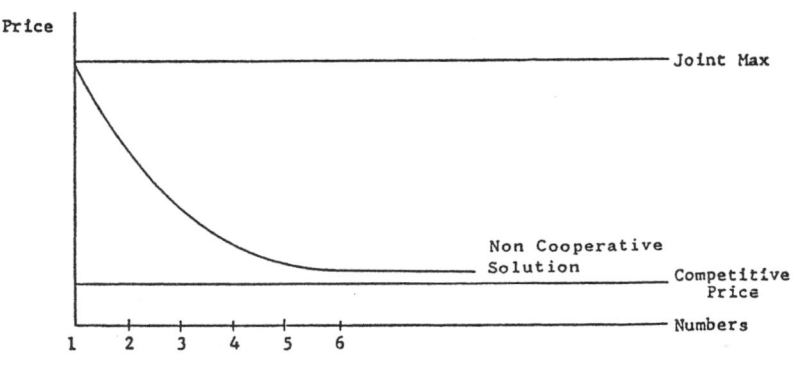

FIGURE 2

It is possible to renormalize the overall size so that we replace the firms by shrunken images of themselves (with the point of minimum average cost for each firm being at a level of output of $1/n$ in the n firm market). However the replication appeared to be intuitively more acceptable.

If the firms had linear costs or no costs then fractionation might appear to be the easiest. This was the case with Cournot's market.

In devising a method to study competition in a closed trading market the natural point of departure was Edgeworth's work. Shapley and Shubik originally elected to push in the direction of sidepayment models because we believed that they offered mathematical tractability and that the insights and results obtained from the analysis of these models would lead to insights, conjectures and methods to handle no sidepayment models.

Essentially what is meant by the existence of a sidepayment medium is that the utility function for each individual contains at least one common commodity for which that component of each individual's utility function is linear and separable. If there are $m + 1$ commo-

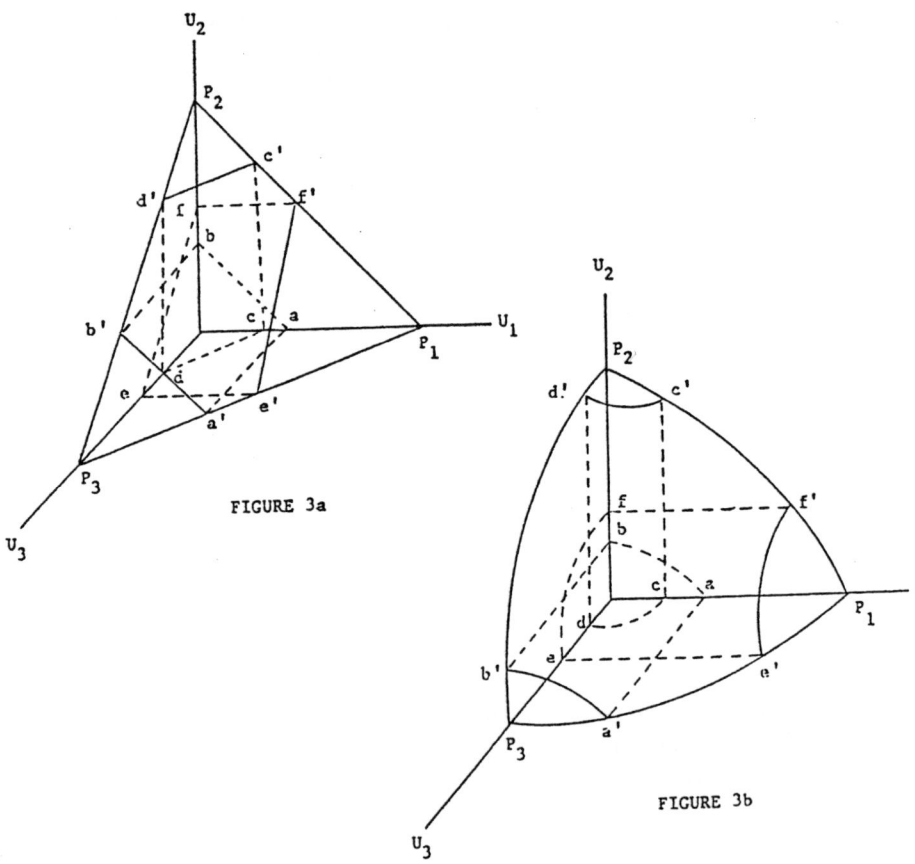

FIGURE 3

dities where the last is a transferable utility or a sidepayment medium then the preferences of an individual i can be represented by a utility function of the form :

$$U_i (x_1, x_2, ..., x_m, x_{m+1}) = \varphi_i (x_1, x_2, ..., x_m) + \lambda_i x_{m+1}$$

Furthermore it is assumed that there exists enough of the $m + 1^{st}$ commodity and that it is sufficiently distributed that the Pareto optimal surface when seen in the n dimensional utility space will be flat, i.e. it will be a hyperplane.

The mathematical methods needed to describe the powers of coalitions of traders are much simpler with sidepayments than without. Figues 3a and 3b show this in a 3-person market.

With sidepayments the overall Pareto optimal surface is flat as is shown by the plane $P_1 P_2 P_3$. Without sidepayments it is curved as is shown by the surface $P_1 P_2 P_3$ in Figure 3b. Similarly the two dimensional Pareto optimal surfaces which show how well pairs of traders can do if they by themselves are in the two cases respectively lines or curves. These are shown by ab (for treaders 1 and 2), cd (for 1 and 3) and ef (for 2 and 3).

3. The Core and Other Cooperative Solutions

3,1. THE REPLICATION RESULTS

The core of a market game is that set of imputations which are undominated by any coalition; i.e. those divisions of goods such that no subcoalition can achieve more by trading separately. In terms of Figures 3a and 3b the core is the set of imputations bounded by $a' b' d' c' f' e'$.

Edgeworth's model of the shrinking of the contract curve was put forth as a model of perfect competition [8].

There is free communication throughout a *normal* competitive field. You might suppose the constituent individuals collected at a point, or connected by telephones — an ideal supposition, but sufficiently approximate to existence or tendency for the purposes of abstract science.

A *perfect* field of competition professes in addition certain properties peculiarly favourable to mathematical calculation; namely, a certain indefinite *multiplicity* and dividedness, analogous to that *infinity* and *infinitesimality* which facilitate so large a portion of Mathematical Physics (consider the theory of Atoms, and all applications of the Differential Calculus). The conditions of a *perfect* field are four; the first pair referrible [1] to the heading *multiplicity* or continuity, the second to *dividedness* or fluidity.

I. Any individual is free to *recontract* with any out of an indefinite number, e.g., in the last example there are an indefinite number of Xs and similarly of Ys.

II. Any individual is free to *contract* (at the same time) with an indefinite number; e.g., any X (and similarly Y) may deal with any number of Ys. This condition combined with the first appears to involve the indefinite divisibility of each *article* of contract (if any X deal with an indefinite number of Ys he must give each an indefinitely small portion of x); which might be erected into a separate condition.

III. Any individual is free to *recontract* with another independently of, *without the consent* being required of, any third party e.g., there is among the Ys (and similarly among the Xs) no *combination* or precontract between two or more contractors that none of them will recontract without the consent of all. Any Y then may accept the offer of any X irrespectively of other is.

IV. Any individual is free to *contract* with another independently of a third party; e.g., in simple exchange each contract is between two only, but *secus* in the entengled contract described in the example (p. 17), where it may be a condition of production that there should be three at least to each bargain.

There will be observed a certain similarity between the relation of the first to the second condition, and that of the third to the fourth. The failure of the first involves the failure of the second, but not *vice versa*; and the third and fourth are similarly related.

A *settlement* is a contract which cannot be varied with the consent of all the parties to it.

A *final settlement* is a settlement which cannot be varied by recontract within the field of competition.

Contract is *indeterminate* when there are an indefinite number of *final settlements*.

He thus suggested recontracting and free communication. Furthermore although the method of analysis of the range of contract was clearly by means of replication, as can be seen from the quotation above he had in mind a continuous distribution of traders.

When I first considered Edgeworth's model in terms of the theory of games, the core had just been suggested as a solution concept by Gillies [10]. I observed the connection between Edgeworth's approach to the emergence of price and the core. However my prime interest was based on two considerations. The first was that the general equilibrium models of Walras [32] and subsequently Arrow and Debreu [1] did not in their mathematical formulations make explicit use of the numbers of competitors whereas Edgeworth's model and the core did.

The second point involved stability. It appeared to me that the approach of studying the stability of a competitive equilibrium point in economics by a displacement and measurement of excess supply and demand together with an adjustment mechanism depending on excess supply and demand was too slavish an imitation of the virtual displacement model in physics and was not necessarily the best model for economic stability. The core on the other hand could be viewed as a set of points which are combinatorically or coalitionally stable and this type of stability is different from the former and more socio-economically oriented.

It was the approach to stability that prompted me to propose the conjecture that the core of a general no sidepayment trading game would « shrink » under replication to Scarf immediately after he had presented a paper on an example of global instability of the competitive equilibrium [14] at Columbia University.

The limited results of Edgeworth [8] and Shubik [26] and the more general results of Scarf [15] all using replication for trading games with a finite number of traders established that the core would shrink and that the competitive equilibrium points are always within the core.

3,2. THE MEASURE THEORY RESULTS

Aumann has offered a different mathematical approach treating small traders as an « ocean » or continuum [2]. This certainly appears to catch the underlying implications of both Walras and Edgeworth.

The result of Aumann is that the core and competitive equilibria coincide. This is a much stronger statement than the statements concerning the shrinking of the core.

It might be thought that the Aumann result provides the appropriate logical underpinnings for the concept of a price system in the competitive market. It clearly does not. The fact that given Aumann's analysis two extremely different concepts of market organization coincide should lead us to suspect that they must be each telling only a partial story.

3,3. WHY THE CORE WAS THE WRONG MODEL OF COMPETITION

The Walrasian, Arrow-Debreu models of competition present a mechanistic maximizing price-taking nonstrategic model of economic man. The strategic power of the individual is removed by assumption and coalitions are ruled out *a priori*.

The core presents a combinatoric view of the market. All coalitions are simultaneously coalitions-in-being. This solution concept provides implicity a *high information state*, costless communication, static, institution-free model of the economy. Unlike the competitive equilibrium model the core can immediately accomodate one form of market imperfection. If a group of traders (or a single trader) has a finite measure a mixed market with some competitive and some oligopolistic components can be defined.

In spite of the greater flexibility of the core solution when information and production are considered new difficulties appear.

From Adam Smith onward the verbal description of the competitive market price has clearly been of a group of competitors each with individual freedom acting independently in an environment with *low* levels of information and communication. The core implicitly models *high* levels of information and communication [3]. As is noted in Section 4 below the model of Cournot appropriately modified provides an alternative to both of the approaches noted above. Furthermore it is a low information and communication strategic model.

When one tries to model production, immediately difficulties are encountered in modeling either decreasing or increasing returns to scale. Are production processes individually or jointly owned ? Without going into detail at this point, the crux of the difficulty appears to be that the production process essentially requires at least a rudimentary dynamic treatment. Even at the simplest level it requires a framework in which elementary institutions called firms play a distinguished role.

The core makes use of the coalitional form representation of the economy [19] whereas a model that adequately portrays production calls for the strategic form [19] and one that adequately portrays information conditions calls for the extensive form [19]. Scarf has managed to provide, as he himself points out, a not completely satisfactory extension of the core model with production [16]. The difficulty is with the model not the mathematics.

3,4. OTHER COOPERATIVE SOLUTIONS

Using the method of replication Shapley and Shubik have been able to show that several other cooperative solutions which differ considerably in concept from either the competitive equilibrium or the core also « approach » the competitive equilibrium as the number of replications is large. These solutions include the value [20] and the bargaining set [22].

---

[3]. The situation is even worse with the competitive equilibrium model which is not suited to handle nonsymmetric information conditions, i.e. where individuals have different amounts of information concerning the economy. See for instance Radner [13].

Unfortunately these solutions have all of the same weaknesses encountered with the core. They are noninstitutional static, high information models, which although they may be of considerable interest in and of themselves, are not necessarily the appropriate models for the study of the competitive process.

4. Replication with Strategic and Extensive Form Models

4,1. REALISM AND PROCESS

Walras may be regarded as providing the mathematical structure for the general equilibrium model of the economy, Edgeworth provided the basis for a cooperative game model of the general equilibrium exchange economy.

In contrast with the above, Cournot provided a mathematical model for competition in a partial equilibrium or open economic system.

In Cournot's model the individuals do not communicate, debate and recontract-they act independently. Cournot shows how the competitive price emerges as the number of competitors increases [6]. As his model was selected with zero costs of production replication or the division of firms are equivalent.

His model was presented in strategic form, i.e. a payoff for each firm is specified as :

$$\pi_i = D_i\, f\, (D_1 + ... + D_n)$$

From the context of the model one could construct a single move extensive form model to indicate that all firms select their levels of production in ignorance of each other's moves.

The criticisms of the Cournot model are primarily of two types : 1°) the realism of the description of market moves (in particular the selection of production as the strategic variable) and 2°) the treatment of information and dynamics.

Various writers have argued that the Cournot model could be made more « realistic » by considering product differentiation, product variation, price as the strategic variable or price and production simultane-ously as strategic variables. These are criticisms concerning relevance and the empirical problems in the modeling of markets.

I believe that the importance of strategic variables varies from market to market as a function of technological and institutional details. The selection of the most important variables cannot be done *a priori* and gives rise to a task in applied economics.

4,2. Institutions and Information

Unlike the core the Cournot model can be converted into a dynamic model virtually directly. Instead of considering a single move model we might consider a series of periods with moves each period and various information conditions.

Different markets and institutions use various processes thus there are sealed bids, auctions, Dutch auctions, two-sided matching procedures used in some stock markets, elaborate distribution and retail systems and so forth. They reflect information differences and evaluation procedures. Buying antiques, houses or bonds differ in the amounts of inspection, evaluation and calculation required of the buyers.

Although Cournot did not concern himself with closed noncooperative models of an exchange economy it is possible to construct such models. Shapley and Shubik [23] and Shubik [28], [30], have built such models and have shown that under replication the noncooperative price approaches the competitive price. It is my belief that these models are far closer to being the appropriate mathematical models of the competitive price system than either the competitive equilibrium or the core.

4,3. Behavioral Solutions

The multiperiod process models of markets call for a complete specification of all positions that the system can attain, as well as a complete specification of the information states. Given the model one can examine it for the existence of noncooperative equilibria; however for games in extensive form there are several types of noncooperative solutions which can be specified and furthermore they melt imperceptibly into behavioral solutions. This is discussed in more detail elsewhere [29]. Here the three major classes of solutions are noted :

1) *Noncooperative Equilibria with State Strategies* : This class contains the Cournot equilibrium, other noncooperative oligopoly models and the competitive equilibrium as a special case ;

2) *Noncooperative Equilibria with Historical Strategies* : This class includes equilibria whose stability depends not merely upon the current state of the system, but the history of the process leading to the current state ;

3) *Behavioral Solutions* : The emphasis is changed from a search for equilibrium to a specification and study of process. Equilibrium

may be reached and may be consistent with the noncooperative solution, but the focus is on process and equilibrium considerations are incidental [4].

5. Further Modeling Problems

5,1. Money and Credit

In trying to model a closed economy as a game in strategic form problems are encountered in defining strategies and payoffs for the individuals. In particular because the system must be defined for all positions of disequilibrium as well as equilibrium if individuals move independently there arise many occasions for which the books do not balance. Unless we postulate a world of perfect trust we need to introduce money and credit to help to « decouple » individual decisions and to simplify market structures. A simple pair of examples shows this.

We first consider a world with two types of traders trading in two commodities; where each type owns one of the commodities. Next we consider trade in more than two commodities.

Suppose traders of Type 1 each have an initial bundle of (A, O) and traders of Type 2 have (O, B). Let prices be quoted in terms of the first commodity. Thus a strategy for trader i of Type 1 might be (p_i, d_i) where p_i is the price he offers for the second commodity and the amount he demands. Similarly trader j of Type 2 has as his strategy (q_j, s_j). The price he asks for the second commodity and the amount he offers. Assume that some market mechanism produces a final price [30].

It is reasonable to impose the conditions that $p_i d_i \leq A$ and $s_j \leq B$. There is no need for a money or credit.

Suppose now that there are $k + 1$ commodities and traders of Type 1 have $(A_1, A, ..., A_{k+1})$ and Type 2 have $(B_1, B_2, ..., B_{k+1})$. For purposes of illustration assume that traders of Type 1 want to be buyers of the first m commodities and sellers of the rest, and for Type 2 it is vice versa.

We might simplify and organize the markets in such a way that instead of having $k(k+1)/2$ markets for every pair of goods the $k + 1^{st}$ good is used as money and there are k markets, each for one good and the money.

If we impose the condition on traders of Type 1 that $\sum_{k=1}^{m} p^k_i d^k_i$

---

4. See for instance Nelson and Winter [12].

$\leq A_{k+1}$ this would be tantamount to granting them no credit for the goods they are selling. This is far too stringent a financial restriction and would result in nonoptimal trade. If we used the limit $\sum_{k=1}^{m} p^k_i d^k_i$ $\leq A_{k+1} + \sum_{r=m+1}^{k} p^r_i s^r_i$ this would be extremely liberal as it amounts to believing that an individual will sell everything he offers at the price he names.

In order to completely define the model we need to postulate a bank or referee who will extend « commercial credit » for the period of trade.

5,2. BANKRUPTCY AND NEGATIVE CLAIMS

If credit is granted then there is the possibility that an individual is unable to pay back that which he owes. In order to cover this eventuality, bankruptcy laws must be specified.

A reasonable way to model insolvency and bankruptcy conditions is to introduce credit as a « paper commodity » which unlike real commodities can take on positive and negative values. Whereas with real commodities it is not particularly naturel or meaningful to consider « minus a ton of wheat », most financial instruments appear on two balance sheets with opposite signs. If we consider the financial instrument as a special form of good the extension into a negative orthant is natural and we can thereby extend our indifference maps to reflect insolvency.

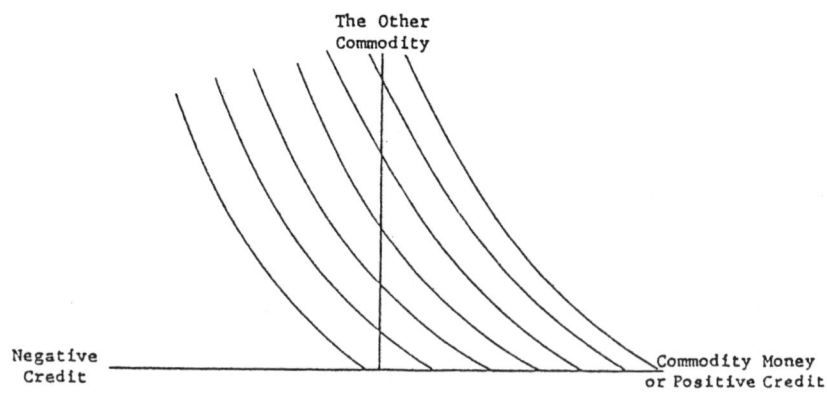

FIGURE 4

Figure 4 shows a simple case of an economy with one commodity, one commodity money and a credit instrument. If we assume that « the bank » or the system pays out any creditor whatever his full claim is in the commodity money then we need only two dimensions, as a positive claim on paper would be the same as ownership of the commodity money. The positions which involve ending up in debt will have the trader's indifference curves shaped by his preferences for the outcomes determined by insolvency and bankruptcy. Shapley and Shubik are investigating this type of extension of the indifference map in considering credit.

In a modern economy financial instruments are as real as bread. They provide the means for defining subtle distinctions in ownership rights and responsibilities. Mass market, low information dynamic models which include goods, services and financial instruments appear to me to offer a far better potential for the understanding of the formation of the price system than do either the competitive equilibrium or core models.

Martin Shubik.

REFERENCES

[1] Arrow K. J. and G. Debreu, « Existence of an Equilibrium for a Competitive Economy », *Econometrica*, 22 (1954), pp. 265-290.

[2] Aumann R. J., « Markets with a Continuum of Traders », *Econometrica*, 32 (1964), pp. 39-50.

[3] Bertrand J., « Théorie mathématique de la richesse sociale » (review), *Journal des Savants*, Paris, sept. 1883, pp. 499-508.

[4] Boehm Bawerk E. von, *Positive Theory of Capital* (translated by William Smart), New York, G. E. Steckert, 1923 (original 1891).

[5] Chamberlin E. H., *The Theory of Monopolistic Competition*, Cambridge, Harvard University Press, 6th ed., 1950.

[6] Cournot A. A., *Researches into the Mathematical Principles of the Theory of Wealth* (translated by N. T. Bacon from French, 1838), New York, A. M. Kelley, 1960, ch. vii, Section 43.

[7] Debreu G., *Theory of Value*, New York, John Wiley, 1959.

[8] Edgeworth F. Y., *Mathematical Psychics*, London, Kegan Paul, 1881, p. 40.

[9] Edgeworth F. Y., *Papers Relating to Political Economy, I*, London, Macmillan, 1925, pp. 111-142.

[10] Gillies D. B., « Solutions to General Non-Zero Sum Games », *Annals of Mathematics : study 40* (1959), pp. 47-85.

[11] NASH J. F. Jr., « Non Cooperative Games », *Annals of Mathematics*, LIV (Sept. 1951), pp. 286-295.

[12] NELSON R. and S. WINTER, « Neoclassical Versus Evolutionary Theories of Growth : Critique and Prospectus » Institute of Public Policy Studies, University of Michigan, August 1973.

[13] RADNER R., « Equilibrium under Uncertainty », *Econometrica*, 36 (1968), pp. 31-58.

[14] SCARF H. E., « Some Examples of Global Instability of the Competitive Equilibrium », *International Economic Review*, I (1960).

[15] SCARF H.E., « The Core of an N Person Game », *Econometrica*, 35 (1967), pp. 50-69 and also DEBREU G. and H. E. SCARF, « A Limit Theorem on the Core of an Economy », *International Economic Review*, 4 (1963), pp. 235-246.

[16] SCARF H. E., « On the Existence of a Cooperative Solution for a General Class of n-Person Games », *Journal of Economic Theory*, 3, 2, 1971.

[17] SHAPLEY L. S., « Markets as Cooperative Games », p. 629, Rand Corporation, Santa Monica, California, 1955.

[18] SHAPLEY L. S., « Values of Games with Infinitely Many Players », *Recent Advances in Game Theory*, Princeton University Conference, Princeton N.J., 113-118.

[19] SHAPLEY L. S. and M. SHUBIK, « Competitive Welfare and the Theory of Games » (unpublished manuscripts), chapters available as RAND reports.

[20] SHAPLEY L. S. and SHUBIK M., « Pure Competition, Coalitional Power and Fair Division », *International Economic Review*, 10, 3 (1969).

[21] SHAPLEY L. S. and SHUBIK M., « The Assignment Game I : The Core », *International Journal of Game Theory*, 1, 2 (1972), pp. 73-94.

[22] SHAPLEY L. S. and SHUBIK M., « Convergence of the Bargaining Set for Differential Games » (mimeographed), 1972.

[23] SHAPLEY L. S. and SHUBIK M., « Some Strategic Models Related to General Equilibrium » (mimeographed), February 1973.

[24] SHAPLEY L. S. and SHUBIK M., « Price Strategy Oligopoly with Product Variation », *Kyklos*, 22 (1969).

[25] SHUBIK M., *Strategy and Market Structure*, New York, John Wiley, 1959.

[26] SHUBIK M., « Edgeworth Market Games » in *Contributions to the Theory of Games IV*, edited by R. D. Luce and A. W. Tucker, Princeton, Princeton University Press, 1959, pp. 267-278, also Seminar Notes CASBS, 1955.

[27] SHUBIK M., « Fiat Money and Noncooperative Equilibrium in a Closed Economy », *International Journal of Game Theory*, 1, 4 (1972), pp. 243-268.

[28] SHUBIK M., « Commodity Money, Oligopoly, Credit and Bankruptcy in a General Equilibrium Model », *Western Economic Journal*, 10, 4 (1972), pp. 24-38.

[29] SHUBIK M., « The General Equilibrium Model is the Wrong Model for the Reconciliation of Micro and Macroeconomic Theory », CFDP 365, November 1973.

[30] SHUBIK M., « A Trading Model to Avoid Tatonnement Metaphysics », CFDP 368, February 1974.

[31] SHUBIK M. and W.WHITT, "Fiat Money in an Economy with One Nondurable Good and No Credit" in *Topics in Differential Games,* edited by A. Blaquiere, North Holland, 1973.

[32] WALRAS L., *Element of Pure Economics,* Allen and Unwin, 1954, original 1926 (translated by W. Jaffe).

[15]

An Example of a Trading Economy with Three Competitive Equilibria

This brief note is presented merely as a convenience for those who wish to see what an actual numerical example of a smooth trading economy with multiple equilibria looks like when depicted in an Edgeworth box diagram. We present a two-trader two-commodity economy in terms of a fanciful exchange between two kinds of money. The example is robust, in that its qualitative features would survive small perturbations in the data.

The tourists.—Ivan has R 40 in his pocket and wants some dollars; John has $50 to spare and would be happy to exchange some of it for rubles. Their utility functions (x = rubles, y = dollars) are

$$\begin{cases} u^1(x, y) = x + 100(1 - e^{-y/10}) & \text{(Ivan, in rubles)} \\ u^2(x, y) = y + 110(1 - e^{-x/10}) & \text{(John, in dollars)}. \end{cases} \quad (1)$$

Note that these functions are not only concave and smooth (C^∞) but additively separable, with one good entering linearly in each. It is well known (but virtually ignored in many textbook treatments of competitive uniqueness)[1] that the competitive equilibrium is unique if the *same* good is linear and separable in all utility functions, provided only that this good is in sufficient supply and the preference sets are smooth (C^1) and strictly convex, as they are here. The present example shows that this "transferable utility" or "welfare maximization" approach to uniqueness does not allow even a modest tinkering with the hypotheses.

In figure 1 the indifference curves are indicated by the broken lines. The locus of points of tangency is the straight line $D^1 D^2$, given by

$$y = x + 50 - 10 \log 110 = x + 2.995. \quad (2)$$

Work supported by NSF grant SOC71-03779 AO2 and ONR contract N00014-76-C-0085, respectively.

[1] See, e.g., Arrow and Hahn 1971, chap. 9.

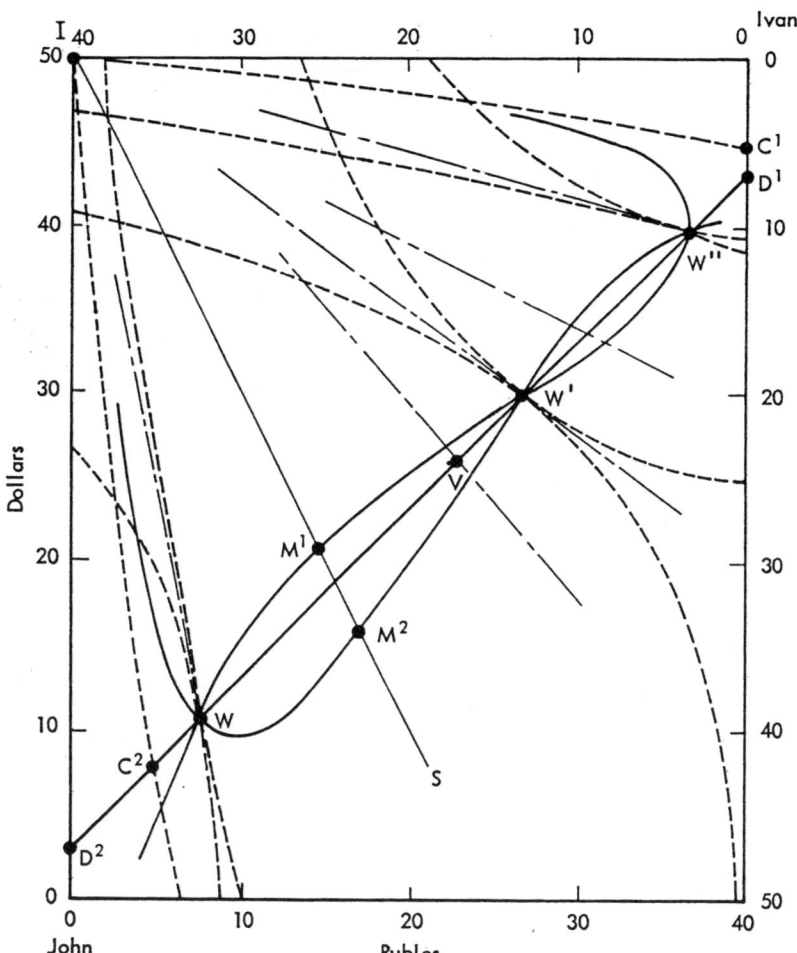

Fig. 1.—Three competitive equilibria

Edgeworth's "contract curve" C^1C^2 runs along this line and a short piece of the boundary. The conditions for a competitive allocation reduce by elementary calculus to the following transcendental equation:

$$x(1 + 11e^{-x/10}) = 10 \log 110, \qquad (3)$$

which has three roots in the region of interest. These lead to the three solutions indicated by W, W', W'' in figure 1 and given numerically in table 1. Their relation to the two response curves (solid lines) is also shown in figure 1.[2]

[2] To illustrate the definition of "response curve," suppose the price ray IS is given exogenously. Then Ivan's best trade is M^1, John's M^2.

TABLE 1
Numerical Data for Figure 1

	Allocation (To John)*		Exchange Ratio	Utility Payoffs		
	Rubles	Dollars	(Dollars: Rubles)	Ivan	John	
I	0.00	50.00	...	40.00	50.00	Initial point
C^1	40.00	44.89	0.20:1 (m) / 0.13:1 (a)	40.00	152.88	Endpoints of core
C^2	4.83	7.83	6.79:1 (m) / 8.73:1 (a)	133.69	50.00	
W	7.74	10.74	5.07:1	130.29	70.01	
W'	26.83	29.82	0.75:1	99.88	132.30	Competitive solutions
W''	36.78	39.77	0.28:1	67.27	146.99	
V	23.00	25.99	1.10:1 (m) / 1.04:1 (a)	107.94	124.96	Game value

Note.—(m) = marginal, (a) = average.
* For Ivan, subtract from (40, 50).

If we take a contract point between W and W', the direction of common tangency (dot-dash line) passes above the initial point I; if we take one between W' and W'', it passes below I. This means that the equilibrium prices associated with W' are dynamically unstable, in the sense that raising the price of either good would create a positive excess demand for that good. This in turn (in a suitable dynamic model) would tend to drive that price up still further. The two other solutions, W and W'', are dynamically stable (see Gale's article [1963], where another simple example of nonuniqueness will be found).

In table 1 we have also indicated the core and value solutions of the trading game in order to suggest outcomes alternative to those of the competitive equilibrium (see e.g., Shapley and Shubik 1969).

L. S. Shapley

RAND Corporation

M. Shubik

Yale University

References

Arrow, K. J., and Hahn, F. H. *General Competitive Analysis*. San Francisco: Holden-Day, 1971.
Gale, D. "A Note on Global Instability of Competitive Equilibrium." *Naval Res. Logistics Q.* 10 (March 1963): 81–87.
Shapley, L. S., and Shubik, M. "Pure Competition, Coalitional Power, and Fair Division." *Internat. Econ. Rev.* 10 (October 1969): 337–62.

[16]

On concepts of efficiency

This is a brief expository article designed to indicate how, by adding extra, natural considerations, a concept which originally appears to be crystal clear can quickly pose some deep problems in definition and in policymaking.

Efficiency as viewed from four viewpoints

'Efficiency' has a multiplicity of meanings and interpretations to different individuals. When applied to major social problems such as the supply of energy it is desirable to understand that there are at least four basically different viewpoints from which efficiency is judged. These viewpoints are sufficiently diverse that in public decisionmaking on basic issues such as the energy supply it is important to always keep in mind who the different groupings are; how and why their views of efficiency differ.

The four broad groups of individuals are:

1. Administrators, bureaucrats and politicians;
2. Engineers and physical scientists;
3. Economists and other 'soft scientists'

and last, but by no means least, a category that is frequently mystical and vague called:

4. The Public.

These groups are by no means always easy to identify in pure form. A politician may have been trained as an engineer, may be a member of the Sierra Club and may find that the site of his summer home has been selected as the most efficient location for a nuclear energy site.

In the processes which characterize an ongoing society, each individual may play several roles; hence the taxonomic division into four groups as suggested here is at best an extremely crude first-order approximation to help us make some, but by no means all, the distinctions we may need to make in understanding the process of decisionmaking in energy plant siting.

Before discussing efficiency from the different viewpoints noted above, a general discussion of efficiency is given; this can then serve as a basis for understanding [121] why different groups may emphasize different aspects of the same broad general problem.

On efficiency

The key items in understanding efficiency are as follows:

1. Are there one or more groups of individuals involved? If there are more than one, do they have identical or differing interests?
2. Does an individual or group know clearly what it wants?
3. Do people perceive clearly what they are willing to pay in direct involvement in time and money in order to attain their goals?
4. Do people understand enough of the technical details and the alternatives available that they feel confident that they can make a reasoned decision?
5. If a group is going to delegate the decisionmaking either implicitly or explicitly to others does it trust the honesty, concern and competence of the fiduciary decisionmakers?

These five points seem to be simple enough, but their implications can be complex.

Simple economic efficiency

The idealization of efficient economic decisionmaking by a single individual is a far cry from reality, but it is a useful base case to start from. A simple example is used to illustrate the point. Consider a single individual who owns his own business manufacturing french fried potatoes. We assume that he has only his own capital invested, that he knows all about the technology of manufacturing the french fries and that his goal is to maximize his net income subject to working not more than sixty hours a week.

The inputs into the business beyond his own time and money are the machinery he buys, the labor, potatoes, oil and a few other ingredients he requires. He can, for example, choose a labor-intensive technology using a little more machinery than knives and vats, or a capital intensive process using machinery to peel, slice and fry the potatoes. If he is an intelligent, technically informed economic man, his choice of inputs and technology will depend upon relative costs and technical efficiencies. If potatoes and labor are cheap he will employ labor to peel the potatoes crudely. If the potatoes were expensive then he would require that the potato peelers be more careful and 'waste less' of the potato. The word 'waste' in the sentence above really does not describe the economist's views of the alternatives he considers. If potatoes were cheap enough and disposal of unused potatoes were relatively cheap it might pay him to throw away a great amount of potato. This is not waste in an economic sense. It becomes economic waste and inefficiency if the price of potatoes is so high that the amount saved by careful peeling more than offsets the cost of extra labor.

Many of us have been brought up to believe that wasting food is a sin. Thus although the manufacturer of potato chips might find that with cheap potatoes, economic efficiency dictates that he throw away a great amount of edible potato with the peelings, our upbringing revolts against this procedure. The gesture of saving food may [122] be regarded by some as more important than the abstract idea that in doing so one is actually raising the costs of food production by being economically inefficient.

This simple example may be summarized more abstractly as follows: A single economic entity is economically efficient from its point of view if its economic organization and process selected are such that given its goals and knowledge there is no alternative organization which will yield it a higher payoff when all costs and outputs are taken into account.

This concept of efficiency is introspective (i.e., it is not explicitly concerned with the goals of others) and its takes technology as given (i.e., it is not concerned directly with technological efficiencies, but takes them as a datum).

Most economic problems are not as simple as this one. Even if an economic entity has a well defined goal, if it is not big enough, the odds are that its actions will influence its environment so that it must take into account the possibility that actions which enable it to achieve its goals may be harmful to others.

Technological efficiency
It may well be that our french fries manufacturer discovers a machine which peels potatoes with ½% of waste as compared with 10% of waste produced by a careful hand peeler. The definition of waste used here could be a purely technological one such as the percentage of edible potato adhering to the peel that is thrown away. Using this definition it is clear that the machine is technologically far more efficient than human labor. If, however, the machine is sufficiently expensive and labor is sufficiently cheap, buying the machinery will not be economically efficient.

Multiperson economic efficiency
Suppose that there are many individuals in an economy, each with different individually and jointly owned resources and each with somewhat different but well defined goals. Can we extend the concept of economic efficiency that sufficed for our french fries manufacturer to a group of individuals with differing goals and resources? In a relatively weak and limited way we can.

The essence of an economic society is that although individuals may have differing goals, there is no way in which an economic society can split into two groups with absolutely opposing goals. It is only in extreme conditions of war or in games such as chess that what one side loses, the other side gains. The underlying concept of an economy with production, trade and exchange is that by cooperation it is possible to make all individuals better off than they would be without cooperation. This concept has very little to do with fairness of equity. Suppose, for example, that separately two individuals could do nothing, but together they could make a 10 pound cake, and they both like cake. Economic efficiency tells us that they should make the cake rather than fail to cooperate. It does not tell us how they should divide the cake.

Given that everyone knows what they want and understands the technology and the alternatives with which they are faced, a society may be regarded as being [123] *economically efficient* if there is no way to reorganize which can make any individual better off without making at least one other individual worse off.

In particular, a society will not be efficient if it turns out that Group A could subsidize, compensate, pay or reward Group B to go along with some scheme that would leave them better off and no one else worse off. In practice, in many activities, societies fail to achieve economic efficiency because they lose the cooperation of those whose ox is being gored, by failing to pay them sufficient compensation.

Long-run efficiency

Returning to our french fries manufacturer, we suggested that his goal was to maximize net profits subject to a limitation on not working more than sixty hours a week. This problem was not quite fully defined, as simple as it was. We failed to specify the time horizon. If his goals were extremely short term he might wish to maximize the net revenues he can obtain in the near future even at the cost of wrecking his machinery and eventually losing customers. In contrast he might have an extremely low discount rate or might even be more interested in building for the future than consuming in the present.

Especially when we are dealing with major capital investments, the discount factor is a key variable in determining what will be economically efficient. If you know that you are going out of operation next year, although technical efficiency could be ensured by maintenance, it may not make economic sense to pay for maintenance. In some instances danger might increase with lack of maintenance, in which case it may be economically better to shut down early than pay high maintenance.

Long-term goals and desires of a society as a whole are extremely hard to characterize. Once investment programs extend to twenty years or more we are no longer planning for ourselves but for future generations.

Efficiency under Uncertainty

In general, even in the most orderly of societies the future is by no means certain. Even if an individual or an organization has well defined goals they must reflect their attitude towards risk. In some rare instances risk may be evaluated statistically; for example when you gamble at Las Vegas the odds at a fair roulette table can be calculated. When a population is large enough and regular enough, some odds can be calculated with fair accuracy as is exemplified by some calculations in life insurance. In general, however, many of the aspects of uncertainty involve low probability or infrequent events. Different people and organizations manifest a considerable variety of reactions to risk. Perhaps one of the important considerations that enters into risk behavior is whether the individual making the decisions is risking his own money or life or the lives and money of others.

If you paid a high fire insurance policy to protect your house and your house never burned down was the purchase of that policy economically efficient? There is no simple answer; but the question cannot be avoided. How much redundancy and peak capacity provisions should be built into a system is not any easy question to answer but it is a fundamental question to ask. [124]

There is virtually no major industrial process in existence which does not involve some degree of danger involving injury or even death. People are massacred by automobiles; major dams can and have broken killing hundreds; nuclear reactors could leak radiation or explode. But reaction to hard-to-measure risk has qualitative as well as quantitative aspects. Heart disease may be more deadly than cancer but cancer is more frightening. It does not matter if 'the wise men' tell the public that the burning of coal for new power is more dangerous than using nuclear energy if the public knows that they do not know how to evaluate the risk themselves, do not really understand the nature of this type of danger and do not trust 'the wise men'.

A fluctuating demand for french fried potatoes may cause a businessman some inventory problems and the price he may have to pay in order to avoid too many unsatisfied customers is that some french fries may go soggy and be thrown away. But this minor bout with uncertainty is not the same as a major power failure, national generating power shortage, or a nuclear explosion.

Perception, evaluation and communication
An old wisecrack goes: 'Are you trying to solve the problem or are you part of the problem?' In a dynamic world with myriads of strangers trying to cooperate and coordinate their actions to fulfill different goals, the answer is both. The social and political processes are the major part of the mechanism we have for perceiving, setting and evaluating overall goals; for operationally defining how much society as a whole is willing to allocate to avoid risk in different activities. The processes themselves are costly to run. The wisdom of nuclear scientists, engineers and economists are all vital inputs to the process; but they are not substitutes for the process. An understanding of what constitutes economic or technological efficiency, given that society as a whole knows what it wants and understands the risks involved, is a necessary but not a sufficient body of knowledge for a society to operate with.

Efficiency, process and participation

Most people are simultaneously idealists and realists; heros and cowards. The scope of everyone's perceptions is limited by their occupations. The politicians' goals are an intermix of their desires to do what they perceive to be right and to be reelected. The administrators and bureaucrats also want to do what they perceive to be right but their constituency is not the electorate. It consists of the politicians and the technicians and scientists, economists and lawyers they must interact with. Their criteria must include their own promotions and the protection of their jobs, and the protection and promotion of their staff and associates. Their measures of efficiency must include number of documents processed, jobs finished, permits granted, disasters avoided and so forth.

The engineers see a somewhat different picture; their concept of efficiency include meeting schedules; solving concrete technical problems, meeting production standards and in a relatively narrow societal context delivering what they promised. Scientists at a more basic level may value solutions to relatively abstract challenges more than they value the meeting of deadlines. To many of them the public is a nuisance which [125] gets in the way of solving basic problems. Even worse the public tends either to believe too little or too much of many scientific pronouncements. The praise of a scientific peer group or the publication of a new result is far closer a measure of efficiency than most interactions with the bureaucracy or the public at large.

The economists, ecologists, and others in applied social science march to yet another drummer. They may or may not comment on social process as a whole, but even if they do, the accountability is with the bureaucrats, administrators and politicians. If yet another scheme that they suggested goes up in flames, the odds are that they can go back to the drawing board. Their responsibility and measures of efficiency for their own performance are not one with the administrators and politicians.

The last set of actors is that mysterious mass called the public. The public is all of us, but in different roles. In general unless a specific member of the public has cast himself in the role of a public defender, or has a piece of property threatened by eminent domain, lives within ten or twenty miles of a new plant site or is actively concerned with some special interest that is directly affected, he has little interest or time to spend on worrying about how power is supplied or if there will be a surplus supply of plutonium. He does however have some direct concerns and he is quite able to supply his own list of efficiency measures. It is reasonable to assume that they will include:

1. How much will power cost me to heat my household and run my car?
2. Am I going to be taxed or is someone else going to be taxed for extra power development costs?
3. Are nuclear energy power plants going to radiate me to death or blow me up? How can I get these questions answered in a way that I believe the answers and trust the judgment, honesty and concern of those answering the questions?

Many members of the public may also be concerned somewhat with other problems of pollution and fear of nuclear war and a few may actively wish to know what are the alternatives to nuclear power and how those alternatives would influence their standard of living.

A useful measure of efficiency for those concerned directly with sources and the delivery of power is how well we can answer these questions to ourselves and how well can these answers be disseminated to the public as a whole in a manner that promotes belief, good faith and confidence. [126]

INFORMATION CONDITIONS, COMMUNICATION AND GENERAL EQUILIBRIUM*†

PRADEEP DUBEY AND MARTIN SHUBIK

Cowles Foundation for Research in Economics

It is shown that if the information sets of one game are a refinement of the information sets of the other, then the set of pure strategy equilibrium points of the game with less information is contained within the set of pure strategy equilibrium points of the game with more information.

1. Introduction. In this paper we demonstrate a relationship between games which have identical moves but different levels of information. Specifically, it is shown that if the information sets of one game are a refinement of the information sets of the other, then the set of pure strategy equilibrium points of the game with less information is contained within the set of pure strategy equilibrium points of the game with more information.

This result is proved in §2. In §3 we discuss the implications of the result for strategic market games, i.e., for games with an economic structure and specific price formation mechanisms [1], [6], [7].

2. Equilibrium points and information in games in extensive form. Let Γ and $\hat{\Gamma}$ be games in extensive form [3] on the player-set $N = \{1, \ldots, n\}$. We will say that $\hat{\Gamma}$ is a *refinement* of Γ (or Γ is a *coarsening* of $\hat{\Gamma}$), and denote it by $\Gamma < \hat{\Gamma}$, if Γ is obtainable from $\hat{\Gamma}$ *only*[1] by forming partitions of the information sets in $\hat{\Gamma}$.

Let $\Gamma < \hat{\Gamma}$. Though the strategy sets S^i and $\hat{S}^i$ of i in Γ and $\hat{\Gamma}$ are formally different, there is a natural inclusion $S^i \subset \hat{S}^i$. A strategy s^i in S^i is identified with $\hat{s}^i$ in $\hat{S}^i$, where $\hat{s}^i$ is as follows: the move chosen by $\hat{s}^i$ at any information set $\hat{I}_t^i$ of i in $\hat{\Gamma}$ is the same as the move chosen by s^i at I_j^i, where I_j^i is the *unique* information set of i in Γ for which $\hat{I}_t^i \subset I_j^i$.

For any game Γ we will denote by $\eta(\Gamma)$ the set of all its *pure-strategy* noncooperative equilibrium points.

Our aim in this section[2] is the following straightforward but striking result.

---

*Received November 24, 1977; revised March 1, 1980.
AMS 1980 subject classification. Primary 90D12.
IAOR 1973 subject classification. Main: Games.
OR/MS Index 1978 subject classification. Primary 236 Games/group decisions, noncooperative.
Key words. Noncooperative games, information, equilibrium points.
†This work relates to Department of the Navy Contract N00014-77-C-0518 issued by the Office of Naval Research under Contract Authority NR 047-006. However the content does not necessarily reflect the position or the policy of the Department of the Navy or the Government, and no official endorsement should be inferred.

The United States Government has at least a royalty-free, nonexclusive and irrevocable license throughout the world for Government purposes to publish, translate, reproduce, deliver, perform, dispose of, and to authorize others so to do, all or any portion of this work.

[1] All the elements of the extensive game (the tree, player partitions, etc.) are held fixed, and only the information sets are varied.
[2] We assume here that Γ has no chance moves. (See, however, Remark 2.) Also, while the length of Γ is finite, we permit the number of moves at any node to be infinite, in contrast with [3]–indeed this is required to be so for the application in §3. (The length of Γ is the supremum of lengths of all plays, i.e., paths from the initial move (root) to a terminal node.)

PROPOSITION. *Suppose* $\Gamma < \hat{\Gamma}$. *Then* $\eta(\Gamma) \subset \eta(\hat{\Gamma})$.

PROOF. If $\eta(\Gamma) = \emptyset$ there is nothing to prove. Let $s = (s^1, \ldots, s^n)$ be in $\eta(\Gamma)$. Put $\hat{s} = (\hat{s}^1, \ldots, \hat{s}^n)$ where $\hat{s}^i$ corresponds to s^i as described earlier. We will show that $\hat{s}$ is in $\eta(\hat{\Gamma})$.

First observe that s and $\hat{s}$ lead to the same play P. Now if $\hat{s}$ is not in $\eta(\hat{\Gamma})$ there is some player i who can improve his payoff by deviating from $\hat{s}^i$ to $\hat{s}'^i$, provided that the others keep their strategies fixed according to $\hat{s}$. Take the path P defined by $(\hat{s}^1, \ldots, \hat{s}^{i-1}, \hat{s}^i, \hat{s}^{i+1}, \ldots, \hat{s}^n)$, and let $\hat{I}_1^i, \ldots, \hat{I}_l^i$ be the information sets of i in $\hat{\Gamma}$ which contain nodes of P. Consider $I_{j1}^i, \ldots, I_{jl}^i$ where I_{jl}^i is the (unique) information set of i in Γ such that $\hat{I}_l^i \subset I_{jl}^i$. Since no two nodes of a play can lie in the same information set, the sets $\{I_{jl}^i : t = 1, \ldots, l\}$ are all disjoint. Construct the strategy s'^i for i in the game Γ as follows: the choice made by s'^i at I_{jl}^i is the same as the choice made by $\hat{s}'^i$ at $\hat{I}_t^i$; the choice made by s'^i at information sets other than $I_{j1}^i, \ldots, I_{jl}^i$ is arbitrary. Clearly the play defined by $(s^1, \ldots, s^{i-1}, s'^i, s^{i+1}, \ldots, s^n)$ is also P. Thus s is not in $\eta(\Gamma)$ since i can improve his payoff by deviating to s'^i, a contradiction.

REMARK 1. This proposition is not true for mixed strategies. Consider the game of matching pennies, where Player 1 wins if they match. If both players move simultaneously the payoff matrix is 2×2 (Figure 1a); each player has two moves, which coincide with his strategies. The only noncooperative equilibrium is in mixed strategies where each player uses a mixture of $(\frac{1}{2}, \frac{1}{2})$ over his two pure strategies. The expected payoff to each is zero. A refinement of the information sets in this game is given if we assume that Player 2 is informed of Player 1's move before Player 2 is called upon to move.

In Figure 1b we observe that the noncooperative equilibrium in the 2×4 matrix game is given where Player 2 uses his strategy C. The notation $(1,2; 2,1)$ can be read as the sentence: "If Player 1 chooses his first strategy choose move 2; if he chooses his second strategy choose move 1."

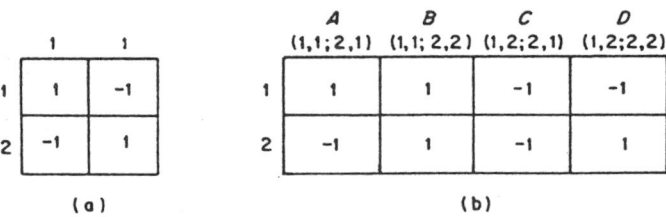

FIGURE 1

REMARK 2. If there are chance moves in the game, and if we vary information of the traders regarding *each others' moves only, while their information about chance moves is held fixed and identical*, then the proposition continues to hold (with the obvious modifications in the proof). However, the following example shows that the result breaks down outside of this case.[3] Consider the game in Figure 2 where players' information about chance moves is kept fixed but not identical. The game is zero-sum, the stated payoff being that to Player 1. All the nodes not enclosed explicitly by information sets form singleton information sets. The two games differ only in 1's information about 2's move. An equilibrium point in the coarse game is for 1 to go right in both cases, whereas 2 goes left on the left node and right on the right node; the payoff is 3. This is not an equilibrium point in the refined game because 1 will always

---

[3]We are grateful to R. J. Aumann for this example which improved upon an earlier example of ours.

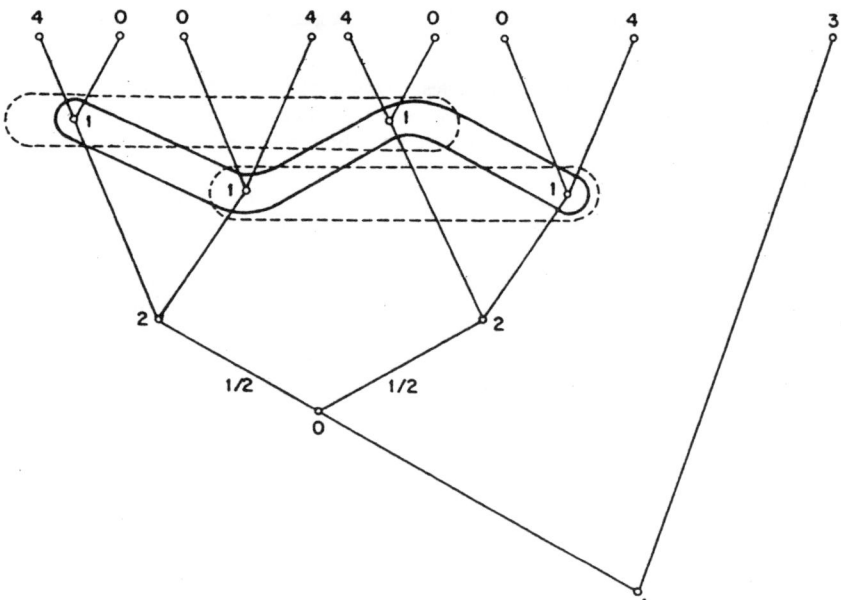

FIGURE 2

know what 2 has done and so can get 4. When the players' information about chance moves is allowed to vary, then a counterexample is even easier; for example, see §5.2 of Ponssard's paper [2]. This is similar to Hirschleifer's famous mutual crop insurance example [4], where reliable weather forecasts may actually be harmful; i.e. adding information may destroy beneficial equilibrium points.

REMARK 3. The result also breaks down if one substitutes "perfect equilibrium" [5] for "equilibrium," as shown by the example in Figure 3. In the coarse game, a perfect equilibrium point is for both players always to go left; in the refined game this is still an equilibrium point, but it is not perfect. In fact, here the inclusion goes in the opposite direction: the refined game has fewer perfect equilibria than the coarse one. It might be conjectured that this is always the case, but that is not true either. Indeed, consider a 2 × 2 matrix game with no pure strategy equilibria. Now give one player

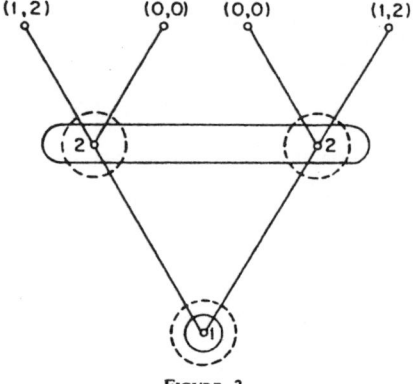

FIGURE 3

perfect information (converting it to a 2 × 4 matrix game). This always has at least one perfect equilibrium.

3. Equilibrium points in market games. One approach to the competitive equilibrium of a market is to represent a market as a game with a trading method fully defined and to show that the competitive equilibria may be obtained as noncooperative equilibria of this game. There are many ways of doing this, (e.g., [1], [6], [7], [8]). A problem with these models is that it is not clear to what extent the conclusions depend on the precise details of the informational structure of the game (i.e. who knows what at which stage).

For example, [1] describes a bid-offer game, which we call Γ^*. Each player decides how much of his endowment of each good to offer for sale, and how much of each good he wants to buy (his *bid* for that good). In Γ^*, all bids and offers are made simultaneously, so that really nobody knows anything about other players' moves at the time he makes his. In reality the various players' bids and offers may be made at different times, so that players may have partial information about other bids and offers at the time they make theirs. We will call such games *information variants* of Γ^*.

The game Γ^* has many representations in extensive form, all of them equivalent. What is common to them all is that whenever a player is called upon to make a move, he knows *nothing* about what has previously been done; each of them is an extensive game in its coarsest form.[4] We have shown that–in an appropriate sense and under appropriate conditions–the N. E.'s of Γ^* "converge" to the C. E.'s of the market if the market itself "approaches" a nonatomic market. In this sense, the N. E.'s of Γ^* may be considered "associated" with the C. E.'s of the market.

Any information variant Γ of Γ^* may be obtained by refining information sets in some representation of Γ^*. But then for all such Γ, it follows from our proposition that $\eta(\Gamma^*) \subset \eta(\Gamma)$. In words: the *C. E.-associated noncooperative equilibria are the only noncooperative equilibria that are common to all information variants of the market game.*

Similar results hold for any of the noncooperative game models of markets mentioned above.

Acknowledgment. We would like to thank R. J. Aumann for his tough editing and helpful comments.

---
[4] Strictly speaking, players in an extensive game in its coarsest form do not even remember what they themselves previously did. However, it can be shown that the pure strategy N. E.'s do not change if such a game is modified by having each player remember his own previous moves.

References

[1] Dubey, P. and Shubik, M. (1978). The Noncooperative Equilibria of a Closed Trading Economy with Market Supply and Bidding Strategies. *J. Economic Theory.* **17** 1–20.
[2] Hirschleifer. (1971). The Private and Social Value of Information. *Amer. Econom. Rev.* **61** 561–574.
[3] Kuhn, H. (1953). Extensive Games and the Problem of Information. *Ann. of Math. Studies,* Number 28. Princeton University Press, Princeton, N. J., 193–215.
[4] Ponssard, J. P. (1976). On the Concept of the Value of Information in Competitive Situations. *Management Sci.* **22** 739–747.
[5] Selten, R. (1975). Reexamination of the Perfectness Concept for Equilibrium Points in Extensive Games. *Internat. J. Game Theory.* **4** 25–55.
[6] Shapley, L. S. and Shubik, M. (1977). Trade Using One Commodity as a Means of Payment. *J. Political Economy.* 937–968.
[7] Shubik, M. (1972). Commodity Money, Oligopoly, Credit and Bankruptcy in a General Equilibrium Model. *Western Econom. J.* **10** 24–38.
[8] ———. (1976). A Trading Model to Avoid Tatonnement Metaphysics. *Studies in Game Theory and Mathematical Economics, Bidding and Auctioning for Procurement and Allocation.* Y. Amihud, ed. NYU Press, New York, 129–142.

COWLES FOUNDATION FOR RESEARCH IN ECONOMICS, BOX 2125, YALE STATION, NEW HAVEN, CONNECTICUT 06520.

THE MANY APPROACHES TO THE STUDY OF MONOPOLISTIC COMPETITION*

Martin SHUBIK

Yale University, New Haven, CT 06520, USA

What are the key questions in the study of monopolistic competition, oligopoly theory and allied topics dealing with competition among the few? A specification and discussion of some of these questions is given. In particular both problems and their resolution in studying oligopoly in a closed economy and the reconciliation of cooperative and non-cooperative models of oligopolistic competition with general equilibrium theory is discussed. Some problems with dynamics are noted. In particular the limitations of the non-cooperative equilibrium solution are considered. Questions are raised concerning the domain of application of perfect equilibria, the treatment of subjective probabilities and non-symmetric information.

1. Dynamics or statics

In the concluding remarks in the opening chapter of *Theory of Games and Economic Behavior* von Neumann and Morgenstern (1944) note:

'We repeat most emphatically that our theory is thoroughly static. A dynamic theory would unquestionably be more complete and, therefore, preferable. But there is ample evidence that it is futile to try to build one as long as the static side is not thoroughly understood... A static theory deals with equilibria... — the relationship between statics and dynamics — may be generically different from that of the classical physical theories.'

They proceeded to develop a cooperative theory based upon the characteristic function of a game. Although in their developments they sketched out a procedure by which one could obtain the characteristic function by starting with the details of the extensive form, both in the development of solution theory and in application, the characteristic function can be regarded as given initially to the economist. In essence, by starting with the characteristic function, the economist has assumed away details of process, has hidden express consideration of the rules of the game and the institutions of the economy which serve as the carriers of process. In return for this simplifi-

---

*This work relates to Department of the Navy contract N00014-77-C-5018 issued by the Office of Naval Research under Contract Authority NR 047-006. However, the content does not necessarily reflect the position or the policy of the Department of the Navy or the Government, and no official endorsement should be inferred.

cation it becomes feasible to concentrate on the study of the combinatorics of potential coalitions.

A different equilibrium approach is the one formalized generally by Nash (1951) and discussed in an economic context far earlier by Cournot (1897). The non-cooperative equilibrium has played a dominant role in much of the work on oligopoly theory and associated topics.

Although the non-cooperative equilibrium solution pervades many of the mathematical, verbal and diagrammatic treatments of oligopoly, two distinct themes are present. They are explicit one period equilibrium analyses, often fully mathematized and implicit or explicit multistage models of competition where the dynamics of action and reaction are often described verbally or diagrammatically. In either instance the game description required for the application of the non-cooperative equilibrium calls for the strategic or extensive form (or an infinite multistage) representation of the game. In all of these instances the rules must be specified. A full definition of the game calls for the description of a process model. Thus, although the cooperative game solutions and the non-cooperative equilibrium can all be regarded as equilibrium solutions, the former depend on models with the strategic structure supressed, or at best implicit, whereas the latter calls for the strategic or extensive form.

Some time ago, in attempting to describe the study of the dynamics of oligopolistic competition, I suggested the title 'Mathematical institutional economics' [Shubik (1959)]. The reason for this paradoxical title is that if an economic environment and players are modeled as a game in extensive or strategic form, the institutions of the economy are reflected in the rules.

Much effort in modern industrial organization, corporate planning and competitive analysis has been aimed at the study of dynamics yet in many of the models and much of the writing the importance of lapsed time as a factor is hardly developed.

The economy dwells in the polity and the polity exists in the context of a society. The rules of one game which may appear to be immutable at one point of time may be modified in a larger context in the space of a few months, years or generations. There are games within the game. In actuality the rules of a single game which are reflected in the laws and customs of society as well as current tastes and technology are not immutable. They are subject to challenge by the players (we can get away with this), reinterpretation by the lawyers and judges and changes by the legislation or public at large.

It is suggested here that there are many highly different purposes for the study and application of theories of oligopolistic competition, monopolistic competition and allied topics. Before they can be criticized, or before suggestions can be made as to where we go from here, the different purposes and questions must be set in context.

2. What are the questions?

2.1. A disclaimer on monopolistic and oligopolistic competition

Taxonomies are frequently highly useful in structuring thoughts. The phrase 'monopolistic competition' suggested by Chamberlin (1933) called attention to the key aspect of differentiation of product (and firm) in competition among the few. Whether Robinson (1933) really meant the same or not is, in my estimation, a topic best left to biographers and historians of economic thought. In the usage followed here, monopolistic competition covers oligopolistic competition and imperfect competition. The essential features being few competitors with strategic power, possibly differentiation, entry barriers and a host of other 'imperfections' such as humans with finite memories and abilities, indivisible goods, imperfect laws, less than infinitely fast markets, and so forth.

2.2. Nine or more related topics for investigation

Nine topics are noted here. No attempt has been made to be exhaustive or to produce completely mutually exclusive categories. Those selected are suggestive of much of the work currently in progress. The topics are listed and then discussed briefly:

(1) Static one period partial equilibrium oligopoly theory.
(2) The reconciliation of general equilibrium and static oligopoly theory.
(3) The new industrial organization.
(4) Regulation, contracts, law and economics.
(5) The structure of the firm, information, incentives and agency.
(6) Empirical institutional studies.
(7) Corporate strategic analysis.
(8) Behavioral theories of the firm and competition.
(9) Experimental gaming.

(1) Static one-period partial equilibrium oligopoly theory. For those who like their oligopoly theory straight there is a broad selection of well defined mathematical models of games in strategic form which have been built and still remain to be built. The original duopoly models of Cournot, Bertrand, Edgeworth and Hotelling can all be regarded as games in strategic form with pure or mixed strategy non-cooperative equilibria. This is also true for much of the work of Chamberlin, Stackelberg and Mrs. Robinson. But even with Cournot and certainly with Chamberlin, Stackelberg, Fellner and the writings on 'kinked oligopoly demand' the verbal argument is oriented towards dynamics.

The straight mathematical economist has many nice clean questions which can be asked in the context of a well defined model of a game in strategic form. An exploration of price, quantity or price and quantity as strategic variables; an investigation of quality differentiation or differentiation via the selection of location in a world with costly transportation all are worth noting.

Although entry, advertising, innovation and purchase of capacity all are clearly relevant to the understanding of oligopolistic competition and although they can be modeled as though they were one shot games in strategic form [see, for example, Nti and Shubik (1981a), Schmalensee (1983)]; in essence, there are many features of the items noted which call for considering dynamics.

The verbal treatments of Chamberlin's large and small groups, Stackelberg's reaction curves; Bain's barriers to entry all are able to tell a better story than the formal games in strategic form. But the cost of the more plausible explanation comes at the price of a less precise analysis. In a verbal treatment it is easy to slur over details such as the non-uniqueness of equilibrium points in even the simplest of entry models [see, for example, Nti and Shubik (1981b), Shubik (1983)]. This applies to contestable markets [Baumol, Panzer and Willig (1982)] where the assumption of zero entry and exit costs in general destroys uniqueness of equilibria in the one shot game [see Shubik (1959, p. 132)].

There is by now a large literature on spacial competition following Hotelling's original paper. This is attested to by the special issue of the *Journal of Industrial Economics* (Sept./Dec., 1982) devoted to this topic.

Questions concerning the relative importance of different strategic variables are made more precise by constructing a strategic form game. And the analysis of the game provides a useful way to carry out a sensitivity analysis.

(2) The reconciliation of general equilibrium and static oligopoly theory. The Arrow–Debreu models of the Walrasian system are closed and are not process models. Even the simplest of games in strategic form may be interpreted as a process model. It tells us what the feasible set of strategies is for each individual and it specifies what the outcome of the game will be regardless of equilibrium considerations.

Until recently almost all of the work on oligopoly theory has been done assuming a partial equilibrium context. Triffin (1940) tried somewhat prematurely to embed the monopolistic competition model into a general equilibrium system, but did not succeed. In the past ten years considerable progress has been made in doing so; this is discussed in section 3 below.

Among the more important questions are:

(1) For a finite number of firms are the non-cooperative equilibria generally inefficient?

(2) Can we construct Cournot and Bertrand–Edgeworth models which converge to the general equilibrium model as the number of competitors grows large?
(3) How do we describe the degree of differentiation among products as the number of monopolistic competitors each with a different commodity becomes large?
(4) How do we describe entry in a closed model?
(5) Can we model and solve a closed model of monopolistic competition with transportation costs on the sphere?

The recent survey of Hart (1983a) deals with many of the latest developments in oligopoly theory. The comments here are designed to supplement rather than duplicate his.

(3) The new industrial organization. The new industrial organization is an amalgam of good empirically oriented microeconomics combined with operations research, management science and modern marketing theory. Here a few of the types of questions it seeks to answer are noted.

(1) What are useful measures of concentration?
(2) Can we measure oligopolistic inefficiency?
(3) What is the effect of the need to search on pricing?
(4) How optimal are the various rules of thumb used in different industries?
(5) How do production, capacity, inventorying and pricing conditions interact in various industries?
(6) What are the barriers to entry?

The thrust of the subject is towards applied and somewhat mathematical economics with some bias in the direction of concern with regulation and societal welfare; but on the whole with much of the work not directly espousing a regulatory or an individual firm's point of view. The text of Scherer (1980) provides an excellent example of the approach, the type of questions and an increased concern with quantification.

From the viewpoint of the mathematical economist all of the questions are in terms of partial equilibrium and most are formulated as one person optimization problems. When competition is dealt with explicitly, it is, in general, as some form of non-cooperative equilibrium analysis ranging from the one-shot equilibrium to an equilibrium in a differential game [see Case (1979) for instance].

Although there is extensive literature on bargaining and on cooperative game solutions, the formal application of cooperative solutions to industrial organization is slim.

(4) Regulation, contracts, law and economics, (5) The structure of the firm,

incentives and agency, and (6) *Empirical institutional studies*. Possibly growing faster than the new industrial organization has been the burgeoning literature on law and economics especially in the United States where it has provided economic arguments for both regulation and deregulation of industry. Journals such as the *Journal of Law and Economics* and the (now reincarnated) *Bell Journal of Economics* provide examples.

Tied in closely with much of the above has been a concern for the internal structure of the firm, especially a recognition of limits on information and control. Furthermore, there is now a growing appreciation of the problem of the design of incentives in a hierarchical system and the occasion of need for a hierarchical system as an alternative to a price system. The work of Williamson offers a perceptive and primarily verbal discussion of the organization of the firm [see Williamson (1975)]; the work in agency theory [see, for example, Ross (1973)], non-symmetric information [see, for example, Green (1983)], and contract theory [see, for example, Grossman and Hart (1984)] is considerably more mathematical — the mathematical formulation of these problems generally involving the Bayesian updating of subjective probabilities in a sequential game utilizing a specialized non-cooperative equilibrium solution such as the alternative to a perfect equilibrium suggested by Kreps and Wilson (1982).

The questions asked are of considerable concern to society:

(1) What are the criteria to be used to decide that a firm is too big?
(2) When is regulation (deregulation) called for?
(3) How are the incentives to be designed to keep fiduciaries acting in the best interests of their constituents?
(4) In what sense can the best interests of a heterogeneous group of stockholders be defined?
(5) What is the economic basis for contracts?

Yet although much wisdom starts with knowing what are interesting and good questions, it is desirable to be able to answer them. In doing so, we must distinguish between the development of a general theory from which answers to these questions can be derived and the proliferation of *ad hoc* models designed to bolster the plausibility of the testimony of expert witnesses.

I suspect that prior to the construction of an adequate general theory many *ad hoc* models need to be built. At least at this time the type of verbal theorizing offered by Williamson combined with a detailed specific industry study and *ad hoc*[1] econometric investigations have far more to offer the courts and society in general than do the theoretical writings in oligopoly theory.

[1] I wish to stress that I use *ad hoc* in a positive and *not* in a negative sense. It frequently stands for knowing your business in detail.

(7) *Corporate strategic analysis.* In the past thirty years there has been an explosion in the utilization of corporate strategic planning procedures. Samples of the publications which contain an intermix of organizational considerations, management science, operations research, computer simulation and economic theory are given by the works of Ansoff (1965), Lorange (1980), Naylor (1980), Porter (1980) and Shubik (1983).

Leaving aside the organizational aspects, (which are often critical for the actual manager) from the viewpoint of the economic theorist, much of corporate strategic analysis may be regarded as industrial organization theory turned on its head. Instead it looks at firm behavior from the viewpoint of the firm. The type of questions asked are:

(1) How can I optimally enter market x?
(2) What determines the decision to acquire an existing firm or start up fresh in a new market?
(3) What is the optimal form of integration?
(4) What is the optimal way to exit from a declining market?
(5) What are the tradeoffs between market expansion and antitrust costs?

Neither corporate vice presidents nor MBA students in a hurry to join a strategic planning consulting firm appear to have much interest in the existence or non-existence of pure strategy perfect non-cooperative equilibria in Bertrand–Edgeworth oligopolistic structures. We must ask is it because we do not yet know how to teach the correct basic oligopoly theory to practitioners or could it be that the theory is not yet sufficiently developed that except for a few concepts such as best response or threat, it is hardly of consequence in daily application?

In the United States there is an important and growing market for the services of the academic as a consultant and expert witness. The experience gained in being paid to work on real specific problems is undoubtedly of great value in building up a body of insight into the workings of actual industry. But it appears to be premature to judge whether this has yet had a significant influence on a theory of monopolistic competition. Perhaps the gap between *ad hoc* dynamics and a predominantly general static equilibrium theory is too large to be bridged in any great generality at this time.

(8) *Behavioral theories of the firm and competition and* (9) *Experimental gaming.* The last two items on the list are explicitly process oriented and hence institutional. The process is constrained by the rules of the game, be they technological, legal or societal. The process is manifested in the behavior of mechanisms and the mechanisms are the institutions of the economy, polity and society.

The seminal work of Simon [for a broad sampling, see Simon (1979), March and Simon (1958), Cyert and March (1963)] and other members of

Carnegie Institute of Technology together with the building of the Carnegie Technology business game appeared to signal the development of a fundamentally new approach where the economist's narrow (and possibly irrational) penchant for rational behavior (well defined only in a highly restricted domain) was challenged and was to be replaced.

Although this work was started around thirty years ago (Simon's paper 'A Behavioral Model of Rational Choice' appeared in 1955), the operational concept of 'satisficing' as contrasted with optimizing still remains illusive. Yet in spite of the will-o'-the-wisp aspects of satisficing it is broadly felt by practitioners and consultants that rules of thumb are representative [as Baumol and Quandt (1964) have phrased it] of optimally imperfect decisions.

The book of Nelson and Winter (1982) represents a continuation and extension of the school of the behavioral theory of the firm. It sketches the dynamics of innovation and offers sensitivity studies of simulations of the growth of industries. Yet in spite of many individual insights the central theoretical basis for the rules of thumb, short cuts and behavior strategies does not yet exist. The problem reduces to the development of a theory of behavior of finite capacity machines required to behave in bounded time. And not only do we not have clear questions and answers, many economists are unaware that this is a central problem at the core of the development of modern psychology and artificial intelligence.

These comments are not meant to detract from the work of Nelson and Winter and the growing volume of corporate simulations [see, for example, Rosenkranz (1979)]. It is possible that progress will eventually be made by the concerted efforts of those building specialized process models of many specified firms and industries developing insights into rules of thumb which work well in limited domains and devising explanations as to why they work.

Related to but considerably different from the work in behavioral theory and simulation has been the experimental approach evinced in activity in experimental economics [see Shubik (1975a, b), Smith (1962, 1979, 1982) and Plott (1982)].

In the context of simple simulated markets, some of the virtues of the price system appear to be demonstrable by experiment. The evidence in favor of the non-cooperative equilibrium or for that matter any game theoretic solution concept in general is poor. Institutional structure and the specifics of context and mechanism are far more important than appears to be generally recognized.

3. Some answers: Statics

3.1. Oligopoly in a closed economy

In section 2 a broad sketch of the many approaches to monopolistic competition was provided in order to set the context for the questions and answers noted here without offering spurious generality. In particular, I suggest that several questions of central interest to the mathematical economist are of little concern to those dealing with policy and applications of the study of industrial organization.

How does one model oligopolistic competition in strategic form in a closed model? What is the relationship between the competitive equilibria and non-cooperative equilibria of a closed economic system modeled as a game in strategic form? These two questions can be answered and can cast further light on our understanding of the Walrasian system, prices and non-cooperative equilibria.

Shapley and Shubik (1967), and Gabszewicz and Vial (1972) produced non-symmetric models of games in strategic form where the consumers are regarded as price-takers. Shubik (1973) and Shapley and Shubik (1977) produced symmetric models of exchange modeled as a game in strategic form with markets and the use of a money.[2] These models are members of a general class of *strategic market games* which provide game or mechanism models of a closed economy. There are many technical difficulties to be faced in modeling production, entry, price strategies and product variation, but the loose coupling provided by markets, money and credit rules makes this simpler than it would be otherwise.

The formulation of a game in strategic form is a preliminary step which must be taken before we can well define a non-cooperative solution concept. But the explicit definition of the game does not imply that the non-cooperative equilibrium should be adopted as a solution concept.

3.2. The reconciliation of non-cooperative oligopoly theory and general equilibrium

The non-cooperative equilibrium (NE) solution of Nash is a natural generalization of the Cournot oligopoly equilibrium which can be applied to any game in strategic form. Let there be n players; each player i has a set of strategies S_i and a payoff function $P_i(s_1, s_2, \ldots, s_n)$ where $s_i \in S_i$ is a strategy of i from his set. A non-cooperative equilibrium $(s_1^*, s_2^*, \ldots, s_n^*)$ is such that if i

---

[2]The main thrust of this article is on the study of monopolistic competition, its relationship to general equilibrium and the value of and difficulties with the non-cooperative equilibrium solution. It would detract too far from our major purpose to include a lengthy discussion of the relationship among the theory of money and financial institutions, oligopoly theory and general equilibrium. But it would be remiss to fail to point it out. In essence, the general equilibrium system is not fully defined as a game in strategic form. Little attention is paid to describing the system in disequilibrium. The act of modeling either the Cournot or the Bertrand–Edgeworth models of oligopolistic competition (or variants with product differentiation, entry, etc.) calls for specifying a game in strategic form. A natural simplification which makes this easy to do is to restrict the model to one market per commodity and to invent money, credit and bankruptcy rules. Basically, the introduction of money and financial institutions and the embedding of oligopolistic competition into general equilibrium are closely related problems [Shubik (1984)].

is informed of the strategies of all others [let $s=(s_1, s_2,\ldots,s_n)$ and (s/s_i) be s without s_i] then

$$\max P_i(s^*/s_i) \quad \text{implies that} \quad s_i = s_i^* \quad \text{for all } i, \quad s_i \in S_i.$$

The key property of the NE is that of stability against the optimal response of any individual. This is different from the competitive equilibrium where the individual is assumed to make his decisions based on prices over which he has absolutely no influence. If we assume that there is a continuum of economic agents each of which is of measure zero, then it can be shown that the non-cooperative equilibria of the strategic market game coincide with the competitive equilibria of the related general equilibrium model [see Postlewaite and Schmeidler (1978), Dubey and Shapley (1977), Dubey, Mas-Colell and Shubik (1980)].

Although the formal mathematical models can 'reconcile' oligopoly theory and general equilibrium theory via strategic market games with a continuum of economic agents, it is my opinion that the difficulties with this reconciliation involve the modeling, not the mathematics. In particular, the key concepts behind the competitive market are powerlessness, mass anonymity and irrelevance of strategic information. The basic factors central to oligopolistic competition are the power of few players and the importance of strategic information as well as the low likelihood of anonymity.

When there are at least two competitors of each type, the competitive equilibria of an economy represented as a strategic market game with price competition (Bertrand–Edgeworth) can be shown to be non-cooperative equilibria [Dubey and Shubik (1980), Dubey (1982)] immediately. Thus, for a fixed finite number of products and firms, the Bertrand–Edgeworth strategic market games appear to be more competitive than the Cournot games.

But with, say, only two or three competitors in a market even though the model and the logic of the mathematical analysis may be accepted, a new problem arises. Do we accept the solution concept? If there are only two or three large players who are known to each other, why should they play non-cooperatively? We return to this in section 3.3.

If all firms are genuinely monopolistic competitors, each sells a differentiated product, then Bertrand–Edgeworth results cannot be obtained [see Hart (1983a–c)]. The Bertrand–Edgeworth strategic market games provide a striking exception to the general results of Dubey and Rogawski (1982) who have shown the generic inefficiency of non-cooperative equilibria for games with a finite number of players.

Novshek and Sonnenschein (1978) have been able to introduce entry into a closed model. By considering firms with a slight non-convexity in production, that is, small relative to the overall market, epsilon-non-cooperative equilibria can be obtained. This provides an endogenous determination of the size of industry.

In a closed system, as everyone has to be somewhere, the concept of entry is at least one that involves comparative statics, if not dynamics. We begin with everyone somewhere, change a parameter and see who moves in or out of production.

3.3. The reconciliation of cooperative oligopoly theory and general equilibrium

The two major divisions of game theory solutions are cooperative and non-cooperative. The former are associated with games in coalitional form and the latter with games in strategic or extensive form. The core, value, nucleolus, bargaining set and stable set solutions are the major solutions which are applied to games in coalitional form. They have been explored extensively in application to market games [for a summary and review, see Shubik (1984)]. For the most part the results which show convergence of large finite game solutions for the core and value to the competitive equilibria, or which show equivalence results for games with a continuum of agents, are in reference to exchange economies. When items such as production and manager run firms are considered, the modeling problems in the coalitional form become considerable and are not yet settled.

Aumann (1973), Shitovitz (1973, 1974) and others have considered exchange economies with an ocean of traders and one or several atoms (i.e., large traders). Results concerning advantageous and disadvantageous outcomes in the core have been obtained. But without going into any discussion of these results an overall comment concerning the models and solutions is made. At the formal mathematical level important connections between the core and value of strategic market games and the related competitive equilibria have been made. These results confirm our intuition that if there are many competitors and none is large, details concerning strategies and institutions do not matter. But among few producers the investigations of the game in strategic form show that the results in general depend upon the specifics of the strategy sets. In the current investigations this detail has been ignored. If we merely regard *market games* as a mathematical structure, study their core (value) properties and compare them with the related exchange economy, we may say that formally general equilibrium and oligopoly theory have been reconciled. I suggest, however, that the characteristic function (in either its sidepayment or no-sidepayment versions) is inadequate for the substantive modeling of monopolistic or oligopolistic competition. If we wish to consider games in coalitional form to study oligopolistic competition, then we need to derive the coalitional form from a strategic form.

4. Some problems: Dynamics

The essence of monopolistic or oligopolistic competition involves few firms with some form of communication or signaling in a multistage game.

Although there are some nice and relevant problems for the theorist to clean up in showing logical consistency between strategic market games and other closed economic models, these are not the major problems at this time.

The general equilibrium model was not suited for the study of oligopoly theory or of dynamics. It was and is a parsimonious construct designed to answer different questions. The strategic market game is more general than the general equilibrium model in the sense that the full strategy set for each agent must be defined.

It is possible that there are still many questions concerning oligopoly which can be answered using a game in strategic form but for the most part the development of further analysis appears to call for models of games in extensive form or in another form which is adequate to represent the information and moves and other sufficient details needed to consider dynamics.

Unfortunately, even for the game in strategic form and certainly for the extensive form the intuitive appeal of the non-cooperative equilibrium point as a solution is far less than it might appear from an examination of a Cournot duopoly model with a unique symmetric non-cooperative equilibrium.

4.1. Variations on non-cooperative equilibria

The story told to support the competitive equilibrium is one of individual maximization under certainty or under uncertainty with individuals as risk neutral or risk averse and able to obtain any mix of insurance needed via futures markets (or adequate stock markets in a more complex model). As all prices are given, there is no strategic problem and no regret in having to reconcile ex ante and ex post strategic rational behavior. These features do not generalize to games in strategic or extensive form with a finite number of players.

The competitive equilibrium (under the appropriate conditions) can be proved to be Pareto optimal. It is, however, not generally unique. When there is no uniqueness unless one wishes to subscribe to a selection process such as that suggested by Harsanyi (1982), there is no endogenous way of selecting among the equilibria.

The argument for the competitive equilibrium can be phrased both normatively and behavioristically. Given prices people should optimize and as a first approximation given prices and more or less complete markets they do try to optimize.

When, however, we move to the non-cooperative equilibrium the normative argument is considerably weakened, the empirical evidence from oligopolistic studies is not clear and the experimental evidence with games in strategic form shows only weak support of the best response property of the noncooperative equilibrium as the only property considered in determining individual 'rational behavior' in a non-constant sum $n(>1)$ person game.

The great virtues of competitive markets are that they are faceless and in theory uninfluenced by individual behavior. The sociology, psychology, social psychology and politics of the situation is killed by the mechanism. The strength of the price system in the context of the mass anonymous market is that the other factors of society do not matter. A major contribution of game theory has demonstrated that many different solutions lead to the same outcomes; this tells us that if the structure of a market is appropriate one can be extremely crude in assumptions made about motivation. The core, value, nucleolus and non-cooperative equilibrium theories will all lead to competitive prices. Information leaks will not matter and free communication among groups will not influence the outcome.

When we assume that there are only few large actors present, all of the different game theoretic solutions give different predictions. Details count and the economist cannot ignore the need to justify choice among solution concepts.

We confine our comments to the non-cooperative equilibrium (NE) and present a brief list of possible properties or desiderata. They are discussed in some detail elsewhere [Shubik (1981)]. We may desire:

(1) Existence: sometimes there are no pure strategy NEs [see also Dierker and Grodal (1982)].
(2) Uniqueness: in general the NEs are not unique.
(3) Symmetry: there may be non-symmetric NEs for symmetric games.
(4) Value: the different NEs do not necessarily have the same value.
(5) Pareto optimality: in general the NEs are not Pareto optimal.

Sensitivity analysis properties are important. We might wish to consider sensitivity to:

(1) perturbation in payoffs,
(2) perturbation in moves and strategies (trembling hand),
(3) perturbation in information.

In general, games in strategic or extensive form do not display smooth changes in the NEs against perturbation in payoffs or small errors in the choice of moves. Furthermore, for games with a finite number of players, changes in information may make considerable differences. This is illustrated by the Cournot duopoly where players 1 and 2 move simultaneously, where 1 moves first and where 2 moves first.

Selten introduced the concept of the perfect non-cooperative equilibrium for games in extensive form. He subsequently introduced the idea of 'trembling-hand perfection' to take care of situations in which some part of the game tree might never be reached. Kreps and Wilson (1982) have offered the concept of sequentially rational non-cooperative equilibrium. At first sight these modifications appear to offer an important restriction on the set of non-cooperative equilibria and might be regarded by some as particularly 'rational'. The argument being made in essence is based upon a backward induction which totally ignores threats as 'irrational', but by the same token we could regard any code of behavior which fails to conform to local individual short run optimization as irrational.

I suggest that the strength of the perfect equilibrium analysis lies precisely away from the study of oligopoly or reputation or agency but when there is a continuum of agents. It is then that threats do not matter. This supplies the justification for treating a multistage strategic market game with a continuum of traders as though each faced a one-person dynamic program with the market providing prices.

When numbers are few, there is no justification to excluding open threats as less rational or reasonable than subgame local optimization [see Shubik (1959, ch. 10) and Anderson (1984)].

4.2. Exogenous uncertainty and subjective probability

Bayes' theorem must be regarded as a contribution to pure logic. Once the assumptions regarding the model are accepted, the way new information should be utilized to update current expectations should be in accordance with the Bayes theorem. But where the initial subjective probabilities came from and how we completely revise our cognitive maps of a situation are not part of the logic, but a basic part of how we structure our view of the change in the causal model invoked by new information. Schumpeter's description and Shackle's concept of 'potential surprise' are both examples of this. The introduction of new products such as the fastner Velcro or the 1-2-3 spreadsheet provide examples where once the additional information was available the whole cognitive maps of the competitors were changed.

A development of this topic would lead us far from the consideration of monopolistic competition alone. Two key survey references are noted. They are those of Abelson and Levi (1983) and Machina (1983). Before we invest too much intellectual capital in Bayesian games, it is desirable to consider the scope of reasonable modeling. The conceptualization of new product introduction and consumer preference formation for new products is based at best on tentative knowledge.

4.3. Non-symmetric information

There has been a considerable growth in the literature on decisionmaking with non-symmetric information. The surveys of Radner (1982) and Green (1983) cover much of the work. My comment here is merely cautionary.

When the number of competitors is few, the conclusions of the rational expectations literature may be false [see Dubey, Geanakoplos and Shubik (1982)]. But leaving that aside, from the game theoretic viewpoint in order to mount the theoretical apparatus of the non-cooperative equilibrium, we have to accept not merely the non-cooperative equilibrium as a solution concept, we must add on top Bayesian updating and some form of subgame perfection. Although I do not subscribe completely to either the kinked oligopoly curve analysis or the tit-for-tat strategy discussed by Axelrod (1984), both of these appear to me to naturally reflect long run threats simply and more plausibly than the variants of perfect equilibria.

Are the solutions on variations of sequential equilibria to be regarded as normative or behavioral? Or are they meant to be some blend? If so, what is the blend? Furthermore, what is the justification for any belief that the failure to carry out short term expensive threats is globally rational?

5. Future developments

It is my contention that as soon as we become concerned with economic dynamics we must consider mechanisms. But mechanisms implicitly define institutions. Thus, in a basic way further developments of mathematical economics in the study of competition must recognize that the rules of the game and institutional structure are highly related even at the highest level of abstraction. Hence, the search for minimal economic institutions is called for. A way to go about this is to build the simplest playable game in which the essence of the function appears. Thus, we must be concerned with the minimal representation of the manager-run-firm, bank, consumer, market and insurance company (among others).

I suggest that gaming has at least three functions. They are in experimentation, teaching and as an aid to model building and verification in theory construction. In particular, in the construction of games portraying oligopolistic markets the game construction methodology forces the economist to make explicit the assumptions concerning the legal and socio-political environment. Thus, for example questions of economic efficiency are asked in the appropriate context of legal and socio-political feasibility and costs.

Applied economics never abandoned institutional knowledge. The courts will survive quite adequately with whatever blend of *ad hoc* study and theory is available. The legal process is pragmatically oriented. It is more concerned with rationalizing what has happened (as good or bad) than with developing any theory of rational behavior. Thus, the economic theorist (as contrasted with the consultant in industrial organization, antitrust or corporate planning) has no burning applied problem to be solved immediately. We can ask where are the new paths to be followed. I suggest four topics, three of which are general to economic and game theoretic analysis and the fourth is specific to the theory of oligopolistic competition.

The four topics are:

(1) The study of subjective probability and risk behavior including the concept of threat.
(2) The construction of games with players with limited capacity.
(3) The development of experimental gaming to test current and new solutions.
(4) The development of oligopoly models combining explicit financial and capital goods structure with stress upon the role of time lags.

The first two involve interacting with psychology, social psychology and artificial intelligence. The third calls for a mixture of experimentation and theory and an explicit program to not merely check non-cooperative behavior but to suggest and consider alternatives to the non-cooperative equilibrium. Specifically, I suggest that a coincidence of properties (such as symmetry, Pareto optimality, uniqueness and best response) all improve the plausibility of a prediction.

The fourth item appears to me at this time to offer the most profitable avenue to combine institutional insight, operations research and economic theory. Although, in general, non-cooperative equilibria may be profuse and the dynamics hard to describe, if a specific structure is imposed caused, say, by time lags in buying new capacity and bringing it into production; inventory, production and delivery lags; lags in financing; organizational lags and costs in getting into or out of business and other specific factors, then the dynamics will be highly constrained [see Shubik and Levitan (1980)]. The important work of Bain (1956) was to a great extent successful due to his introduction of time associated facts of life relevant to barriers to entry. Explicit teaching and experimental game models utilizing computer simulations rich enough to incorporate capital structure yet simple enough to be consistent with analytical models merit further development.

References

Abelson, R.P. and A. Levi, 1983, Decision making and decision theory, in: G. Lindzey and E. Aronson, eds., Handbook of social psychology (Addison-Wesley, Reading, MA).
Anderson, R.M., 1984, Quick response equilibrium, Paper IP 323, May (Institute of Business and Economic Research, University of California, Berkeley, CA).
Ansoff, H.I., 1965, Corporate Strategy (McGraw-Hill, New York).
Aumann, R.J., 1973, Disadvantageous monopolies, Journal of Economic Theory 6, 1–11.
Axelrod, R., 1984, The evolution of cooperation (Basic Books, New York).
Bain, J.S., 1956, Barriers to new competition, their character and consequences in manufacturing industries (Harvard University Press, Cambridge, MA).
Baumol, W.J. and R.E. Quandt, 1964, Rules of thumb and optimally imperfect decision, American Economic Review 14, no. 2, part 1, 23–46.
Baumol, W.J., J.C. Panzar and R.D. Willig, 1982, Contestible markets and the theory of industrial structure (Harcourt Brace Jovanich, New York).
Case, J.H., 1979, Economics and the competitive process (New York University Press, New York).

Chamberlin, E.H., 1933, The theory of monopolistic competition (Harvard University Press, Cambridge, MA).
Cournot, A.A., 1897, Researches into the mathematical principles of the theory of wealth (translated from French; original 1838) (Macmillan, New York).
Cyert, R. and J. March, 1983, A behavioral theory of the firm (Prentice-Hall, Englewood Cliffs, NJ).
Dierker, H. and B. Grodal, 1982, Nonexistence of Cournot Walras equilibrium in a general equilibrium model with two oligopolists, Discussion paper no. 95 (University of Bonn, Bonn).
Dubey, P., 1982, Price quantity strategic market games, Econometrica 50, 111–126.
Dubey, P. and J.D. Rogawski, 1982, Inefficiency of Nash equilibria in a private goods economy, Cowles Foundation discussion paper no. 631 (Yale University, New Haven, CT).
Dubey, P. and L.S. Shapley, 1977, Noncooperative exchange with a continuum of traders, Cowles Foundation discussion paper no. 447 (Yale University, New Haven, CT).
Dubey, P. and M. Shubik, 1980, A strategic market game with price and quantity strategies, Zeitschrift für National Ekonomie 40, nos. 1–2, 25–34.
Dubey, P., J. Geanakoplos and M. Shubik, 1982, Revelation of information in strategic market games: A critique of rational expectations, Cowles Foundation discussion paper no. 643 (Yale University, New Haven, CT).
Dubey, P., A. Mas-Colell and M. Shubik, 1980, Efficiency properties of strategic market games: An axiomatic approach, Journal of Economic Theory 22, no. 2, 339–362.
Gabszewicz, J.J. and J.P. Vial, 1972, Oligopoly à la Cournot in a general equilibrium analysis, Journal of Economic Theory 4, 381–400.
Green, J., 1983, Differential information, the market, and incentive compatibility, presented at Yrjo Jahnsson Symposium (Helsinki).
Grossman, S.J. and O. Hart, 1984, The costs and benefits of ownership: A theory of vertical integration, Mimeo.
Harsanyi, J.C., 1982, Solutions for some bargaining games under the Harsanyi-Selten solution theory 1: Theoretical preliminaries; 2: Analysis of specific bargaining games, Mathematical Social Sciences 3, 179–191.
Hart, O.D., 1983a, Imperfect competition in general equilibrium: An overview of recent work, presented at Yrjo Jahnsson Symposium (Helsinki).
Hart, O.D., 1983b, Monopolistic competition in the spirit of Chamberlain: A general model, paper no. 82 (International Center for Economics and Related Disciplines, London School of Economics, London).
Hart, O.D., 1983c, Monopolistic competition in the spirit of Chamberlain: Special results, paper no. 85 (International Center for Economics and Related Disciplines, London School of Economics, London).
Kreps, D.N. and R. Wilson, 1982, Sequential equilibria, Econometrica 50, no. 4, 863–894.
Lorange, P., 1980, Corporate planning (Prentice-Hall, Englewood Cliffs, NJ).
Machina, M.J., 1983, The economic theory of individual behavior toward risk: Theory, evidence and new directions, Technical report 433, Oct. (Center for Research on Organizational Efficiency, Stanford University, Stanford, CA).
March, J.G. and H.A. Simon, 1958, Organizations (Wiley, New York).
Nash, J.F., Jr., 1951, Noncoopoerative games, Annals of Mathematics 54, 289–295.
Naylor, T.H., 1980, Strategic planning management (Planning Executives Institute, Oxford, OH).
Nelson, R.R. and S.G. Winter, 1982, An evolutionary theory of economic change (Harvard University Press, Cambridge, MA).
Novshek, W. and H. Sonnenschein, 1978, Cournot and Walras equilibria, Journal of Economic Theory 19, 223–266.
Nti, K. and M. Shubik, 1981a, Noncooperative oligopoly with entry, Journal of Economic Theory 24, no. 2, 187–204.
Nti, K. and M. Shubik, 1981b, Duopoly with differentiated products and entry barriers, Southern Economic Journal 48, no. 1, 179–186.
Plott, C.R., 1982, Industrial organization theory and experimental economics, Journal of Economic Literature 20, 1485–1527.
Porter, M.E., 1980, Competitive strategy (Free Press, New York).
Postlewaite, A. and D. Schmeidler, 1978, Approximate efficiency of non-Walrasian Nash equilibria, Econometrica 46, 127–135.

Radner, R., 1982, Equilibrium under uncertainty, in: K. J. Arrow and M. D. Intriligator, eds., Handbook of Mathematical Economics, Vol. 2, ch. 20 (North-Holland, Amsterdam).
Robinson, J., 1933, The economics of imperfect competition (Macmillan, London).
Rosenkranz, F., 1979, An introduction to corporate modeling (Duke University Press, Durham, NC).
Ross, S., 1973, The economic theory of agency, American Economic Review 63, 134–139.
Scherer, F.M., 1980, Industrial market structure and economic performance, 2nd ed. (Rand-McNally, Chicago, IL).
Schmalensee, R., 1983, Advertising in entry deterrence: An exploratory model, Journal of Political Economy 4, 636–653.
Shapley, L.S. and M. Shubik, 1967, Concepts and theories of pure competition, in: M. Shubik, ed., Essays in mathematical economics in honor of Oskar Morgenstern (Princeton University Press, Princeton, NJ) 63–79.
Shapley, L.S. and M. Shubik, 1977, Trade using one commodity as a means of payment, Journal of Political Economy 85, 937–968.
Shitovitz, B., 1973, Oligopoly in markets with a continuum of traders, Econometrica 41, 467–501.
Shitovitz, B., 1974, On some problems arising in markets with some large traders and a continuum of small traders, Journal of Economic Theory 8, 458–470.
Shubik, M., 1959, Strategy and market structure (Wiley, New York).
Shubik, M., 1973, Commodity money, oligopoly, credit and bankruptcy in a general equilibrium model, Western Economic Journal IX, no. 3, 24–38.
Shubik, M., 1975a, Games for society, business and war (Elsevier, Amsterdam).
Shubik, M., 1975b, The uses and methods of gaming (Elsevier, New York).
Shubik, M., 1981, Perfect or robust noncooperative equilibrium: A search for the philosopher's stone?, in: Essays in game theory and mathematical economics in honor of Oskar Morgenstern (Bibliographisches Institut, Mannheim).
Shubik, M., 1983, The strategic audit: A game theoretic approach to corporate competitive strategy, Managerial and Decision Economics 4, no. 3, 160–171.
Shubik, M., 1984, Game theory in the social sciences, Vol. II (MIT Press, Cambridge, MA).
Shubik, M. and R.E. Levitan, 1980, Market structure and behavior (Harvard University Press, Cambridge, MA).
Simon, H.A., 1955, A behavioral model of rational choice, Quarterly Journal of Economics 69, 99–118.
Simon, H.A., 1979, Models of thought (Yale University Press, New Haven, CT).
Smith, V.L., 1962, Experimental studies of competitive market behavior, Journal of Political Economy 70, 111–137.
Smith, V.L., ed., 1979, Research in experimental economics, Vol. 1 (JAI Press, Greenwich, CT).
Smith, V.L., ed., 1982, Research in experimental economics, Vol. 2 (JAI Press, Greenwich, CT).
Triffin, R., 1940, Monopolistic competition and general equilibrium theory (Harvard University Press, Cambridge, MA).
Von Neumann, J. and O. Morgenstern, 1944, Theory of games and economic behavior (Princeton University Press, Princeton, NJ).
Williamson, O.E., 1975, Markets and hierarchies: Analysis and anti-trust implications (Macmillan, London).

PART II

OLIGOPOLY

[19]

A COMPARISON OF TREATMENTS OF A DUOPOLY SITUATION[1]

By J. P. Mayberry, J. F. Nash, and M. Shubik

A specific duopoly situation is treated here from the viewpoint of several theories of duopoly in order to obtain and compare the numerical results. Thus one is able to compare, in a sense, the behavior of producers operating under free competition with their behavior in collusion.

I. INTRODUCTION

The problem of duopoly has been discussed at length in the literature on restricted competition, together with the related problems of bilateral monopoly, oligopoly, and, in general, economic situations involving a small number of important participants. There are several theories applicable to certain aspects of these situations. The most recent of these theories springs from the work done in the theory of games [6].

The purpose of this paper is to take a simple model of two firms in competition, with explicit cost functions and an explicit demand function, and to examine the behavior of the firms on the basis of each of several theories. We assume there is no collusion among the buyers, so that the demand function remains fixed and describes the action of the market. Each theory discussed here, except the "contract curve" of F. Edgeworth, gives a uniquely determined pair of production rates, and all the others, with the exception of the von Neumann and Morgenstern solution, determine the profit made by each of the two producers. The graphs in Section VII show the production rates and profits for the various solutions, and will serve to compare the effect of the different formulations on the behavior of the firms.

II. HISTORICAL REMARKS

A. Cournot and J. Bertrand offered solutions to the duopoly problem, each of which consisted of definite outputs and a price; their solutions differ essentially. Edgeworth modified Bertrand's work, and suggested that one would expect to find a price oscillation in the case of duopoly. H. Stackelberg developed a very complex indifference-map method whereby he produced, among other solutions, the Cournot solution.

Edgeworth applied an indifference-curve method to bilateral monopoly and obtained his famous contract curve. This is not a solution in the same sense as that of Cournot, as it does not prescribe the profit made

---

[1] The authors wish to acknowledge the financial support of The RAND Corporation and ONR contracts N6 onr-27009 and N6 onr-27011 and to express an indebtedness for some illuminating conversations with L. S. Shapley.

141

by the individual monopolists. It is, like the von Neumann and Morgenstern solution, a solution only in the sense that it restricts the possibilities and not in the sense that it determines the outcome uniquely.

The problems of bilateral monopoly and duopoly, when consumer coalitions are excluded, are quite similar from a game theory viewpoint. Each may be regarded as a two-person nonzero-sum game [6]. This game theoretical approach, with its notion of strategy and, in particular, mixed strategy, provides a means for clarifying some of the concepts involved in previous approaches to these problems.

The older approach often depended on the assumption of a specific "conjectural behavior" [2]. For example, one obtains the Cournot solution by presuming that each producer chooses his new production rate on the assumption that his competitor's production rate will remain fixed. The solution is then that situation where the producers' policies do not impel them to any changes in their rates of production. The great difficulty with these hypothetical rules of behavior is their multiplicity; in general, too, they require the producers to act in a rather shortsighted manner. In other words, if producer A could count on producer B behaving according to the hypothesis, he could generally do better for himself by departing from this pattern.

III. SOLUTIONS TREATED IN THIS PAPER

We consider the following solutions: (1) the efficient point solution, (2) the Edgeworth contract curve, (3) the Cournot solution, (4) the von Neumann and Morgenstern solution, (5) the cooperative game with side-payments, and (6) the cooperative game without side-payments.

Certain general assumptions are made in all these solutions. We assume that the duopolists are intelligent men, attempting to maximize their individual utilities. These utilities are assumed measurable in the von Neumann and Morgenstern sense—i.e., they are assumed to be determinate up to a linear transformation. However, individual utilities are not necessarily comparable, and it is not in general meaningful to say "the utility of a dollar to A is greater than the utility of a dollar to B." Since it is necessary to assume some function for the utility of money to each firm, we make the simplest assumption, and take both these functions to be linear. Then the profit in dollars provides a valid utility-function for each firm that we use, in preference to any linear transformation thereof, because of its simplicity.

We assume that there is complete information [1]. We assume that the duopolists produce the same product. We ignore advertising, which could be included with no substantial change in the theory, because our purpose is to illustrate these diverse solutions for a simplified case. We seek a solution that will remain constant, and therefore we exclude the possibility that production (and therefore price) might vary in time.

IV. DESCRIPTION OF THE PHYSICAL SITUATION

We assume that the cost for one firm depends only on the rate of production for that firm. As simple illustrative functions which first decrease, then increase, we take

$$\gamma_1 = 4 - q_1 + q_1^2$$
$$\gamma_2 = 5 - q_2 + q_2^2$$

to be the average-cost functions for the firms, where q_1 and q_2 are the amounts produced in one unit of time by firms one and two, respectively.

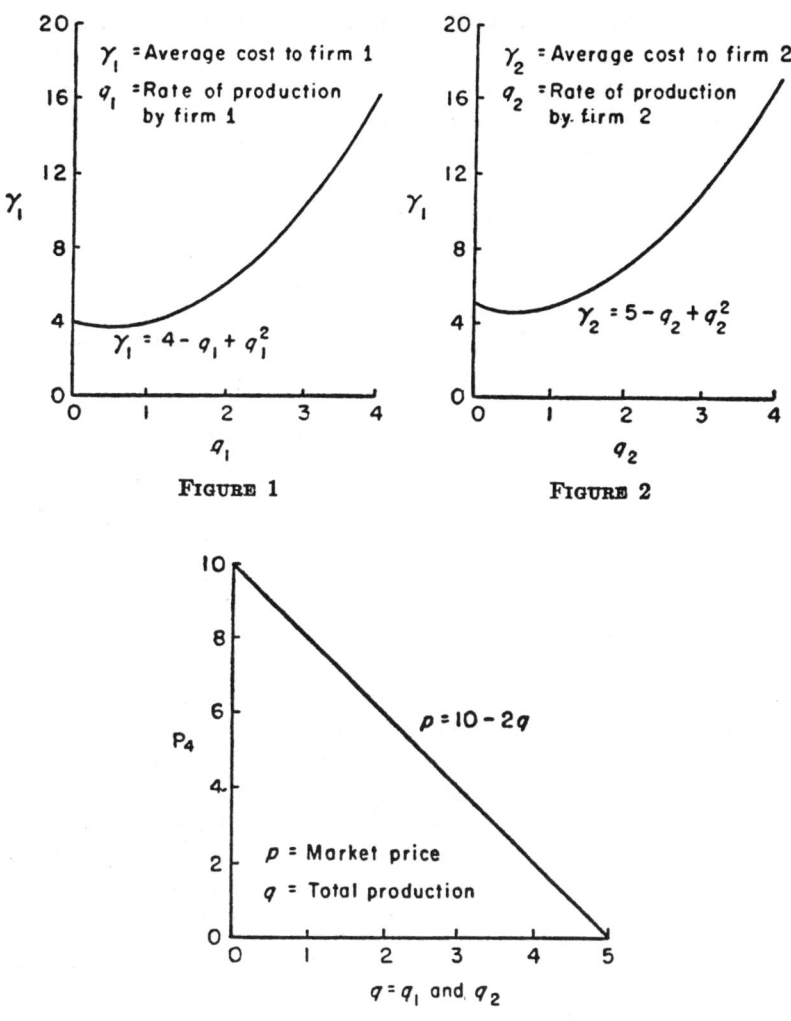

FIGURE 1

FIGURE 2

FIGURE 3

Since the duopolists have been assumed to produce the same good, it is reasonable to take a demand curve (interpreted as giving price as a function of quantity) in which only the total production $q = q_1 + q_2$ enters, and not the individual productions q_1 and q_2. We assume for this function the form

$$p = 10 - 2(q_1 + q_2) = 10 - 2q,$$

where p is the price when total production is $(q_1 + q_2)$.

These functions are graphed in Figures 1, 2, and 3.

V. DESCRIPTION OF THE VARIOUS SOLUTIONS

1. *The efficient-point solution.* The concept of an efficient point was obviously not intended for application to duopoly problems, where the producers are presumably acting from selfish motives. Actually, it cannot be applied in any straightforward way, because it is a total economy concept and a duopoly situation is a fragment of the economy.

However, rules of operation have been invented to lead to efficient production without centralized administration. These rules [3] tell a producer how to adjust his production rate to meet varying costs and prices.

Thus we can define at least an analogous concept for duopoly by supposing that the producers operate according to such rules. This gives what might be called an altruistic production schedule.

The essential efficiency rule [3] is that each producer behave as if the market price were constant and try to maximize profits. So if his marginal cost is less than the market price, he increases production until they are equalized. This gives us the equations

$$\frac{\partial(q_i \gamma_i)}{\partial q_i} = p, \qquad (i = 1, 2).$$

2. *The Edgeworth contract curve.* The condition that a point lie on the Edgeworth contract curve is that it be impossible for both players to improve their situation simultaneously. In other words, the corresponding point in the (P_1, P_2) plane must lie on the upper-right boundary of the set of attainable pairs of profits. [Note that P_1 and P_2, the profits for the two firms, may be expressed explicitly; $P_i = q_i(p - \gamma_i)$, $(i = 1, 2)$.] In our example, since the boundary of the set of attainable points in the (P_1, P_2) plane slopes down to the right at each point, the Edgeworth contract curve is precisely this boundary, which is characterized by the Jacobian condition (see Section VI.6),

$$\frac{\partial(P_1, P_2)}{\partial(q_1, q_2)} = 0.$$

3. *The Cournot solution.* This solution has been discussed in detail elsewhere [2], hence we mention only its main feature. Each producer behaves as if the other will not change his output. The solution may be obtained by solving

$$\frac{\partial P_i}{\partial q_i} = 0 \qquad (i = 1, 2).$$

4. *The von Neumann and Morgenstern solution.* The duopoly problem may be set up as a two-person nonzero-sum game. Space does not permit the development of the theory here, but the reader may refer to the *Theory of Games and Economic Behavior* [6]. The economic interpretation of the result which they obtain is that the two firms cooperate in their policy against the market, and act in such a way as to maximize joint profits.[2] Then they settle between themselves by means of a side-payment. The amount of the payment is not determinate (in general), but is limited by the amounts which the firms could assure for themselves regardless of the competitor's actions. [In computing its minimum level, each firm must assume that the competitor will disregard the result for him (the latter), and act so as to minimize the former's outcome.]

The production rates will satisfy

$$\frac{\partial}{\partial q_i}(qp - q_1\gamma_1 - q_2\gamma_2) = 0 \qquad (i = 1, 2).$$

5. *The cooperative game with side-payment.* Here the final mode of behavior is to produce at the same rates as in the von Neumann and Morgenstern case; however the side-payment will now be uniquely determined by the threat potentialities of the firms. The best threat for each player is that production rate which has the greatest value as a club held over the other's head. The threat production rate of firm one will be such that the maximum value obtainable by firm two for the quantity $(P_2 - P_1)$ is minimized. This explanation requires amplification in case the threats are mixed strategies, but in our example the optimal threats are pure strategies.

6. *The cooperative game without side-payments.* If it is not possible (perhaps for legal reasons) for the producers to make side-payments, the cooperative game solution will in general give different production rates, and total profit will not be maximized; for the production rates must now bear the full burden of adjusting the profit distribution. On

---

[2] In the more general case, where neither firm need have a linear utility for money, the definition of the von Neumann and Morgenstern solution may not be possible; these authors assume that utility is transferable.

the graphs in Section VII, points relevant to the non-side-payment case are marked *NSP*.

In another paper, one of the present authors [5] has shown how to analyze the producers' bargaining positions in terms of the threats they may exert on one another. The result of this analysis is a solution that gives the utility of the situation to each participant. This theory analyzes the threat potentialities from what is believed to be a more complete viewpoint than that of von Neumann and Morgenstern, for they consider a threat only in terms of its effect on the threatened player, whereas here the effects on both players are considered. They are justified in this since they do not attempt to determine the utility of the situation to a participant, but merely to determine a worst and a best outcome for him.

VI. DETERMINATION OF NUMERICAL RESULTS

1. *The efficient point.* The conditions which hold at this point are

$$\frac{\partial(q_1 \gamma_1)}{\partial q_1} = \frac{\partial(q_2 \gamma_2)}{\partial q_2} = p.$$

These equations yield

$$3q_1^2 + 2q_2 = 6 \quad \text{and} \quad 3q_2^2 + 2q_1 = 5.$$

From these, one obtains by elimination

$$27q_2^4 - 90q_2^2 + 8q_2 + 51 = 0,$$

which was solved by Newton's method.

2. *The Edgeworth contract curve.* The Jacobian equation given in Section V.2 gives the equation of the contract curve as

$$6q_1^3 + (9q_2^2 + 6q_2 - 11)q_1^2 + (6q_2^2 + 4q_2 - 22)q_1 \\ + (6q_2^3 - 14q_2^2 - 22q_2 + 30) = 0.$$

A few points enabled the curve to be plotted.

3. *The Cournot solution.* The conditions

$$\frac{\partial P_1}{\partial q_1} = \frac{\partial P_2}{\partial q_2} = 0$$

are equivalent to

$$3q_1^2 + 2q_1 + 2q_2 - 6 = 3q_2^2 + 2q_2 + 2q_1 - 5 = 0;$$

whence we deduce

$$27q_2^4 + 36q_2^3 - 90q_2^2 - 60q_2 + 71 = 0,$$

and solve, as before, by Newton's method.

4. The von Neumann and Morgenstern solution. The conditions

$$0 = \frac{\partial}{\partial q_1}(P_1 + P_2) = \frac{\partial}{\partial q_2}(P_1 + P_2)$$

yield

$$3q_1^2 + 2q_1 + 4q_2 - 6 = 3q_2^2 + 2q_2 + 4q_1 - 5 = 0;$$

whence the solution was calculated by elimination and successive approximations.

5. The cooperative game with side-payments.
The final production rates are identical with those for the von Neumann and Morgenstern solution, but the threat rates of production must also be evaluated in order to determine the magnitude of the side-payment. The conditions satisfied by the threat production rates are

$$\frac{\partial}{\partial q_1}(P_1 - P_2) = \frac{\partial}{\partial q_2}(P_1 - P_2) = 0,$$

which are equivalent to

$$3q_1^2 + 2q_1 - 6 = 3q_2^2 + 2q_2 - 5 = 0.$$

These equations also were solved by Newton's method.

6. The cooperative game without side-payments.
We outline the method of obtaining a solution if it can be found in terms of pure strategies, which is the case in our example.

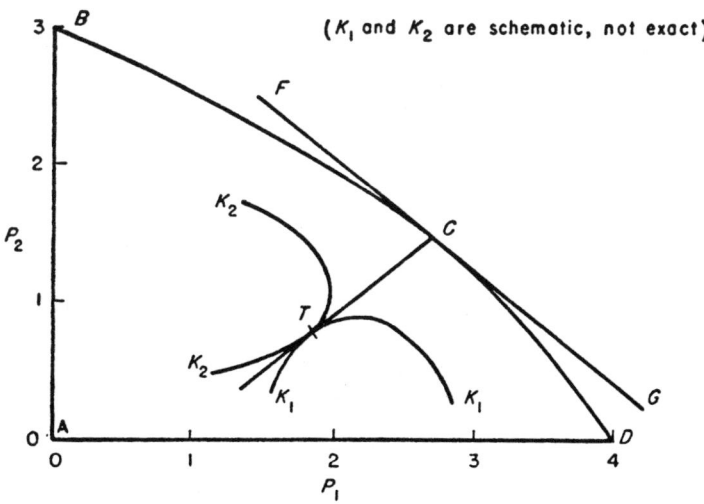

FIGURE 4—Illustration for description of NSP solution

First, the set of attainable utility pairs is determined. In our example, P_1 and P_2 serve as utilities, so it is the attainable set in the (P_1, P_2)

plane that is of interest, and especially the upper-right boundary of this set. This set is shown in Figure 4 as the region $ABCD$. The condition satisfied by production rates (q_1, q_2) giving a point on that boundary is

$$(1) \qquad \begin{vmatrix} \dfrac{\partial P_1}{\partial q_1} & \dfrac{\partial P_2}{\partial q_1} \\ \dfrac{\partial P_1}{\partial q_2} & \dfrac{\partial P_2}{\partial q_2} \end{vmatrix} = 0.$$

In Figure 4, BCD is the relevant boundary of the attainable region in the (P_1, P_2) plane, T is the threat point, C is the final point, FCG is the tangent to BCD at C, and TC is the line through C such that the slope of TC is the negative of the slope of FCG. The coordinates of T show the profits resulting if the threat production rates q_1^T and q_2^T are in force. K_1 shows the pairs of profits resulting if q_2 varies while q_1 remains constant and equal to q_1^T. K_2 shows the analogous curve for q_1 varying while q_2 remains equal to q_2^T. The condition for the threat point T and the boundary point C to make up a solution is that K_1 lie entirely below TC and K_2 lie entirely above TC. However, since these curves have derivatives, we must have K_1 and K_2 tangent at T, so that the above determinant must also vanish at the threat point. Thus there are two branches of this curve where equation (1) holds; T lies on the one, C on the other.

The slope of FCG is

$$\frac{\dfrac{\partial P_2}{\partial q_1}}{\dfrac{\partial P_1}{\partial q_1}} = \frac{\dfrac{\partial P_2}{\partial q_2}}{\dfrac{\partial P_1}{\partial q_2}},$$

and the slope of TC is

$$\frac{P_2^C - P_2^T}{P_1^C - P_1^T}.$$

Now the slope of TC must be the slope of K_1 and K_2 at T, so, defining

$$\frac{\dfrac{\partial P_2}{\partial q_i}}{\dfrac{\partial P_1}{\partial q_i}} = D_i,$$

we have the following equations:

$$-D_1^C = -D_2^C = \frac{P_2^C - P_2^T}{P_1^C - P_1^T} = D_1^T = D_2^T.$$

These four equations in the four unknowns q_1^C, q_2^C, q_1^T, and q_2^T, were solved by graphical methods of successive approximation.

VII. NUMERICAL RESULTS

1. *Table and bar graphs.* The table presents the various numerical results. The Edgeworth solution cannot be included in such a table, as none of the quantities q_1, q_2, P_1, P_2, q, and p, are determined by the contract curve. The von Neumann and Morgenstern solution is omitted because those quantities which are determinate in their solution (viz., q_1, q_2, and p) are the same as the corresponding quantities in the side-payment case. The P_1 and P_2 quoted in the table for the side-payment case are the profits after the side-payment has been made. The unadjusted values are: $P_1 = 3.1327$, $P_2 = 1.0664$.

TABLE

Case	q_1	q_2	P_1	P_2	$P_1 + P_2$	p
Efficient point	1.1716	0.9411	1.8437	0.7812	2.6249	5.7747
Cournot	0.9386	0.7400	2.5346	1.3581	3.8927	6.6428
Non-side-payment	0.7812	0.5817	2.6913	1.4644	4.1557	7.2742
NSP threat	1.1708	0.9419	1.8436	0.7811	2.6247	5.8873
Side-payment	0.9161	0.4125	2.6299	1.5692	4.1991	7.3428
Side-payment threat	1.1196	1.0000	1.8214	0.7607	2.5821	5.7607

The values of production, profits, and price are also tabulated for the two threat points.

The bar graphs (Figures 5–10) exhibit much of this information in graphical form. The shaded portions of P_1 and P_2 for the side-payment case represent the amount of the side-payment, which is 0.5028, paid by firm one to firm two.

2. *Comprehensive graphs.* Figures 11 and 12 present graphically most of the numerical results of this paper. Figure 11 shows the quantities produced under various conditions, and Figure 12 shows the corresponding profits which accrue to the two firms.

Several aspects of Figures 11 and 12 are noteworthy. In Figure 11 the "threat curve" lies, as would be expected, entirely outside the "contract curve." E_1 and E_2 are points which bound the portion of the threat curve where P_1 and P_2 are both positive. (The images in the (P_1, P_2) plane of these points are marked with the same letters.)

The point cusp in Figure 11 corresponds to the cusp in Figure 12, which also appears, to a larger scale, in the insert to Figure 12. In spite of the fact that the efficient point, the *NSP* threat, and the side-payment threat, all lie close to it, this cusp has little economic significance or relevance to the duopoly model in general. This assertion is verified by the remark that the cusp depends only on the local properties of the mapping of the (q_1, q_2) plane into the (P_1, P_2) plane, whereas the various

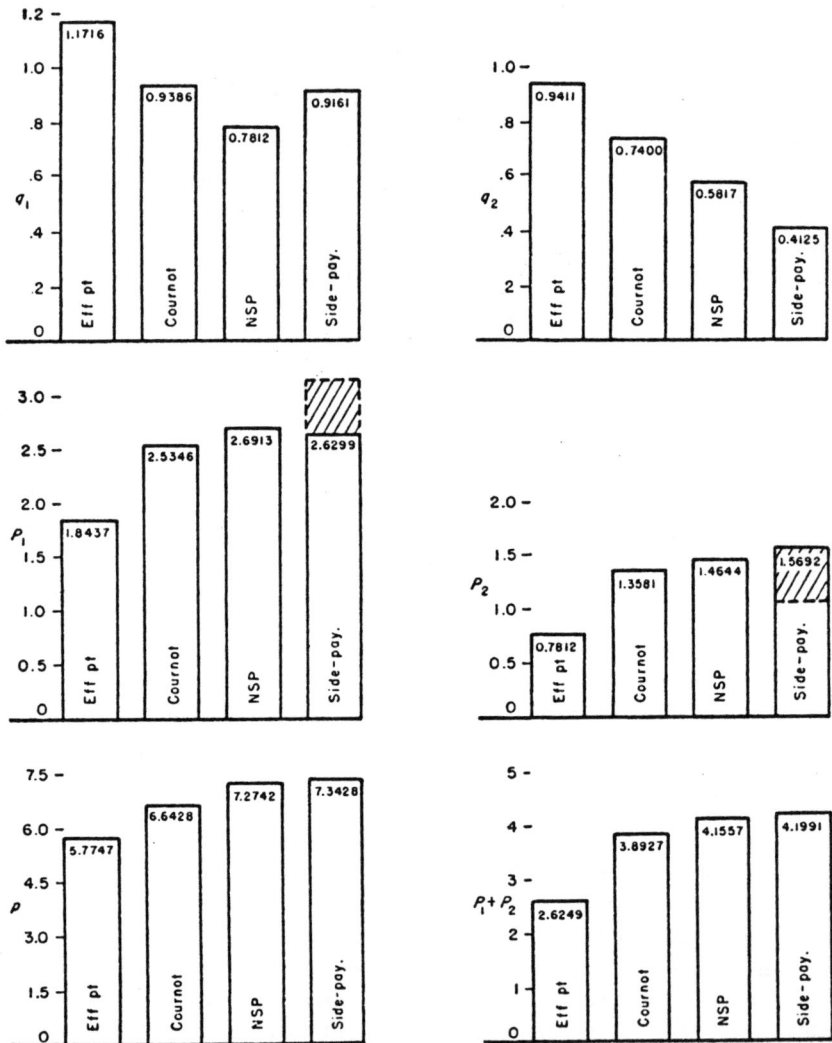

FIGURES 5–10

threat points depend also on the nature of the boundary curve in the (P_1, P_2) plane. Thus the cusp and threat points may be varied independently. It may also be remarked that it is only the behavior of the tangent to the threat curve which is interesting, and the tangent to a curve in the neighborhood of a cusp is not ill-behaved.

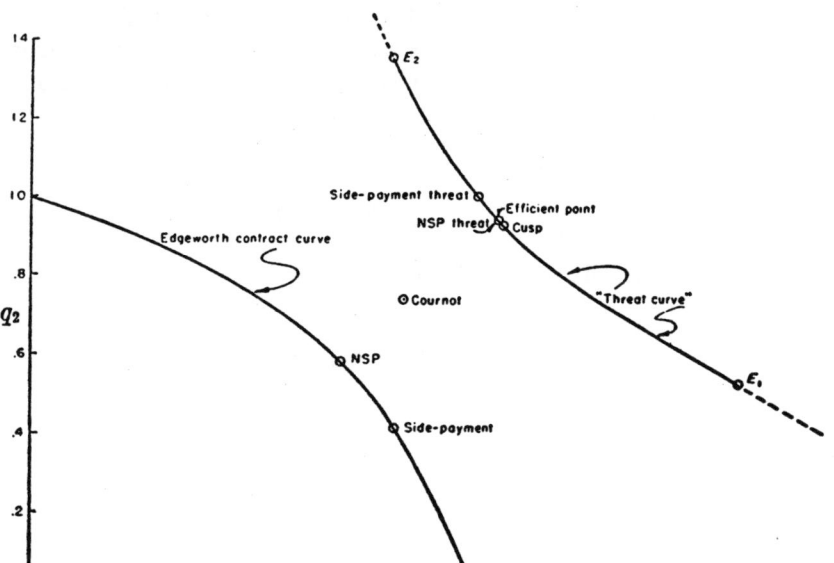

FIGURE 11—Production under various circumstances

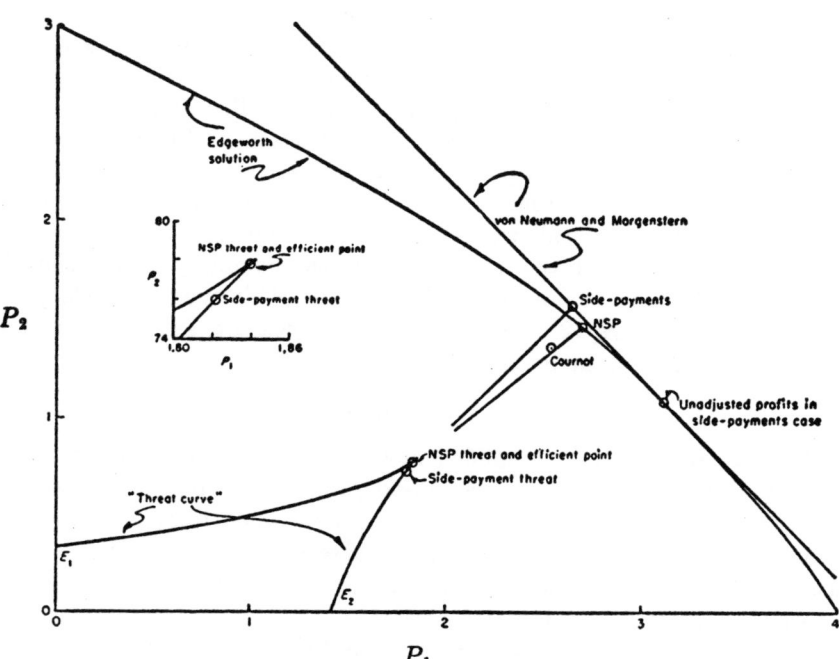

FIGURE 12—Profits obtained under various circumstances

It may be noticed from Figures 11 and 12 that the efficient point lies on the threat curve. This fact is not accidental, but must be true whenever the functions describing the situation are of the form assumed in this paper; i.e., whenever γ_1 depends on q_1 only, γ_2 depends on q_2 only, and p depends on q only. For at the efficient point, we have

$$\frac{\partial}{\partial q_1}(\gamma_1 q_1) = \frac{\partial}{\partial q_2}(\gamma_2 q_2) = p;$$

whereas a point will lie on the threat curve if

$$\frac{\partial P_1}{\partial q_1}\frac{\partial P_2}{\partial q_2} = \frac{\partial P_1}{\partial q_2}\frac{\partial P_2}{\partial q_1}.$$

But for the efficiency point,

$$\frac{\partial P_1}{\partial q_1} = \frac{\partial}{\partial q_1}[q_1(p - \gamma_1)] = p - q_1 p' - \frac{\partial(\gamma_1 q_1)}{\partial q_1} = -q_1 p',$$

and

$$\frac{\partial P_2}{\partial q_1} = \frac{\partial}{\partial q_1}[q_2(p - \gamma_2)] = q_2 p',$$

where p' denotes the derivative of p with respect to q. Since analogous formulas will obtain for the partial derivatives with respect to q_2, the condition for the efficiency point to lie on the threat curve becomes $(-q_1 p')(-q_2 p') = (q_1 p')(q_2 p')$, an obvious identity. Hence the efficiency point will lie on the threat curve, regardless of the shape of the functions γ_1, γ_2, and p, provided only that they are differentiable.

Another fact, which appears from an inspection of Figures 11 and 12, is that the *NSP* threat point lies very close to the efficient point. As with the cusp, this fact must be coincidental, since the efficient point depends only on the local properties of the mapping, while the *NSP* threat point depends on the whole shape of the boundary curve. These two points, though so close as to be indistinguishable even on the large-scale insert to Figure 12, are not actually coincident, as may be seen from the tabulation in Section VII.1.

VIII. CONCLUSIONS

We have seen, in this simplified model, how collusion may tend to restrict production and raise prices and profits. It is noteworthy that these effects are still quite marked in the case when there are restrictions ("laws") against side-payments. It seems, therefore, that such laws or restrictions would naturally result in implicit collusion. The Cournot solution shows that the mere striving for an equilibrium position vis-a-vis one's competitor maximizes neither social product nor profits, for the producers could aid society more if compelled to operate at some "efficient point," and could make a larger profit by collusion (even with antitrust legislation in force).

An interesting phenomenon is observed in the case of implicit collusion. As expected, production is higher and the price lower than when there is open collusion with side-payments; however, the more efficient firm (firm one) actually makes more profit under this arrangement than with open collusion. One might think, *a priori*, that anything which facilitates the collusion should improve the situation of both firms. An example will demonstrate the falsity of this principle. Suppose that A and B can obtain \$10,000 and \$100, respectively, by collaboration and nothing if they do not collaborate, but they cannot make side-payments. Then clearly they will collaborate, and B will be happy to take \$100 in return for collaborating. But if side-payments may be made, B will surely demand that A give him part of the \$10,000 in return for collaborating. Thus A will be better off when side-payments are prohibited.

This example merely exaggerates the phenomenon which appears in our duopoly situation. The efficient producer, firm one, is intrinsically more capable of making profits than firm two. However, firm two has the power to cut one's profits considerably, by increasing his own production. Thus, when side-payments can be made, two can blackmail one into paying two for restraining his (two's) production.

The model employed in this paper has been a very simple one. It would be desirable that more complex models be constructed which would embody such aspects of the problem as incomplete information, nonlinear utility for money, and a more extensive set of strategies for each player. However, it is interesting to note the appearance of several aspects of cartel behavior, even in this simple prototype model. The inefficient firm appears here in the role of blackmailer, whose position is maintained by the damage he might do.

An adequate economic theory of competition involving a small number of firms is yet to be developed. The analysis of the duopoly problem is a step in that direction, and it is to be hoped that the development of game theory apparatus for use in economic analysis may eventually lead to more general results.

Princeton University

REFERENCES

[1] Brems, Hans, "Some Notes on the Structure of the Duopoly Problem," *Nordisk Tidsskrift for Teknisk Økonomi*, Løbe Nr. 37, 1948 (1–4), pp. 41–74.
[2] Fellner, William, *Competition Among the Few*, New York: A. Knopf, 1949, 328 pp.

[3] LANGE, OSCAR, *On the Economic Theory of Socialism*, Minneapolis: The University of Minnesota Press, 1938, 143 pp.
[4] LERNER, A. P., *The Economics of Control; Principles of Welfare Economics*, New York: The Macmillan Company, 1944, 428 pp.
[5] NASH, J. F., "Two Person Cooperative Games," ECONOMETRICA, Vol. 21, January, 1953, pp. 128–140.
[6] VON NEUMANN, J. AND O. MORGENSTERN, *Theory of Games and Economic Behavior*, 2nd ed., Princeton: Princeton University Press, 1947, 641 pp.

[20]

A COMPARISON OF TREATMENTS OF A DUOPOLY PROBLEM (PART II)[1]

By M. Shubik

1. INTRODUCTION

In an earlier investigation,[2] J. F. Nash, J. P. Mayberry, and the present writer compared several treatments of a duopoly problem where the quantity produced by each entrepreneur was taken as the independent variable and the price was assumed to be dependent. In order to continue our study of the strategic possibilities in a duopolistic market, situations involving the treatment of price as the independent variable and, finally, both price and quantity as independent must be examined.

A satisfactory model of a group of firms in competition should take into account the inventory positions, asset positions, and information states of the firms; it should examine the possibility of entry of new firms into competition; and it should attempt to evaluate the effect of the simultaneous application of the many different policy weapons available to each firm, such as price variation, product variation, and advertising. In addition to the above, the analysis should be dynamic. The construction of such a model is well beyond the scope of this article. However, even by limiting our discussion to far less complex models we may be able to extend our knowledge of the nature of oligopolistic competition.

In this paper our discussion will be confined to a duopoly where each firm moves according to the reaction process based on the belief that the opponent's previous price will remain unchanged; a duopoly with alternate moves which are price statements (no probability mixes considered) and to a duopoly treated as a two-person non-zero-sum simultaneous–move game. The first model, which is akin to a model treated by Edgeworth,[3] will be seen to be closely related to the resultant non-zero-sum game.

2. THE ASSUMPTIONS

We assume that the duopolists produce the same product, that there is no advertising, no possibility of outside entry into the market, and that they possess complete information.[4] We abstract, moreover, from such frictions as transportation costs due to different locations. All these conditions, except that of complete

---

[1] The research for this paper was done under Office of Naval Research Contract No. N6onr-27009. I am indebted to L. S. Shapley for aid on the mathematical problems encountered in this paper. A special debt of gratitude is owed to William Vickrey for valuable criticisms and suggestions on rewriting.

[2] J. P. Mayberry, J. F. Nash, and M. Shubik, "A Comparison of Treatments of a Duopoly Situation," *Econometrica*, XXI (January, 1953), pp. 141–154.

[3] F. Y. Edgeworth, *Papers Relating to Political Economy*, Vol. I (London: Macmillan and Co., 1925), pp. 111–142; also *Mathematical Psychics*.

[4] J. von Neumann and O. Morgenstern, *Theory of Games and Economic Behavior* (Princeton: Princeton University Press, 1947), p. 541.

information, may be relaxed. The possibility of entry into the market involves a considerable complication of the model, but yields several interesting results that will be discussed elsewhere.[5] In order to simplify matters as much as possible, let us imagine that the method of marketing for the duopolists is such that each phones in his strategy to a marketing board which is in touch with all the customers and knows their individual demand schedules. This board sells to the customers and remits to the duopolists.

A strategy consists of a statement of a price and the maximum quantity that the player is willing to supply at that price. The marketing board works out how much each of the duopolists will be able to sell (utilizing the goods of the lower-priced firm first) and then it informs the firms how much they are to deliver to the market; i.e., the firms only produce to advance orders from the marketing board after having stated their price and production limits.

In any price variation duopoly we are faced with difficulties that do not exist in the Cournot[6] or any other quantity variation model. These difficulties are respectively the problem of how demand is split when the two firms attempt to sell a homogeneous product at different prices in the same market, and the problem of accounting for the effect of unsold inventories upon a firm's profits. Suppose that the lower-priced firm is unable or unwilling to supply the whole market; then there will be some demand left for the goods of the higher-priced firm. How do we determine this residual demand? It depends upon the way in which the market is split and the customers served. In order to determine the residual demand we must have information concerning which customers were able to purchase the lower-priced goods. Given that this is known and that we have a method for working out the residual demand for the higher-priced firm, it is then perfectly possible that the residual demand will be smaller than the output of the higher-priced firm, in which case that firm will be caught with inventories. The method of stating a strategy given in the preceding paragraph avoids having to deal with this aspect of the problem at this time. (See Section 10 for further discussion.)

3. RECONSTITUTION OF DEMAND

There are three problems that arise in the determination of residual demand in a market after one firm has sold its product but has failed to satisfy all the customers. They are: (1) the reconstitution[7] of individual demand, (2) the reconstitution of aggregate demand, and (3) the regeneration of demand.

(1) The reconstitution of individual demand takes into account the possibility that the individual purchaser may not be satisfied and might wish to buy more of some good even if he has to pay a higher price for the extra units. This could

---

[5] M. Shubik, *Competition and the Theory of Games* (publication in process), Chapter IV.

[6] Augustine A. Cournot, *Researches into the Mathematical Principles of the Theory of Wealth* (New York: The Macmillan Co., 1897).

[7] See: O. Morgenstern, "Demand Theory Reconsidered," *Quarterly Journal of Economics*, LXII (February, 1948), p. 168. The terminology used here is not quite the same as that of Morgenstern.

easily take place under rationing. For the present analysis, however, assume that the individual who buys from the lower-priced duopolist has his demand completely satisfied, but that many individuals do not get a chance to buy from the less expensive firm.

(2) Given this assumption concerning the satisfaction of individual demand, the aggregate reconstituted demand may be obtained by aggregating the demand schedules of the unsatisfied purchasers. It can have a great many shapes, depending upon which group of buyers was satisfied by the lower-priced duopolist. Given all the individual demand schedules to start, there are several conventions whereby we can reconstitute demand for the higher-priced duopolist:

Convention 1. Customers are chosen at random (i.e., if there are k customers, the probability that an individual is served first is $1/k$).

Convention 2. Proportional weightings are given to the desire of each would-be purchaser at the price asked. In other words, if one purchaser would buy N units, he is given N ballots. The sales are made by choosing randomly from all the ballots. Each customer selected is then permitted to buy all he wants at the given price.

Convention 3. Customers are chosen in a manner to maximize effective demand in the reconstituted demand curve (that is, to maximize quantity demanded at the price set by the second firm).

The first two conventions are of the nature of symmetry assumptions and, for the purpose of constructing a complete model without putting in special asymmetric conditions that would imply knowledge of a specific society's trading methods, these appear to be the most reasonable conventions that can be assumed. The third convention is one that would be followed if the marketing board wished to maximize trade. In actual application, this extra parameter of the model would have to be evaluated empirically. Such Veblenesque conventions as that where some customers actually prefer trading with the higher-priced firm might easily fit many societies better than the three others suggested. Greater detail concerning the theory of the reconstitution of demand will soon be available elsewhere.[8]

(3) The third demand problem raised is of a dynamic nature and ties in with inventory policy. Given that an individual is eliminated from the market during a price battle between firms by virtue of his having made a purchase, when is his demand regenerated? In the case of bread, an individual would be out of the market for a day; if we consider consumer durables, however, it may be several years before the purchaser returns to the market. In any attempt to discuss a dynamic model of duopoly this feature of demand must play an important role.

4. SOLUTIONS TREATED IN THIS PAPER

We consider explicitly (1) a modified Edgeworth duopoly, (2) an alternate move duopoly, and (3) a noncooperative price game. We discuss, but do not solve, (4) the noncooperative price-quantity game.

---

[8] M. Shubik, "Reconstitution and Regeneration of Demand," paper delivered at the September, 1954 meetings of the Econometric Society in Montreal.

5. DESCRIPTION OF THE MARKET SITUATION

The same functions are used here as were used in the joint paper with J. F. Nash and J. P. Mayberry.[9] They are the average cost curves of the firm

$$\gamma_1 = 4 - q_1 + q_1^2$$
$$\gamma_2 = 5 - q_2 + q_2^2$$

where q_1 and q_2 are the amounts produced by firms 1 and 2, respectively. We take the collective demand schedule to be

$$p = 10 - 2(q_1 + q_2) = 10 - 2q.$$

As we must reconstitute the market demand, we not only need the collective demand schedule, but must know the individual demands as well. In order to simplify the problem as much as possible, consider the individual demand sched-

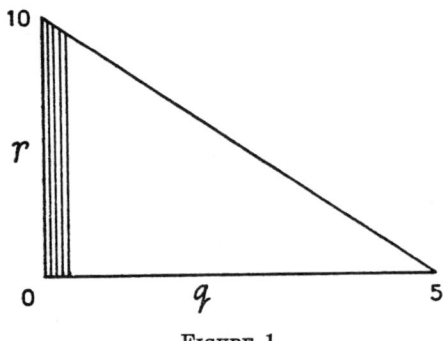

FIGURE 1

ules to be step functions such that one customer is willing to buy a certain amount at price 10 or lower; the next customer is willing to buy the *same* amount at $10 - \epsilon$ or lower, and so on, as is illustrated in Figure 1. With these individual demands, conventions 1 and 2 become the same and can be given a simple geometric interpretation. Suppose that the first player (firm) offers his goods at a lower price than does the second player, but is unable to saturate the market. Then a reconstituted demand for the goods of the second player is the quantity indicated by the horizontal distance between the old demand curve and the line labeled DR in Figure 2, at any price above that of the first player (this actually is the expected or mean reconstituted demand). Although the demand schedules considered here are peculiar (they might represent the demand for some consumer durable of a nature that only one is wanted by any customer), it must be stressed that all phenomena discussed could have been obtained even by taking

[9] J. P. Mayberry, J. F. Nash, and M. Shubik, *loc. cit.*, p. 143.

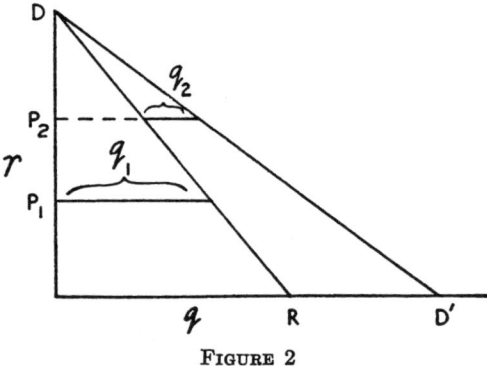

FIGURE 2

a group of customers with identical demand schedules (for instance, each with the same hyperbolic demand schedule).

6. DESCRIPTION OF THE SOLUTION

(1) *The Modified Edgeworth Duopoly.* If we consider that the duopolists act as though each others' prices were going to be the same as they were in the previous period, then we obtain perpetual fluctuations in the market of the type indicated by Edgeworth.[10] This is true of the model presented here, but it is not always true and depends explicitly upon the reconstitution of demand. When the reconstituted demand is very unfavorable, the result of this sort of price variation duopoly may be equilibrium at the efficient-point production rates and price. This will be shown at the end of this subsection.

In order to explain the solutions to the three duopoly models with a minimum of the lengthy computations required, we use the payoff diagram for the second duopolist which is given in Figure 3. P_2 is the profit of the second player and p_2 is the price he charges. The line $Ea'a''a'''T$ is the line of monopolistic profit that the second player can make if he has the market to himself. T is the maximum and is his monopoly profit. E represents the efficient point profit (the figure is drawn approximately to scale). The lines $Ea'M'c'$, $Ea''M''c''$, or $Ea'''M'''c'''$ represent the profits that the second duopolist makes as he varies his price, given that the first duopolist is charging respectively p'_1, p''_1, and p'''_1. When he is the low-price man in the market, his profit grows along the monopoly profit line until his price reaches that of his opponent. As soon as his price goes above that of his opponent, there is a discontinuity in his payoff. (The curve $M'M''M'''$ represents the duopolist's profits if he is always just undercut by his opponent.) As the duopolist's price rises to become considerably higher than that of his opponent, his profits start to rise again until they assume a second maximum, as is indicated by the points c', c'', and c'''.

The range of fluctuation of price in this modified Edgeworth duopoly is given

[10] F. Y. Edgeworth, *op. cit.*

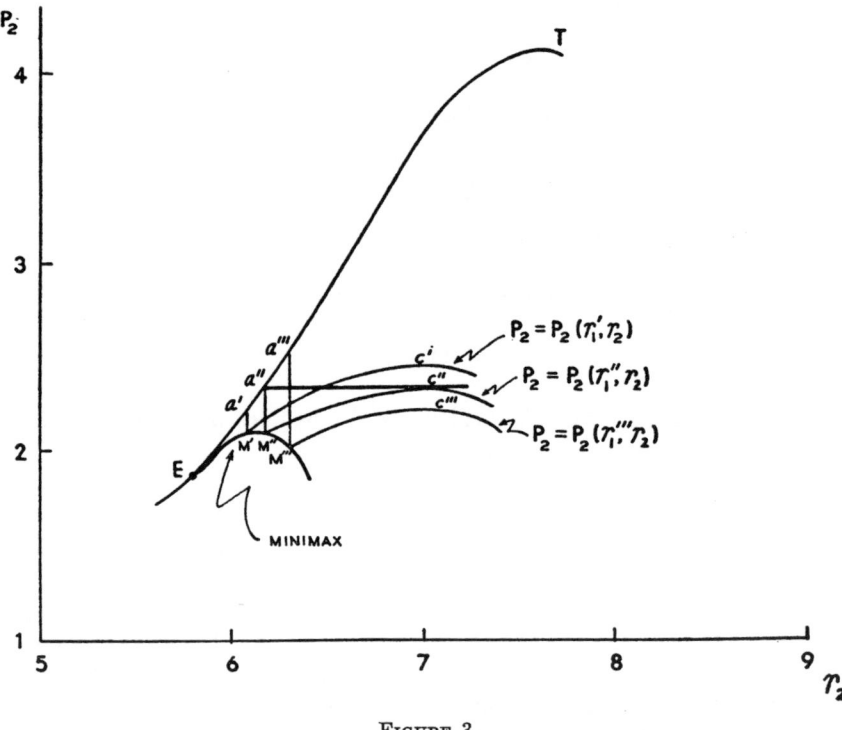

FIGURE 3

by the line $a''c''$. Suppose that the prices charged in the market were above the price p_1'', and that the first duopolist had just undercut the second; then in the next period the second would undercut the first. The first duopolist would retaliate in the following period, (this can be seen from the payoff diagram of the first duopolist which is similar to, but not identical with, Figure 3) and gradually the price would work down to p_1''. Suppose that the second duopolist just undercuts the first by naming p_1''; in the next period the first will retaliate by naming a price slightly lower. If the second assumes that the first man's price is fixed for the next time period, he now raises his price considerably because at that point the maximum obtained by a large raise in price is better than the maximum obtained by just undercutting the first duopolist. This range can be computed generally from one of the two equations

$$q_i\phi_{i\bar{p}}(q_i) - q_i\gamma_i(q_i) = q_i'\bar{p} - q_i'\gamma_i(q_i') \qquad (i = 1, 2)$$

where the $\phi_{i\bar{p}}(q_i)$ is the reconstituted demand function for player i computed on the assumption that the other player's price is $\bar{p}$. q_i is the production rate that maximizes against the reconstituted demand. q_i' is the production rate of the ith firm on the assumption that it is charging price $\bar{p}$ and is not undersold.

The two equations given above determine the "cycling" points for both duopolists, which are the points at which each man is indifferent between being low man in the market or being very high man in the market on the assumption that the other player is charging the low price. For any duopoly of the type described above, the range of fluctuations will not be lower than the lower of the two lower "cycling" points and will not be higher than the higher of the two upper "cycling" points.

This range of fluctuation depends upon the secondary maxima c', c'', etc., which, in turn, are determined by the reconstitution of the demand. If there is no secondary maximum for a given price being charged by the opponent that is larger than the maximum obtained by just undercutting, then the duopolists will cut prices until the efficient point is reached.

LEMMA: Given cost curves that are convex from below (U-shaped) and a demand curve that is monotone decreasing with a second derivative > 0, then the condition required for a pure strategy equilibrium (efficient-point solution) to exist in an Edgeworth duopoly is that

$$(q_i p_E - q_i \gamma_i) \geq (q_i \varphi_{i,p_E}(q_i) - q_i \gamma_i) \qquad \text{for } i = 1, 2$$
$$\text{at efficient} \qquad \text{at } \epsilon \text{ above efficient}$$
$$\text{point price} \qquad \text{point price}$$

where $p = \phi_{i,p_E}(q_i)$ is the reconstituted demand curve for player i on the assumption that the other player charges p_E, which is the price charged at the efficient point.

PROOF: The efficient point is determined by the players acting as though price were a given fixed parameter, i.e., demand were perfectly elastic. At this point their production just saturates the market. Given that one player is charging the efficient point price and is producing his efficient point production, then if the other player can increase his profits this can only be done if it is possible to make more by raising his price.

If a player attempts to raise his price, then the condition given in the lemma is necessary because if it were not true, it would pay a player to depart from the efficient point if he knew that the other was playing his efficient point strategy; hence, the efficient point could not be an equilibrium point.

It is sufficient because above the efficient point $dP_i < 0$ everywhere if the condition holds; hence, the efficient point will be an equilibrium point. This completes the proof of the lemma.

The above amounts to saying that if one player wishes to depart from the efficient point by naming a higher price, he faces a demand which is restricted by a boundary condition which makes it *always* less than the amount he was willing to sell at the efficient point. Whether or not he can make an extra profit depends upon the size of the loss suffered by a cutback in volume as compared with the gain made through the change in unit price and unit cost.

The classes of duopolies which can have the efficient point as their solution include those in which the demand function has a perfectly elastic section about the efficient point; trivially, those in which the plants have very low capacity limits on production and some cases where the efficient point happens to coincide with the minimum average cost. An example of this last type is given below. In this case the stability of the efficient point as an equilibrium depends explicitly upon the reconstitution of the demand.

Consider a symmetric duopoly where the average costs are

$$\gamma_i = q_i^{-.8} + q_i^{.8}$$

(more generally $q_i^{-n} + q_i^n$ where $1/2 < n < 1$).

Demand is given by $\begin{cases} p = 2/q + q/2, & \text{where } q = \sum q_i \text{ and } 0 < q \leqslant 2, \\ p = 2, & \text{for } q > 2. \end{cases}$

The average cost functions are U-shaped in the range $0 < q < 3$ with a unique minimum at $q = 1$. This can be seen by observing that:

$$\gamma_i' = -.8q_i^{-1.8} + .8q_i^{-.2} \text{ is } < 0 \text{ for } q < 1, = 0 \text{ for } q = 1, \text{ and } > 0 \text{ for } q > 1$$

while $\gamma_i'' = 1.44q^{-2.8} - .16q^{-1.2}$ is > 0 for $0 < q < 3$ (the cost curves are U-shaped for a somewhat larger range than $0 < q < 3$, but this is sufficient for this example).

The demand function is monotonic, "downwards sloping to the right." It is continuous as is the marginal revenue function for the duopolists acting jointly.

The efficient point is determined by solving the equations

$$d(q\gamma_i)/dq_i = p$$

or

$$\begin{cases} .2q_1^{-.8} + 1.8q_1^{.8} = \dfrac{2}{q_1 + q_2} + \dfrac{q_1 + q_2}{2} \\ .2q_2^{-.8} + 1.8q_2^{.8} = \dfrac{2}{q_1 + q_1} + \dfrac{q_1 + q_2}{2}. \end{cases}$$

They have a solution of $q_1 = q_2 = 1$ where the price $p = 2$ and the profits made by each firm $P_i = 0$.

Suppose that one firm is committed to producing an output of 1 and charging the price $p = 2$ for it. We wish to examine the action of the other as it tries to maximize against the residual demand. We consider two ways in which the residual demand could be reconstituted. They are (1) that at any price the remaining demand is the original demand minus the production of the firm at the efficient point, or (2) that at any price the remaining demand is half of the original demand. The first convention amounts to stating that the low-priced firm satisfies those customers who would have been willing to pay a high price for the goods. We can write the reconstituted demand as

$$p = \frac{2}{q+1} - \frac{q+1}{2}.$$

In this case, the duopolist must maximize

$$P = q\left[\frac{2}{q+1} + \frac{q+1}{2} - \frac{1}{q^{.8}} - q^{.8}\right].$$

This amounts to the condition that

$$\frac{2}{q+1} - \frac{2q}{(q+1)^2} + q + \frac{1}{2} - .2q^{-.8} - 1.8q^{.8} = 0$$

which is satisfied by $q = 1$. This is a maximum as can be seen by observing that the second derivative of the payoff function is negative at $q = 1$.

The second convention assumes that equal numbers of all types of customers were satisfied by the low-priced firm. We can write the reconstituted demand as

$$p = q^{-1} + q.$$

In this case, the duopolist must maximize

$$P = q[q^{-1} + q - q^{-.8} - q^{.8}].$$

This amounts to the condition that

$$2q - .2q^{-.8} - 1.8q^{.8} = 0.$$

This is satisfied by $q = 1$, but is *not* a maximum as the second derivative is positive here. We can see that the first derivative of the payoff function is always negative for $q < 1$; hence, any maximum that exists must be a boundary maximum. For the purposes of this example, we can restrict the range of q such that $.1 \leqslant q < 3$. If the first duopolist stays at the efficient point, the second raises his price to 10.1, produces $q = .1$, and makes a profit of $P = .3631$. Thus, in this example, the bounds on the range of fluctuation of price are such that $2 < p < 10.1$.[11]

(2) *An Alternate Move Duopoly.* If we replace the reaction conditions by a situation in which one duopolist has to move first, the second one is informed of his move and then declares his own move, the solution is an equilibrium point, depending on the "cycling" points ($a'c'$ in Figure 3) and the minimax points (M in Figure 3). If there is no range of fluctuation, then the efficient point remains the solution. (We note that in this case the minimax points and the efficient point coincide; this amounts to saying that the maximum of the envelope curve $MM'M''$ in Figure 3 is at the efficient point.)

If the Edgeworth price fluctuation range exists, then if the duopolist whose lower "cycling" point (a' in Figure 3) is lower than his opponent's moves first, he names the price just under the other player's lower "cycling" point. In this case, the other player then names his upper "cycling" point and the equilibrium con-

---

[11] We note that in this particular example the Cournot equilibrium and efficient point coincide because the elasticity of demand happens to be infinite just at that point. In the example computed in detail the Cournot value lies above the price game value.

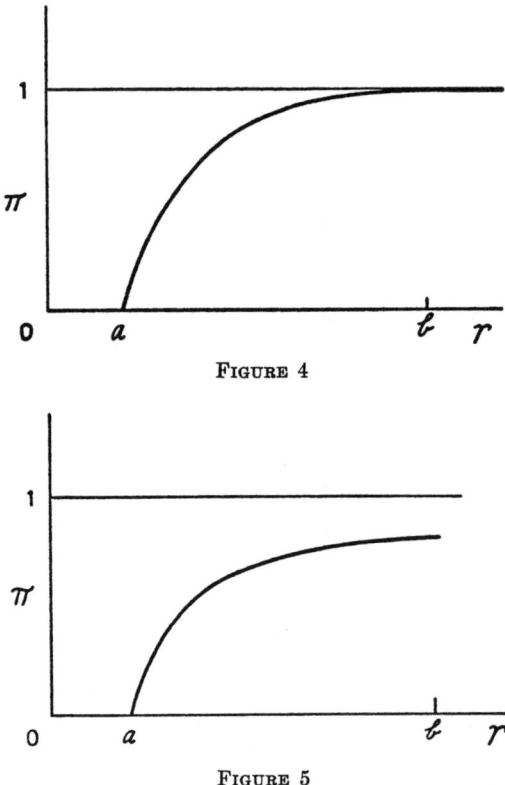

FIGURE 4

FIGURE 5

sists of the two players each at a different price. If the first move goes to the player whose lower "cycling" point is higher than that of his opponent, he names a price just under the bottom of his opponent's "cycling" range or else names his minimax point (whichever gives him the greater profit).

If the Edgeworth price fluctuation range exists for only one of the duopolists, then if the player who has no fluctuation range moves first, he names his efficient-point price and production (which constitutes his minimax strategy[12]), and the other player maximizes against this. If the other player moves first, he names his minimax strategy and is just undercut.

(3) *The Price Game.* We examine a market form in which the two firms, simul-

[12] Two difficulties are faced in the alternate-move duopoly. The first is encountered in the definition of the minimax strategy. This only exists at or above the efficient point if we put in the condition that the player who is low in the market is not permitted to produce more goods than that amount which would maximize his profit at the price he charges. If we allow him to produce uneconomic quantities, then the minimax for any firm will usually be the point where it shuts down and loses overhead costs. The second has to do with how the market is split if both firms charge more than the efficient point price but yet charge the same price. Some convention must be given. In the price game this does not really matter much because such occurrences have probability zero.

taneously and without knowledge of each other's strategies, inform a central marketing board of their price strategies, which then tells them what to produce. We wish to examine this game for the existence of equilibrium points. This model is closely linked to the modified Edgeworth duopoly discussed above. If there is no range of fluctuations present in the Edgeworth duopoly, then the simultaneous-move price game has the efficient point as its solution and this is also the solution to the other models.

The existence of a range of fluctuations in the Edgeworth model implies that if the simultaneous-move price game has an equilibrium point, this point entails the use of at least one mixed strategy. Shapley[13] has shown the existence of a unique equilibrium point for symmetric games of this type. He has also established the general shape of the probability density functions used by the players in the nonsymmetric game presented here as an example. The corresponding cumulative distributions are shown in Figures 4 and 5.

The problem of finding an equilibrium point in such a game devolves into finding two functions $f(p_1)$ and $g(p_2)$ such that

$$\int_a^b P_1(p_1, p_2) \, dg(p_2) \equiv V_1,$$

for all p_1 in the range: $(a \leqslant p_1 \leqslant b)$

$$\int_a^b P_2(p_1, p_2) \, df(p_1) \equiv V_2,$$

for all p_2 in the range: $(a \leqslant p_2 \leqslant b)$

where V_1 and V_2 are the values of the game to the first and second players, respectively. These relations follow directly from the definition of an equilibrium point. In essence, each player attempts to present the other with a strategy that "flattens" his payoff in such a manner that he has an expectation of V_i no matter what strategy he plays (we note that it is possible that the functions $g(p_2)$ and $f(p_1)$ do not exist; this means that the game has no equilibrium point whatsoever).

7. DETERMINATION OF NUMERICAL RESULTS

(1) *The Modified Edgeworth-Bertrand Duopoly.* The conditions that give the range of fluctuation are

$$q_2(10 - Q_1 q_2) - q_2(5 - q_2 + q_2^2) = q_2' p_2 - q_2'(5 - q_2' + q_2'^2)$$

where

$$q_2 = \frac{1 - Q_1 + \sqrt{(Q_1 - 1)^2 + 15}}{3}; \qquad q_2' = \frac{1 + \sqrt{3p_1 - 14}}{3}$$

and Q_1 is defined as $2q^*/(q^* - q_1)$ with

$$q^* = (10 - p_1)/2 \quad \text{and} \quad q_1 = (1 + \sqrt{1 + 3(p_1 - 4)})/3.$$

[13] L. S. Shapley, "Solutions to Bertrand Games" (unpublished).

By successive approximation we find that at $\bar{p}_2 = 6.184$ the high cost producer is indifferent to undercutting or raising his price to $p_2 = 7.358$. He makes the same profits $P_{2L} = P_{2H} = 1.188$ by doing either. His production at the low price is $q_2' = 1.0443$ and at the high price is $q_2 = .4560$. The two prices define the Edgeworth range of fluctuation. Both points depend explicitly upon the way in which the market splits, or, in other words, upon the way in which demand is reconstituted. The profits of player 1 can vary at most between $P_{1L} = 2.339$ and $P_{1L}' = 3.875$ where P_{1L} and P_{1L}' stand for the profits he would make if he just undercut player 2 at the low and high points of the range of fluctuation.

(2) *The Alternate Move Duopoly.* We give only the case where player 1 moves first. He names the price $p_1 = 6.184$ which is the lower "cycling" point of his opponent's range; his opponent names the price $p_2 = 7.358$. The profits of the first player are $P_{1L} = 2.339$ and the second player $P_{2H} = 1.188$.

(3) *The Price Game.* The actual computation of the distribution functions for the strategies involves a considerable amount of numerical integration and requires the use of an electronic computer. We are able, however, to obtain bounds on the value of the game without actually computing these distributions. This is done for player 1 only. Let P_{1U} represent the payoff to player 1 if he is just undercut by the other player. In terms of price we obtain

$$P_{1U} = (.70000p_1 - 5.18333)(.49000p_1^2 - 7.55666p_1 + 25.68358).$$

This has a maximum at $p_1 = 6.073$ where the profit $P_{1U} = 2.176$ and is the lower bound for V_1. It is the pure strategy minimax point for player 1. The equation above is that of the curve $M'M''M'''$ in Figure 3. The minimax value is somewhat greater than that of the efficient point and can be obtained regardless of what the other player does. This establishes it as a lower bound to the value of the game for player 1. It has been proved elsewhere[14] that the value of the game to either player cannot be higher than that given by the highest of the lower ends of the players' range of fluctuations. The highest of the lower ends of the two ranges is at $p = 6.184$ where $P_{1L} = 2.339$.

Figures 4 and 5 indicate the shapes of the cumulative distribution functions used by each player in his strategy.

8. COMPARISON OF NUMERICAL RESULTS

The bar graphs of Figures 6 and 7 give comparisons of the prices charged and of the profits made, respectively, by the first player in four noncollusive solutions to the duopoly problem. For a more detailed comparison of some of these solutions the reader is referred to the earlier paper.[15]

9. STATICS AND DYNAMICS

This paper has dealt primarily in statics, although the modified Edgeworth duopoly was discussed dynamically. The range of fluctuation that we cited as the

---

[14] M. Shubik, *op. cit.*, Chapter II.
[15] J. P. Mayberry, J. F. Nash, and M. Shubik, *loc. cit.*, p. 150.

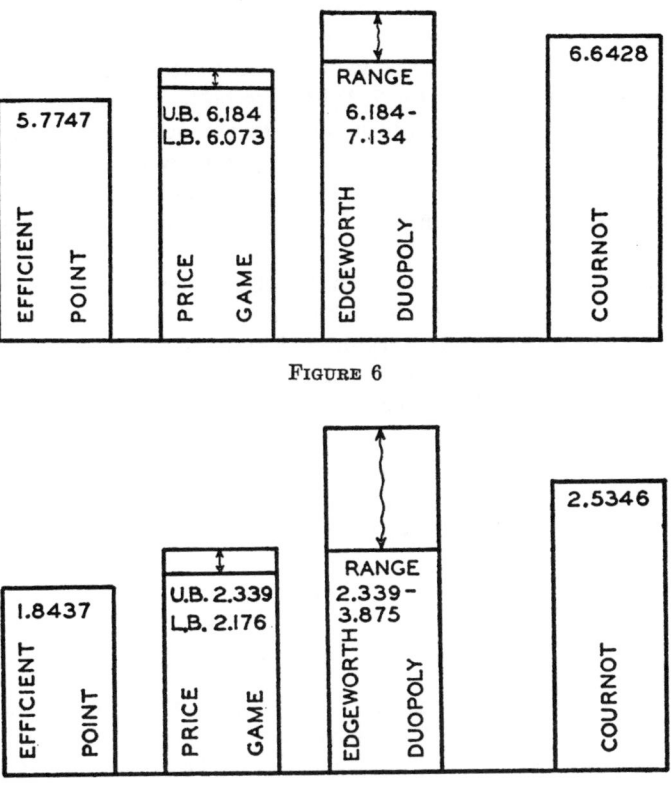

FIGURE 6

FIGURE 7

solution for this model can be described as the range within which all frictionless noncooperative price duopoly solutions must lie, regardless of the order of moves.[16]

10. THE PRICE-QUANTITY GAME

Economists have hardly discussed the possibility that both price and quantity may be independent variables in many situations. The retailer may decide both the quantities of goods to buy and the prices he intends to charge for them, thereby taking the risk of being left with inventory. By specifying the value of inventories left over at the end of trading, we can formulate a static model of a price-quantity duopoly. A strategy consists of phoning in a statement concerning both price and quantity of goods offered to the market. In the analysis in this paper, inventory trouble was avoided by assuming that the marketing board told the duopolists how much to produce after they had named their prices.

---

[16] This does not hold true if we take asset positions into consideration. For instance, it might pay one firm to try to ruin the other if long-run profits are to be considered, in which case it may set a price way below the range and a production rate far higher than would be consistent with short-run economic gain.

The modified Edgeworth duopoly and the alternate-move example can be reformulated to take inventories into account. There is absolutely no difference between the alternate-move price game and the alternate-move price-quantity game. The solutions presented in 7(2) hold for both. The game analogous to the Edgeworth duopoly is one in which each entrepreneur acts on the assumption that the price charged and the quantity produced by the other is fixed. In this case, when he turns out to be wrong, he may be penalized both by being undercut and by being stuck with inventory. In the formulation in this paper we ignored the inventory penalties.

Although the price-quantity game has been completely formulated,[17] no existence proofs for equilibrium points are known as yet, and the computational problems are far worse than those encountered in the price game, even if existence were assured. In essence, the problem is to find two probability distributions $f(p_1, q_1)$ and $g(p_2, q_2)$ over a square such that

$$\int_\alpha^\beta \int_a^b P_1(q_1, q_2, p_1, p_2) \, dg(p_2, q_2) \equiv \bar{V}_1,$$

for all p_1 and q_1 in the ranges: $(a \leqslant p_1 \leqslant b; \alpha \leqslant q_1 \leqslant \beta)$

$$\int_\alpha^\beta \int_a^b P_2(q_1, q_2, p_1, p_2) \, df(p_1, q_1) = \bar{V}_2$$

for all p_2 and q_2 in the ranges: $(a \leqslant p_2 \leqslant b; \alpha \leqslant q_2 \leqslant \beta)$.

The value for this game has the same lower bound as the value for the price game. That is the minimax point which remains unchanged in both cases. It is conjectured that the game's value lies below that of the price game. It is easy to see that in general the equilibrium in the price game, if it is not a pure strategy equilibrium, does not become an equilibrium in the price-quantity game. Suppose that it did; then by the very definition of an equilibrium point, if player 2 knows player 1's probability distribution over his active strategies, then player 2 can compute that he is indifferent to his expectation obtained by playing any of his optimal strategies. In particular, this holds for the lower and upper points of the active strategy range. If the ranges and the distributions over price are the same, then in both games the second player knows that he will make the same amount by playing at the bottom of the range. If he plays at the top of the range, in the price game, as formulated, there is a mechanism which adjusts his production to maximize against the actual price charged by his opponent on the assumption that his opponent produces an amount that maximizes the opponent's income at the prices he charges when he is low man in the market. Thus, there is no inventory problem for either player. In the price-quantity game there is no such mechanism; hence, if the second player were to name a quantity which would maximize his income if the first player were to charge his lowest price, then if the first player charges any other price, the second player will be caught with some inventory and make less than he would have in the price game. In this case the values of

---

[17] M. Shubik, *op. cit.*, Chapter II.

the strategies at the upper bound of the range are different. Thus, the equilibrium in the price game does not in general give an equilibrium in the price-quantity game.

Another way to look at the price game dealt with here is as a price-quantity game in which any producer caught with inventory is allowed to sell it at cost to a surplus product purchasing agency.

11. CONCLUSIONS

The narrow study of the highly idealized models of duopoly has enabled us to formulate and examine some not at all intuitively evident aspects of economic behavior. Previously, we noted the actions for maximizing social product[18] and for maximizing joint profits. We were able to study the phenomenon of implicit collusion and examined the blackmailer role of firms threatening others prior to collusive agreements. The study has now been extended to price and price-quantity warring situations. We observed that in order to be able to set up our simplified models we were forced to consider the nature of the reconstitution of demand and to examine the role of inventories even in a static model. The price-variation game appears to be usually more "competitive" than the Cournot game.[19] The possibility of both price and quantity as independent variables was raised. The price-quantity game has not yet yielded to complete analysis. It is the belief of this writer that the solution of this problem is necessary to our understanding even static duopoly models.

The above models have dealt with the technical features of two firms battling during one time period. In order to construct a duopoly model which begins to approach "reality" we must obviously extend our consideration to multi-period games between firms in which both their technical and financial structures play a role.[20] The static analysis in this paper is offered in full realization that it attempts to do only a small amount of the necessary preliminary spadework prior to the erection of such a theory.

Princeton University

[18] J. P. Mayberry, J. F. Nash, and M. Shubik, *loc. cit.*, p. 144.

[19] It is of interest to note that as the number of competitors increases the price variation duopoly model has its equilibrium converge to the perfect competition equilibrium. This result is known to be true for the Cournot duopoly, where the pure strategy equilibrium points become closer and closer until finally they converge to the perfect competition equilibrium. In the price-variation game, however, the convergence takes place through a series of mixed strategies which in the limit have all of their probability concentrated on the perfect competition equilibrium point. M. Shubik, *op. cit.*, Chapter IV.

[20] It is possible to formulate such a situation as an "economic ruin game." In such a game the payoffs are a discounted stream of income obtained by active firms. There are "ruin conditions" which force firms into liquidation if their assets ever drop below a certain level.

[21]

GAMES OF ECONOMIC SURVIVAL*

Martin Shubik and Gerald L. Thompson †

General Electric Co. and Ohio Wesleyan University

1. INTRODUCTION

The games examined in this paper were originally formulated by one of the authors in order to investigate some problems in economic theory pertaining to the theory of the firm [4]. A corporation has to give consideration both to its prospects for survival and to its ability to pay dividends. Survival <u>per se</u> could be a goal, as could the maximization of the discounted value of the dividend payments. The latter might even involve the eventual ruin of the firm as part of the optimal policy.

In order to portray the relationship between the market and the financial aspects of the firm, we can construct a type of game whose form is closely related to the Gambler's Ruin problems. The fortunes of the firm are divided into a corporate and a withdrawal account. Ruin conditions apply to the former, and dividend payments are made in the latter.

In general, complex dynamic problems are soon encountered. This paper deals only with simple examples involving one and two person dynamic nonzero sum games.

2. ONE-PERSON GAMES OF ECONOMIC SURVIVAL

2.1. General Description of One-Person Games of Economic Survival

A one-person game of economic survival is characterized by the following quantities: a discrete set of times $t = 0, 1, 2, \ldots$; a set of integers a_i, $i = 1, \ldots, n$ that represent the amounts that a player obtains each time he plays; a set of positive numbers p_i, $i = 1, \ldots, n$, with $\Sigma p_i = 1$, where p_i represents the player's probability of obtaining a_i when he plays and where successive plays of the game are assumed to be independent; a corporate account whose size at time t is denoted by $C(t)$; a withdrawal account whose size at time t is denoted by $W(t)$; a discount rate ρ (< 1) which is effective on the withdrawal account only; a ruin payment F that is paid to the player upon ruin or on leaving the game by withdrawing all assets; a ruin level B and a ruin condition that says, if $C(t-1) > B$ and $C(t) \leq B$ the player received the ruin payment F and the game is terminated at time t.

At the beginning of any time period he may transfer any integral part of his corporate account into his withdrawal account; let us indicate such a transfer by w. Such a payment may be interpreted as a dividend payment to stockholders and is nonrecoverable, hence payments

---

*Manuscript received September 12, 1958.

†The authors wish to express their gratitude to Drs. Lloyd Shapley and Alan Hoffman for valuable criticism. Additionally, the authors are indebted to Mr. R. Singleton for several interesting comments and observations, including an alternative proof for the solution to the one-person game.

111

can never be made from the withdrawal to the corporate account. Let the initial amount in the corporate account be $C(0) = x$ (an integer). The goal of the player is to maximize the discounted expected value of his withdrawal account. This completes the description of the game.

Suppose for a moment there were no transfer privileges. Then if we watch the corporate account it varies as if it were a random walk on the line. Thus it moves from a value $C(t)$ at time t to the value $C(t+1) = C(t) + a_i$ at time $t+1$ with probability p_i.

If there are transfer privileges, then, since we have assumed that each time the game is played the probabilities of obtaining payments a_i are independent of previous plays of the game, we see that the transfer payment w should be a function only of the value of the corporate account. Therefore, if transfer payments $w(C(t))$ are included, the corporate account moves from $C(t)$ to $C(t+1) = C(t) + a_i - w(C(t))$ with probability p_i for $i = 1, \cdots, n$.

A financial strategy (or simply, strategy) is a transfer payment function $w(x)$. (The function w is from positive to non-negative integers.) Let W be the set of all financial strategies. Given a strategy w, one can compute the value function $V(x, w)$ that measures the worth to the player of playing the game with initial assets x and strategy w. The value function gives the expected discounted value of the player's withdrawal account.

In order to compute the expected discounted value of the withdrawal account, given initial assets x and withdrawal strategy w, we consider truncated games of length n, that is, games which are played at times $t = 0, 1, \cdots, n$. Let s_n be the outcome of a series of plays of the subgames played each period. This indicates whether the player has obtained the maximum or minimum revenue from the first play of the game, the second, etc., up to the nth play. From this information we can compute the amount $K(s_n, w, t)$ that is added to the player's withdrawal account for each of the times $t = 0, 1, \cdots, n$. Let $q(s_n)$ be the probability that s_n occurs. Then, taking sums and passing to the limit, we have

$$V(x, w) = \lim_{n \to \infty} \sum_{s_n} \left(\sum_{t=0}^{n} K(s_n, w, t) \rho^t \right) q(s_n)$$

as the expected discounted value of the withdrawal account. In order to show that this limit exists, we consider the most favorable game—the one in which the player wins <u>every</u> time he plays. For this game, the most favorable strategy is w^*, for which the player withdraws all his assets above the ruin level as soon as he can. Then

$$V(x, w^*) = (x - B) + \sum_{t=1}^{\infty} \rho^t = (x - B) + \frac{\rho}{1 - \rho}$$

for this game. And, obviously, for any other course of the game and any other strategy w, we have $0 \leq V(x, w) \leq V(x, w^*)$. Since the expression for $V(x, w)$ is a series of positive terms and is bounded, it converges. For an explicit calculation of $V(x, w)$ in the case that w is an optimal strategy, we refer the reader to Section 2.3.

A strategy w is said to <u>dominate</u> strategy w' if

(a) $V(x, w) \geq V(x, w')$ for all $x \geq 1$;
(b) $V(x, w) > V(x, w')$ for some $x \geq 1$.

That undominated strategies exist can be seen from the following argument. A withdrawal strategy w is completely characterized by the smallest integer N such that $w(N) > 0$, since the corporate account increases only by integral values and N acts as a "reflecting barrier." We shall call this N the N-value of w. Now we have $K(s_n, w, t) = 0$ for $n = 1, 2, \ldots, N-x$, so that, letting $V_n(x, w)$ be the discounted amount in the withdrawal account out to time n, we have

$$V_n(x, w) = \sum_{s_n} \left\{ \sum_{t=0}^{n} K(s_n, w, t) \rho^t \right\} q(s_n)$$

$$= \sum_{t=0}^{n} \left\{ \sum_{s_n} K(s_n, w, t) q(s_n) \right\} \rho^t$$

$$= \rho^{N-x} \left\{ \sum_{t=n-x}^{n} \left[\sum_{s_n} K(s_n, w, t) q(s_n) \right] \rho^{t-N+x} \right\}.$$

Now the term in braces on the right is bounded by the same kind of argument that we used to show that $V(x, w)$ was bounded. Since $\rho < 1$, the factor ρ^{N-x} becomes small if x is fixed and N becomes large. Hence all those strategies w whose N-values are sufficiently large are dominated by other strategies with smaller N-values. Since there are only a finite number of withdrawal strategies with bounded N-values, there are only a finite number of values $V(x, w)$ to consider, and hence an undominated strategy w exists.

Let W^o be the set of undominated strategies. In Section 2.3. we shall characterize W^o for a certain kind of one-person game of economic survival.

2.2. A Simple One-Person Survival Game

Consider the game G for which $a_1 = 1$, $a_2 = -1$, $p_1 > 0$, $p_2 > 0$, $p_1 + p_2 = 1$, $B = 0$, and $F = 0$. We shall characterize the undominated strategies for this game and thus obtain information about the result for the general one-person survival game.

LEMMA 1: An undominated w^o in G is a solution of the functional equation*

(1) $\quad V(x, w^o) = \max \left[\rho p_1 V(x-1, w^o) + \rho p_2 V(x+1, w^o) \right], \max_{a \geq 1} \left[a + V(s-a, w^o) \right].$

---

*The optimal financial strategies are not the only solutions of (1). For instance, the function

$$V(x) = \left[\frac{1 + \sqrt{1 - 4\rho^2 p_1 p_2}}{2 \rho p_1} \right]^x$$

satisfies the equation

$$V(x) = \max_{a \geq 0} [\rho p_1 V(x - a - 1) + \rho p_2 V(x - a + 1) + a],$$

and we can regard this function as the solution for a zero withdrawal strategy. But this solution is not economically interesting, since, although $V(x) \to \infty$, the expected discounted value of the withdrawal account for any finite length of time is zero.

PROOF: Equation (1) states merely that the player always has the choice of immediately playing the game or first withdrawing an amount a and then playing. Moreover, he must choose the withdrawal payment a so that his value is maximized. It is obvious that any strategy, not a solution of (1), is dominated by such a solution.

COROLLARY: $V[x, w^o]$ is a monotone strictly increasing function of x.

PROOF: From (1) we have $V(x, w^o) \geq 1 + V(x - 1, w^o)$, hence $V(x, w^o) > V(x - 1, w^o)$.

Lemma 1 shows that the game is a dynamic programming problem in the sense of Bellman [1]. It also brings out a difficulty: namely, suppose that for one or more values of a, both terms of the right-hand side of (1) are equal. An undominated strategy could then choose either term. We determine a unique choice by saying that he will not play the game when the terms are equal. More precisely we give the following definition.

DEFINITION: By a special undominated strategy we shall mean an undominated strategy with the following property: $w^o(x) = 0$ if, and only if,

(2) $$\rho p_1 V(x - 1, w^o) + \rho p_2 V(x + 1, w^o) > \max_{a \geq 1} [a + V(x - a, w^o)].$$

Remark: If $\rho p_1 < 1/2$, the solution to the game is trivial, the player pays out all the resources to his withdrawal account immediately as the game is "unfair" and not in his favor. We exclude this case and henceforth assume $\rho p_1 \geq 1/2$.

LEMMA 2: If w^o is undominated, then there exists an x such that $w^o(x) > 0$.

PROOF: If $w^o(x) = 0$ for all x, then, obviously, $V(x, w^o) = 0$. But, since all assets can immediately be withdrawn, we see that $V(x, w^o) \geq x \geq 0$, which is a contradiction for positive x. Therefore for some x, $w^o(x) > 0$.

LEMMA 3: Let w^o be a special undominated strategy. If z is an x such that $w^o(x) = 0$ and $w^o(x + 1) > 0$, then $w^o(z + 1) = 1$.

PROOF: Suppose, on the contrary, that $w^o(z + 1) = k + 1 > 1$. By Lemma 1 we have

$$V(z + 1, w^o) \geq V(z, w^o) + 1 \geq V(z - 1, w^o) + 2 \geq \ldots \geq V(z - k, w^o) + k.$$

Since the player always withdraws to produce the maximum amount possible, by Lemma 1 we have

$$V(z + 1, w^o) = V(z + 1 - n, w^o) + n,$$

hence all of the above inequalities are actually equalities. And, in particular, $V(z, w^o) = V(z - 1, w^o) + 1$. The definition of a special undominated strategy now shows that $w^o(z) > 0$, contradicting the fact that $w^o(z) = 0$.

LEMMA 4: Let w^o be a special undominated strategy and let z be the smallest x such that $w^o(x+1) > 0$, then for all $x \geq z$ we have $w^o(x) = x - z$.

PROOF: No payment can be greater than $x - z$, for if $w^o(x) = a > x - z$ then $z > x - a$ and

$$V(x - a, w^o) + a \geq V(z, w^o) + (x - z),$$

or

$$V(x - a, w^o) + (z - (x - a)) = V(z - (z - (x - a)), w^o) + (z - (x - a)) \geq V(z, w^o),$$

contradicting the fact that w^o is a special undominated strategy and z is the smallest x such that $w^o(x+1) > 0$. (See (1).)

If there is at least one $x > z$ such that $w^o(x) < x - z$, then there must be an $x' > z$ such that $w^o(x') = 0$, for otherwise (2) would be violated. (For instance, $x' = x - w^o(x)$ has this property.) Let $\bar{z}$ be the least $x > z$ such that $w(x) = 0$. By an argument similar to that of Lemma 2, there is an $x > \bar{z}$ such that $w(x) = 0$ and $w(x+1) > 0$. Call the smallest such z_1. By an argument analogous to Lemma 3, $w(z_1 + 1) = 1$. Then we have the following information about w^o:

$$w^o(x) = \begin{cases} 0 & \text{for } x \leq z \\ x - z & \text{for } z < x < \bar{z} \\ 0 & \text{for } \bar{z} \leq x \leq z_1 \\ 1 & \text{for } x = z_1 + 1 . \end{cases}$$

We shall show that w^o is actually dominated by another strategy w' defined as follows:

$$w'(x) = \begin{cases} 0 & \text{for } x \leq z \text{ or if } x = z_1 \text{ for the first time,} \\ x - z & \text{for (i) } z < x < \bar{z} - 1 \quad \text{if } x \text{ has always been } < z_1, \\ & \quad\ \ (ii) \ z \leq x \leq \bar{z} \quad\quad\ \text{if } x \text{ has ever been } = z_1, \\ 0 & \text{for (i) } \bar{z} - 1 \leq x < z_1 \quad \text{if } x \text{ has always been } < z_1, \\ & \quad\ \ (ii) \ \bar{z} \leq x \leq z_1 \quad\quad\ \text{if } x \text{ has ever been } = z_1. \\ w^o(x) & \text{for } x \geq z_1 + 1 . \end{cases}$$

We shall show that strategy w^o does not do as well as w' by comparing two games, one starting with $C(o) = a$ and using strategy w', and the other starting with $C(o) = a + 1$ and using strategy w^o, where $\bar{z} \leq a < a + 1 \leq z_1$. Let $C'(t)$ be the player's corporate account in the first game and $C^o(t)$ be his corporate account in the second game. We compare these two strategies for the same course of outcomes of individual plays of the game.

Case 1: For some $t \geq 1$, C' reaches z_1 before C^o reaches $\bar{z} - 1$. Then 1 is withdrawn from the C^o account, but the C' remains unchanged so that at time t the two corporate accounts will be equal to z_1. And since w' becomes the same as w^o after reaching z_1, the two corporate accounts will be equal for the rest of the game. Hence we see that $V(a + 1, w^o) - V(a, w') = \rho^t < 1$.

Case 2: For some $t \geq 1$, C^o reaches $\bar{z} - 1$ before C' reaches z_1. Then $\bar{z} - 1 - z$ is withdrawn from C^o and $\bar{z} - 2 - z$ is withdrawn from C' so that at time t the two corporate accounts will be equal (to z); and since w' is the same as w^o when $x \leq z$, the two corporate accounts will remain equal for the rest of the game. Hence we see that $V(a + 1, w^o) - V(a, w') = \rho^t < 1$.

In both of the above cases it is false that $V(a + 1, w^o) \geq V(a, w') + 1$, which must necessarily be the case for an undominated strategy. (One could do better than w^o at $a + 1$ by first paying out 1 and then using w'.) From this contradiction we conclude that $\bar{z}$ does not exist and $w^o(x) = x - z$ for $x > z$, concluding the proof.

THEOREM 1: There is a unique special undominated strategy w^o defined as follows: there exists a $z > 0$ such that $w^o(x) = 0$ if $x < z$ and $w^o(x) = x - z$ if $x \geq z$.

PROOF: Observe from the definition of special undominated strategies that z (the largest x for which $w^o(x) = 0$) is the same for all special undominated strategies. Lemmas 3 and 4 and mathematical induction now show that any two special undominated strategies are equal.

Remarks:
(A) If $\rho = 1$, then this one-person game becomes the classical gambler's ruin game. The unique special undominated strategy for this game would be to initially withdraw all funds and not play the game at all.
(B) The calculation of the value of an example of a game of the type here discussed is carried out in Section 2.3.
(C) If the changes of state are rational numbers a_1 and a_2, where $a_1 > 0$ and $a_2 < 0$, then the changes in the possible fortunes of the player will be multiples of d, where d is the greatest common divisor of a_1 and a_2. We regard the player as ruined whenever his corporate account is less than $|a_2|$. The optimal strategy for the player is as follows: (a) initially pay out an amount $\gamma = x - m|a_2|$ where m is the largest integer such that γ is nonnegative; (b) there is a number z such that, if $x < z$, then $w^o(x) = 0$, and if $x \geq z$, then $w^o(x) = x - z$. (It can be shown in this case that $x - z$ is a multiple of d.)

2.3 The Value of a Simple One-Person Game of Economic Survival

In Section 2.2 we have established the form of the special optimal strategy w for a game with $a_1 = +1$ and $a_2 = -1$, and $F = 0$. The value of a game in which the special optimal strategy is used can be calculated directly as follows: Let $z + 1$ be the first value of the corporate account at which a payment is made to the withdrawal account under the optimal strategy w^o. By setting $P_1 = p$ and $P_2 = q = 1 - p$, the value of the game is the solution to:

(3) $\quad V(x, w^o) = \rho p V(x+1, w^o) + \rho q V(x-1, w^o) \quad \text{for } x \leq z$

subject to:

(4) $\quad V(0, w^o) = 0,$

(5) $\quad V(z+1, w^o) = V(z, w^o) + 1.$

We set $V(x, w^o) = \alpha^x$ in (1), which yields:

$$\alpha = \frac{1 \pm \sqrt{1 - 4\rho^2 pq}}{2\rho p}.$$

Denoting the roots by α_1 and α_2,

$$V(x, w^o) = A_1 \alpha_1^x + A_2 \alpha_2^x.$$

From (4) and (5) we obtain:

$$A_1 = -A_2$$

and

$$A_1 = \frac{1}{\left(\alpha_1^{z+1} - \alpha_2^{z+1}\right) - \left(\alpha_1^z - \alpha_2^z\right)}.$$

The solution may be expressed as:

$$V(x, w^o) = A_1(z) \left\{\alpha_1^x - \alpha_2^x\right\}.$$

The solutions for games with $\rho p > 1/2$ for different values of ρ and p have been calculated and examples are provided in the tables below.

$\rho = .7, p = .99$

x	z		
	1	2	3
1	2.26	1.59	1.10
2		2.29	1.60
3			2.29

$\rho = .995, p = .99$

x	z		
	2	3	4
1	191.34	195.08	195.08
2	194.21	198.02	198.02
3		199.04	199.04

$\rho = .99, p = .55$

x	z		
	<u>7</u>	8	9
1	1.95	1.93	
2	3.57	3.55	
3	4.97	4.94	
4	6.20	6.16	
5	7.33	7.27	
6	8.38	8.32	
7	|9.39|	9.32	
8		10.31	
9			
10			

$\rho = .995, p = .55$

x	z			
	7	8	9	<u>10</u>
1	2.66	2.84	2.97	3.06
2	4.86	5.18	5.42	5.59
3	6.71	7.15	7.48	7.71
4	8.28	8.83	9.24	9.52
5	9.65	10.28	10.75	11.09
6	10.85	11.57	12.10	12.47
7	11.94	12.73	13.32	13.72
8		13.80	14.43	14.87
9			15.48	15.95
10				|16.98|

The numbers in the boxes represent the values of the largest investment that will be made under an optimal policy. The underlined z's are the amounts associated with the optimal policies.

3.1. GENERAL DESCRIPTION OF TWO-PERSON GAMES OF ECONOMIC SURVIVAL

A two-person game of economic survival is characterized by the following quantities: an $m \times n$ matrix $A = \|a_{ij}\|$, a discount rate ρ (<1), ruin payments F_1 and F_2, ruin conditions B_1 and B_2, corporate accounts $C = (C_1(t), C_2(t))$, and withdrawal accounts $W = (W_1(t), W_2(t))$. The role that each of these quantities plays is analogous to those played by the corresponding quantities in the one-person case. The subscript 1 refers to the first player and 2 refers to the second player. Let the initial amounts in the players' corporate accounts be $C_1(0) = x$ and $C_2(0) = y$. We shall seek equilibrium point solutions for a simple class of these games for various values of x and y.

3.2. A Simple Two-Person Survival Game

Here we consider the case in which $B_1 = B_2 = 0$, $F_1 = F_2 = F$, and A is the matrix

$$A = \begin{pmatrix} 1 & -1 \\ -1 & 1 \end{pmatrix}.$$

We shall characterize the equilibrium point solutions of this game for various initial fortunes x and y.

In general it is false to assume that the players will always use optimal strategies for every matrix game A, as will be shown in Section 3.4. In the simple example selected here, however, the unique optimal strategies for each player are to choose each of his alternatives with probability 1/2, and by known martingale theorems [2] there is no optimal way to deviate from these. Hence each player can expect +1 and -1 with equal probability each time the game is played.

Let us compute the values $V(x, y) = (V_1(x, y), V_2(x, y))$ of withdrawal accounts of each player, assuming that they begin with fortunes x and y and that they both <u>always</u> keep their fortunes in their corporate accounts.

We must distinguish between two cases: the first, where at the termination of play the winner obtains only the "prize" F, and the second, where he obtains the prize F and adds the assets in his corporate account to his winnings to bring his total to $x + y + F$. Initially, we examine the first case. However, it is easier to give an economic interpretation to the second. This is done later in Section 3.4.

For the first player the quantity $V_1(x, y)$ satisfies

(6) $$V_1(x, y) = \frac{\rho}{2} V_1(x + 1, y - 1) + \frac{\rho}{2} V_1(x - 1, y + 1).$$

By use of the trial solution $V_1(x, y) = c^x d^y$ where c and d are constants, and the boundary conditions $V_1(x + y, 0) = F$ and $V_2(0, x + y) = 0$, it can be shown that the solution to these difference equations is

(7) $$V_1(x, y) = F \left(\frac{c_1^x - c_2^x}{c_1^{x+y} - c_2^{x+y}} \right),$$

where

(8) $$c_1 = \frac{1 + \sqrt{1 - \rho^2}}{\rho} \quad \text{and} \quad c_2 = \frac{1 - \sqrt{1 - \rho^2}}{\rho} = \frac{1}{c_1}.$$

If $0 < \rho < 1$, then $c_1 > 1$ and $0 < c_2 < 1$. (This is certainly the range of economic interest.) If $x + y$ is relatively large, then the second factor in the denominator of (7) is negligible and we may write

(9) $$V_1(x, y) = \frac{F}{c_1^y} (1 - k^x) \quad \text{where} \quad k = \frac{c_2}{c_1} = c_2^2 \ (k < 1).$$

We use this approximate relation to investigate the solutions. Note that for fixed y the quantity $V_1(x, y)$ is a monotone increasing, but bounded, function. Thus, given a fixed fortune y for player 2, player 1 will not want his capital account to be arbitrarily large, since it would be more profitable for him to withdraw some of this money and invest it elsewhere.

Now suppose the players can withdraw any part of their corporate account, so long as neither player is ruined. But if either player is ruined, the survivor gets F and forfeits the money in his corporate account.

Player 1 will withdraw a dollar whenever his expected value obtained by leaving it in is less than a dollar more than his expected value obtained by withdrawing it—in other words, when (for fixed y) x is such that

(10) $$\frac{F}{c_1^y}(1 - k^x) + 1 \geq \frac{F}{c_1^y}(1 - k^{x+1}).$$

After some algebraic manipulations we take logarithms and obtain a relation of the form

(11) $$x = b - \frac{1}{2} y,$$

where b is a positive constant.

Since we have made an approximation in the course of deriving the very simple formula (11), we must expect our results from now on to be quantitatively inaccurate. However, for large discount rates (small discount factors) the error due to the approximations is small, so that our conclusions are qualitatively correct.

Carrying out the analogous procedure for player 2 we find that, given a fortune x of player 1, the maximum fortune y that player 2 would keep in the game is given by the equation

(12) $$y = b - \frac{1}{2} x,$$

where the constant b is the same as in equation (11).

We are now in a position to analyze the equilibrium points in the game.

3.3. Equilibrium Points in the Two-Person Game

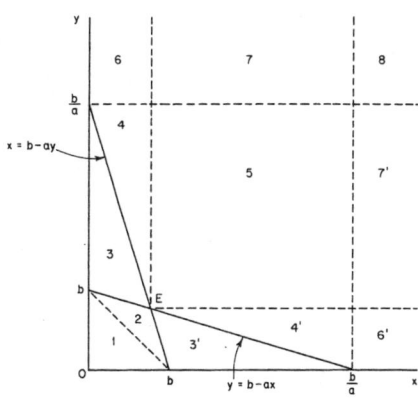

Figure 1

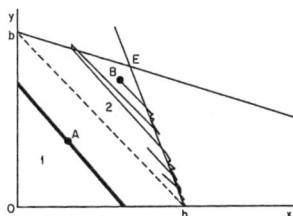

Figure 2

In Figure 1 we have sketched the graphs of equations (11) and (12) and have numbered different areas of the plane that lead to essentially different behavior, described by a pair of strategies in equilibrium.

In area 1 neither player has more money that he would like to have, based on the most pessimistic assumption that his opponent will not withdraw money from the game. It is impossible for them to move to a point at which either player would prefer to withdraw money rather than keep it in. Hence, if the initial fortunes of the players specify a point A in the interior of 1, as the game progresses their fortunes move along a negative 45° line (see Figure 2) until the game terminates at one of the axes at which point one of the players is ruined. The equilibrium points also have maxmin properties in area 1.

If the initial fortunes specify a point B in the interior of area 2, then again neither player wishes to withdraw money. However, with probability 1, the play of the game will proceed until one of the players makes a withdrawal. This is shown in Figure 2 by the crossing of the negative 45° line with one of the lines of equation (11) or (12). After one of the players withdraws part of his fortune, the point representing the joint fortunes in their corporate accounts moves back into area 2, and the game progresses until it eventually terminates at (0, b) or (b, 0). A possible path of play is illustrated in Figure 2.

Let A be an initial fortune point in area 3 (see Figure 3). Here the second player has more money than he needs under a minmax assumption concerning his opponent's behavior. He withdraws down to the point A' which is in area 2. The game then proceeds as if it started at A' in area 2.

In Figure 3, point B is in area 4 where both players are overcapitalized. Two possible equilibrium paths are as follows: first, player 2 may withdraw capital so that the fortune move to point B' which is in area 2 and the game proceeds from there; or player 1 may withdraw to point B'', after which player 2 withdraws to B''' in area 2 and the game proceeds from there. Other equilibrium paths between B' and B''' are also possible.

Point C in Figure 3 represents a point in which player 2 has enough capital to make it rational for player 1 to withdraw from the game on a minmax assumption concerning player 2's behavior. This is shown at C''. If, however, player 1 retained his capital, player 2 would withdraw down to C'. There is mixed strategy equilibrium involving the decisions of each player to keep in all his funds and the decision of player 1 to leave the game and player 2 to reduce investment. There are also many other equilibrium points.

The equilibrium point solutions in areas 3', 4', and 6' are analogous to those in areas 3, 4, and 6, but with the roles of the two players interchanged.

Point A in area 5 of Figure 4 represents a point at which both players are overcapitalized. There are two obvious equilibrium points: in one player 2 withdraws to A' and then player 1 withdraws to A'''; in the other, player 1 withdraws to A'' and then player 2 withdraws to A''''. There are many other equilibrium points. In particular, E is enforceable as follows: player 2 announces that he will withdraw to E' and, even though overcapitalized, will remain

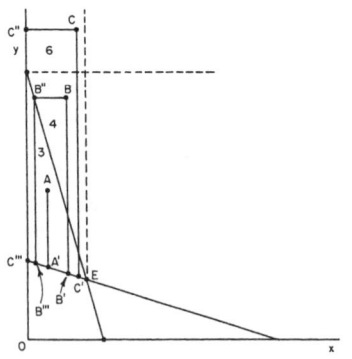

Figure 3

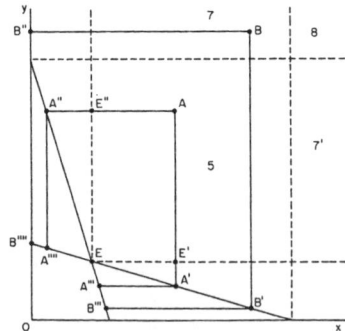

Figure 4

there, regardless of what player 1 does; it is then rational for player 1 to withdraw to E, and the game proceeds from there. There is an analogous equilibrium point in which 1 withdraws to E" and then 2 withdraws to E.

At point B in area 7 of Figure 4, player 2 is so overcapitalized that player 1 should quit entirely. If he does, the fortune point moves to B''. On the other hand, if player 2 first withdraws his capital, the fortune point moves to B'; then player 1 should withdraw to B''' and the game proceeds in area 2; there is a mixed strategy equilibrium involving these choices. The equilibrium point E is also enforceable in the same manner as stated in the paragraph above. Many other equilibrium points are possible. The behavior in area 7' is analogous to that in 7, with the roles of the players interchanged.

In area 8 both players are heavily overcapitalized and there are many equilibrium points, including E.

The analysis is most satisfactory in areas 1, 2, 3, and 3', since unique equilibrium points are obtained and these have maxmin properties. In other areas, the multiplicity of equilibrium points means that the outcome of the game will be settled by means of threats and counterthreats, in which case the equilibrium point analysis is not particularly conclusive. Experimental investigation of the course of play in areas 5 to 8 would be of interest.

3.4. An Economic Interpretation of the One- and Two-Person Games

The one-person game serves to demonstrate the "safety value" of liquid assets in a fluctuating market. It is analogous to an inventory problem. The penalty of being "out of stock" is the ruin of the firm.

The next step in the investigation of models with greater realism involves having the rewards from the random walk depend upon the asset position of the firm. In the model in Section 2.2, the only use of the money in the corporate account was as protection against ruin.

It is possible to give a mathematical formulation for the policy of a firm which wishes to achieve and maintain a given level of safety prior to paying dividends. A discussion and examination of the effects of different policies is given elsewhere [4].

The two-person game has an economic analogue a market in which there is overcapacity with two firms present, but in which one firm could make a profit if it were the sole survivor. The zero-sum matrix indicates that if one firm makes a profit, the other must lose. For example, the only way in which business may be obtained is by obtaining the competitor's accounts. The prize, F, is the discounted value of the income obtainable by the surviving firm in the market. When one firm is ruined the value of the other is F plus its corporate assets at that time. In order to express this we would have had to express:

$$V(x + y, 0) = x + y + F.$$

If $x + y \lll F$, then the same qualitative results as above hold, with the linear equations (11) and (12) being replaced by nonlinear equations whose graphs are convex upward curves similar to the straight lines of Figure 1.

If cooperation and side payments were permitted, the solution would be for the industry to "rationalize" immediately, with one firm exiting but being paid off by the other.

An examination of the solution shows that the richer both firms are, the worse off they may be!!!

$$V(x, x) \geq V(x + k, x + k).$$

This phenomenon is caused by the prolongation of the length of the struggle for survival. If two financially weak firms compete, one will soon fail, leaving the other a lucrative market. If they are stronger, it will take a much longer time to reach a prosperous state, unless there is collusion and rationalization.

3.4. Comments and Further Problems

It is evident that the investigation of the game with:

$$A = \begin{pmatrix} 1 & -1 \\ -1 & 1 \end{pmatrix} \text{ holds for } A = \begin{pmatrix} c & -c \\ -c & c \end{pmatrix}$$

by a simple change of units.

If the matrix A is 2×2 and the one-period zero-sum game on A has a unique mixed strategy solution, then the players will always use optimal strategies in the matrix game each period. For A larger than 2×2, this is not necessarily so. A simple example demonstrates this:

Let
$$A = \begin{pmatrix} 1 & -1 \\ -1 & 1 \\ 10 - \epsilon & -10 \\ -10 & 10 - \epsilon \end{pmatrix}$$

The strategy $(\frac{1}{2}, \frac{1}{2}, 0, 0)$ dominates $(0, 0, \frac{1}{2}, \frac{1}{2})$ in the one period game but not necessarily in the dynamic game in which the duration of play as well as the expected gain per period is of importance.

A further discussion and economic interpretation of this type of game including cases where the one-period "subgame" is nonzero sum is given elsewhere [4].

BIBLIOGRAPHY

[1] Bellman, Richard, <u>Dynamic Programming</u>, Princeton University Press, 1957

[2] Doob, J. L., <u>Stochastic Processes</u>, Wiley, 1953

[3] Feller, W., <u>The Theory of Probability</u>, Wiley, 1950

[4] Shubik, Martin, <u>Strategy and Market Structure</u>, Wiley, 1958

* * *

[22]

PRICE STRATEGY OLIGOPOLY WITH PRODUCT VARIATION*

I. THE MARKET WITH UNDIFFERENTIATED PRODUCTS

Suppose that two firms each with a constant average cost c face a market for an undifferentiated product. For simplicity we consider a model with a linear demand, constant average costs and given capacities for the firms; however our remarks are more general.

In the classical writings on duopoly the various authors observed[1] it is possible to formulate a duopoly model where the firms use either quantity or price as their strategy. The use of quantity as a strategy leads to the well known COURNOT model[2]. It has been suggested that price is a far more natural variable for the firm to use in its strategy. Both BERTRAND and EDGEWORTH[3] investigated the simple price model with the sellers selling an undifferentiated product. They obtained different results and the difference can be explained in terms of different assumptions concerning capacity restrictions (or equivalently rising costs).

A great amount of the difficulty in dealing with oligopolistic price strategy models comes in describing the demand conditions that exist when various prices are quoted. This is especially striking when the product are totally undifferentiated; however even with differentiation, oligopoly contingent demand curves[4] may have kinks and bends which make them difficult to analyze.

CHAMBERLIN suggested in his theory of monopolistic competition[5] that one should treat each competitor as though he were a

* Research undertaken by the Cowles Commission for Research in Economics under Contract Nonr-3055 (01) with the Office of Naval Research.

1. A. A. COURNOT, *Researches into the Mathematical Principles of the Theory of Wealth*, New York, Macmillan, 1897.

F. Y. EDGEWORTH, *Papers Relating to Political Economy I*, London: Macmillan, 1925.

2. A. A. COURNOT, *op. cit.*

3. See M. SHUBIK, *Strategy and Market Structure*, New York, Wiley, 1959, Chs. 4 and 5.

4. Ibid., Ch. 5.

5. E. H. CHAMBERLIN, *The Theory of Monopolistic Competition*, Cambridge, Harvard University Press, 6th ed., 1950.

30

PRICE STRATEGY OLIGOPOLY WITH PRODUCT VARIATION

monopolist in the sense that each individual provides a slightly differentiated product or service when compared with any other even though the other may purportedly be selling the same good. In this paper we mathematically formulate an example of firms in monopolistic competition and investigate noncooperative behavior as the number of firms increases and as the level of product variation is decreased.

II. THE BERTRAND-EDGEWORTH EXAMPLES

We consider two firms each selling the same product to a market whose aggregate demand can be represented by $p = a - bq$. We wish to describe the demand if more than one price exists. This can be done by considering an aggregate consumer with a quadratic utility function:

$$u(q) = aq - \frac{b}{2}q^2 \quad \text{where} \quad q = q_1 + q_2 \qquad (1)$$

It is easy to see that the consumer in this instance will always buy from the firm with the lower price, as long as it has sufficient supplies. If it runs out of supplies then the constrained consumer maximization is such that it might make purchases from the other firm as well. It follows immediately that if each firm has sufficient capacity to supply the whole market at any price (at or above costs), then the firm with the lower price will take all. In this case the noncooperative equilibrium, will, as was observed by BERTRAND, be the competitive equilibrium where each firm sets price equal to marginal cost (in this case marginal cost equals average cost).

EDGEWORTH noted that if capacities were limited the consumers might also buy from the higher priced firm and instead of settling down immediately to the competitive price, price in the market would be indeterminate and would tend to fluctuate over a range. This has been described as the EDGEWORTH cycle[6].

At what capacity level does the nature of the solution change, and how is the solution affected by the presence of more competitors? A diagramatic and algebraic investigation of these questions is presented.

6. SHUBIK, M., op. cit. Ch. 5.

Let each firm have an average cost c and a capacity k. *Figures 1 and 2 represent conditions where respectively*

$$k = \frac{a-c}{b} \quad \text{and} \quad k = \frac{a-c}{2b}$$

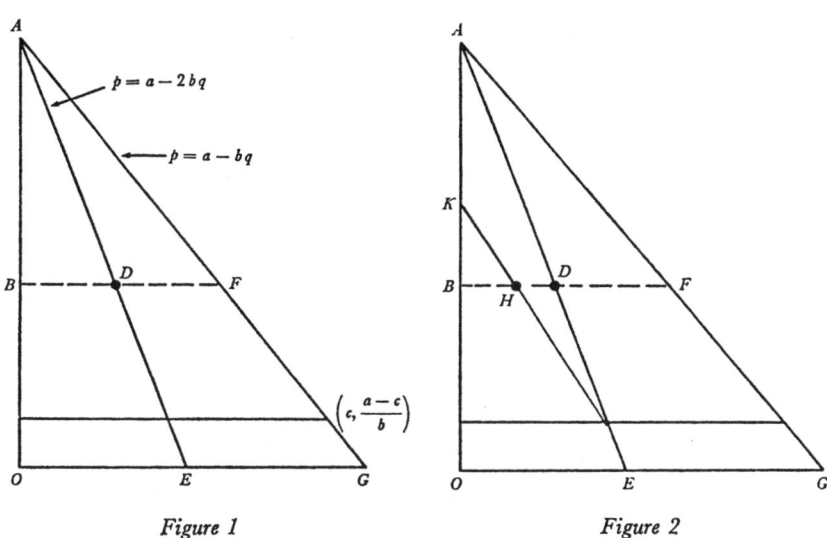

Figure 1

Figure 2

In *Figure 1* the line AG describes the overall market demand when the firms charge the same price. The line AE is the demand on the individual firm when both charge the same price. Suppose that one firm charges a price represented by OB. We may draw the contingent demand curve faced by the other firm as it varies its price. First consider what it will sell if it charges a price higher than OB. It will sell nothing as is indicated by the segment AB. The reason is evident, because at any price above c, the other firm has enough capacity to satisfy the whole market. If both firms name the same price, the market will be split at D. If the second firm undercuts the first then its demand will be given by FG. Hence, given the price of the first firm, the contingent demand for the second as it varies its price from $p = a$ to $p = 0$ is given by the broken curve $AB\,D\,FG$. For every price of the first firm there will be a contingent demand defined for the second firm. Furthermore as long as the first firm

PRICE STRATEGY OLIGOPOLY WITH PRODUCT VARIATION

prices at or above cost, if the second firm names a higher price than the first there will be no market left.

In *Figure 2* we consider the effect of a capacity limitation. We note that together they have enough capacity to satisfy the market at price equals cost, hence the efficient solution is not affected by capacity restrictions. This is not so for the structure of contingent demand. This change is critical for the change in noncooperative behavior.

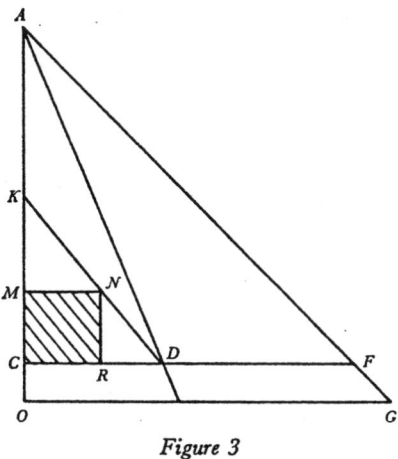

Figure 3

Suppose that the first firm charges the price given by OB. At that price the consumer are willing to buy BF, however the firm has only enough capacity to supply HF leaving an unsatisfied demand of BH. Suppose that the second firm is charging a higher price than the first; its demand may not be completely wiped out. In this instance there will be demand left up to the price OK as is indicated by the segment of the contingent demand curve given by HK. As the price of the second firm varies across the whole range the total contingent demand curve is given by AK KH D and FG. This possibility that there may be some demand left for the higher priced firm destroys the stability of the efficient point as a noncooperative equilibrium. This can be seen from *Figure 3*. Suppose that the first firm sets its price equal to cost. If it had enough capacity to saturate the market then there would be nothing left for the other at any higher price. The argument holds symmetrically hence the efficient point is also a noncooperative equilibrium point inasmuch as given that either knows that the other is charging $p_i = c$ neither is motivated to change his price.

When we limit capacity to $k = (a-c)/2b$ (or the amount given by DF in *Figure 3*) we may not only show why this equilibrium is destroyed but can even demonstrate how far the second firm would raise its price given the information that the first was charging c. The contingent demand curve, given that the first firm charges c is shown by AKD and FG. The best move for the second firm is to maximize monopolistically by charging a price OM. It obtains a profit indicated by the rectangle $MNCR$.

Any capacity less than $k = (a-c)/b$ causes the destruction of the equilibrium point at $p = c$.

III. CONTINGENT DEMAND AND PRODUCT DIFFERENTIATION

The discontinuity in the contingent demand curves was a natural result of the assumptions of the absolute identity of products and rational consumer behavior. This should disappear if we introduce product differentiation. A way to do so which yields us a model sufficiently simple to permit a diagramatic and algebraic treatment is by introducing an extra term into the previously used utility function. This term reflects a degree of differentiation among the goods. In spite of its relative simplicity the model is rich enough to illustrate the interesting features of oligopolistic demand and to serve as the basis for the study of the change in oligopolistic behavior as the number of competitors is increased.

The aggregate utility function used is given by:

$$U = aq - \frac{b}{n}q^2 - \varepsilon\left[\frac{\sum q_i^2}{n} - \frac{q^2}{n^2}\right] - \sum_{i=1}^{n} p_i q_i \quad \text{where} \quad q = \sum_{i=1}^{n} q_i \quad (2)$$

This contains the parameter n in various terms. The reason for its inclusion in the utility function will be explained when we investigate oligopolistic behavior involving markets with increasing numbers of firms and customers. For our discussion of contingent demand we limit ourselves to $n = 2$. This gives us:

PRICE STRATEGY OLIGOPOLY WITH PRODUCT VARIATION

$$U = aq - \frac{b}{2}q^2 - \frac{\varepsilon}{4}(q_1 - q_2)^2 - p_1 q_1 - p_2 q_2 \qquad (3)$$

where ε is the parameter which controls the degree of product differentiation. If it is zero, then the products are perfect substitutes. The higher it is, the more complementary they become and the more the third term dominates the utility function. As the model we are investigating is an open partial model of the economy we ignore income effects as a first approximation and hence may take into account the budget constraint of the consumers by subtracting the terms $p_1 q_1$ and $p_2 q_2$ directly in the utility function.

Using equation (3) and the condition for consumer optimization we may solve for consumer demand in terms of prices. This gives us:

$$\frac{\partial U}{\partial q_1} = a - bq - \frac{\varepsilon}{2}(q_1 - q_2) - p_1 = 0 \qquad (4)$$

$$\frac{\partial U}{\partial q_2} = a - bq - \frac{\varepsilon}{2}(q_2 - q_1) - p_2 = 0 \qquad (5)$$

From (4) and (5) by addition and subtraction (6) and (7) are obtained:

$$2a - 2bq = p_1 + p_2 \qquad (6)$$

$$-\varepsilon(q_1 - q_2) = p_1 - p_2 \qquad (7)$$

These yield

$$q_1 = \frac{2a - p_1 - p_2}{4b} - \frac{p_1 - p_2}{2\varepsilon} \qquad (8)$$

or

$$\boxed{q_1 = \frac{2a - \left(1 + \frac{2b}{\varepsilon}\right)p_1 - \left(1 - \frac{2b}{\varepsilon}\right)p_2}{4b}}$$

and similarly for q_2. This solution will hold only if

$$\varepsilon \geq \frac{2b(p_1 - p_2)}{2a - p_1 - p_2} \qquad (9)$$

Condition (9) is obtained directly from (8) by setting $q_1 = 0$.

LLOYD SHAPLEY AND MARTIN SHUBIK

If $p_1 > p_2$ and

$$0 \leq \varepsilon \leq \frac{2b(p_1 - p_2)}{2a - p_1 - p_2} \qquad (10)$$

then $q_1 = 0$ and from (5) we obtain:

$$q_2 = \frac{a - p_2}{b + \frac{\varepsilon}{2}} \qquad (11)$$

thus we have a continuous contingent demand with two 'kinks' as is shown in *Figure 4*. The labels are for the case where p_1 is fixed and p_2 varies.

Figure 4

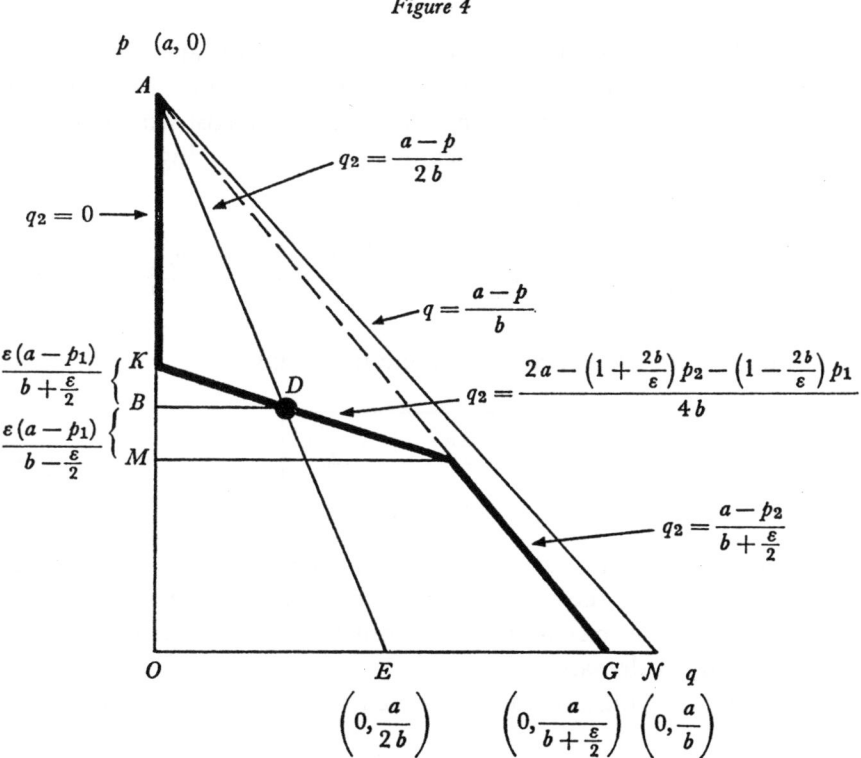

The contingent demand is given by $AKDFG$. It is of interest to note that because the goods are not perfect substitutes the low priced firm will never obtain 'all of the market' in the sense of the amount

PRICE STRATEGY OLIGOPOLY WITH PRODUCT VARIATION

ON which would be sold if both charged the same low price. The gap GN is a measure of the lack of substitutability. The line AN is an 'apples and oranges' addition of the amounts sold by both when they charge the same price. The line AE gives the amount sold by one firm on the assumption that the other is charging the same price. The equations for the three line segments AK, KF and FG which make up the contingent demand are given. At the point F the second firm has completely priced the first out of the market.

IV. THE NONCOOPERATIVE PRICE DUOPOLY WITHOUT CAPACITY CONSTRAINTS

When the firms are equal and capacity constraints are not tight we are able to solve for the noncooperative equilibrium in the market by observing that a symmetric equilibrium will exist which enables us to write down payoff or revenue functions involving demands as given in equation (8). Thus

$$P_i = (p_i - c)\left(\frac{2a - \left(1 + \frac{2b}{\varepsilon}\right)p_i - \left(1 - \frac{2b}{\varepsilon}\right)p_j}{4b}\right) \qquad (12)$$

can be written and maximized analytically giving:

$$\boxed{p_i = \frac{2a + c\left(1 + \frac{2b}{\varepsilon}\right)}{3 + \frac{2b}{\varepsilon}}} \qquad (13)$$

A straightforward check that this is an equilibrium point is obtained by setting the price of one firm at that given by (13) and checking for the maximum for the other using all three segments of his contingent demand.

When $\varepsilon = 2b$ the point G is moved to E and (13) simplifies to

$$p_i = \frac{a+c}{2} \qquad (14)$$

each may charge his monopoly price.

As ε approaches zero p_i approaches c which is the efficient point solution for undifferentiated gods.

The solution given by equation (13) is not the efficient point solution for this monopolistic market. If the economy were being run for the benefit of the consumer the condition that price equals cost would still prevail. It is of interest to note however that as differentiation is removed even with only two competitors the noncooperative equilibrium (without capacity constraints) approaches the efficient solution as indicated by the BERTRAND model.

The noncooperative equilibrium can be illustrated by means of

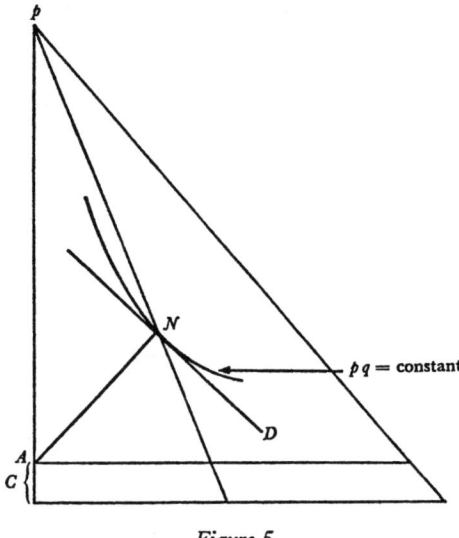

Figure 5

a diagram as is shown in *Figure 5* above. N is the equilibrium point. The curve '$pq = \text{constant}$' is the highest isoprofit curve attainable when on the contingent demand curve ND. The slope of AN is the negative of ND.

V. THE NONCOOPERATIVE PRICE OLIGOPOLY WITHOUT CAPACITY CONSTRAINTS

In order to appreciate the effect of increasing the number of competitors in such a way that we can study the change in economic

PRICE STRATEGY OLIGOPOLY WITH PRODUCT VARIATION

power as numbers increase we introduce the parameter n into the utility function. This has the effect that the size of the market faced by any individual firm, if all are charging the same price will be the same regardless of the number of competitors. The absolute economic size of all firms remains constant but their size relative to the sum of all markets decreases.

Rewriting the utility function for the n person market

$$U = aq - \frac{b}{n}q^2 - \varepsilon\left[\frac{\sum q_i^2}{n} - \frac{q^2}{n^2}\right] - \sum_{i=1}^{n} p_i q_i \qquad (15)$$

we differentiate to obtain demand conditions and solve for demands in terms of all prices.

$$\frac{\partial U}{\partial q_i} = a - \frac{2bq}{n} - \frac{2\varepsilon}{n}q_i + \frac{2\varepsilon}{n^2}q - p_i = 0 \qquad (16)$$

or

$$p_i = a - \frac{2b}{n}q - \frac{2\varepsilon}{n}q_i + \frac{2\varepsilon}{n^2}q \qquad i = 1, \ldots, n \qquad (17)$$

Summing all of the equations (17) and dividing by n we obtain the avarage price:

$$\bar{p} = \frac{1}{n}\sum p_i = a - \frac{2b}{n}q - \frac{2\varepsilon}{n^2}q + \frac{2\varepsilon}{n^2}q = a - \frac{2b}{n}q \qquad (18)$$

Hence

$$p_i - \bar{p} = -\frac{2\varepsilon}{n}(q_i - \bar{q}) \quad \text{where} \quad \bar{q} = q/n \qquad (19)$$

Using (18) we may write

$$\bar{q} = \frac{n(a - \bar{p})}{2nb} = \frac{a - \bar{p}}{2b}$$

From (19)

$$q_i - \bar{q} = \frac{n}{2\varepsilon}(p_i - \bar{p})$$

hence

$$\boxed{q_i = \frac{a - \bar{p}}{2b} - \frac{n}{2\varepsilon}(p_i - \bar{p})} \qquad (20)$$

The revenue for the ith firm may be expressed as:

$$P_i = (p_i - c) q_i \tag{21}$$

Taking derivatives, setting them equal to zero and then setting all $p_i = p$ we may solve for the symmetric noncooperative equilibrium.

$$\frac{d}{dp_i}[(p_i - c) q_i] = (p_i - c)\left[-\frac{1}{2nb} - \frac{n}{2\varepsilon} + \frac{1}{2\varepsilon}\right]$$
$$+ \frac{a - \bar{p}}{2b} - \frac{n}{2\varepsilon}(p_i - \bar{p}) = 0 \tag{22}$$

or

$$(p - c)\left[\frac{-\varepsilon - n(n-1)b}{2\varepsilon nb}\right] + \frac{a - p}{2b} = 0$$

giving

$$\boxed{p = \frac{n(n-1)bc + \varepsilon(na + c)}{n(n-1)b + \varepsilon(n+1)}} \tag{23}$$

As ε approaches zero p approaches c as we have already noted in the case of duopoly. Here however we have a further convergence result. As $n \to \infty$ we observe that $p \to c$. In other words as the number of competitors increases even though they are selling differentiated products the noncooperative equilibrium approaches the efficient point solution where price equals costs.

The results will be qualitatively the same for firms with increasing marginal costs provided that they are not so steep as to have the effect of capacity limitation.

VI. PRICE OLIGOPOLY WITH CAPACITY CONSTRAINTS

It was possible to demonstrate in Section II that in price competition capacity is of critical importance in preserving the stability of equilibrium. If the capacity of the competitor with the lowest price is limited this will increase the market to the others. It is evident that this effect will be present for differentiated as well as undifferentiated competitors.

PRICE STRATEGY OLIGOPOLY WITH PRODUCT VARIATION

The condition needed to preserve the equilibrium in the market for an undifferentiated good is that each firm has enough capacity to supply the whole market as price equals cost. In this section we investigate the conditions needed when the firms sell products differentiated as in the demand structure given in equation (20) as derived from the utility conditions given in (15).

Let each firm have a capacity k, suppose that all except the first are selling at capacitiy, then from (17) we may write:

$$p_1 = a - \frac{2(n-1)}{n}bk + (n-1)\frac{2\varepsilon}{n^2}k - \frac{2}{n}\left[b + \varepsilon - \frac{\varepsilon}{n}\right]q_1 \quad (24)$$

The optimal price and production for the first firm are determined by setting

$$\frac{d[(p_1-c)q_1]}{dp_1} = 0$$

which gives:

$$q_1 = \frac{a - c + \frac{2(n-1)}{n}\left(\frac{\varepsilon}{n} - b\right)k}{\frac{4}{n}\left[b + \varepsilon\left(\frac{n-1}{n}\right)\right]} \quad (25)$$

The profit at (p_1, q_1) is:

$$(p_1 - c)q_1 = \frac{\left[a - c + 2\frac{(n-1)}{n}\left(\frac{\varepsilon}{n} - b\right)k\right]^2}{\frac{8}{n}\left[b + \varepsilon\left(\frac{n-1}{n}\right)\right]} \quad (26)$$

From (23) we may write the profit at the noncooperative equilibrium as:

$$P = \frac{\varepsilon n(a-c)^2[n(n-1)b + \varepsilon]}{2b[\varepsilon(n+1) + n(n-1)b]^2} \quad (27)$$

By setting (26) = (27) and solving for k we can determine the critical capacity below which the equilibrium is destroyed. We call the critical capacity k_n to stress its dependence upon n.

$$k_n = \frac{n^2(a-c)}{2(n-1)(nb-\varepsilon)}\left\{1 - \frac{2}{n}\frac{\sqrt{\varepsilon n[bn + \varepsilon(n-1)][n(n-1)b + \varepsilon]}}{[\varepsilon(n+1) + n(n-1)b]}\right\} \quad (28)$$

41

For $\varepsilon = 0$ we obtain
$$k_n = \frac{n^2(a-c)}{2(n-1)nb}$$

This gives
$$k_2 = \frac{a-c}{b} \quad \text{and} \quad k_\infty = \frac{a-c}{2b}$$

this shows that with undifferentiated products the duopolists need an excess capacity so large that each could individually saturate the whole market, if a pure strategy equilibrium is to be preserved. As the numbers grow each needs only in the limit his efficient production capacity.

For any fixed ε as $n \to \infty$ we have:
$$k \to \frac{a-c}{2b} + O\left(\frac{1}{n}\right) \to \frac{a-c}{2b}$$

For duopoly in general (28) gives:
$$k_2 = \frac{2(a-c)}{(2b-\varepsilon)}\left\{1 - \left(\frac{2b+\varepsilon}{2b+3\varepsilon}\right)\sqrt{2\varepsilon}\right\}$$

We can show the destruction of the pure strategy noncooperative equilibrium by a diagram as in shown in *Figure 6*. In *Figure 4* the contingent demand with no capacity constraint was *AKDFG*; this has a kink at *F*. When capacity constraints come into effect an added 'kink' is introduced as is shown in *Figure 6* where the contingent demand is *AUWNFG*. The point *K* is the capacity limit of the lower priced firm, which causes the kink at *W* and the branch of demand *UW*.

When the isoprofit curve becomes tangent to the contingent demand at two points the pure strategy equilibrium is destroyed. The profit from raising price will be as high as maintaining the same price as the competitor. The profits are shown in *Figure 6* by areas *TVMC* and *RNSC*. *N* is the pure strategy noncooperative equilibrium that would prevail if capacity were plentiful and *V* is the point at which profit possibilities destroy the equilibrium. It can only exist if there is the added kink at *W*, which in turn depends upon capacity limitations.

PRICE STRATEGY OLIGOPOLY WITH PRODUCT VARIATION

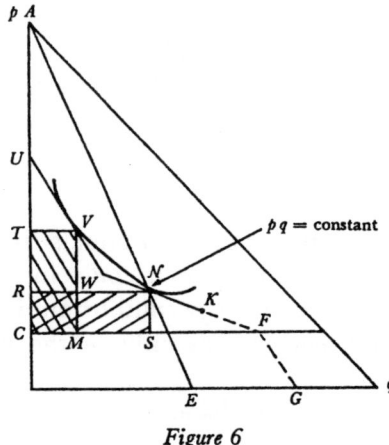

Figure 6

The interpretation of these results is straightforward. In a price oligopoly there will be price instability unless there is sufficient overcapacity. For any level of capacity of $k > (a-c)/2b$ for a fixed ε the introduction of more competitors will eventually remove the price instability, i.e. there will be a specific $n = n(k, \varepsilon)$ beyond which there will be a pure strategy equilibrium in the market, and as n increases this equilibrium approaches the efficient solution.

In many oligopolistic markets with few firms we expect some overcapacity however not necessarily enough to prevent price instability. In larger markets we may expect that the overcapacity is sufficient to lead to stability. Although we have presented our analysis in terms of a specific example, our results appear to be general for any economically reasonable symmetric model.

RAND Corporation, Santa Monica   LLOYD SHAPLEY
Yale University, New Haven   MARTIN SHUBIK

SUMMARY

This article presents a mathematical model for monopolistic price competition among firms with differentiated products. The conditions which distinguish the CHAMBERLINIAN analysis from that of EDGEWORTH are examined. Product differentiation is not sufficient to guarantee the existence of a stable non-cooperative equilibrium point in terms of price. The relationship among the degree of product variation, amount of capacity, and stability are examined. Explicit formulas are obtained for the non-cooperative equilibrium and the capacity conditions for its existence. The behavior of the model as the degree of differentiation approaches zero and as competition becomes large is examined. This connects the analysis of oligopoly with the analysis of pure competition.

ZUSAMMENFASSUNG

Der Artikel bietet ein mathematisches Modell zum monopolistischen Preiswettbewerb zwischen Unternehmungen mit differenzierten Produkten. Die Bedingungen, die CHAMBERLIN's Analyse von EDGEWORTH's Arbeiten unterscheiden, werden geprüft. Produktdifferenzierung genügt nicht, um die Existenz eines stabilen Gleichgewichts der Preise ohne Absprache zu sichern. Die Autoren untersuchen die Beziehungen zwischen dem Ausmass der Produktvariation, der Kapazität und der Stabilität. Formeln für ein nicht-kooperatives Gleichgewicht und die Kapazitätsbedingungen seines Bestehens werden entwickelt. Die Reaktion des Modells wird kontrolliert, wenn sich der Differenzierungsgrad Null nähert, und wenn der Wettbewerb zunimmt. Damit wird die Analyse des Oligopols mit der Analyse des reinen Wettbewerbs verbunden.

RÉSUMÉ

Cet article contient un modèle mathématique traitant de la compétition monopolistique sur le secteur des prix entre des entreprises fournissant différent produits. On y examine les points sur lesquels l'analyse de CHAMBERLIN diffère de celle d'EDGEWORTH. La différentiation des produits ne suffit pas à assurer un équilibre stable des prix sans coopération. Les auteurs examinent les relations entre le degré de la variation des produits, de la capacité et de la stabilité. Ils obtiennent des formules pour un équilibre non coopératif et les conditions de capacité pour son existence. On contrôle la réaction du modèle pour le cas où de degré de différentiation s'approche de zéro et où la compétition augmente. De cette manière, on obtient une connexion de l'analyse de l'oligopole et de celle de la compétition libre.

[23]

PRICE STRATEGY OLIGOPOLY: LIMITING BEHAVIOR WITH PRODUCT DIFFERENTIATION

MARTIN SHUBIK
Yale University

In a previous paper Shapley and Shubik [3] set up a model of an oligopolistic market with n firms selling (symmetrically) differentiated products. They solved this model for the noncooperative equilibrium where price is the strategic variable used by each firm. The behavior of the noncooperative equilibrium was examined under the condition that the number of firms in competition became arbitrarily large, and it was shown that under certain circumstances the noncooperative price approached the price that would prevail under pure competition. In that paper, however, the implications of the model concerning the nature of product differentiation and the different ways in which the concept of "the number of firms in a market becomes large" were not discussed. In this paper they are examined.

I. LIMIT PROCESSES: FIRMS, CUSTOMERS, CAPACITY AND PRODUCT DIFFERENTIATION

"How many is many in a market?" is an old question in the study of the control of industry. With the exception of Cournot [1] and Edgeworth [2] until quite recently this question has been scarcely asked in a precise manner by economic theorists.

A way in which the effect of numbers can be examined precisely in an economic model is to set up a formal mathematical description of the economic activity in which the number of participants appears explicitly, or implicitly in its effect upon certain of the variables or parameters; then investigate the behavior of the model as the number of participants becomes arbitrarily large.

In general there are many different ways in which the model of a market can be constructed and examined for behavior in the limit. The different ways of proceeding to a limit correspond to somewhat different economic phenomena which may not be easy to isolate and distinguish in a verbal or diagrammatic treatment of the subject.

When dealing with oligopolistic competition we make considerable use of symmetry, and in this paper several other simplifications are introduced in order to study a specific model to obtain some simple contrasting results.

The setting-up of the appropriate model of an oligopolistic market involving n single-product firms calls for the specification of the number of firms, the number of customers, the capacities and costs of the firms and the nature of the differentiation amongst the products. Complete symmetry among the firms is assumed. They are each assumed to manufacture and sell one product to the market, they have identical overheads and production costs and their products are symmetrically differentiated. Another way of stating this is that when all the firms charge the same price, the cross-elasticity between any two firms in the market will be the same as that between any other two firms.

If we assume U-shaped average costs of production, then immediately we are confronted with a problem in comparing markets with different numbers of firms. Intuitively, when the economist talks about the effect of more competitors in the market he usually has in mind a comparison between two markets in which the demand or consumer side has remained constant, but the few competitors have been replaced by more competitors who are in some sense "similar to but smaller than" the previous few competitors. The concepts of "size" and "similarity" are not easy to define precisely when the firms have U-shaped average costs. One way to avoid this difficulty yet still be able to comment in a useful manner upon the growth in the number of competitors, is by considering a sequence of different markets in which the costs to the individual producer always remain the same; however, the overall size of market demand grows at the same speed as does the supply, due to the presence of new competitors. In other words, although in the sequence of models the market grows larger, if all the firms charge the same price, no matter how many firms there are, each will sell an amount that is independent of the number of competitors. This way of going to the limit leaves the size and cost conditions of the firms the same, but erodes their market share. Hence this is a useful model to examine the question of how market power is related to market share.

Another way of setting up a limit process would be to hold the overall size of the market fixed, but to chop up the firms. We note that this is difficult to do when we assume U-shaped costs. If however we were to assume that the firms produced under conditions of constant variable costs, c (with no fixed costs) and with a capacity limit k, then it is easy to examine a set of markets where we successively replace few large firms by more firms with the same average costs but with smaller individual capacities.

When fixed average cost and capacities are assumed, then a little reflection will show that the two ways of considering markets with many competitors are equivalent. We can either hold the overall market size fixed and chop up the firms; or we can introduce more firms of the same size and expand the overall size of the market. In the examples in Section II fixed average costs and capacities are assumed, and the latter limiting process is used to examine the effect of numbers.

Before any market models are specified and investigated, the role of product differentiation and numbers of firms must be discussed. Consider a duopolistic market and suppose that we had a measure that indicated the degree to which the products of the firms were differentiated substitutes. One such measure might be as follows: consider that one firm sets its price arbitrarily high while the other sets its price equal to cost, the measure is the ratio of the amount sold by the lower-priced firm to the amount that it would have sold, had both firms each set price equal to cost. An example is provided in Figure 1. If both firms had charged $p=c$ then the amount sold by the first is indicated by AB and the amount sold by the second by BG, where $BG = AB$ as the firms are selling symmetrically related substitutes. If the second firm raises its price to a level so high that it sells nothing, the first will still not pick up all of the amount BG because the products are differentiated. In Figure 1 the final demand for the first firm is indicated by AH, hence the extra demand obtained is BH and the measure that has been suggested is BH/BG. When the firms are selling the identical product the value of this measure is 1, and when they are selling completely isolated products the value of the measure is 0.

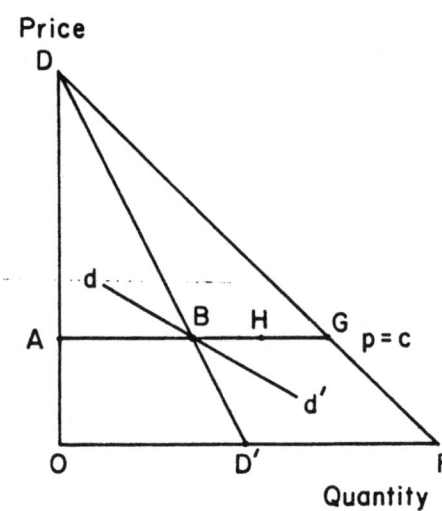

Figure 1

Suppose that the overall size of the market were doubled, and that two more firms selling differentiated products were in the market. Once more if all charged $p = c$, each would sell an amount equal to AB in Figure 1. If one firm were to change its price slightly with the others remaining fixed, it would pick up or lose demand along a contingent demand curve indicated by dd' in Figure 1. Does the slope of this curve remain constant as the number of firms increases or does the slope become flatter? One cannot state which from only a priori considerations. When the market is larger and there are more competing substitutes available, do the substitutes become more competitive or do they stay the same "distance" apart? Are four family cars, in a market twice the size, closer substitutes for each other than two family cars in the original market?

When a market is growing and a few more substitute products appear, it is not unreasonable that the substitutes remain apart. When the number of substitutes becomes large, it is likely that in some sense they become closer substitutes for each other. In Section II examples of both assumptions are examined. In actual markets there is probably an upper bound of between 5 to 10 perceived different substitutes, beyond which further substitutes pack together.

More powerful mathematical methods than those used here might dispense with the type of replication process used here to compare games of different sizes. This might be achieved by using games with a continuum of players and games with a continuum of commodities. However, staying with the somewhat simpler formulation, there are at least four models that might be considered as are indicated below:

Model	Number of firms	Size of market	Capacity	Product differentiation
(1)	n	$O(n)$	k	fixed
(2)	n	$O(n)$	k	$O(1/n)$
(3)	n	fixed	k/n	fixed
(4)	n	fixed	k/n	$O(1/n)$

In the first example the market grows in proportion to the number of firms, but the "distance" of substitutability between the growing number of substitutes remains the same. In the second example the added substitutes are packing closer together as $1/n$. The two other cases are basically the same, however the firms are being cut up and their capacity shrinks while the overall size of the market stays the same.

II. THE NONCOOPERATIVE EQUILIBRIUM AND SUBSTITUTABILITY

Following the models (1) and (2) suggested above, we consider two aggregate consumer utility functions of the form

(1) $$U(q) = aq - \frac{b}{n}q^2 - \frac{\epsilon}{n}\left[\Sigma q_i^2 - \frac{q^2}{n}\right] \quad \text{or}$$

(2) $$U(q) = aq - \frac{b}{n}q^2 - \frac{\epsilon'}{n}\left[\Sigma q_i^2 - \frac{q^2}{n}\right]$$

where $q = \sum_{i=1}^{n} q_i$ and $\epsilon' = \epsilon n$.

These utility functions are obviously only defined for the appropriate range of the q_i where the marginal utilities are not negative. In the first case the product differentiation comes through the presence of the variance term controlled by ϵ. The customer likes variety. As the number of commodities increases they become progressively closer substitutes. In the second case the introduction of the extra term of n in $\epsilon' = \epsilon n$ helps to keep the products apart.

It is a relatively straightforward calculation to write down demand in terms of price, to write down the payoff function for each firm in terms of price and to solve for the noncooperative equilibrium with price as the independent variable. This has been done by Shapley and Shubik [3]. Stating only the results we have

(3) $$q_i = \frac{a - \bar{p}}{2b} - \frac{n}{2\epsilon}(p_i - \bar{p})$$

(4) $$P = (p_i - c) q_i$$

and the noncooperative equilibrium price is given by

(5) $$p_i = \frac{n(n-1)bc + \epsilon(na + c)}{n(n-1)bc + \epsilon(n + 1)}$$

As $n \to \infty$, $p_i \to c$. This is the price that would prevail at a competitive equilibrium. Suppose, however, that ϵ is replaced by $\epsilon' = \epsilon n$. The expression for the price prevailing at the noncooperative equilibrium, given in (5) is now modified and becomes

(6) $$p_i = \frac{(n-1)bc + \epsilon(na + c)}{(n-1)b + \epsilon(n + 1)}$$

As $n \to \infty$, $p_i \to \frac{bc + \epsilon a}{b + \epsilon}$.

Even though price will decrease with more competitors, it does *not* approach that of the competitive equilibrium, but a term involving ϵ remains, indicating that there is always a measure of monopoly power present.

The first model is probably more believable than the second. However, it still involves making the assumptions about substitutability explicit. In a world of monopolies where all products are gross substitutes and where the differentiation among them is continuous, then as numbers grow the noncooperative price approaches that of pure competition. If the substitutes do not pack together, this is not so.

In most markets the truth probably lies in between the two models indicated here. Richard Levitan, in a discussion, has suggested a model in which it is assumed that there is an extra parameter representing the maximum number of different substitutes that a group of customers will consider to be fundamentally different. Beyond this number, if new firms come into the market we may expect that the distinction amongst further substitutes lessens. In such a market the limit is not the competitive equilibrium, but depends upon the extra parameter.

How many competitors constitute "many competitors" in an oligopolistic market? Taking the model of demand based upon (1) we observe that the joint maximum or monopolistic price is given by

(7) $$p_i = \frac{a+c}{2}$$

We might use as a measure the number of firms for which the non-cooperative equilibrium price (given in (5)) comes within 5% of the competitive equilibrium price where the range of 100% is determined by the difference between the monopolistic price and the competitive equilibrium, i.e.,

(8) $$\frac{n(n-1)bc + \epsilon(na+c)}{n(n-1)b + \epsilon(n+1)} \leq c + .05\left(\frac{a-c}{2}\right) \quad \text{or}$$

(9) $$\frac{\epsilon n}{n(n-1)b + \epsilon(n+1)} \leq .025$$

Regarding (9) as an equation, this is a quadratic in n with ϵ and b as parameters. When $\epsilon = 0$ the products are identical; when $\epsilon = b$ the products become completely isolated from each other. A few representative values for ϵ are:

ϵ	n
b	40
$b/5$	9
$b/10$	5
$b/20$	3

The first value represents products that are rather distant as substitutes. The others show closer substitutes. It can be seen that the number of firms needed for "many" drops extremely quickly if the initial degree of substitutability is high and the new products crowd together.

Using the model of demand based upon (2), then from (6) we can calculate how close the limiting state with many competitors is to the competitive equilibrium. Calling this distance K, then

(10) $$\frac{bc + \epsilon a}{b + \epsilon} = c + K\left[\frac{a-c}{2}\right]$$

Hence

$$(11) \quad K = 2\epsilon/(b + \epsilon)$$

Thus unless $\epsilon < b/40$ in this model oligopolistic price never reaches to within the 5% bandwidth above competitive price.

Although the examples calculated in this paper were specific, the basic model and the results appear to be reasonably general. In particular, when we consider the increase in numbers of competitors in an oligopolistic market with differentiated products, there are many different models that are relevant, each reflecting different conditions on substitutability.

Leaving aside the many other features that characterize an oligopolistic market (indivisibilities, increasing returns to scale, etc.), "many" is determined by the market and cost parameters as well as by the number of firms. With the market structure used here convergence to competition was $O(1/n)$, if substitutes were becoming closer. Otherwise, although price may still decrease with new entrants, it does not approach the competitive price as a limit.

With different market structures we may expect virtually any speed of convergence, although it seems to be reasonable to consider the linear case as a manageable approximation to many markets.

"Many" appears to be a number between 2 and 10 for the most part, unless for special reasons, even with new entrants, the firms find it possible to maintain their distance in product differentiation.

REFERENCES

1. A. A. Cournot, *Researches into the Mathematical Principles of the Theory of Wealth*. New York 1897.
2. F. Y. Edgeworth, *Mathematical Psychics*. London 1881.
3. L. Shapley and M. Shubik, "Price Strategy Oligopoly with Product Variation," *Kyklos*, Fasc. 1, 1969, 22, 30-44.

Price Variation Duopoly with Differentiated Products and Random Demand*

RICHARD LEVITAN

*IBM Thomas J. Watson Research Center
Yorktown Heights, New York 10598*

AND

MARTIN SHUBIK

Department of Economics, Yale University, New Haven, Connecticut

Received January 9, 1970

1. INTRODUCTION

It is shown in this paper that under the appropriate conditions the introduction of a random component to demand in a duopolistic (or more generally oligopolistic) market has the competitive effect of *increasing* stability in the sense that the market without a random component may have no noncooperative equilibrium point (in pure strategies), whereas the market with a random component has a noncooperative equilibrium point.

This result is related to the previous results of Shubik [1], Levitan [2], and Shapley and Shubik [3] on duopolistic demand and competition. These are summarized below.

Suppose two firms with identical constant average costs are supplying identical goods to a market. As was noted by Bertrand [4], if the firms employ price as a strategic variable and if they have no capacity limits then after a process of price-cutting, an equilibrium will be established with prices equal to cost. Edgeworth [5] analyzing the competition between two firms selling the identical product with rising average costs (or equivalently capacity constraints) observed that no (pure strategy) equilibrium need exist and that price will fluctuate in a range which we may call the Edgeworth cycle.

---

* Research undertaken by the Cowles Commission for Research in Economics under Contract Nonr-3055(01) with the Office of Naval Research. This paper is a revision of an original version.

It does not seem to be realistic or reasonable to expect prices to bounce up and down incessantly over a wide range in an otherwise static situation. It has been argued that at least Edgeworth's analysis was a step closer to reality than that of Cournot inasmuch as price rather than quantity offered for sale is a more natural independent variable for oligopolists. We feel that neither price nor quantity strategy models are satisfactory but that both price and quantity should be treated as simultaneous independent variables and that inventory shortages and stockouts need to be considered.

Chamberlin added a considerable amount of richness and relevance to the analysis of oligopolistic markets by introducing product differentiation [6]. He examined the equilibrium conditions for a group of firms whose products were symmetrically differentiated and suggested that by means of price policy they would achieve a noncooperative equilibrium. Figure 1 illustrates this equilibrium for two firms, each with the same constant average costs of production, each selling a symmetrically differentiated product and each facing a linear demand when they charge the same price.

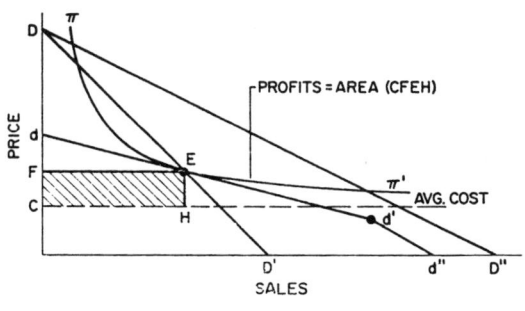

FIG. 1

The line DD' is precisely the same as in Chamberlin's analysis [6] and indicates how much an individual will sell on the assumption that the other charges the same price. The line DD'' may be drawn and interpreted as the total amount sold in both markets when each firm charges the same price. Since they are selling differentiated products, we are adding "oranges and apples" when we draw this curve; quantity actually has two dimensions, not one. Nevertheless, as the products are symmetric, it is useful to draw DD'' provided it is correctly interpreted.

The line segments Dd, dEd', and $d'd''$ describe the contingent demand for one firm as it varies its price from D to zero given that the other maintains its price at the level of E. These line segments are the complete description of the dd' curve of Chamberlin. Their properties and shapes have been investigated by Shubik [7], Levitan [2], and Shapley and Shubik [3]. The point d is that point at which the higher priced firm has priced itself out of its market. The point d' is the point at which the lower priced firm will have priced its competitor out of his market. The point d'' is the maximum demand attainable for any firm in its market regardless of the price charged by the other. This easy to see when the meaning of the segment $d''D''$ and the slope of dd' are considered. They both serve as measures of the lack of substitutability between the two commodities. Suppose that we smoothed away the product differentiation so that the firms sell closer and closer substitutes until in the limit they sell identical products. The line dd' will become more and more horizontal as the differentiation diminishes. Furthermore, the distance $d''D''$ will shrink until in the limit d'' approaches D'' and the distance is zero. The interpretation is direct. If the two firms sell the identical product, a slight difference in price will give all of the demand to the lower priced firm. If their products are the same, then all of the demand will be given by DD''. When there is product differentiation the amount available to any firm will be less than DD''.

The point E is the Chamberlinian noncooperative equilibrium point. Behavioralistically, it is *exactly* the same as the Cournot, Bertrand, Edgeworth or Nash equilibria. The economic analysis may be regarded as more satisfactory than that in other models inasmuch as product differentiation is more realistic. The location of E is determined by examining the family of isoprofit curves and selecting a point on the intersection of DD' with an isoprofit curve such that the curve is tangent to dd' at that point. It is easy to see that the tangency condition guarantees that it is not profitable to undercut or to raise price if both firms are at E.

If both firms were being run solely for the good of the public they would set price equal to marginal cost which in this case is equal to average cost. The rectangle $CHEF$ shows the "monopoly profit" attained by this noncooperative monopolistic competition. As the product differentiation is lessened, the distance EH shrinks until (as was noted by Bertrand) with identical products price will equal costs and there will be no profit.

Shapley and Shubik [3] showed that Chamberlin's analysis is incomplete. The mere introduction of product differentiation does not destroy the instability encountered by Edgeworth. If average costs are increasing sufficiently; if there are low enough capacity constraints or if there are inventory carrying costs, the Chamberlinian monopolistic competition equilibrium will not exist. Any of these conditions cause an extra "kink" or bend to appear in the segment of the contingent demand denoted by SS', as is shown in Fig. 2. In this diagram, the capacity constraint is just severe enough to make the kink SS' exactly tangent to isoprofit curve $\pi\pi'$

through E. For any smaller capacity, the point S' will be closer to E and SS' will meet a higher isoprofit curve. For example, in Fig. 3 there is no capacity excess beyond the abscissa of E, hence, S' coincides with E. The curve SS' meets $\pi_1\pi_1'$, a higher profit locus.

The meaning of the additional kink and segment SS' is that although the firm would lose demand along DD' by raising price, this demand loss is based upon the assumption that the competitor can supply any increased demand going his way. If he is unable to do so (owing to capacity limits or inventory shortage) or unwilling to do so (owing to high costs), then the loss in demand will not be so severe to the higher-priced firm. Instead of proceeding along $S'd$, it becomes SS'. This cuts the isoprofit curve $\pi\pi'$ and destroys the equilibrium.

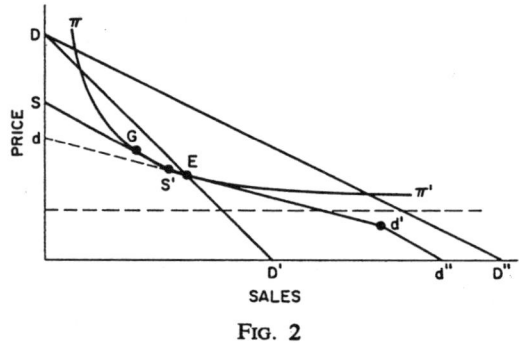

Fig. 2

2. The Inventory Problem

It is easy to see that if we introduce inventory costs, the noncooperative equilibrium must be destroyed. Let production costs be the same as in Fig. 1 and 2, and let there be a positive cost of holding inventory in excess of the sales of the current period; and assume further that the firm can supply consumers in the current period only from previously ordered production. The point E can no longer be an equilibrium point as is shown in Fig. 3. If it were, then no firm would hold more inventory than exactly enough to satisfy demand at that point. However, this is the equivalent of a very severe capacity limitation, so that the contingent demand faced by a firm raising its price is ES rather than Ed. This means that the higher-priced firm is virtually a monopolist in its price range as the lower-priced firm has inventory sufficient only for E.

Following the desires expressed by many for more economically reasonable and realistic models in the study of oligopoly theory we maintain that the addition of inventory costs is a step closer to realism and relevance, yet it apparently helps to destroy equilibrium. We shall show that the addition of a further quite realistic complication helps to restore the existence of the equilibrium. This complication is a degree of uncertainty in expected demand. The introduction of uncertainty may have a stabilizing effect.

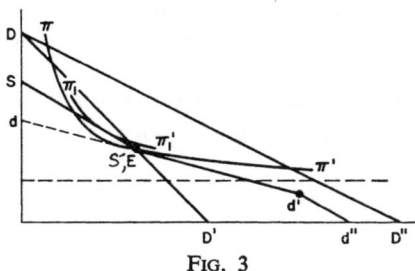

Fig. 3

3. An Oligopolistic Market with Random Demand

The general condition needed for stability is that the minimum inventories that the firms are induced to keep by consideration of demand fluctuations at a potential equilibrium point provide sufficient extra stock for each to make it unprofitable for either to change his price. The test for the condition is straightforward: Solve the model for a pure strategy equilibrium by using the first-order conditions for a local maximum; then verify that no higher payoff can be achieved by having one firm change its price or quantity globally (i.e., over the whole range) given that the other firm is constrained to charging the price and supplying the quantity that is the candidate for the equilibrium.

In practice, the stability will depend specifically upon the forms of the demand, production cost and inventory carrying cost functions. In this paper we are unable to offer a general characterization of the functional forms which satisfy the conditions but confine ourselves to the calculation of an example which establishes the validity of our assertion.

3.1. *The Model*

Our notation is as follows:

Π = profit for the first (or distinguished) firm;
p = price;
q = demand;
x = supply;
ρ = inventory carrying cost per unit.

The same notation with a bar, e.g., $\bar{p}$, stands for the variables of the other firm. We assume production costs are zero.[1] Profits are given by:

$$\Pi = p \min (q, x) - p \max (0, x - q). \tag{1}$$

To obtain the demand we follow the calculations[2] for contingent demand illustrated in Fig. 1 and 2 as shown in Section 1 and developed mathematically by Shubik [1], Levitan [2], and Shapley and Shubik [3]. Here, however, we introduce a random component. The constant term is replaced by a random variable.

Let us give a derivation of the curve $DSS'd'd''$ shown in Fig. 2. We define the (normalized) ordinary linear demand curve,

$$q(p, \bar{p}) = \epsilon - p - \gamma(p - \bar{p}), \tag{2}$$

which describes the demand for the distinguished firm when there are no stockouts or "price-outs."

The parameter γ may be interpreted as a measure of the substitutability of the products of the two firms. $\gamma = 0$ describes a model of two independent monopolists; $\gamma = +\infty$, the case of firms selling perfect substitutes.

Equation (2) is represented by the segment $S'd'$ in Fig. 2. The segment $d'd''$ describes the demand when the distinguished firm has priced its competitor out of the market. The demand on this segment is given by

$$\hat{q}(p) = \frac{(1 + 2\gamma)(\epsilon - p)}{1 + \gamma}, \tag{3}$$

which is derived by imputing to the nondistinguished firm that price, $(\epsilon - \gamma p)/(1 + \gamma)$ which causes its demand to be exactly zero. Finally, the segment SS' gives the demand of the distinguished firm when the nondistinguished firm's demand exceeds its supply $\bar{x}$. The demand on SS' is given by

$$\hat{q}(p, \bar{x}) = \frac{(1 + 2\gamma)(\epsilon - p) - \gamma \bar{x}}{1 + \gamma}. \tag{4}$$

This function is derived similarly by substituting for $\bar{p}$ the price $(\epsilon - \bar{x} - \gamma p)/(1 - \gamma)$, which makes the second firm's demand equal to its

---

[1] This assumption is made for the sake of economy of notation in the sequel. All results are easily generalized to the case of constant costs by substituting $p - c$ for p in the demand and profit functions.

[2] This is a somewhat special calculation. There are many special problems involved in the computation of contingent demand. These involve the aggregation of individual preferences and details such as priorities and queuing when an item is in short supply. A discussion of the details is given in the first two references following.

supply. This is based on an institutional assumption that the market behaves as if all customers have uniform tastes and do not exhibit a significant income effect when they receive an opportunity to purchase a restricted quantity at intramarginal prices.

It is easy to see that the demand function whose graph is $DSS'\,d'd''$ is equal to

$$q_T(p, \bar{p}, \bar{x}) = \max(0,\, \hat{q}(p, \bar{x}),\, \min(q(p, \bar{p}),\, \hat{q}(p))),$$

and the actual sales of the distinguished firm are given by

$$s = \min(q_T, x) = \min(x,\, \max(0,\, \hat{q},\, \min(q, \hat{q}))).$$

In the sequel, we introduce the random component into the demand by considering ϵ to be a random variable.

The calculations required to derive the conditions for a pure strategy equilibrium are straightforward but basically tedious. The difficulty is that the decision space $\{p, x, \bar{p}, \bar{x}\}$ is subdivided into six regions[3] in which the expected sales and profits have distinct formulas. The elucidation of these regions is given in the Appendix. The first derivatives of profit with respect to decision variables are continuous at the boundaries of these regions but the second derivatives are not. The conditions for a symmetric pure strategy equilibrium imply that the decisions must be at the intersection of four of these six regions, where $p = \bar{p}$ and $x = \bar{x}$. At this intersection we can show:

$$\frac{\partial \Pi}{\partial p} = \int_p^{p+x} q\, dF + x(1 - F(p + x)) - (1 + \gamma)(p + \rho)(F(p + x) - F(p)) \tag{7}$$

$$\frac{\partial \Pi}{\partial x} = (p + \rho)(1 - F(p + x)) - \rho, \tag{8}$$

where $F(g) =$ probability that $\epsilon \leqslant g$, and henceforth let $\Pi = E(\Pi)$. In general, we expect that the equations $\partial \Pi / \partial p = 0$ and $\partial \Pi / \partial x = 0$ can be solved to yield a unique pair $(\hat{p}, \hat{x})$. Then it remains to verify that $p = \bar{p} = \hat{p}$ and $x = \bar{x} = \hat{x}$ indeed yield a pure strategy equilibrium. That is, that

$$\Pi(\hat{p}, \hat{x}, \hat{p}, \hat{x}) \geqslant \Pi(p, x, \hat{p}, \hat{x}) \quad \text{for all} \quad (p, x). \tag{9}$$

Typically, the local conditions for a maximum may be met and still the maximum may not be global. We do not have a general set of specifica-

[3] We give here an abbreviated account. The interested reader is referred to Ref. [8] for further details.

tions on parameters and distributions which imply that $(\hat{p}, \hat{x})$ is a symmetric equilibrium. Therefore, we have resorted to the special (tractable) case of the rectangularly distributed ϵ.

3.2. The Case of Uniformly Distributed Demand

We assume now that ϵ is distributed uniformly between a and $a + \Delta$. Sometimes it seems natural to use as a parameter the mean $\bar{\epsilon} = a + \Delta/2$. We may now combine this assumption with (7) and (8) which, together with the first-order conditions give us

$$\hat{x} = \frac{\Delta p}{p + \rho} + a - \hat{p} \tag{10}$$

and

$$\frac{\partial \Pi}{\partial p} = \Phi(p) = \int_{p}^{\Delta p/(p+\rho)+a} (\epsilon - p) f(\epsilon) \, d\epsilon + \left(\frac{\Delta p}{p + \rho} + a - p\right) \frac{\rho}{p + \rho}$$
$$- (\rho - p)(1 + \gamma)\left(\frac{p}{\rho + p} - F(p)\right) = 0, \tag{11}$$

where

$$f(y) = \frac{1}{\Delta}, \quad \text{if} \quad y \in [a, a + \Delta],$$
$$= 0, \quad \text{if} \quad y \notin [a, a + \Delta].$$

We can show that this equation has a unique solution $\hat{p}$, whose properties depend on the value of

$$\Phi(a) = \frac{\Delta}{2} - \frac{\Delta \rho^2}{2(a + \rho)^2} = -(1 + \gamma) a. \tag{12}$$

If $\Phi(a) > 0$, $\hat{p} \in (a, a + \Delta)$ and is expressible as

$$\hat{p} = \frac{a + \Delta + \rho + \sqrt{(a + \Delta + \rho)^2 + 4(3 + 2\gamma) \Delta \rho}}{2(3 + 2\gamma)} - \rho. \tag{13}$$

If $\Phi(a) = 0$, then $\hat{p} = a$, and finally if $\Phi(a) < 0$, then $\hat{p}$ is the unique positive root, less than a, of the cubic expression:

$$2\rho^2 a + 2\rho(\Delta + 2a - \rho(2 + \gamma)) p + (\Delta + 2a - 4\rho(2 + \gamma)) p^2 - 2(2 + \gamma) p^3. \tag{14}$$

In the special limiting case, where $\rho = 0$, the solution assumes a simple form if $\Phi(a) = \Delta/2 - (1+\gamma)a \geq 0$,

$$\hat{p} = \frac{a + \Delta}{3 + 2\gamma}, \tag{15}$$

$$\hat{x} = \frac{2(1+\gamma)}{3 + 2\gamma}(a + \Delta). \tag{16}$$

And if $\Phi(a) < 0$, then

$$\hat{p} = \frac{\bar{\epsilon}}{2 + \gamma}, \tag{17}$$

$$\hat{x} = \frac{1+\gamma}{2+\gamma}\bar{\epsilon} + \frac{\Delta}{2}. \tag{18}$$

We now discuss the conditions for our strategy $(\hat{p}, \hat{x})$ to give a symmetric equilibrium pair that satisfies inequality (9). The necessary second-order local conditions, because of the discontinuity, are somewhat complicated and not sufficient. Therefore, in order to describe sufficient conditions, we resorted to the expedient of computationally scanning the parameter space and tabulating the boundary of the set in which $(\hat{p}, \hat{x})$ was found to be an equilibrium strategy.

Our task was a great deal simplified by the fact that $\partial \Pi/\partial x$ is a monotone decreasing piecewise linear function of x. Therefore, for each specification of $p, \bar{p}$, and $\bar{x}$, there exists an optimal value of x which we call $\hat{x}(p, \bar{p}, \bar{x})$ or, for brevity, $\hat{x}(p)$.

We can show that

$$\hat{x}(p, \bar{p}, \bar{x}) = \max\left(0, \frac{1+2\gamma}{1+\gamma}\left(\frac{\Delta p}{p+\rho} + a - p\right)\right)$$

$$\text{if } (1+\gamma)\bar{p} - \gamma p > \frac{\Delta p}{p+\rho} + a,$$

$$= \max\left(0, \frac{\Delta p}{p+\rho} + a - (1+\gamma)p + \gamma\bar{p}\right), \tag{19}$$

$$\text{if } (1+\gamma)\bar{p} - \gamma p \leq \frac{\Delta p}{p+\rho} + a \leq (1+\gamma)\bar{p} - \gamma p + \bar{x},$$

$$= \max\left(0, \frac{1+2\gamma}{1+\gamma}\left(\frac{\Delta p}{p+\rho} + a - p\right) - \frac{\gamma}{1+\gamma}\bar{x}\right),$$

$$\text{if } (1+\gamma)\bar{p} - \gamma p + \bar{x} \leq \frac{\Delta p}{p+\rho} + a.$$

Since $\Pi(p, \hat{x}(p), \bar{p}, \bar{x}) \geqslant \Pi((p, x, \bar{p}, \bar{x})$ for all $x > 0$, the question of the optimality of $\Pi(\hat{p}, \hat{x}, \hat{p}, \hat{x})$ is a question of whether the maximum of the function of one variable $\Pi(p) = \Pi(p, \hat{x}(p), \hat{p}, \hat{x})$ occurs at $p = \hat{p}$, since it is easily seen that $\hat{x} = \hat{x}(\hat{p})$.[4]

In order to carry out the scan of the parameter space we wrote a computer program which, for any value of the parameters $\bar{\epsilon}$, Δ, γ, and ρ first computes $\hat{p}$ from formula (13), if $\Phi(a) \geqslant 0$, and if $\Phi(a) < 0$, by Newton's method applied to (14). Next, $\hat{x}$ is calculated and from it the function $\Pi(p)$ is computed and searched for its maximum value. If the maximum occurs at $\hat{p}$, then $(\hat{p}, \hat{x})$ gives an equilibrium. In Fig. 4, we gave sample curves showing $\Pi(p)$ and $\hat{x}(p)$ in a case where Π has a secondary maximum exactly equal to $\Pi(p)$.

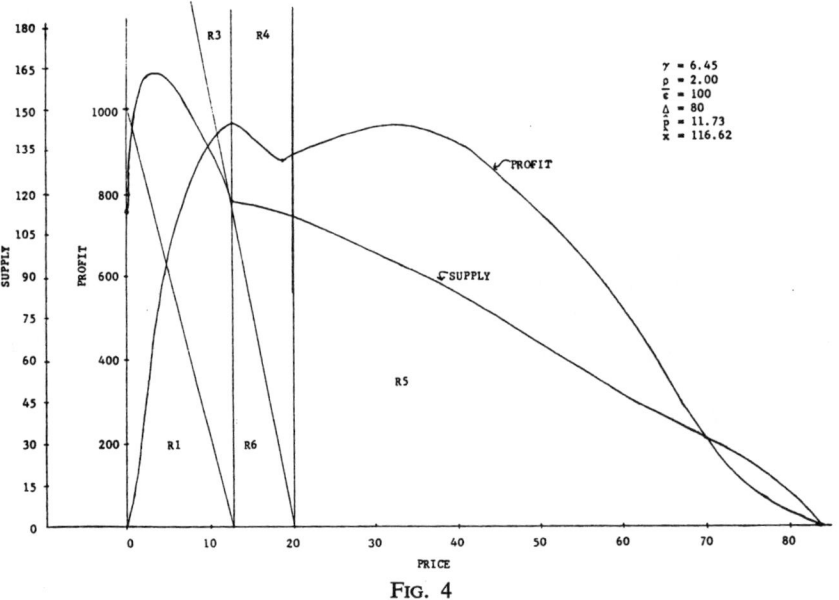

FIG. 4

It was discovered computationally that if there is no equilibrium at a set of values $\bar{\epsilon}_0$, Δ_0, ρ_0, and γ_0, then for $\gamma > \gamma_0$ there is also no equilibrium. Similarly, if at $\bar{\epsilon}_0$, Δ_0, ρ_0, and γ_0, there is an equilibrium, there is also an equilibrium if $\gamma < \gamma_0$. Thus, it is sensible to define and compute a maximal γ compatible with equilibrium. Since the model is homogeneous in $\bar{\epsilon}$, Δ and ρ, we have computed by a search procedure and tabulated in Table I the values of maximal γ as a function of $\rho/\bar{\epsilon}$ and $\Delta/\bar{\epsilon}$.

[4] By the notation $\Pi(\hat{p}, \hat{x}, \hat{p}, \hat{x})$ we mean the expected profit when $(p, x) = (\hat{p}, \bar{x}) = (\hat{p}, \hat{x})$.

TABLE I
Maximal Gamma for Pure Strategy Equilibrium

	$\Delta/\bar{\epsilon}$									
$p/\bar{\epsilon}$	0.025	0.05	0.10	0.20	0.30	0.40	0.50	0.60	0.70	0.80	0.90
0.0	0.71	1.02	1.55	2.54	3.61	4.84	6.30	7.50	8.31	8.83	9.15
0.010	0.70	1.01	1.51	2.43	3.38	4.40	5.53	6.47	7.02	7.32	7.44
0.020	0.69	0.99	1.47	2.33	3.19	4.08	5.01	5.81	6.25	6.45	6.50
0.030	0.69	0.97	1.44	2.25	3.04	3.83	4.63	5.34	5.71	5.86	5.86
0.040	0.68	0.96	1.41	2.18	2.91	3.62	4.33	4.97	5.31	5.42	5.39
0.050	0.67	0.94	1.38	2.12	2.80	3.45	4.09	4.67	4.98	5.07	5.03
0.060	0.66	0.93	1.35	2.06	2.70	3.30	3.88	4.43	4.72	4.80	4.74
0.070	0.65	0.92	1.33	2.01	2.61	3.18	3.71	4.21	4.50	4.56	4.50
0.080	0.65	0.90	1.30	1.96	2.53	3.06	3.56	4.03	4.31	4.37	4.29
0.090	0.64	0.89	1.28	1.91	2.46	2.96	3.43	3.86	4.14	4.20	4.12
0.100	0.63	0.88	1.26	1.87	2.40	2.88	3.31	3.72	4.00	4.05	3.97
0.110	0.63	0.87	1.24	1.84	2.34	2.79	3.21	3.59	3.86	3.92	3.84
0.120	0.62	0.86	1.23	1.80	2.29	2.72	3.11	3.47	3.75	3.80	3.72
0.130	0.61	0.86	1.21	1.77	2.24	2.65	3.03	3.37	3.64	3.70	3.61
0.140	0.61	0.84	1.19	1.75	2.19	2.59	2.95	3.27	3.54	3.60	3.52
0.150	0.60	0.83	1.18	1.71	2.15	2.54	2.88	3.19	3.45	3.52	3.43
0.200	0.58	0.79	1.11	1.59	1.97	2.30	2.59	2.85	3.07	3.18	3.11
0.300	0.54	0.73	1.01	1.42	1.73	2.00	2.23	2.42	2.59	2.73	2.74
0.400	0.51	0.68	0.94	1.30	1.57	1.80	2.00	2.16	2.30	2.41	2.50
0.500	0.48	0.65	0.88	1.21	1.46	1.66	1.84	1.98	2.10	2.20	2.28
0.600	0.46	0.62	0.83	1.14	1.37	1.56	1.72	1.85	1.96	2.05	2.13
0.700	0.44	0.59	0.80	1.08	1.30	1.47	1.62	1.74	1.85	1.94	2.01
0.800	0.43	0.57	0.76	1.03	1.24	1.40	1.54	1.66	1.76	1.84	1.91
0.900	0.41	0.55	0.74	0.99	1.19	1.34	1.48	1.59	1.68	1.76	1.83
1.000	0.40	0.53	0.71	0.96	1.14	1.29	1.42	1.53	1.62	1.70	1.76
1.100	0.39	0.52	0.69	0.93	1.10	1.25	1.37	1.48	1.57	1.64	1.71
1.200	0.38	0.50	0.67	0.90	1.07	1.21	1.33	1.43	1.52	1.59	1.66
1.300	0.37	0.49	0.65	0.87	1.04	1.17	1.29	1.39	1.47	1.55	1.61
1.400	0.36	0.48	0.64	0.85	1.01	1.14	1.25	1.35	1.43	1.51	1.57
1.500	0.36	0.47	0.62	0.83	0.99	1.11	1.22	1.32	1.40	1.47	1.53

Table continued

TABLE I (*continued*)

	$\Delta/\bar{\epsilon}$									
$p/\bar{\epsilon}$	1.00	1.10	1.20	1.30	1.40	1.50	1.60	1.70	1.80	1.90	2.00
0.0	9.30	9.33	9.27	9.13	8.95	8.72	8.46	8.17	7.88	7.88	7.88
0.010	7.44	7.35	7.20	7.01	6.78	6.54	6.28	6.01	5.88	5.88	5.88
0.020	6.43	6.30	6.12	5.91	5.68	5.43	5.18	4.92	4.88	4.88	4.87
0.030	5.77	5.61	5.42	5.20	4.96	4.71	4.46	4.24	4.24	4.23	4.23
0.040	5.28	5.11	4.91	4.69	4.44	4.20	3.95	3.78	3.78	3.77	3.77
0.050	4.90	4.72	4.51	4.28	4.04	3.80	3.55	3.43	3.42	3.41	3.41
0.060	4.60	4.41	4.20	3.96	3.72	3.48	3.23	3.15	3.14	3.13	3.13
0.070	4.35	4.16	3.94	3.70	3.46	3.21	2.97	2.91	2.90	2.90	2.89
0.80	4.14	3.94	3.72	3.48	3.24	2.99	2.74	2.72	2.71	2.70	2.69
0.090	3.96	3.76	3.53	3.29	3.94	2.79	2.56	2.55	2.54	2.53	2.52
0.100	3.81	3.60	3.37	3.13	2.88	2.63	2.41	2.40	2.39	2.38	2.38
0.110	3.67	3.46	3.23	2.98	2.73	2.48	2.28	2.27	2.26	2.25	2.24
0.120	3.55	3.34	3.10	2.85	2.60	2.34	2.16	2.15	2.14	2.13	2.13
0.130	3.45	3.23	2.99	2.74	2.48	2.23	2.06	2.05	2.04	2.03	2.02
0.140	3.35	3.13	2.89	2.64	2.38	2.12	1.97	1.95	1.94	1.93	1.93
0.150	3.26	3.05	2.80	2.54	2.28	2.02	1.88	1.87	1.86	1.85	1.84
0.200	2.94	2.72	2.46	2.19	1.92	1.64	1.54	1.53	1.52	1.51	1.50
0.300	2.60	2.37	2.10	1.81	1.50	1.18	1.13	1.11	1.10	1.08	1.07
0.400	2.43	2.23	1.96	1.63	1.30	0.94	0.88	0.86	0.84	0.82	0.81
0.500	2.32	2.18	1.93	1.60	1.24	0.83	0.70	0.68	0.66	0.65	0.63
0.600	2.12	2.16	1.96	1.64	1.26	0.82	0.58	0.56	0.53	0.52	0.50
0.700	2.06	2.09	2.00	1.73	1.35	0.89	0.49	0.46	0.44	0.42	0.40
0.800	1.96	2.00	2.01	1.83	1.49	1.03	0.42	0.39	0.36	0.34	0.32
0.900	1.88	1.92	1.94	1.90	1.64	1.21	0.60	0.33	0.30	0.27	0.25
1.000	1.82	1.86	1.88	1.89	1.78	1.42	0.85	0.28	0.25	0.22	0.20
1.100	1.76	1.80	1.83	1.85	1.84	1.62	1.12	0.24	0.21	0.18	0.15
1.200	1.71	1.75	1.78	1.80	1.81	1.76	1.40	0.68	0.18	0.15	0.11
1.300	1.66	1.71	1.74	1.76	1.78	1.78	1.63	1.07	0.15	0.12	0.08
1.400	1.62	1.67	1.70	1.73	1.74	1.75	1.74	1.42	0.13	0.10	0.06
1.500	1.58	1.63	1.67	1.70	1.72	1.72	1.72	1.66	1.02	0.08	0.04

These values may be thought of as the maximal amount of product similarity or strategic interlinkage compatible with equilibrium.[5]

The result that very little similarity is tolerable when Δ is small corresponds to our prior intuition. However, we were surprised by and have no explanation for the fact that in general as Δ increased toward $2\bar{\epsilon}$ the maximal γ decreases significantly.

3.3. A Sensitivity Analysis

We shall conclude our analysis of the stochastic duopoly model by examining the sensitivity of equilibrium price to changes in the values of the parameters. Since the result is quite apparent, we shall state without proof that $\bar{p}$ is a decreasing function of γ and an increasing function of $\bar{\epsilon}$. We have been left to deal with changes in the parameters ρ and Δ. The analysis differs depending on whether $\hat{p}$, the equilibrium price is greater than or less than or equal to a, the lower bound of the random variable ϵ.

In the case where $\hat{p} \geq a$, we have from Eq. (13)

$$\hat{p} = \frac{\bar{\epsilon} + \rho + \Delta/2 + \sqrt{(\bar{\epsilon} + \rho + \Delta/2)^2 + 4(3 + 2\gamma)\Delta\rho}}{2(3 + 2\gamma)} - \rho. \quad (20)$$

It is immediately apparent that $\hat{p}$ is an increasing function of Δ. The behavior of $\hat{p}$ as a function of ρ is more ambiguous. For large ρ, $\hat{p}$ behaves like

$$\frac{\bar{\epsilon} + \rho + \Delta/2}{3 + 2\gamma} - \rho,$$

which decreases linearly with ρ. However, at $\rho = 0$, we have

$$\frac{\partial \hat{p}}{\partial \rho} = -\frac{1}{3 + 2\gamma} + \frac{\Delta}{\bar{\epsilon} + \Delta/2} - 1. \quad (21)$$

Since $\hat{p} = (\bar{\epsilon} + \Delta/2)/(3 + 2\gamma) \geq a = \bar{\epsilon} - \Delta/2$, which implies that $1/(3 + 2\gamma) \geq (\bar{\epsilon} + \Delta/2)/(\bar{\epsilon} - \Delta/2)$, we have

$$\frac{\partial p}{\partial \rho} \geq \frac{\bar{\epsilon} - \Delta/2}{\bar{\epsilon} + \Delta/2} + \frac{\Delta}{\bar{\epsilon} + \Delta/2} - 1 = 0. \quad (22)$$

Thus, we have that $\hat{p}$ increases with ρ for small ρ and then decreases for large ρ, until the regime changes and $\hat{p} < a$.

[5] The entries corresponding to the row $\rho = 0$, are in some sense fallacious in that the computation is carried out with x defined as in (19). However, since if $\Delta = 0$, any $x > \hat{x}$ so computed is also optimal because the derivative is zero for larger x. For sufficiently large x, the maximum of $\Pi(p)$ occurs at $\hat{p}$ regardless of the magnitude of γ. Thus this row of the table is only to be regarded as the limit for small values of ρ.

Finally, we examine the sensitivities in the case where $\hat{p} < a$. Here we do not have a closed form solution for $\hat{p}$; but since we know that $\hat{p}$ is a zero of Φ at a point where $\partial \Phi / \partial p < 0$, we can compute $\partial \hat{p} / \partial \Delta$ and $\partial \hat{p} / \partial \rho$, respectively, as

$$-\frac{\partial \Phi / \partial \Delta}{\partial \Phi / \partial p} \quad \text{and} \quad -\frac{\partial \Phi / \partial \rho}{\partial \Phi / \partial p}.$$

We have, from Eq. (11), when $\hat{p} < a$,

$$\Phi = \bar{\epsilon} - (2 + \gamma) p - \frac{\Delta \rho^2}{2(p + \rho)^2}, \tag{23}$$

which immediately implies that $\partial \Phi / \partial \Delta \leqslant 0$ and hence $\partial \hat{p} / \partial \Delta \leqslant 0$. Finally,

$$\frac{\partial \Phi}{\partial \rho} = -\frac{\Delta p \rho}{(p + \rho)^3} \leqslant 0; \tag{24}$$

and in this regime $\hat{p}$ is a decreasing function of both Δ and ρ.

It is very difficult to supply a very convincing intuitive rationale for these somewhat inconsistent results except to state that the effect of inventory penalty and demand variance on equilibrium price does not seem very clear cut. It may turn out that the signs of their particular derivatives may turn out to be artifacts of the form chosen for the demand function and the distribution of its error term. The principal value here would seem to be a warning against too facile theorizing in this field.

4. Conclusions

We have shown that if one improves the realism of the Chamberlinian model by adding inventory carrying costs and making the assumptions that production takes time, then the pure strategy noncooperative equilibrium postulated by Chamberlin *never* exists. Instability of the type suggested by Edgeworth is all that remains.

By introducing a *further* complication into the model, the equilibrium may be restored. This too is a step towards realism. It is the introduction of a random element to overall market size.

We leave as open problems the generalizations and the extension of our work to the *n*-person symmetric and nonsymmetric cases. Our experience with models of this variety indicates that it is a safe conjecture that our results go through for the nonsymmetric market model we have developed elsewhere. Although our results are mathematically inelegant, our conjecture is of substantive and theoretical interest. While our proof of the existence of the equilibrium under uncertainty uses a specially simple example, in order to obtain a more general result, a more powerful and different type of mathematical approach is undoubtedly needed.

APPENDIX

The calculation of expected sales involves integrals over various ranges of the functions q, $\hat{q}$, and $\hat{\hat{q}}$. The possible degeneracy of some of these intervals and the precise specification of the interval depends on the values p, $\bar{p}$, x, and $\bar{x}$. If we are given $(\bar{p}, \bar{x})$, the set of decision pairs $\{(p, x)\}$ has six distinct regions where the value of $E(\Pi)$ has distinct formulas. Figure A1 illustrates these six regions.

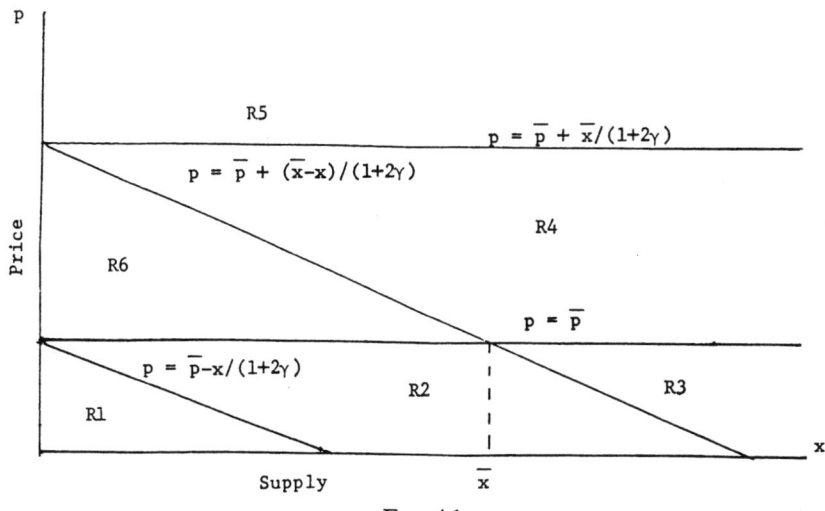

FIG. A1

We define the functions:

$$\epsilon_1 = (1 + \gamma) p - \gamma \bar{p},$$
$$\epsilon_2 = p + \frac{\gamma}{1 + 2\gamma} \bar{x},$$
$$\epsilon_3 = p + \frac{1 + \gamma}{1 + 2\gamma} x,$$
$$\epsilon_4 = (1 + \gamma) \bar{p} - \gamma p,$$
$$\epsilon_5 = (1 + \gamma) \bar{p} - \gamma p + \bar{x},$$
$$\epsilon_6 = (1 + \gamma) p - \gamma \bar{p} + x,$$
$$\epsilon_7 = p + \frac{\gamma \bar{x} + (1 + \gamma) x}{1 + 2\gamma}.$$

We now list, for each region, the definition, sales as a function of the random variable ϵ and expected profit.

In R_1:
$$p < \bar{p} - \frac{x}{1 + 2\gamma},$$

$$\begin{aligned} s &= x, & \text{if } \epsilon \geqslant \epsilon_3, \\ &= \hat{q}, & \text{if } p \leqslant \epsilon \leqslant \epsilon_3, \\ &= 0, & \text{if } \epsilon \leqslant p \end{aligned}$$

and

$$E(\Pi) = (p + \rho)\left[\int_p^{\epsilon_3} \hat{q}\, dF + (1 - F(\epsilon_3))\right] - \rho x.$$

In R_2:
$$\min\left(\bar{p}, \bar{p} + \frac{\bar{x} - x}{1 + 2\gamma}\right) > p > \bar{p} - \frac{x}{1 + 2\gamma},$$

$$\begin{aligned} s &= x, & \text{if } \epsilon \geqslant \epsilon_6, \\ &= q, & \text{if } \epsilon_4 \leqslant \epsilon \leqslant \epsilon_6, \\ &= \hat{q}, & \text{if } p \leqslant \epsilon \leqslant \epsilon_4, \\ &= 0, & \text{if } \epsilon \leqslant p, \end{aligned}$$

and

$$E(\Pi) = (p + \rho)\left[\int_p^{\epsilon_4} \hat{q}\, dF + \int_{\epsilon_4}^{\epsilon_6} q\, dF + (1 - F(\epsilon_6))\right] - \rho x.$$

In R_3:
$$\bar{p} + \frac{\bar{x} - x}{1 + 2\gamma} < p < \bar{p}.$$

$$\begin{aligned} s &= x, & \text{if } \epsilon \geqslant \epsilon_7, \\ &= \hat{q}, & \text{if } \epsilon_5 \leqslant \epsilon \leqslant \epsilon_7, \\ &= q, & \text{if } \epsilon_4 \leqslant \epsilon \leqslant \epsilon_5, \\ &= \hat{q}, & \text{if } p \leqslant \epsilon \leqslant \epsilon_4, \\ &= 0, & \text{if } \epsilon \leqslant p, \end{aligned}$$

and

$$E(\Pi) = (p + \rho)\left[\int_p^{\epsilon_4} \hat{q}\, dF + \int_{\epsilon_4}^{\epsilon_5} q\, dF + \int_{\epsilon_5}^{\epsilon_7} \hat{q}\, dF + x(1 - F(\epsilon_7))\right] - \rho x.$$

In R_4:
$$\max\left(\bar{p} + \frac{\bar{x} - x}{1 + 2\gamma}, \bar{p}\right) \leqslant p \leqslant \bar{p} + \frac{\bar{x}}{1 + 2\gamma},$$

$$\begin{aligned} s &= x, & \text{if } \epsilon \geqslant \epsilon_7, \\ &= \hat{q}, & \text{if } \epsilon_5 \leqslant \epsilon \leqslant \epsilon_7, \\ &= q, & \text{if } \epsilon_1 \leqslant \epsilon \leqslant \epsilon_5, \\ &= 0, & \text{if } \epsilon \leqslant \epsilon_1, \end{aligned}$$

and
$$E(\Pi) = (p+\rho)\left[\int_{\epsilon_1}^{\epsilon_5} q dF + \int_{\epsilon_5}^{\epsilon_7} \hat{q} dF + x(1 - F(\epsilon_7))\right] - \rho x.$$

In R_5:
$$p \geqslant \bar{p} + \frac{\bar{x}}{1+2\gamma},$$
$$s = x, \quad \text{if} \quad \epsilon \geqslant \epsilon_7,$$
$$= \hat{q}, \quad \text{if} \quad \epsilon_2 \leqslant \epsilon \leqslant \epsilon_7,$$
$$= 0, \quad \text{if} \quad \epsilon \leqslant \epsilon_2,$$

and
$$E(\Pi) = (p+\rho)\left[\int_{\epsilon_2}^{\epsilon_7} \hat{q} dF + x(1 - F(\epsilon_7))\right] - \rho x.$$

Finally, in R_6:
$$\bar{p} < p < \bar{p} + \frac{\bar{x}-x}{1+2\gamma},$$
$$s = x, \quad \text{if} \quad \epsilon > \epsilon_6,$$
$$= q, \quad \text{if} \quad \epsilon_1 \leqslant \epsilon \leqslant \epsilon_6,$$
$$= 0, \quad \text{if} \quad \epsilon \leqslant \epsilon_1,$$

and
$$E(\Pi) = (p+\rho)\left[\int_{\epsilon_1}^{\epsilon_6} q dF + x(1 - F(\epsilon_6))\right] - \rho x.$$

Reference

1. M. Shubik, "Strategy and Market Structure," Chaps. 4 and 5, Wiley, New York, 1959.
2. R. E. Levitan, Demand in an oligopolistic market and the theory of rationing, IBM Watson Research Center Report 1545, Yorktown Hights, New York, January 21, 1966.
3. L. S. Shapley and M. Shubik, Price strategy oligopoly with product variation, Kyklos 1 (1969).
4. J. Bertrand, Théorie mathématique de la richesse sociale, *J. des Savants Paris* (September 1883), 488–503.
5. F. Y. Edgeworth, "Papers Relating to Political Economy," Vol. I, pp. 111–142, MacMillan, London, 1925.
6. E. H. Chamberlin, "The Theory of Monopolistic Competition," (6th ed.) Harvard University Press, Cambridge, Mass. 1950.
7. M. Shubik, A further comparison of some models of duopoly, *West. Econ. J.* 6 (1968), 260–276.
8. R. E. Levitan and M. Shubik, Price variation duopoly with differentiated products and random demand, Cowles Foundation Discussion Paper #270, Yale University, New Haven and IBM Research Center Report 2433, Yorktown Heights, New York, 1969.

Printed in Belgium by the St. Catherine Press Ltd., Tempelhof 37, Bruges

PRICE DUOPOLY AND CAPACITY CONSTRAINTS*

By Richard Levitan and Martin Shubik[1]

1. INTRODUCTION

IN THIS PAPER we examine an extremely simple model of a duopoly situation in which the two firms compete with price as the strategic variable and in which the firms are limited by capacity constraints. In terms of this model we shall review some of the important developments of duopoly theory concerned with the existence of equilibrium. Such a market as Edgeworth [4] showed that it does not, in general, have an equilibrium. We have found that our model, however, has a rather simply described equilibrium in mixed strategies.

Beckmann [1] in 1965 gave a mixed strategy equilibrium, as a solution of an integral equation for a similar model suggested by Shubik [7]. The difference between the two models is the mode by which demand gets redistributed in case shortage occur. Our model has a much simpler solution and has the appealing property that, as we let the capacities vary, we get at one end of an interval as a limiting case the Cournot quantity strategy equilibrium and, at the other, the Bertrand price strategy equilibrium.

While we do not claim any particular realism for this model we present it both as a useful exposition of a game theoretic attack on the duopoly problem and as a valid point of departure for the study of model with more realism. These issues are discussed in the concluding section.

2. A SMALL DUOPOLY MODEL AND A HISTORICAL INTRODUCTION

Our model of duopoly is starkly simplified but it will serve to illustrate the issues we are considering. Let two firms be selling identical goods in a market where demand is a linear function of price. We shall assume, for simplicity that production cost is zero.[2]

We shall assume that the firms have identical capacities. In the Appendix we shall analyze the extreme asymmetric case where only one firm is capacity-limited.

The firms compete by independently selecting a price to charge. In order to make the model complete we shall have to include an assumption about what happens when the low price firm does not have enough capacity to satisfy the whole market at his price. We assume that the market behaves as if the buyers

---

* Manuscript received December 13, 1970; revised February 5, 1971.
[1] Research undertaken by the Cowles Commission for Research in Economics under Contract Nonr-3055(01) with the Office of Naval Research. This research was also partially supported by a grant from the Ford Foundation. The authors wish to thank the referee for his careful reviewing and correcting of a large number of errors in the original version of this paper.
[2] However, all results of this analysis can be generalized to the case of constant unit rates by considering the price in the analysis as the unit margin over cost.

111

all have uniform access to the sellers and further, that if their purchases at the low price is restricted, then their total demand is the maximum of the demand at the high price and the available amount at the low price. This amounts to the assumption of zero income effect on consumption.

Our last assumption is that in the event of equal prices the market is split evenly.

Let the market demand be:

(1) $$q = a - p.$$

Capacities are given by k_1 and k_2. To start with, we assume $k_1 = k_2 = k$; i.e., the firms have equal capacity.

We assume that $k \leq a$; i.e., a firm has no more capacity than enough to supply the whole market.

Suppose the firms charge prices p_i and p_j. We assume that demand is as follows:

(2) $$q_i = \begin{cases} a - p_i & \text{if } p_i < p_j \\ \frac{1}{2}(a - p_i) & \text{if } p_i = p_j \\ a - k - p_i & \text{if } p_i > p_j \end{cases}.$$

Three historical analyses of duopoly can now be illustrated quickly. The first is that of Cournot [3]. Suppose that each firm had $k \geq a/3$ and named *quantity* not price as its strategic variable. The relation determining price is given by:

(3) $$p = a - (q_1 + q_2) \quad \text{or} \quad p = a - q,$$

where

$$q = q_1 + q_2,$$

hence the payoff to Firm 1 is:

(4) $$\Pi_1 = q_1(a - q_1 - q_2).$$

Taking derivatives of (4) and a like expression for Π_2 and setting them equal to zero we obtain

(5) $$q_1 = q_2 = a/3; \; p = a/3 \quad \text{and} \quad \Pi_1 = \Pi_2 = a^2/9.$$

These describe the Cournot noncooperative equilibrium. The price and total output in the Cournot market are shown at point C in Figure 1. Point J gives the price and output if the firms act together as a monopolist.

(6) $$q_1 = q_2 = a/4; \; p = a/2 \quad \text{and} \quad \Pi_1 = \Pi_2 = a^2/8.$$

Suppose that the firms were competing via price. Furthermore, suppose that each had enough capacity to satisfy the whole market at any price. Bertrand [2] argued that the noncooperative equilibrium would fall to $p = 0$ as the firms would keep undercutting each other. This is shown at the point E in Figure 1.

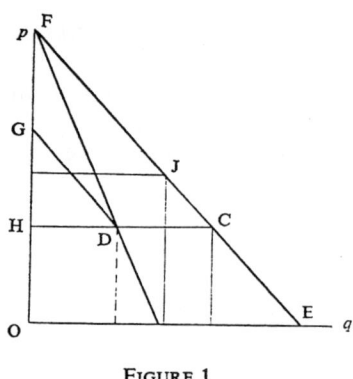

FIGURE 1

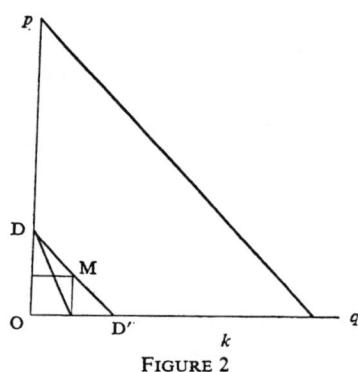

FIGURE 2

(7) $\qquad q_1 = q_2 = a/2; p = 0 \quad \text{and} \quad \Pi_1 = \Pi_2 = 0.$

Edgeworth [4] introduced the possibility of limited capacity. The case for $k = 3a/4$ is illustrated in Figure 2. Suppose that one firm were charging $p_i = 0$ it can only sell up to capacity k. This would leave the other firm a demand given by DD'. It would pay to seek monopoly profit against this contingent demand by raising price to M as shown in Figure 2. Subsequently, if we are willing to follow a loose dynamic argument the other firm may raise its next price to just under that at M and a period of price-cutting may follow. Thus Edgeworth suggested that there was a range over which price might be expected to fluctuate. This range will depend upon the capacity k.

It is relatively simple to solve for the upper and lower bounds of this range as follows. Let the prices at the bottom and top of the range be respectively $\bar{p}$ and $\hat{p}$. Suppose that the price of one firm is indefinitely close to the bottom of the range and that the other firm has the choice of picking the price at the bottom of the range or raising its price to p. It will be indifferent if profits are equal, i.e., if:

(8) $\qquad k\bar{p} = (a - k - \hat{p})\hat{p}.$

Fortunately, in this simple example, the profit to the high-priced firm depends only on its price and the capacity of the other hence we know that:

(9) $\qquad \hat{p} = \dfrac{a-k}{2},$

(this is illustrated by the point M when $k = 3a/4$ in Figure 2). Thus:

(10) $\qquad \bar{p} = \dfrac{1}{k}\left(\dfrac{a-k}{2}\right)^2$

and

(11) $\qquad \bar{\Pi} = \hat{\Pi} = \left(\dfrac{a-k}{2}\right)^2.$

We note if (as would be the case in a competitive industry) there was no excess

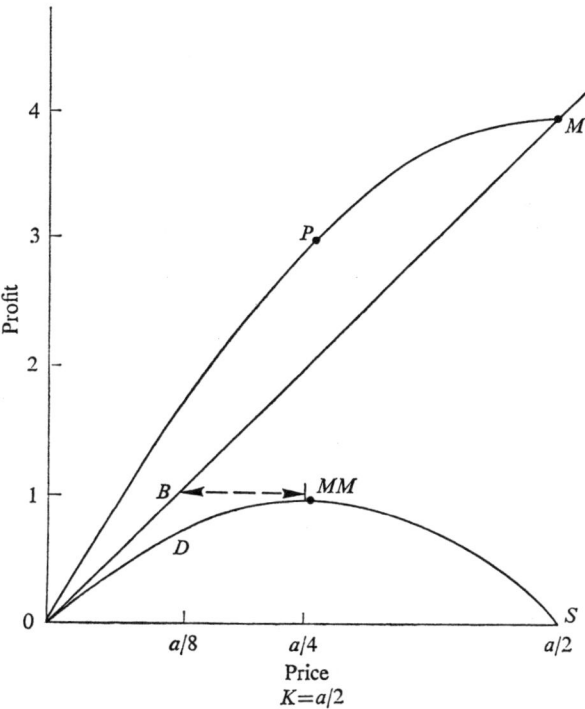

FIGURE 3

or shortage of capacity, then $k_1 = k_2 = a/2$ and the range and profits would be:

(12) $$\bar{p} = \frac{a}{8}, \; \hat{p} = \frac{a}{4} \quad \text{and} \quad \bar{\Pi} = \hat{\Pi} = \frac{a^2}{16}.$$

We now come to a result that at first is surprising, then obvious. Suppose that we continue to shrink capacity in this market. When $k_1 = k_2 = a/3$ we see from (9) and (10) that $\bar{p} = \hat{p} = a/3$, in other words the range of fluctuations shrinks to a point and a pure strategy noncooperative equilibrium reappears. This occurs precisely at the value of the Cournot equilibrium point for the following reason. If the lower priced firm is at capacity at $a/3$, this leaves $2a/3$ of the market for the other who will act as a monopolist and produce $a/3$.

In Figures 3 and 4, the Edgeworth range and the reason why it disappears at $k = a/3$ are illustrated. We first consider $k = a/2$. This case is shown in Figure 3. Each firm has just enough capacity to satisfy the whole market at monopoly price $p = a/2$. If the firms had a capacity of $k = a$, then the curve OPM would show the growth of revenues to the lower priced firm as it raises price, always being able to satisfy total market demand. Because for prices $p < a/2$, a capacity of $k = a/2$ is not sufficient to satisfy the market, the growth of revenue is given

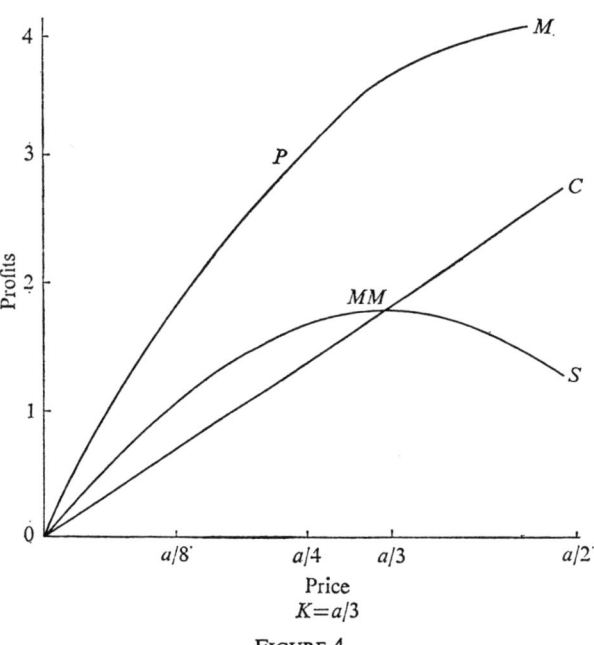

FIGURE 4

by the line *OM* which intersects the curve *OPM* at *M*.

The curve *OMMS* shows the change in revenues as the higher priced firm increases its price. It has more than enough capacity to satisfy any demand that is left for it. Its revenues reach a maximum at the point *MM* and decline to 0 if it continues to increase its price to $p = a/2$.

Consider the lower priced firm charging $p = a/8$ and the higher priced firm charging a price a shade higher. The former will make a profit indicated by *B* and the latter will make a profit shown at *D*. Suppose that the firm with the higher price has an opportunity to change his price. If he cuts his price to just below the other his profit will be approximately *B*. If he raises his price to $p = a/4$ his profit, which is shown at *MM*, will be as high as at *B*.

The Edgeworth range is given by *B* and *MM*. Furthermore if a firm pessimistically assumes that it will be undercut it should set its price $p = a/4$. The point *MM* is also the $\max_{p_j} \min_{p_i} \Pi_j$. It is the security level for either firm.[3]

Turning to Figure 4, the curve *OPM* has the same meaning as in Figure 3 the line *OMMC* is related to the line *OM*, but here we observe that the individual firm no longer has enough capacity to satisfy the market at the monopoly price without capacity limits. At that price revenue is shown at *C*. We now note that this line goes through *MM*. But the Edgeworth cycle is determined by the horizontal distance from *MM* to *OC* which is now zero.

---

[3] In a more general model this is not the case. The relationships are somewhat more complicated. For further discussion and an example, see Shubik [10].

If one firm adopts its maximum strategy the optimum reply for the other is to also adopt its maximum strategy hence they are in equilibrium. Thus when $k = a/3$ the price and quantity noncooperative equilibria are the same. This holds true in the range $0 \leq k \leq a/3$ where both firms will produce to capacity and price will be $\bar{p} = a - 2k$.

When the firms together have a total capacity less than a, the efficient point or competitive equilibrium[4] is no longer at $p = 0$ but becomes $p = a - 2k$. This price can be interpreted as the shadow price for the worth of an increment of new capacity.

Our results can be summed up in Table 1:

TABLE 1

	Cournot	Edgeworth-Bertrand	Efficient Point
$a \leq k$	$a/3$	0	0
$a/2 < k < a$	$a/3$	fluctuation	0
$a/3 < k \leq a/2$	$a/3$	"	$a - 2k$
$0 < k \leq a/3$	$a - 2k$	$a - 2k$	$a - 2k$

The entries in the table are prices for the appropriate solution. Instability in price competition is bounded from below by extreme excess capacity (if we regard $k_1 + k_2 = a$ as the "correct" amount of capacity then there is 100% excess). It is also bounded above by a 1/3 shortage of capacity. Beyond this point capacity is so tight that both firms need not worry about undercutting.

3. MIXED STRATEGY SOLUTION TO THE PRICE GAME

We now take the range $a/3 < k < a$ and investigate the nature of the mixed strategy in the price game. The mixed strategy could be interpreted as providing an indication of the distribution of prices in an unstable market.

A mixed strategy equilibrium of a two-person game consists of a pair of probability distributions over the respective strategy spaces with the property that for each player any strategy chosen with positive probability must be optimal against the other player's probability mixture.

In the present example we will avoid a complicated argument and merely assume that the equilibrium strategy involves a single interval, $[p_l, p_h]$ with positive probability density and no lumping of probabilities at any point except perhaps at p_h.[5]

Since we have assumed no lumps except at the highest price and except in the case where both players sell capacity when they both charge their highest price, which is ruled out here, the probability of a tie is zero and we can write

---

[4] Without entry it is not quite correct to refer to the solution of Bertrand as the competitive equilibrium. It is better described as the efficient solution which assigns a shadow price to the value of capacity.

[5] The interested reader is referred to Reference [5] for a rigorous argument for these assumptions in the case of a sealed bid auction.

the sales of Firm i as a function of both prices as

$$x_i = \begin{cases} \min(k, a - p_i) & \text{if } p_i < p_j \\ \max(0, \min(k, a - k - p_i)) & \text{if } p_i > p_j \end{cases},$$

and expected sales as

(14) $\quad E(x_i) = (1 - \phi_j(p_i)) \min(k, a - p_i) + \phi_j(p_i) \max(0, \min(k, a - k - p_i))$

where $\phi_j(p)$ is the distribution function of the price of player j.

Expected profit, of course, is

(15) $\quad\quad\quad\quad\quad\quad\quad\quad \Pi_i = p_i E(x_i) .$

The reason why our model is solved with such facility is the fact that sales are a step function of the competitor's price and the functional equations (14) and (15) involve no integrals.

We now assert that $p_h < a - k$. Suppose $p_h \geq a - k$. Then

(16) $\quad\quad\quad\quad\quad\quad \Pi_i(p_h) = p_h(1 - \phi_j(p_h))(a - p_h)$

which is equal to zero unless j has a lumped probability at p_h. But both firms cannot sell k at $a - k$ so this is ruled out.

Further, from equation (10) and its associated argument, it will never pay a firm to charge less than $\bar{p} = (1/k)((a - k)/2)^2 \geq a - 2k$. Thus we can rewrite (14) as

(17) $\quad\quad\quad\quad\quad\quad \Pi_i(p) = p[(1 - \phi_j(p))k + \phi_j(p)(a - k - p)]$

or

(18) $\quad\quad\quad\quad\quad\quad \dfrac{\Pi_i(p)}{p} = k - \phi_j(p)[p + 2k - a] .$

Solving (18) for ϕ_j, one obtains

(19) $\quad\quad\quad\quad\quad\quad \phi_j(p) = \dfrac{k - \pi_j/p}{p + 2k - a} .$

Since Π_i is constant on $[p_l, p_h]$, (19) gives the mixed strategy equilibrium once Π_i, p_l and p_h are evaluated.

It is simple to show that $p_h = (a - k)/2$. If a firm is charged p_h he knows that he will be surely undercut, and charging other than $(a - k)/2$ will not be optimal. Thus $\Pi_i = \Pi_i(p_h) = ((a - k)/2)^2$, and further, since $\Pi_i = \Pi_i(p_l) = k \cdot p_l$ we have $p_l = (1/k)((a - k)/2)^2$ and the upper and lower bounds for the mixed strategy coincide with those for the Edgeworth fluctuation or cycle (this is not generally true).

Substituting the value of $\Pi_i = ((a - k)/2)^2$ into (19) we obtain the cumulative probability function:

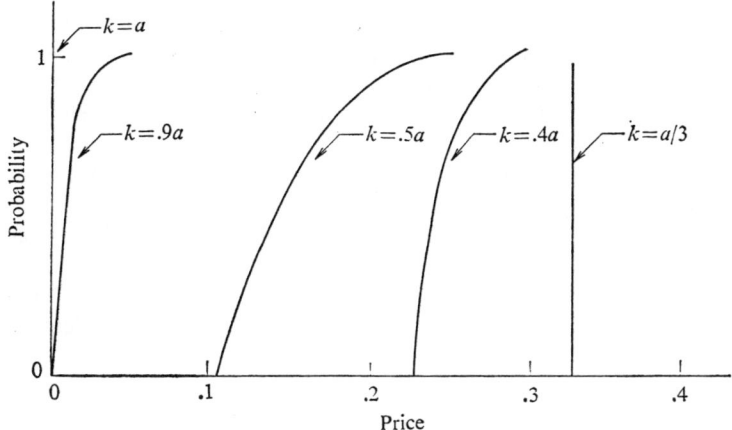

FIGURE 5

(20)
$$\phi_j(p) = \frac{kp - \left(\frac{a-k}{2}\right)^2}{p(p + 2k - a)}.$$

The graph in Figure 5 shows how the distribution changes for the values $k = .9a$, $.5a$ and $.4a$. It is easy to check the limiting values of $k = a$ and $a/3$ in equation (19).

What happens when capacities are different? If $k_1 \geq a$ and $k_2 \geq a$ or $k_1 \leq a/3$ and $k_2 \leq a/3$ a pure strategy exists. When the capacities are unequal but not in the ranges noted there is a mixed strategy solution.

The mixed strategy is no longer continuous when the firms have unequal capacities but the firm with the larger capacity selects the upper point in its bidding range with a finite probability. The solution appears to be more of a mathematical curiosum than of economic interest. An example with $k_1 = a$ and $k_2 < a$ is given in the Appendix.

4. CONCLUDING REMARKS: REALISM AND GENERALITY

In Section 2 we encountered the surprising result that as capacity was shrunk the pure strategy equilibrium reappears at the Cournot equilibrium point. This is not general; it will be determined by the type of contingent demand structure that is postulated. For example, Beckmann using a contingent demand method originally suggested by Shubik [8], does not obtain a pure strategy equilibrium at the Cournot point [1].

The determination of the reappearance of the pure strategy equilibrium depends upon the value of $\partial \Pi_i/\partial p_i$ at the point of potential equilibrium. The effect of moving capacity into the range $k_1 + k_2 \leq 2a/3$ puts a constraint on the derivative in the price-cutting direction. If a pure strategy equilibrium exists then both firms will be producing at capacity.

The test to see if capacity production is in the equilibrium comes in the direction of raising price. In Figure 1 the contingent demand at the point D is GD which has an elasticity of 1 at D hence there is no motivation to move price up.

Depending upon the method used for the calculation of contingent demand, as capacity is varied the slope of the contingent demand for each firm (when both are at capacity and the price is such that the market just clears) may change. This point will be an equilibrium point if the elasticity of demand along the contingent demand curve is 1.

We have picked the most pessimistic method of calculating contingent demand. The high priced part of the demand curve is satisfied first. This gives a parallel translation of the demand inward as the shape of the contingent demand. In this case the Cournot point must be the point of demarcation for the reappearance of the pure strategy.

The actual shape of contingent demand cannot be specified generally from *a priori* reasoning. It will depend upon priorities in service of customers and details of the summation of individual demand in specific markets. It is an important and complicated marketing problem which needs specific empirical investigation and model building.

We have discussed capacity constraints in this paper. Inventories are often used as a means for avoiding the short term effect of capacity constraints. We deal with the introduction of inventories elsewhere [6].

Unequal capacities complicate the analysis but do not appear to introduce any particularly interesting new phenomena.

Is price the right variable? Furthermore is the simultaneous move noncooperative game the right model? In general we would argue that price may not be the most important variable. Furthermore the one period noncooperative game is a gross oversimplification of oligopolistic competition. However, this model does represent an extreme case and as such merits investigation. What strategic variable is important depends upon the specifics of the market. Furthermore, the manner in which an economic weapon can be used also depends upon the market. In some markets price can be moved almost instantaneously; in others it may be highly inflexible.

A dynamic theory of oligopolistic competition needs to take into account both technological and institutional detail which enable us to give structure to the strategic possibilities for each firm and to the nature of threats.

The introduction of product differentiation does not appear to add any new qualitative results. The capacity limits on the conditions for the existence of a pure strategy equilibrium will change but beyond that the phenomena encountered will be qualitatively the same as without product differentiation.

It has been shown elsewhere that as the number of firms is increased in the market, if $\sum_{i=1}^{n} k_i \geq a$ and $k_i \geq a/n$ then the value of the game approaches that of the competitive equilibrium and the probability distribution on prices shifts towards the lower end of the range as n increases [9].

IBM Thomas J. Watson Research Center, U.S.A. and
Yale University, U.S.A.

APPENDIX

A Price Game with Unequal Capacities

The payoff to Firm i may be expressed as:

(20) $\Pi_i = p\{(1 - \Phi_j) \min(k_i, a - p) + \Phi_j \max(0, \min(k_i, a - k_j - p))\}$,

which may be written as

(21) $\Pi_i = p\{\min(k_i, a - p) - \Phi_j \min(k_i, a - p, \max(0, k_i + k_j - a + p), \max(k_j, a - k_i - p))\}$.

We solve for the special case: $k_1 = a$ and $k_2 = k < a$. The payoffs to the two firms are:

(22) $\qquad \Pi_1 = p[a - p - \Phi_2 \min(a - p, k)]$,

(23) $\qquad \Pi_2 = p[\min(a - p, k)](1 - \Phi_1)$.

We see immediately from (23) that Φ_1 cannot take the value 1 for an active strategy in the continuous range, as the value of Π_2 would then be zero.

At the lower end of the range of active strategies we may assume that the capacity constraint is effective on the firm with limited capacity, hence $a - p_l \geq k$ and

(24) $\qquad \Pi_2 = p_l k$

where p_l is the lowest active price. This follows immediately from (23).

At the highest price in the range, p_h, the condition $a - p_h \geq k$ must hold, or $p_h \leq a - k$.

From (22) we have:

(25) $\qquad \Pi_1 = p_h[a - p_h - k]$.

At p_h, the derivative $\partial \Pi_1/\partial p$ must be nonpositive, as this is the end of the range of active strategies.

(26) $\qquad \dfrac{\partial \Pi_1}{\partial p} = a - p_h - k + p_h(-1) \leq 0$,

hence

(27) $\qquad p_h \geq \dfrac{a - k}{2}$.

From (22) we may write:

(28) $\qquad \Phi_2' = \dfrac{a - p - \Pi_1/p}{\min(a - p, k)} = \dfrac{a - p - \Pi_1/p}{k}$,

this must have a positive derivative.

(29) $\qquad \Phi_2 = \dfrac{1}{k}\{-1 + \Pi_1/p^2\}$,

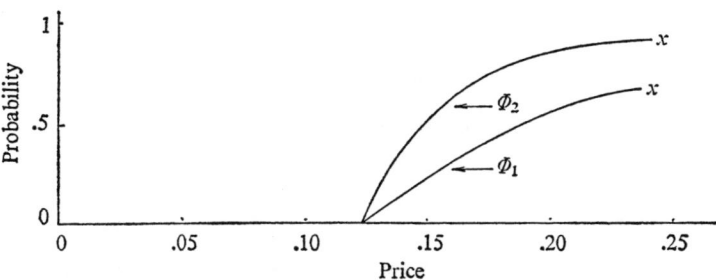

FIGURE 6

hence

$$\frac{\Pi_1}{p^2} \geq 1$$

and from (25) $p_h^2 \leq \Pi_1 = p_h(a - p_h - k)$; hence $p_h \leq (a-k)/2$ and from (27)

(30) $$p_h = \frac{a-k}{2}.$$

Given the top of the range the values and the cumulative density functions Φ_1 and Φ_2 become easy to calculate. We modify our notation replacing k by $k = \theta a$ where $0 < \theta < 1$.

From (30) and (25):

(31) $$\Pi_1 = \left(\frac{a-k}{2}\right)^2 = \left(\frac{a(1-\theta)}{2}\right)^2 = p_l(a - p_l),$$

from which

(32) $$p_l = \frac{a}{2}\{1 - \sqrt{\theta(2-\theta)}\}.$$

We check to verify that $p_h - p_l > 0$

$$p_h - p_l = \frac{a}{2}\{\sqrt{\theta(2-\theta)} - \theta\} > 0, \qquad \text{for } \theta < 1.$$

From (23) we have:

(33) $$\Phi_1 = 1 - \frac{a(1 - \sqrt{\theta(2-\theta)})}{2p},$$

and from (28)

(34) $$\Phi_2 = \frac{a - p - \frac{a^2(1-\theta)^2}{4p}}{\theta a}.$$

In Figure 6 we illustrate the distributions for the case where $k = a/2$.

REFERENCES

[1] BECKMANN, M. J. (with the assistance of Dieter Hochstadter), "Edgeworth-Bertrand Duopoly Revisited," *Operations Research-Verfahren, III*, edited by Rudolf Henn, Sonderdruck, Verlag, Anton Hain, Meisenheim, 1965.
[2] BERTRAND, J., "Theorie Mathematique de la Richesse Sociale," (Review), *Journal des Savants* (Paris, 1883), 499–508.
[3] COURNOT, A. A., *Researches into the Mathematical Principles of the Theory of Wealth* (New York: Macmillan, 1897), 79–80.
[4] EDGEWORTH, F. Y., *Papers Relating to Political Economy, I* (London: Macmillan, 1925), 111–142.
[5] GRIESMER, J. H., R. E. LEVITAN, AND M. SHUBIK, "Toward a Study of Bidding Processes, Part IV: Games with Unknown Costs," *Naval Research Logistics Quarterly*, XIV (December, 1967), 415–433.
[6] LEVITAN, R. L. AND M. SHUBIK, "Duopoly with Price and Quantity as Strategic Variables," in process.
[7] SHUBIK, M., *Strategy and Market Structure* (New York: Wiley, 1959), Chapter 6.
[8] _____, *Strategy and Market Structure*, Chapter 5.
[9] _____, *Strategy and Market Structure*, Chapter 6.
[10] _____, "A Comparison of Treatments of a Duopoly Problem (Part II)," *Econometrica*, XXIII (October, 1955), 417–431.

Duopoly with Price and Quantity as Strategic Variables

By *R. Levitan*, New York[1]), and *M. Shubik*, New Haven[2])

Abstract: In this paper we provide an explicit mixed strategy equilibrium solution for an oligopoly game. In the specification of the model, we assume that each firm has to make a decision on the production level while it names its prices, and we introduce a fixed unit cost for unsold inventory. Hence, both price and quantity appear as strategic variables.

1. Introduction

This paper can be considered as a continuation of our effort initiated in *Levitan/Shubik* [1970] to find explicit mixed strategy solutions to duopoly games. In *Levitan/Shubik* [1970], for example, price was treated as the only decision while production was considered to be under a capacity constraint. In this paper both price and quantity are decision variables, but there is no capacity limitation. In both papers, demand is assumed to be linear and nonstochastic, and products are assumed to be identical.

It turns out that while we consider duopoly with a linear demand function, and identical inventory carrying cost, our solution can be easily generalized to the case of olipoly with nonlinear demand, and asymmetric inventory costs. Further, the solution is not sensitive to the model of demand redistribution when one firm's product is in short supply.

In Section 2, we review our assumptions concerning interrelated demand and specify the model. In Section 3, we state the mixed strategy solution to the price quantity game, leaving the detailed proof for the Appendix. In Section 4, we examine some alternative specifications of the duopoly game model, including a dynamic sequential game and a minorant version where one firm must make its choices first. In Section 5, we extend our solution to the n firm situation where penalty costs may be asymmetric. In the concluding section, we discuss the realism of the model and suggest extensions.

2. Specification of Duopoly Model

In Figure 1 the line AD represents the overall market demand for the product. If one firm charges a lower price than the other we expect that it will obtain all of the

---

[1]) Dr. *R. Levitan*, IBM Thomas H. Watson Research Center, Yorktown Heights, New York.
[2]) Prof. *M. Shubik*, Yale University, Department of Economics, Box 2125, Yale Station, New Haven, Connecticut 06520.

demand as they are selling identical products. If it can satisfy this demand then there is nothing left for its competitor. Suppose, however, that it runs out of inventory before

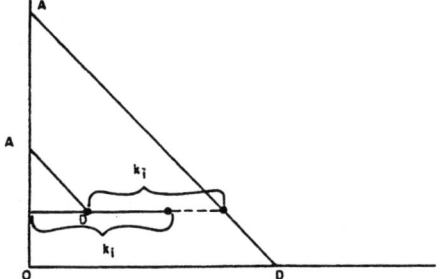

Fig. 1

it has satisfied all demand. What is left for its competitor? Looking at a relatively pessimistic possibility (from the viewpoint of the second firm) the demand to him is given by the line $A'D$ which is obtained by shifting the original line AD in towards the left by a distance k_j where k_j is the level of inventories held by the lower priced player. Obviously if k_j were sufficient to supply all of the demand there would be nothing left for the individual with the higher price.

Mathematically we may describe demand under all circumstances as given by:

$$d_i = \begin{cases} a - p_i & \text{when } p_i < p_j \text{ for } 0 \leq p \leq a \\ (a - p_i)/2 & \text{when } p_i = p_j \\ \max(a - k_j - p_i, 0) & \text{when } p_i < p_j. \end{cases}$$

Suppose that the firms had no capacity limitations and there were no costs to carrying inventories. For ease of argument we have selected an overall demand function of the form $d = a - p$ and furthermore we assume that there are no costs of production, i.e., $c_i = 0$. Neither of these restrictions is critical to the argument and they will be relaxed later.

If both firms had no capacity limitations and no inventory carrying costs then they each could always carry enough inventory to supply the whole market at any price. In particular if both were to carry an inventory of size $q_1 \geq a$ and $q_2 \geq a$ then, as with the unlimited capacity model of *Bertrand* [1883], there would be a pure strategy equilibrium in the market at $p = 0$ with the firms splitting the market. In this case this is precisely the same price as the competitive equilibrium.

The profits made by each firm at the competitive equilibrium are zero. What happens to the equilibrium and the profits if we introduce an inventory carrying cost? Before we can investigate this we must completely specify our model.

Two firms compete in a market in which they are selling identical products. They each simultaneously select a price and an inventory level for the period ahead. Production takes time, hence the inventory level selected is tantamount to a short term capacity constraint. All of the inventories are available for sale at the start of the period. There is an inventory cost of λ charged on each unit of inventory left in stock at the end of the period. There are several different ways of attacking an inventory cost which reflect the financing, physical handling and excess stock aspects of inventories.

Given inventory costs which vary as the size of the inventory, then even when there are no capacity limitations on the firms there can be no pure strategy equilibrium. This is true even if the products they are selling are differentiated. In particular this means that neither the *Bertrand* case nor the Chamberlinian large group equilibrium can exist if there are inventory carrying costs. The reason is quite simple. Assume the contrary. In which case with undifferentiated products both firms would be charging $p = 0$ and each would supply half of the market. If this were truly an equilibrium then each would cut down his inventories so that he had just enough to supply his half of the market. But if one firm knows that the other is charging $p = 0$ and has a supply of only $a/2$ it will face a contingent demand of $d = (a/2) - p$ if it raises its price. Hence, it will pay it to raise its price to $p = a/4$ and there will be no equilibrium.

We have shown that the pure strategy equilibrium is destroyed by the introduction of inventory costs. Furthermore the effect of the inventory costs is to shade prices in an upward direction. What does it do to the profits of the firms? Is there a mixed strategy equilibrium to this game with price and quantity as simultaneous independent variables?

3. Mixed-Strategy Solution to the Price-Quantity Game

Two firms simultaneously name prices p_1 and p_2 for their products and select their supplies for the period at q_1 and q_2. The payoff (shown for Firm 1) is given by:

$$\Pi_1 = \begin{cases} p_1 \min [q_1, a - p_1] & \text{if } p_1 < p_2 \\ p_1 \max [0, \min (q_1, a - q_2 - p_1)] - \lambda \max [0, q_1 - (a - q_2 - p_1)] & \text{if } p_1 > p_2 \end{cases}$$

The case in which $p_1 = p_2$ is not formally written down. It plays no role in the determination of the mixed strategy equilibrium. For other purposes one might wish to define it. An *ad hoc* convention must be given. A reasonable one is that if they charge the same price they split the market. If one cannot supply his share then the other obtains the excess demand.

There is no pure strategy equilibrium. We assume that there is a mixed strategy equilibrium.

Let $F_i(p,q) = $ probability that $p_i \leq p$ and $q_i \leq q$

In the sequel for ease of notation we shall refer to the price and inventory of the distinguished Firm i as p and q instead of p_i and q_i except where confusion might arise.

The profit of Firm i is:

$$\Pi_i = (p + \lambda) \{\min(a - p, q) [1 - F_j(p, a) + F_j(p, a - p - q)]$$
$$+ \int_{\substack{0 \leq x \leq p \\ a-p-q \leq y \leq a-p}} (a - p - y) \, dF_j(x, y)\} - \lambda q.$$

Assuming that $0 \leq q \leq a - p$

$$\frac{\partial \Pi_i}{\partial q} = (p + \lambda) \{1 - F_j(p, a) + F_j(p, a - p - q)\} - \lambda.$$

At $p = p_e$, the lowest price

$$F_j(p, a) = F_j(p, a - p - q) = 0.$$

Hence, $\dfrac{\partial \Pi_i}{\partial q} = (p_e + \lambda) - \lambda = p_e \geq 0.$

Let us assume that $F_j(p, q)$ assigns no positive probability to any price, we wish to examine two possibilities:

Case 1: $p_e = 0$ and $\Pi_i = 0$.

Case 2: $p_e > 0$ and $\Pi_i > 0$.

We assert that if Case 1 were to hold, $q = q(p) = a - p$. This would give:

$$\Pi_i = 0 = (a - p) \{(p + \lambda)(1 + F_j(p, a)) - \lambda\}$$

or

$$F_j(p, a) = \frac{p}{p + \lambda}; \text{ for } 0 \leq p < a.$$

Thus we have an equilibrium point of the form:

$$F_j(p, q) = \Phi_j(p) = \frac{p}{p + \lambda} \text{ and } q = a - p, \text{ for } 0 \leq p < a,$$

and, as follows from $p_e = 0$ being an active strategy, $\Pi_i = 0$.

It is shown in the Appendix that this is the only equilibrium point for Case 1 with $p_e = 0$. In the Appendix, Case 2 is examined and it is shown that there can be no solution if $p_e > 0$ thus the equilibrium point is unique. We now turn to examining its economic properties and meaning.

The solution is unique for all serious "moves" i.e., those moves which involve a positive supply. In our linear case, at equilibrium the players each make a non-serious move with probability $\lambda/(a + \lambda)$. When making a non-serious move, in our model the price has no effect and can have any value. Making a non-serious move corresponds to staying out of the market. This is reasonably consistent with economic sense. As the inventory carrying cost λ grows so does the probability that a firm stays away from the market. In this case its method of nonparticipation is to have nothing for sale.

Figure 2 shows the shape of $\phi_j(p)$ for several values. Suppose that $a = 10$ and $\lambda = .1, 1$ or 10. In the neighborhood of $\lambda = 0$, both firms are almost always ready to

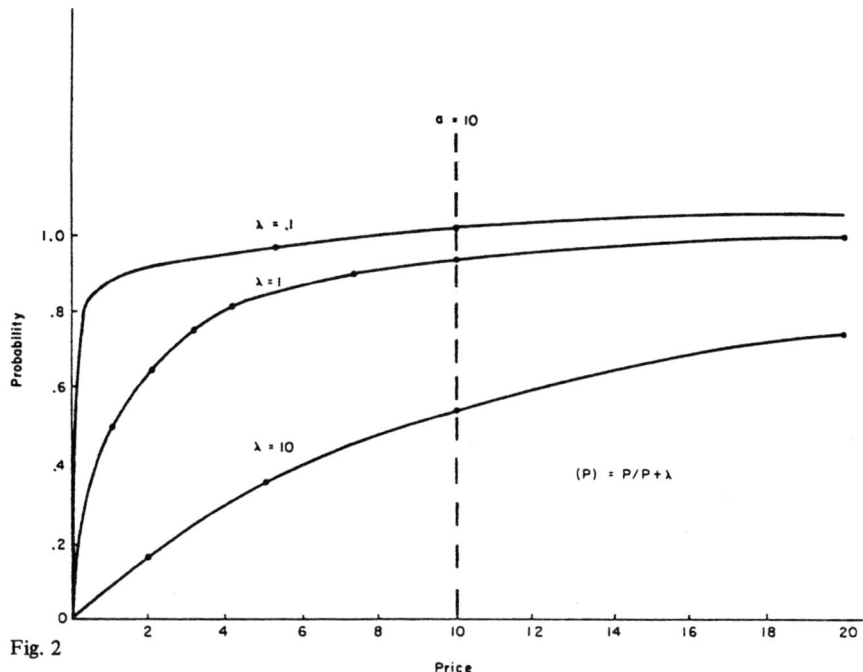

Fig. 2

supply the whole market at a price close to zero. As λ increases, the chance that the market is not served increases until at around $\lambda = 40$ there is over a 50 % chance that the market will not even be supplied. An inventory carrying cost this high is, in general, unreasonable hence one would expect to see risk being cut down. The range of λ that would cover most goods is close to zero.

It is easy to observe that the introduction of an inventory carrying cost in this instance, while it does not lower the payoffs to the firms, considerably lowers consumer welfare. This follows immediately from observing that for any finite λ there is a finite possibility that the customers will not even be served. Furthermore as can be seen from Figure 2 as the inventory carrying cost as increased not only do the customers have a smaller chance of being served, but if they are served the chances that it will be for a small amount at a high price are increasing. This is consistent with the observation that the firms have to earn enough to cover expected inventory carrying costs.

One could introduce identical constant costs of production into the model by merely making a transformation on price, replacing p by $(p + c)$ where c is the cost of a unit of production. We have not worked out the more difficult case of unequal production costs.

The equilibrium point described above is virtually independent of the method used for calculating contingent demand. This is because if a firm decides to produce anything it will produce enough to saturate the market if it charges the lower price. The higher firm is not going to sell anything anyhow hence the type of reconstitution is irrelevant. Furthermore, the equilibrium holds for a monotonically decreasing demand function. The proof of its uniqueness becomes more difficult and has not been done.

4. The Modified Edgeworth Cycle, Maxmin and the Minorant Game

Suppose that the firms were in some sort of dynamic market where they could undercut each other sequentially. The maxmin point for the firm moving first is to announce a price of $p = 0$ and a production of $q = a$. The best reply for the other is to produce nothing. This does appear to be dynamically stable as long as the second firm is around to undercut the first if it departs from its price policy. However, modeling difficulties now obviously appear in specifying the details of production and entry into competition. It is unlikely that there is going to be a nonproducing firm-in-being always present and always in a position to move immediately against any action of the production firm.

The minorant game has the pure strategy maxmin point at $p = 0$. This is an equilibrium point in this game; but it is not the only equilibrium point. There is another at which both firms make profits.

Suppose that Firm 1 must commit itself first. It selects a price and level of production such that when Firm 2 is informed it is slightly more profitable for Firm 2 to charge a higher price, taking advantage of the contingent demand that has been for it, than to undercut Firm 1. This calculation obviously depends upon the specific form of contingent demand. We use that as described in Section 2.

If Firm 1 charges p_1 and q_1 then the calculation made by Firm 2 to decide if it should undercut or name a higher price is given by:

$$p(a-p) = \left(\frac{a-q_1}{2}\right)^2$$

hence

$$q_1 = a - 2\sqrt{p(a-p)}.$$

Firm 1 may regard its problem as an attempt to maximize $p_1 q_1$ subject to $q_1 = a - 2\sqrt{p_1(a-p_1)}$. We obtain a maximum at $p_1 = .16a$ and $q_1 = .266a$ [3]). Hence $p_2 = .367a$ and $q_2 = .367a$.

It is of interest to note that this equilibrium is independent of the parameter λ. This we should expect, as no inventories are left over. Profits at this equilibrium are:
$\Pi_1 = .0426a^2$ and $\Pi_2 = .1342a^2$.

The possibility of a cycle and the existence of an equilibrium with a positive value, both appear to depend delicately upon the formulation of the model. Furthermore they stress its limitations, in the sense that although the λ is a parameter connected with inventories, it is more of a discount parameter that one might apply to having to mark down goods left at the end of the season than a straight per unit charge. It is still necessary to investigate other forms of charges on inventory. An inventory charge of say $c/2$ on all units (implying an average time in stock of $1/2$ a period) can be treated as an addition to cost of production.

---

[3]) More accurately the maximization condition gives $16p^3 - 24ap^2 + 10a^2 p - a^3 = 0$.

5. An n-Firm Asymmetric Model

Suppose that there are n firms in the market. Each firm has an inventory cost of λ_i. We shall assume that an equilibrium point with all firms obtaining zero profit exists and then attempt to calculate it. Instead of using the function $F_i(p_i, q_i)$ we assume that firms supply the whole market at any price they charge and let

$$\psi_i(p) = 1 - F_i(p, a).$$

The profit of Firm i can be written as:

$$\Pi_i = d_i(p)\left((p + \lambda_i) \prod_{j \neq i} \psi_j(p) - \lambda_i\right) = 0$$

where $d(p)$ is the market demand at price p.

From which it follows that

$$[A] \quad \prod_{j \neq i} \psi_j(p) = \frac{\lambda_i}{p + \lambda_i};$$

multiplying by ψ_i we obtain:

$$\prod_{j=1}^{n} \psi_j(p) = \frac{\lambda_i \psi_i}{p + \lambda_i} \text{ for } i = 1, \ldots, n.$$

Multiply these n equations together, and we obtain

$$\prod_{j=1}^{n} \psi_j(p) = \prod_{i=1}^{n} \frac{\lambda_i}{p + \lambda_i} \prod_{i=1}^{n} \psi_i(p);$$

or

$$\prod_{j=1}^{n} \psi_j(p) = \prod_{i=1}^{n} \left(\frac{\lambda_i}{p + \lambda_i}\right)^{1/(n-1)}$$

From Equation [A] this becomes

$$\frac{\lambda_i \psi_i}{p + \lambda_i} = \prod_{i=1}^{n} \left(\frac{\lambda_i}{p + \lambda_i}\right)^{1/(n-1)}, \text{ or}$$

$$[B] \quad \psi_i(p) = \frac{p + \lambda_i}{\lambda_i} \prod_{i=1}^{n} \left(\frac{\lambda_i}{p + \lambda_i}\right)^{1/(n-1)}$$

Suppose that all $\lambda_i = \lambda$. This becomes:

$$\psi_i(p) = \frac{p + \lambda}{\lambda}\left(\frac{\lambda}{p + \lambda}\right)^{n/(n-1)} = \left(\frac{\lambda}{p + \lambda}\right)^{1/(n-1)}$$

For $n = 2$, $\psi_i(p) = \lambda/(p+\lambda)$ hence

$$F_i(p, a) = 1 - \psi_i(p) = \frac{p + \lambda - \lambda}{p + \lambda} = \frac{p}{p + \lambda}$$

and this checks with the results in Section 2.

As n becomes large when all $\lambda_i = \lambda$ we observe that

$$\left(\frac{\lambda}{p+\lambda}\right)^{1/(n-1)} \to 1 \text{ for a bounded } p.$$

Hence, $F(p, a) \to 0$ for any positive price, p. We obtain a result that at first might appear to be paradoxical. If there are many firms in the market the possibilities for being undercut are enormous. As they increase, a protective strategy is to almost always price yourself out of the market. When a firm does not price itself out, the tendency will be to offer very little at a high price in order to be able to recoup expected losses from being caught with unsold stock.

6. Concluding Remarks

Pure duopoly theory has always smacked somewhat of esoterica. Yet if we are willing to note explicitly the limitations of our results it is probably well worth pursuing. Sometimes in an extremely simple model it is possible to demonstrate that effects which have been regarded as personal or social idiosyncracies are manifested without needing recourse to sociopsychological or other explanations.

In parts of the fashion industry many competitors produce few high priced items which are no marked down during the season. After the season is over a firm may take its losses by destroying or remaindering its inventory. This type of behavior is manifested in this model.

The model is obviously far too simple for any satisfactory application. We need to know how to handle dynamic and above all information conditions. Especially when firms are few in number a simple single period model is inadequate because it cannot reflect threat conditions. This is illustrated strikingly here when we observe the intuitively unsatisfactory equilibrium point in the nonsymmetric two firm market. Form [B] we obtain

$$\psi_1(p) = \frac{\lambda_2}{p + \lambda_2} \text{ and } \psi_2(p) = \frac{\lambda_1}{p + \lambda_1}.$$

Suppose $\lambda_1 = 0$ and $\lambda_2 > 1$ this gives $\psi_2(p) = 0$ or $F_2(p, a) = 1$ immediately. The equilibrium suggested is such that the firm with inventory costs supply the whole market at the competitive equilibrium price while the other sells nothing. This is not particularly reasonable. Furthermore we note that if we changed the information conditions somewhat the solution would change drastically in favor of the more efficient firm.

Duopoly models in which both price and quantity are treated as independent variables simultaneously are difficult to analyze. Among the open problems are firms with different costs and the treatment of differentiated products.

Appendix

Here we shall sketch a proof the principal results in this paper, the existence and uniqueness of the given mixed strategy equilibrium. We shall assume without proof that a mixed strategy equilibrium must have no price with a lumped probability except perhaps at the maximum of the range of active prices. Also we shall assume that the set of prices for which the probability density is positive is an interval. The proof of these propositions is quite tedious; those interested in such an exercise may consult Levitan/Shubik [1978]. We shall make a further assumption which is quite innocuous but which makes it possible to define the equilibrium of serious offers uniquely. This assumption is that if a firm names a price for which there is a positive market demand then it must offer a positive quantity for sale.

We define $F_i(p, q)$ as the probability that Firm i charges a price less than or equal to p and supplies a quantity less than or equal q. Since a is an upper bound on sales, optimality gives $F(p, \infty) = F(p, a)$ for any choice, p, such that $0 \leq p \leq a$.

Suppose Player i supplies $q \leq a - p$. He will undercut his competitor and sell q with probability $1 - F_j(p, a)$; and if he himself is undercut he might still sell q due to j's shortage with probability $F_j(p, a - p - q)$. If his competitor undercuts him and supplies y in the interval $[a - p - q, a - p]$ his sales will be $a - p - y$ with probability density $dF(x, y)$ given that $x \leq p$.

Hence, the expected profit of firm i is given by

$$\Pi_i = \Pi_i(p, q) = (p + \lambda) q (1 - F_j(p, a) + F_j(p, a - p - q))$$
$$+ \int (a - p - y) dF_j(x, y) - \lambda q,$$
$$a - p - q \leq y \leq a - p$$
$$x \leq p$$

subject to $0 \leq q \leq a - p$.

Suppose that $\Pi_i > 0$. Consider p_l the minimum price for which the firms have a positive probability density. Now $\Pi_i(p_l, q) = (p_l + \lambda) q - \lambda q = p_l q$ has its maximum at $(a - p_l)$ since $p_l > 0$ is implied by $\Pi_i > 0$. Thus $\Pi_i = p_l(a - p_l)$. Now, for all (p, q), defining $\Pi_i'(p, a)$ as

$$\frac{\partial}{\partial q} \Pi_i = (p + \lambda)(1 - F_j(p, a) + F_j(p, a - p - q)) - \lambda$$

is a non-increasing function of q, and $\partial/\partial q \ \Pi_i(p, a - p)$ is continuous and $\partial/\partial q \ \Pi_i(p_l, a - p_l) = p_l > 0$. Hence, there exists a maximal interval (p_l, p^*) in which $a - p$ is the unique optimal supply. Thus for $p \in (p_l, p^*)$

$$\Pi_i = (a-p)\,[(p+\lambda)\,(1 - F_j\,(p,a)) - \lambda], \text{ and}$$

$$F_j\,(p,a) = \frac{p - (\Pi_i / (a-p))}{p + \lambda}, \; p_l \leq p < p^*.$$

Since the above function is decreasing near a, $p^* < a$, and also in (p_l, p^*), $F_j\,(p^*, a) < 1$. Firm j cannot lump his remaining probability at p^*. If this were the case i would always undercut p^* and p^* would give j a nonpositive profit. Thus $F_j\,(p, a)$ is continuous at p^* and for $p > p^*$, $a - p$ is no longer uniquely optimal; for $p > p^*$, $\partial/\partial q \; \Pi_i\,(p, a-p) \leq 0$, while for $p < p^*$, $\partial/\partial q \; \Pi_i\,(p, a-p) \geq 0$. By continuity $\partial/\partial q \; \Pi_i\,(p^*, a-p^*) = 0$. However, $(a-p^*)\,\partial/\partial q \; \Pi_i\,(p^*, a-p^*)$ is the limit of $\Pi_i\,(p, a-p)$ as p approaches p^* from below. This implies $\Pi_i = 0$, a contradiction and we have $\Pi_i \leq 0$.

The case $\Pi_i < 0$ is ruled out by the fact that $q = 0$ is an available strategy which gives $\Pi_i = 0$. Hence $\Pi_i = 0$ and $p_l = 0$ is the remaining case.

We assert that $F_i\,(p, q) = 0$ for all $p > 0$ and $0 < q < a - p$, $i = 1, 2$. Suppose it is not true; then for some (p_0, q_0), $F_1\,(p_0, q_0) > 0$ with $p_0 > 0$ and $0 < q_0 < a - p_0$.

For any optimal q and $j = 1, 2$:
$\Pi'_1\,(p_0, q) = \partial/\partial q \; \Pi_j\,(p_0, q) = (p_0 + \lambda)\,(1 - F_j\,(p_0, a) + F_i\,(p_0, a - p_0 - q)) - \lambda = 0$. But $\Pi_j\,(p_0, q) = q\,\Pi'_1\,(p_0, q) + (p + \lambda)\int (a-p-y)\,dF\,(x, y) = 0$. Hence the integral must equal zero and $\Pi'_j\,(p_0, q)$ must be non-positive for all positive q or else a positive profit is attainable.

Now by our assumption about serious offers there exists a (p_1, q_1) such that $p_1 > 0$, $0 < q_1 < a - p_1$ and $F_1\,(p, q)$ is strictly increasing with both p and q in every neighborhood of (p_1, q_1). Hence $F_1\,(p_1, q_1 + \Delta) > F_1\,(p_1, q_1 - \Delta)$ for all $\Delta > 0$ and $\Pi'_1\,(p_1, q_1) = 0$.

Thus for $p_1 \leq p < a - q_1$, letting $0 \geq \Pi'_2\,(p, a - q_1 - p - \Delta) =$
$= (p + \lambda)\,(1 - F_1\,(p, a) + F_1\,(p_1, q_1 + \Delta)) - \lambda > (p + \lambda)\,(1 - F_1\,(p, a) + F_1\,(p_1, q_1 - \Delta)) - \lambda = \Pi'_2\,(p, a - q_1 - p + \Delta)$.

Consequently, for prices in $[p_1, a - p_1)$, $a - q_1 - p$ is an upper bound for the supply of Firm 2. Specifically, for $p_1 \leq p \leq a_1 - q_1$ and $q \geq a - q_1 - p_1$

$$F_2\,(p, q) = F_2\,(p_1, q).$$

Finally, for $0 < \Delta < a - q_1 - p_1$

$$\Pi'_1\,(p_1 + \Delta, q_1 - \Delta) = (p + \lambda + \Delta)\,[1 - F_2\,(p_1 + \Delta, a) +$$
$$+ F_2\,(p_1 + \Delta, a - p_1 - q_1)] - \lambda$$
$$= (p + \lambda + \Delta)\,[1 - F_2\,(p_1, a) + F_2\,(p_1, a - p_1 - q_1)] - \lambda$$
$$= \frac{\Delta\lambda}{p + \lambda} + \Pi'_1\,(p_1, q_1) > 0.$$

This contradicts $\Pi'_1 \leq 0$ and the assertion is proved. The case remains that $F_j\,(p, a - p - q) = 0$ for all $a - p > q > 0$, and we have

$$\frac{\partial}{\partial q} \Pi_i = (p + \lambda)(1 - F_j(p, a)) - \lambda = 0,$$

or

$$F_j(p, a) = \frac{p}{p + \lambda}.$$

This gives

$$\Pi_i(p, q) = q \left[(p + \lambda)\left(1 - \frac{p}{p + \lambda}\right) - \lambda \right] = 0$$

for all $0 \leq q \leq a - q$ which completes the proof.

References

Bertrand, J.: Theorie mathematique de la richesse sociale. (review) Journal des Savants (Paris: September 1883), 499–508.

Levitan, R.E., and M. Shubik: Price Duopoly and Capacity Constraints. Cowles Foundation Discussion Paper No. 287, February 10, 1970.

— : Duopoly with Price and Quantity as Strategic Variables. Cowles Foundation Discussion Paper 289, July 17, 1970 (Revised 1978).

Received May, 1974
(revised version May, 1976)

# Noncooperative Oligopoly with Entry*

KOFI O. NTI

School of Business, Washington University, St. Louis, Missouri 63130

AND

MARTIN SHUBIK

*Cowles Foundation for Research in Economics,
Yale University, New Haven, Connecticut 06520*

Received May 21, 1979; revised April 7, 1980

1. INTRODUCTION

The literature on oligopolistic competition with entry is substantial (Nti and Shubik [11]). However, with the exception of Shubik [13], Friedman [4], Spence [14], Dubey and Shubik [2], and a few others, most of the theoretical analyses on entry tend to ignore strategic issues. The well-known results of Kamien and Schwartz [7], Gaskins [6], and others, have generally focussed on how a group of oligopolists may collectively control entry by their pricing policies. But established firms need not act collectively. They may engage in intra-oligopolistic competition and, above all, they may perceive entry threats differently. In addition, potential entrants in oligopolistic industries are often big firms whose interests and strategic options need to be explicitly considered. Hence, strategic issues, involving varying degrees of cooperation and antagonism, are critical to the entry problem.

Furthermore, strategic oligopoly with entry is pertinent to research on Cournot–Walras equilibrium; it provides an entry-inclusive partial equilibrium building block to supplement general equilibrium models that integrate oligopolistic production and competitive exchange markets (Gabszewicz and Vial [5], Novshek and Sonnenschein [9]).

In this paper, competition between several established firms in an oligopolistic industry with a finite number of potential entrants is modelled as a noncooperative game with quantity and entry decisions as strategic variables. The varieties of disadvantages facing potential entrants are

* Partially supported by the Office of Naval Research under contract N0014-77-C-0518. Parts of this paper appeared in Dr. Nti's Ph. D. thesis, Yale University, 1977.

187

aggregated into single period fixed cost barriers. The existence of a noncooperative equilibrium solution is proved. At equilibrium, no active firm has an incentive to change production or exit and no firm outside the industry has an incentive to enter. Apart from potential entrants who may randomize their entry decisions, the equilibrium solution involves no randomization in the outputs of any firm. The solutions are characterized and their dependence on entry costs are determined. The solutions partition entry costs into several regions which have natural interpretations regarding the ease or difficulty of entry and the use of threats or exclusionary tactics to enforce certain outcomes.

The model is formulated in Section 2. Cournot oligopoly without entry is analyzed in Section 3. And in Sections 4 and 5, the main results proving and characterizing the noncooperative equilibrium solution for the game with one potential entrant are established and discussed. Section 6 extends the results to the situation with several asymmetric potential entrants. We conclude with a brief review and discussion of the results of the paper.

2. The Model

Consider a one product oligopolistic industry with n established firms and k potential entrants. The sole decision variable of an established firm is the quantity of the product it should offer for sale. A potential entrant, on the other hand, must simultaneously decide whether it should actually enter into competition and the quantity of the product it should offer for sale. All decisions are made independently.

We study a highly symmetric model as this serves to illustrate the qualitative results we wish to show with analytical simplicity.

The price, p, of the commodity is determined by the relation

$$p = \phi(q), \qquad (1)$$

where q is the total quantity of the commodity offered for sale. All firms, established as well as potential entrants, face identical variable cost function $c(q_i)$, where q_i is the output of the ith firm. Established firms have no fixed costs but each potential entrant incurs a non-negative fixed cost D if it actually enters into competition and offers any quantity of the product for sale. The objective is to identify and characterize noncooperative equilibrium solutions for the problem under the following assumptions:

Assumption 1. *The variable cost function $c(q_i)$ is twice differentiable with*

$$c'(q_i) > 0, \, c''(q_i) \geqslant 0. \qquad (2)$$

ASSUMPTION 2. *For all total outputs, q, offered for sale by all market participants, the demand function $\phi(q)$ is decreasing for $0 \leq q \leq Q$. Also $\phi(0) > 0$ and there exists Q for which $\phi(Q) = 0$ and $\phi(q) = 0$ for all $q \geq Q$. Furthermore, $\phi(q)$ is twice differentiable and satisfies*

$$\phi'(q) < 0, \phi''(q) \leq 0 \quad \text{for all} \quad 0 \leq q \leq Q. \tag{3}$$

We also assume that all firms have equal capacity and that the firms in the industry have enough capacity to supply the Cournot equilibrium requirements. This avoids having to discuss boundary solutions.

Assumption 1 states that each firm's variable cost is a non-decreasing function of its output and its marginal cost exhibits increasing returns. Assumption 2 describes a negatively sloping demand function with decreasing first derivative. These two assumptions are typical of cost and demand functions in the theory of the firm. Together, they imply concave profit functions and help establish the existence of noncooperative equilibrium for classical oligopoly without potential entrants. In the framework of symmetric oligopoly, following Roberts and Sonnenschein [12], it is possible to dispense with implied assumptions of concavity (and differentiability) but to do so here would only raise technical complexities in subsequent optimizations, especially when we attempt to characterize equilibrium solutions for the problem with entry.

Before considering the entry problem, we need some results for the Cournot oligopoly problem without potential entrants.

3. Cournot Oligopoly without Potential Entrants

Consider n identical oligopolists competing with quantity as strategic variables under the assumptions of the model.

Suppose the ith firm offers an output q_i for sale, $i = 1,..., n$. Then the profit, π_j, of the jth firm is given by

$$\pi_j = q_j \phi \left(\sum_{i=1}^{n} q_i \right) - c(q_j), \quad j = 1, 2,..., n. \tag{4}$$

In general, a strategy for the ith firm may involve randomization over all non-negative outputs. A noncooperative equilibrium solution will then be a set of n strategies, one for each firm, with the property that no firm acting alone can improve its expected profit if it uses a strategy other than its part of the equilibrium solution.

The following theorem asserts that without loss of generality we may restrict our search for noncooperative equilibrium solutions to pure strategies. That is, no firm needs to randomize its production decision and we can look for an n-tuple of output $(q_1^*, q_2^*,..., q_n^*)$ such that

$$\pi_j^* = q_j^* \phi \left(\sum_{i=1}^{n} q_i^* \right) - c(q_j^*)$$

$$\geqslant q_j \phi \left(\sum_{i \neq j}^{n} q_i^* + q_j \right) - c(q_j), \quad \text{for all } q_j \geqslant 0, j = 1, 2,..., n. \quad (5)$$

To establish the existence of a n-tuple of noncooperative equilibrium production levels we need a theorem of Nikaido and Isoda [8].

THEOREM 1 (Nikaido and Isoda). *If the following conditions hold for an n-player game*

(i) *The set of decisions, Σ_i, for each player is a bounded, closed and convex subset of Euclidean space R^k.*

(ii) *The payoff function $\pi_i(q_1,..., q_n)$ is concave in q_i, for fixed $q_1,..., q_{i-1}, q_{i+1},..., q_n$, $i = 1, 2,..., n$, $q_i \in \Sigma_i$.*

(iii) *$\pi_i(q_1,..., q_n)$ is continuous in $q_1,..., q_n$.*

Then there exists an n-tuple $(q_1^*,..., q_n^*)$ with $q_i^* \in \Sigma_i$, $i = 1,..., n$ which is a noncooperative equilibrium solution for the game.

Under the assumptions of the Cournot oligopoly problem above, it is clear no firm needs to produce more than Q units of the commodity. Hence condition (i) is satisfied. Also the second partial derivative of π_i with respect to q_i is nonpositive for all i so the concavity requirements in condition (ii) are met. It is also clear π_i is continuous in all variables. Hence:

COROLLARY 1. *Under the assumptions of the model, there exists a noncooperative equilibrium solution of the form $(q_1^*,..., q_n^*)$ with $0 \leqslant q_i^* \leqslant Q$, $i = 1, 2,..., n$, to the Cournot oligopoly problem.*

A characterization of the pure strategy noncooperative equilibrium solution, which we state as a theorem, is given in Burger [1, pp. 49–52].

THEOREM 2. *Under assumptions 1 and 2 the n-firm Cournot oligopoly problem has exactly one noncooperative equilibrium point $(q^*,..., q^*)$ where*

(a) *$q^* = 0$ if $c'(0) \geqslant \phi(0)$ or*

(b) *q^* is the only root of the equation*

$$H_n(q) = \phi(nq) + q\phi'(nq) - c'(q) = 0 \quad (6)$$

on the interval $0 < q < Q/n$ if $c'(0) < \phi(0)$.

The condition $c'(0) < \phi(0)$, for which there exists a positive output for each firm at the noncooperative equilibrium solution, only says that marginal cost at zero output is less than the market price. This is obviously a necessary condition for a firm to send a positive output to the market. On the other hand, if marginal cost at zero output exceeds the market price, then it is unprofitable for any firm to produce. From henceforth, we will rule out this trivial situation and assume each active firm has an incentive to send a positive output to the market.

ASSUMPTION 3. *The functions $\phi(\cdot)$ and $c(\cdot)$ satisfy $\phi(0) > c'(0)$.*

The noncooperative equilibrium solutions to the Cournot oligopoly problem may be parametrized with respect to the number of active firms. Let

$$H_n(\xi) = \phi(n\xi) + \xi\phi'(n\xi) - c'(\xi). \tag{7}$$

Then $H'_n(\xi) < 0$ for $0 < \xi < Q/n$. And the unique noncooperative equilibrium output, ξ_n, for each firm in the n-firm oligopoly game satisfies

$$H_n(\xi_n) = 0 \quad \text{with} \quad 0 < \xi_n < Q/n. \tag{8}$$

It is also straightforward to establish

THEOREM 3. *For $m > n$, let ξ_m and ξ_n be the noncooperative equilibrium outputs for the m-firm and n-firm Cournot oligopolies respectively. Also let π^m and π^n be the respective profits of the firms.*
Then

(a) $\xi_n > \xi_m$,
(b) $\pi^n > \pi^m$,
(c) $n\xi_n < m\xi_m$.

Theorem 3 states that the individual firms in the n-firm oligopoly problem produce more and earn more than their counterparts in the m-firm problem but the total output in the n-firm problem is less than the total output in the m-firm problem.

4. Cournot Oligopoly with One Potential Entrant

Continuing with the entry model we now assume there are n established firms and one potential entrant. The potential entrant is designated the 0th firm and the established firms are labelled from 1 through n.

Define δ by

$$\delta = 0 \quad \text{if potential entrant stays out.}$$
$$ = 1 \quad \text{if potential entrant enters.}$$

Then the profits may in general be written as

$$\pi_j^\delta = q_j \phi \left(\sum_{i=1}^n q_i + \delta q_0 \right) - c(q_j), \quad j = 1,\dots, n, \qquad (9)$$

and

$$\pi_0^\delta = \delta \left[q_0 \phi \left(\sum_{i=1}^n q_i + \delta q_0 \right) - c(q_0) - D \right]. \qquad (10)$$

A strategy for the jth active firm is a distribution function $F_j(\xi)$ over all nonnegative outputs, where $F_j(\xi) = \text{Prob}\{q_j \leqslant \xi\}$, $j = 1,\dots, n$. Similarly, a strategy for the potential entrant is a probability distribution $F_0(\xi)$ over all nonnegative outputs if it enters and a discrete probability distribution h over its entry decision, where

$$F_0(\xi) = \text{Prob}\{q_0 \leqslant \xi \mid \delta = 1\}$$

and

$$h_\delta = \text{Prob}\{\delta = v\}, \quad v = 0, 1.$$

The expected profit of the firms under the strategies F_0, F_j, h_δ are

$$E\pi_j^\delta = \sum_{\delta=0}^1 \left[\int \prod_{i=1}^n dF_i(q_i) \pi_j^\delta \right] h_\delta, \quad j = 1,\dots, n, \qquad (11)$$

and

$$E\pi_0^\delta = \sum_{\delta=0}^1 \left[\int \prod_{i=1}^n dF_i(q_i) \pi_0^\delta \right] h_\delta. \qquad (12)$$

A noncooperative equilibrium strategy for the problem is, in general, a set of $n+2$ probability distributions $\{F_j^*\}$, $j = 0,\dots, n$, h_δ^* satisfying

$$\sum_{\delta=0}^1 \int \left[\prod_{i=1}^n dF_i^*(q_i) \pi_j^\delta \right] h_\delta^*$$
$$\geqslant \sum_{\delta=0}^1 \int \left[\prod_{\substack{i=1 \\ i \neq j}}^n dF_i^*(q_i) \, dF_j(q_j) \pi_j^\delta \right] h_\delta^* \qquad (13)$$

for $j = 1,..., n$, and all distribution functions F_j. And

$$\sum_{\delta=0}^{1} \left[\int \prod_{i=1}^{n} dF_i^*(q_i) \, dF_0^*(q_0) \pi_0^\delta \right] h_\delta^*$$
$$\geq \sum_{\delta=0}^{1} \left[\int \prod_{i=1}^{n} dF_i^*(q_i) \, dF_0(q_0) \pi_0^\delta \right] h_\delta \qquad (14)$$

for all distribution functions F_0 and probability distribution h_δ.

We will show that without loss of generality, the search for noncooperative equilibrium solutions may be restricted to pure production strategies for all firms and a possible randomization in the entry decision of the potential entrant.

That is, if we let δ be the probability of entry with $0 \leq \delta \leq 1$ and compute the expected profit from (9) and (10) as follows:

$$\pi_0^\delta = \delta \left[q_0 \phi \left(\sum_{i=1}^{n} q_i + q_0 \right) - c(q_0) - D \right] \qquad (15)$$

and

$$\pi_j^\delta = (1-\delta) \left[q_j \phi \left(\sum_{i=1}^{n} q_i \right) - c(q_j) \right]$$
$$+ \delta \left[q_j \phi \left(\sum_{i=1}^{n} q_i + q_0 \right) - c(q_j) \right], \quad j = 1,..., n. \qquad (16)$$

Then there exists an $n+2$-tuple $(q_1^*,..., q_n^*, q_0^*, \delta^*)$ with $0 \leq \delta^* \leq 1$ and $0 \leq q_i^* \leq Q$, $i = 0, 1,..., n$, such that

$$\pi_j^{\delta*}(q_1^*,..., q_{j-1}^*, q_j^*, q_{j+1}^*,..., q_n^*, q_0^*)$$
$$\geq \pi_j^{\delta*}(q_1^*,..., q_{j-1}^*, q_j, q_{j+1}^* \cdots q_n^*, q_0^*) \qquad \text{for all} \quad q_j, \, j = 1,..., n \quad (17)$$

and

$$\pi_0^{\delta*}(q_1^*,..., q_n^*, q_0^*) \geq \pi_0^\delta(q_1^*,..., q_n^*, q_0) \qquad \text{for all} \quad \delta \text{ and } q_0. \qquad (18)$$

Observations. (a) The function π_j^δ and π_0^δ are continuous in δ, $q_0, q_1,..., q_n$.

(b) The feasible decision set for each firm is convex and compact, $0 \leq \delta \leq 1$, $0 \leq q_i \leq Q$, $j = 0,..., n$.

(c) For fixed δ, $q_0, q_1,..., q_{j-1}, q_{j+1},..., q_n$ the function π_j^δ is concave in q_j, $j = 1,..., n$.

(d) For fixed $q_1,..., q_n$, the function π_0^δ is concave *separately* in q_0 and δ but not *jointly* in δ and q_0.

For fixed $\delta, q_0, q_1,..., q_{j-1}, q_{j+1},..., q_n$ let

$$B_j(\delta, q_{-j}) = \{q_j^* \mid \pi_j^\delta(q_0,..., q_j^*,..., q_n) \geq \pi_j^\delta(q_0,..., q_j,..., q_n)\} \quad \text{for } j = 1, 2,..., n. \quad (19)$$

And for fixed $q_1, q_2,..., q_n$ let

$$B_0(q_{-0}) = \{(\delta^*, q_0^*) \mid \pi_0^{\delta*}(q_0^*, q_1,..., q_n) \geq \pi_0^\delta(q_0,..., q_1,..., q_n)\} \quad (20)$$

Observations. (e) For fixed $\delta, q_0, q_1,..., q_{j-1}, q_{j+1},..., q_n$, $B_j(\delta, q_{-j})$ is the set of optimal responses for the jth firm. The set is nonempty since $\pi_j^\delta(q_0, q_1,..., q_n)$ is concave in q_j and the decision set is closed and convex. Actually the second order differentiability conditions on the cost and demand functions $c(\cdot)$ and $\phi(\cdot)$ imply $\pi_j^\delta(q_0, q_1,..., q_n)$ is strictly concave in q_j and so the jth firm has a unique optimal response.

(f) The set $B_0(q_{-0})$ is the set of optimal responses for the potential entrant for fixed $q_1,..., q_n$. The set $B_0(q_{-0})$ is nonempty, not necessarily convex but it is contractible.

Let $I = \{t \mid 0 \leq t \leq 1\}$.

DEFINITION. A subset Z of Euclidean space R^n is *contractible* (deformable) into a point $z_0 \in Z$, if there exists a continuous function $h(t, z)$ mapping $I \times Z$ into Z such that for all $z \in Z$, $h(0, z) = z$ and $h(1, z) = z_0$.

We list without proof some standard results on contractible sets.[1]

(i) If a set Z is contractible to $z_0 \in Z$, then Z is contractible on any other point $z \in Z$.

(ii) The Cartesian product of a finite number of contractible sets in contractible.

(iii) A convex set is contractible.

THEOREM 4. *The set $B_0(q_{-0})$ is contractible.*

Proof. For fixed $q_1, q_2,..., q_n$, the set $B_0(q_{-0})$ is defined by

$$B_0(q_{-0}) = \left\{(\delta^*, q_0^*) \mid \delta^* \left[q_0^* \phi \left(\sum_{i=1}^n q_i + q_0^*\right) - c(q_0) - D\right]\right.$$
$$= \max_{\substack{0 \leq \delta \leq 1 \\ 0 \leq q_0 \leq Q}} \delta \left[q_0 \phi \left(\sum_{i=1}^n q_i + q_0\right) - c(q_0) - D\right]\right\}. \quad (21)$$

---

[1] See, for example, Murray Eisenberg, "Topology" p. 372, Holt, Rinehart & Winston, New York: 1974.

Since $q_0\phi(\sum_{i=1}^{n} q_i + q_0) - c(q_0) - D$ is (strictly) convex in q_0, there exists a (unique) point $\bar{q}_0$ satisfying

$$\bar{q}_0 \phi \left(\sum_{i=1}^{n} q_i + \bar{q}_0 \right) - c(\bar{q}_0) - D$$

$$\geqslant q_0 \phi \left(\sum_{i=1}^{n} q_i + q_0 \right) - c(q_0) - D \quad \text{for all } 0 \leqslant q_0 \leqslant Q. \quad (22)$$

The value of $\bar{q}_0\phi(\sum_{i=1}^{n} q_i + \bar{q}_0) - c(\bar{q}_0) - D$ determines the nature of the set $B_0(q_{-0})$.

Case 1. $\bar{q}_0 \phi(\sum_{i=1}^{n} q_i + \bar{q}_0) - c(\bar{q}_0) - D < 0$.

In this case, the optimum response for the potential entrant is attained by setting $\delta^* = 0$, $0 \leqslant q_0^* \leqslant Q$. That is $B_0(q_{-0}) = \{(\delta^*, q_0^*) \mid \delta^* = 0, 0 \leqslant q_0^* \leqslant Q\}$.

Case 2. $\bar{q}_0 \phi(\sum_{i=1}^{n} q_i + \bar{q}_0) - c(\bar{q}_0) - D > 0$.

In this case, the optimal response for the potential entrant is attained on setting $\delta^* = 1$, $q_0^* = \bar{q}_0$. That is, $B_0(q_{-0}) = \{(\delta^*, q_0^*) \mid \delta^* = 1, q_0^* = \bar{q}_0\}$.

Case 3. $\bar{q}_0\phi(\sum_{i=1}^{n} q_i + \bar{q}_0) - c(\bar{q}_0) - D = 0$.

Here the optimal response for the potential entrant is attained by either setting $\delta^* = 0$, $0 \leqslant q_0^* \leqslant Q$ or by setting $0 \leqslant \delta^* \leqslant 1$, $q_0^* = \bar{q}_0$. That is, $B_0(q_{-0}) = \{\delta^* = 0, 0 \leqslant q_0^* \leqslant Q\} \cup \{0 \leqslant \delta^* \leqslant 1, q_0^* = \bar{q}_0\}$.

Clearly for fixed $q_1,\ldots, q_n$ the set $B_0(q_{-0})$ may fall into any of the three cases identified above.

Figure 1 is a sketch of the possible shapes of $B_0(q_{-0})$ corresponding to these three cases.

In Cases 1 and 3, the optimal response sets $B_0(q_{-0})$ are convex and so contractible. And in Case 2, the inverted T-shaped set in Fig. 1 is also contractible. To see this, define the map $h(t, q) : I \times B_0(q_{-0}) \to B_0(q_{-0})$ by

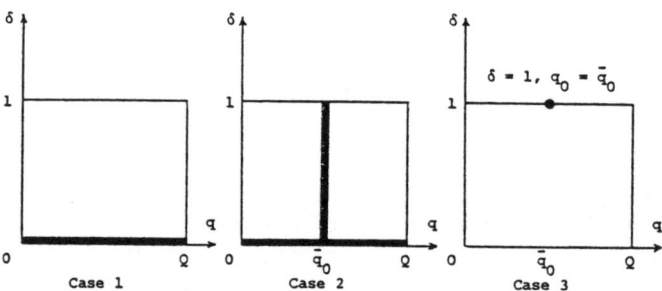

FIG. 1. Optimal response sets for the entrant.

$h(t, q) = t\bar{q}_0 + (1-t)q$. Then $h(t, q) \in B_0(q_{-0})$ for $0 \leq q \leq Q$. Also $h(0, q) = q$ and $h(1, q) \times \bar{q}_0$.

Hence $B_0(q_{-0})$ is contractible in all cases. Q.E.D.

Since the optimal response sets $B_j(\delta, q_{-j})$ for the established firms are convex, they are contractible and hence

COROLLARY 2. *The sets $B_j(\delta, q_{-j})$, $j = 1,...,n$ and $B_0(q_{-0})$ defined in (20) and (21) are contractible.*

A generalization of the Kakutani fixed point theorem due to Eilenberg and Montgomery [3] may now be applied to the entry problem.

THEOREM 5 (Eilenberg and Montgomery). *Let Z be a contractible polyhedron and $g: Z \to Z$ be a semicontinuous point to set mapping such that for every $z \in Z$, the set $g(z)$ is contractible. Then there exists z^* such that $z^* \in g(z^*)$.*

THEOREM 6. *There exists a noncooperative equilibrium solution $(\delta^*, q_0^*, q_1^*,..., q_n^*)$ to the Cournot oligopoly problem with one potential entrant.*

Proof. Define the function g by $g: (\delta, q_0, q_1,..., q_n) \to B_0(q_{-0}) \times B_1(\delta, q_{-1}) \times \cdots \times B_n(\delta, q_{-n})$. Then g is a semi-coninuous mapping of a contractible (convex) set into a contractible set. Hence by Theorem 5, g has a fixed point. That is, there is a fixed point $\delta^*, q_0^*, q_1^*,..., q_n^*$ of the point to set mapping g satisfying

$$(\delta^*, q_0^*, q_1^*,..., q_n^*) \in B_0(q^*_{-0}) \times B_1(\delta^*, q^*_{-1}) \times \cdots \times B_n(\delta^*, q^*_{-n}). \quad (23)$$

The relation (23) just translates into

$$\pi_j^{\delta*}(q_0^*, q_1^*,..., q_n^*) \geq \pi_j^{\delta*}(q_0^*, q_1^*,..., q_{j-1}^*, q_j, q_{j+1}^*,..., q_n^*)$$

$$\text{for all} \quad 0 \leq q_j \leq Q, \quad j = 1, 2,..., n. \quad (24)$$

and

$$\pi_0^{\delta*}(q_0^*, q_1^*,..., q_n^*) > \pi_0^{\delta}(q_0, q_1^*,..., q_n^*)$$

$$\text{for all} \quad 0 \leq \delta \leq 1, \text{ and } \quad 0 \leq q_0 \leq Q, \quad (25)$$

which is the result we are seeking.

The proof of Theorem 6 can also be based on the Kakutani fixed point theorem if the decision set of the entrant is redefined by means of a homeomorphic transformation so its optimal response set becomes convex.[2]

---

[2] Pradeep Dubey, private communication.

5. Structure of the Noncooperative Equilibrium Solutions

Suppose $(\delta^*, q_0^*, q_1^*, ..., q_n^*)$ is a noncooperative equilibrium solution to the entry game of the form identified in Theorem 6. That is, the maxima of

$$\pi_j^{\delta*} = (1 - \delta^*) \left[q_j \phi \left(\sum_{i \neq j}^{n} q_i^* + q_j \right) - c(q_j) \right]$$
$$+ \delta^* \left[q_j \phi \left(\sum_{i \neq j}^{n} q_i^* + q_j + q_0^* \right) - c(q_j) \right] \quad (26)$$

and

$$\pi_0^{\delta} = \delta \left[q_0 \phi \left(\sum_{i=1}^{n} q_i^* + q_0 \right) - c(q_0) - D \right] \quad (27)$$

are attained at $q_j^*, j = 1,..., n$ and (δ^*, q_0^*), respectively.

If $q_0^* \phi(\sum_{i=1}^{n} q_i^* + q_0^*) - c(q_0^*) - D > 0$, then $\delta^* = 1$, otherwise $\pi_0^{\delta*}$ would not be maximal for the potential entrant. But if $\delta^* = 1$, then the maximization problems reduce to

$$\pi_j^{\delta*} = \max_{0 \leq q_j \leq Q} q_j \phi \left(\sum_{i \neq j}^{n} q_i^* + q_0^* + q_j \right) - c(q_j), \quad j = 1,..., n, \quad (28)$$

and

$$\pi_0^{\delta*} = \max_{0 \leq q_0 \leq Q} q_0 \phi \left(\sum_{i=1}^{n} q_i^* + q_0 \right) - c(q_0) - D, \quad (29)$$

which, apart from the fixed cost D in (29), is just the $n + 1$-firm oligopoly problem. But D does not affect the maximization, so the noncooperative equilibrium output for this problem is the unique output, ξ_{n+1}, of the $n + 1$-firm oligopoly problem defined as in Eqs. (7) and (8).

Hence $\delta^* = 1$, $q_0^* = q_1^* = \cdots = q_n^* = \xi_{n+1}$ form a noncooperative equilibrium solution to the entry game so far as

$$D < \xi_{n+1} \phi[(n + 1)\xi_{n+1}] - c(\xi_{n+1}) = \pi^{n+1} = D_1. \quad (30)$$

On the other hand, if

$$q_0^* \phi \left(\sum_{i=1}^{n} q_i^* + q_0^* \right) - c(q_0^*) - D < 0. \quad (31)$$

Then $\delta^* = 0$, otherwise $\pi_0^{\delta *}$ would not be maximal for the entrant. But for this situation, the maximization problem reduces to

$$\pi_j^{\delta *} = \max_{0 < q_j < Q} \left[q_j \phi \left(\sum_{i \neq j}^{n} q_i^* + q_j \right) - c(q_j) \right], \quad j = 1, \ldots, n, \quad (32)$$

which is just the maximization problem for the n-firm Cournot oligopoly. Therefore, the output of the established firms is the unique output, ξ_n, of the n-firm oligopoly problem where ξ_n is defined as in (7) and (8). Hence $\delta^* = 0$, $q_1^* = q_2^* = \cdots = q_n^* = \xi_n$ form a noncooperative equilibrium solution to the problem so far as

$$q_0^* \phi(n\xi_n + q_0^*) - c(q_0^*) < D. \quad (33)$$

In particular, the result still holds if

$$D > \max_{0 < q_0 < Q} [q_0 \phi(n\xi_n + q_0) - c(q_0)].$$

And it is straightforward to show that there exists a unique q_0^+, with $0 < q_0^+ < Q - n\xi_n$, which attains the maximum in (33).

And so, for all

$$D > D_2 = q_0^+ \phi(n\xi_n + q_0^+) - c(q_0^+),$$

there exists a noncooperative equilibrium solution to the entry problem with

$$\delta^* = 0, q_1^* = q_2^* = \cdots = q_n^* = \xi_n.$$

Finally, if the equilibrium $\delta^*, q_0^*, q_1^*, q_2^*, \ldots, q_n^*$ has $0 < \delta^* < 1$, then

$$q_0^* \phi \left(\sum_{i=1}^{n} q_i^* + q_0^* \right) - c(q_0^*) = D,$$

otherwise, the potential entrant can increase (decrease) δ^* towards 1(0) to improve its profit.

Hence for $0 < \delta^* < 1$, the differentiability conditions for maximum imply the folllowing set of equations:

$$(1 - \delta^*) \left[q_j^* \phi' \left(\sum_{i=1}^{n} q_i^* \right) - c'(q_j^*) + \phi \left(\sum_{i=1}^{n} q_i^* \right) \right]$$

$$+ \delta^* \left[q_j^* \phi' \left(\sum_{i=1}^{n} q_i^* + q_0^* \right) + \phi \left(\sum_{i=1}^{n} q_i^* + q_0^* \right) - c'(q_j^*) \right] = 0,$$

$$j = 1, \ldots, n, \quad (34a)$$

$$q_0^* \phi' \left(\sum_{i=1}^n q_i^* + q_0^* \right) + \phi \left(\sum_{i=1}^n q_i + q_0^* \right) - c'(q_0^*) = 0 \qquad (34b)$$

$$q_0^* \phi \left(\sum_{i=1}^n q_i^* + q_0^* \right) - c(q_0^*) = D. \qquad (34c)$$

From the symmetry in the problem, the conditions for equilibrium may further be simplified to

$$\delta^* = \frac{q_a \phi'(nq_a) + \phi(nq_a) - c(q_a)}{q_a[\phi(nq_a) - \phi'(nq_a + q_0^*)] + [\phi(nq_a) - \phi(nq_a + q_0^*)]}, \qquad (35a)$$

$$q_0^* \phi'(nq_a + q_0^*) + \phi(nq_a + q_0^*) - c'(q_0^*) = 0, \qquad (35b)$$

$$q_0^* \phi(nq_a + q_0^*) - c(q_0^*) = D, \qquad (35c)$$

where $q_1^* = q_2^* = \cdots = q_n^* = a_a$.

Note that the denominator in (35a) is positive because $\phi(\cdot)$ is nondecreasing and $\phi''(\cdot) \leq 0$. Also, from the assumed range of δ^* together with the properties of $H_n(\cdot)$ and other assumptions of the model, it is shown in [10] that

$$q_0^+ < q_0^* < \xi_{n+1} < q_a < \xi_n$$

and that

$$D_2 < D < D_1, \quad \text{if} \quad 0 < \delta^* < 1.$$

The results on the structure of the noncooperative equilibrium solutions are summarized into a theorem.

THEOREM 7. *Let ξ_n and ξ_{n+1} be the noncooperative equilibrium solutions to the n firm and $n+1$ firm Cournot oligopoly problem with no potential entrants. Also define D_1, D_2 and q_0^+ by*

$$D_1 = \pi^{n+1} = \xi_{n+1} \phi(n\xi_{n+1} + \xi_{n+1}) - c(\xi_{n+1}), \qquad (36)$$

$$D_2 = q_0^+ \phi(n\xi_n + q_0^+) - c(q_0^+) = \max_{0 \leq q \leq Q} [q\phi(n\xi_n + q) - c(q)]. \qquad (37)$$

Then the noncooperative equilibrium solution $(\delta^*, q_0^*, q_1^*, \ldots, q_n^*)$ for the Cournot oligopoly probem with one potential entrant may be characterized with respect to entry costs, D, as follows:

Case 1. *If $D < D_1$, then $\delta^* = 1$, $q_0^* = q_1^* = \cdots = q_n^* = \xi_{n+1}$ is a noncooperative equilibrium solution.*

Case 2. If $D > D_2$, then $-\delta^* = 0$, $q_1^* = q_2^* = \cdots = q_n^* = \xi_n$ is a noncooperative equilibrium solution.

Case 3. If $D_2 < D < D_1$, then in addition to the pure oligopoly solutions identified above there also exists a randomized ebtry solution with $0 < \delta^* < 1$, $q_1^* = q_2^* = \cdots = q_n^* = q_a$ and q_0^* satisfying the set of equations

$$\delta^* = \frac{q_a \phi'(nq_a) + \phi(nq_a) - c'(q_a)}{q_a[\phi'(nq_a) - \phi'(nq_a + q_0^*)] + [\phi(nq_a) - \phi(nq_a + q_0^*)]}, \quad (38a)$$

$$q_0^* \phi'(nq_a + q_0^*) + \phi(nq_a + q_0^*) - c'(q_0^*) = 0, \quad (38b)$$

$$q_0^* \phi(nq_a + q_0^*) - c(q_0^*) = D, \quad (38c)$$

with $q_0^+ < q_0^* < \xi_{n+1} < q_a < \xi_n$.

Figure 2 is a sketch of the equilibrium solutions characterized in Theorem 7.

The oligopoly problem with entry has a three band equilibrium solution. For entry costs less than D_2, the $(n+1)$-firm Cournot output solves the problem while for entry costs greater than D_1, the n-firm Cournot output solves it. In other words, if entry costs are moderately low there are no real impediments to entry but once entry costs exceed D_1, entry is forestalled. For intermediate entry costs, a randomized entry solution or any of the pure oligopoly outcomes may be realized.

The pure oligopoly solutions in the middle band are enforceable as threat strategies, one by the potential entrant which announces its intention to enter at the $(n+1)$-firm oligopoly output, and the other is enforceable by any of the established firms which makes known its intention to produces at the n-firm oligopoly output. The randomized entry equilibrium captures the effects of uncertain entry on the output decisions of established firms. They are forced to reduce their outputs somewhat below what would otherwise be optimal for n noncooperative oligopolists in competition. But the restriction of output serves a dual purpose. It acts as a hedge against overproduction if entry should occur. Yet the restricted output is large enough to discourage entry at the $(n+1)$-firm oligopoly output.

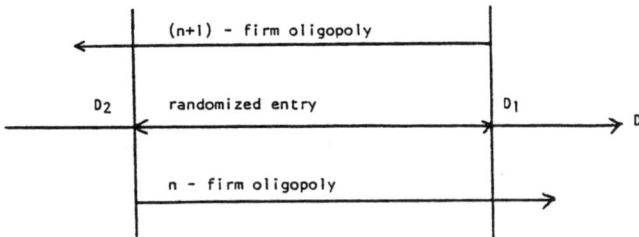

FIG. 2. Variation of equilibrium solution with entry cost.

In the middle band we have three noncooperative solutions whose characteristics can be described qualitatively in terms of threats implicit in the behavior. No entry occurs when the entrant believes that the firms in the industry will stand and fight; entry occurs when those in, believe that new entry is a foregone conclusion and they must adjust to it, and the third case is where all are uncertain.

6. Cournot Oligopoly with Finite Potential Entrants

For the model under consideration assume there are n established firms and k potential entrants. Established firms are labelled by the index 1 through n and potential entrants by the index $n+1$ through $n+k$. The variable cost functions $c_j(\cdot)$, $j=1,...,n+k$ individually satisfy Assumption 1. In addition, potential entrants face different entry costs. That is, the entry cost of the jth potential entrant is D_{n+j}. Then we have the following general existence theorem.

THEOREM 8. *Under the assumptions of the model, the general Cournot oligopoly problem with n established firms and k potential entrants facing asymmetric entry costs has a noncooperative equilibrium solution of the form $(\delta_1^*,..., \delta_k^*, q_1^*,..., q_j^*,..., q_n^*, q_{n+1}^*,..., q_{n+k}^*)$, where $0 \leq \delta_j^* \leq 1$, $j=1,...,k$, and $0 \leq q_i^* \leq Q$, $i=1,...,n+k$. q_i^* is the output of the ith established firm and δ_j^*, q_{n+j}^*, respectively, are the entry probability and output of the jth potential entrant.*

Proof. Due to combinatorial reasons, the notation in the proof of the general problem with k potential entrants could get unnecessarily cumbersome. Consequently, we illustrate the proof with only two potential entrants.

Suppose the ith established firm produces q_i, and the jth potential entrant enters with probability δ_j and produces q_{n+j} if it enters; then the expected profits are given by

$$\pi_i = (1-\delta_1)(1-\delta_2)\left[q_i\phi\left(\sum_{j=1}^n q_j\right) - c_i(q_i)\right]$$
$$+ \delta_1(1-\delta_2)\left[q_i\phi\left(\sum_{j=1}^{n+1} q_j\right) - c_i(q_i)\right]$$
$$+ \delta_2(1-\delta_1)\left[q_i\phi\left(\sum_{j=1}^n q_j + q_{n+2}\right) - c_i(q_i)\right]$$
$$+ \delta_1\delta_2\left[q_i\phi\left(\sum_{j=1}^{n+2} q_j\right) - c(q_i)\right], \quad i=1,...,n, \tag{39a}$$

$$\pi_{n+1} = \delta_1(1-\delta_2)\left[q_i\phi\left(\sum_{j=1}^n q_j + q_{n+1}\right) - c_{n+1}(q_{n+1}) - D_{n+1}\right]$$

$$+ \delta_1\delta_2\left[q_{n+1}\phi\left(\sum_{j=1}^n q_j + q_{n+1} + q_{n+2}\right) - c_{n+1}(q_{n+1}) - D_{n+1}\right], \quad (39b)$$

$$\pi_{n+2} = \delta_2(1-\delta_1)\left[q_{n+2}\phi\left(\sum_{j=1}^n q_j + q_{n+2}\right) - c_{n+2}(q_{n+2}) - D_{n+2}\right]$$

$$+ \delta_1\delta_2\left[q_{n+2}\phi\left(\sum_{j=1}^n q_j + q_{n+1} + q_{n+2}\right) - c_{n+2}(q_{n+2}) - D_{n+2}\right]. \quad (39c)$$

The expression for π_i, $i = 1,\ldots,n$, π_{n+1} and π_{n+2} are all continuous functions of $q_1,\ldots,q_n, \delta_1, \delta_2, q_{n+1}, q_{n+2}$. Also for fixed $\delta_1, \delta_2, q_1, \ldots, q_{i-1}, q_{i+1},\ldots,q_n, q_{n+1}, q_{n+2}$, π_i is a concave function of q_i, $i = 1,\ldots,n$. This follows because π_i is a convex combination of four strictly concave functions of q_i for fixed $\delta_1, \delta_2, q_1,\ldots,q_{i-1},q_{i+1},\ldots,q_n, q_{n+1}, q_{n+2}$. Similarly the expressions for π_{n+j} are separately concave in δ_j and q_{n+j} if all other variables other than δ_j and q_{n+j} are fixed, $j = 1, 2$. To see this just rewrite (39b), for example, as

$$\pi_{n+1} = \delta_1\left\{(1-\delta_2)\left[q_{n+1}\phi\left(\sum_{j=1}^n q_j + q_{n+1}\right) - c_{n+1}(q_{n+1}) - D_{n+1}\right]\right.$$

$$\left.+ \delta_2\left[q_{n+1}\phi\left(\sum_{j=1}^n q_j + q_{n+1} + q_{n+2}\right) - c_{n+1}(q_{n+1}) - D_{n+1}\right]\right\}$$

and observe that the expression in the braces is a convex combination of concave functions of q_{n+1} for fixed $\delta_2, q_1,\ldots,q_n, q_{n+2}$, so the expression in braces is concave in q_{n+1} if these variables are fixed. Now to prove the theorem, just observe that the same technique used in the proof of Theorem 6 is applicable. Q.E.D.

In general the solution identified in Theorem 8 depends on $D_{n+1},\ldots,D_{n+k}$ and all the other data of the problem such as $\phi(\cdot)$, n, k, $c_j(\cdot)$ and Q. The equilibrium solutions also partition entry costs into several bands, but some bands may coalesce depending on the data of the problem.

7. Conclusion

We have presented a model of oligopoly which explicitly considers entry barriers and strategic behavior by potential entrants and established firms. Two main results were obtained.

First, we established the existence of a noncooperative equilibrium. At equilibrium, no active firm has an incentive to change production or exit from competition and no firm outside the industry has an incentive to enter. The equilibrium consists of pure strategy outputs for all firms with a possibility that potential entrants may randomize their entry decisions. Our existence proof relied on the Eilenberg–Montgomery fixed point theorem rather than that of Kakutani because optimal response sets are not necessarily convex if potential entrants face non-zero entry costs.

Second, we characterized the equilibrium solutions for a symmetric n-firm oligopoly with one potential entrant and related the equilibrium outcomes to levels of entry costs. Entry costs were endogenously partitioned into three regions; low, moderate and high. When entry costs are low, all firms becomes active as $n + 1$ Cournot oligopolists and when entry costs are high, the potential entrant is excluded, leaving the established firms to operate as n Cournot oligopolists. Uncertainties and subjective factors appear to play a critical role when entry costs are moderate. In this regime, the solutions capture some subtleties in perceptions and behavior. One solution corresponds to a situation where all firms are unsure of each others' intentions and all are cautious; entry is randomized and established firms reduce their outputs. However, if established firms can sufficiently communicate their unwillingness to accommodate entry, an n-firm Cournot solution is enforcible. Likewise, a potential entrant may be accommodated if it can convince the other firms of its intention to penetrate the industry. Thus, the use of threats and other exclusionary tactics become extremely important when entry costs are moderate. The more general problem with finite potential entrants may be analyzed and interpreted in a similar way.

A benefit of the approach used in this paper is that it permits an analysis of strategic behavior in the face of entry barriers. Further insights into the entry problem may be forthcoming if other barriers to entry, such as scale economies and product differentiation, are studied from a game theoretic perspective.

Acknowledgments

We are indebted to Matthew J. Sobel for several suggestions and to Pradeep Dubey for a sketch of the mathematical approach to entry used here.

References

1. E. Burger, "Introduction to the Theory of Games," Prentice-Hall, Englewood Cliffs, N. J., 1963.
2. P. Dubey and M. Shubik, Entry and exit in a closed economic model, unpublished manuscript available from authors, 1978.

3. S. EILENBERG AND D. MONTGOMERY, Fixed point theorems for multivalued transformations, *Amer. J. Math.* **68** (1945), 214–222.
4. J. W. FRIEDMAN, "Oligopoly and the Theory of Games," North-Holland, Amsterdam, 1977.
5. J. J. GABSZEWICZ AND J. P. VIAL, Oligopoly "a la Cournot" in a general equilibrium analysis, *J. Econ. Theory* **4** (1972), 381–400.
6. D. W. GASKINS, JR., Dynamic limit pricing: Optimal pricing and uncertain entry, *J. Econ. Theory* **3** (1971), 306–322.
7. M. I. KAMIEN AND N. L. SCHWARTZ, Limit pricing and uncertain entry, *Econometrica* **39** (1971), 441–454.
8. H. NIKAIDO AND K. ISODA, Notes on noncooperative convex games, *Pacific J. Math.* **5** (1955), 807–815.
9. W. NOVSHEK AND H. SONNENSCHEIN, Cournot and Walras equilibrium, *J. Econ. Theory* **19** (1978), 223–266.
10. K. O. NTI, "Oligopoly with Potential Entrants," Ph. D. Dissertation, Yale University, December, 1977.
11. K. O. NTI AND M. SHUBIK, "Entry in Oligopoly Theory: A Survey," S. O. M. Working Paper No. 17, Series A, Yale University, 1979.
12. J. ROBERTS AND H. SONNENSCHEIN, On the existence of Cournot equilibrium without concave profit functions, *J. Econ. Theory* **13** (1976), 112–117.
13. M. SHUBIK, "Strategy and Market Structure," Wiley, New York, 1959.
14. A. M. SPENCE, Entry, capacity, investment and oligopolistic pricing, *Bell. J. Econ.* **8** (1977), 534–544.

PART III

GAMING

SOME EXPERIMENTAL NON-ZERO SUM GAMES WITH LACK OF INFORMATION ABOUT THE RULES*†

MARTIN SHUBIK

Cowles Foundation for Research in Economics, Yale University

A discussion of the relationship between the theory of games and experimental gaming is presented. This includes comments on games of indefinite length and lack of knowledge concerning the rules. Six simple games are constructed and examined in the light of four different solution concepts. These games were used in an experiment with a class of Yale seniors as the subjects. The results of these experiments are discussed. They appear to lend weight to the non-cooperative equilibrium concept of solution.

1. The Formalization of a Game

Two disciplines, Game Theory and Experimental Gaming have grown up in the past few years. Unfortunately a confusion exists which often causes the misinterpretation of results and constructs of the (relatively mathematical) methodology of Game Theory with the results of Experimental Gaming. Game Theory has given great impetus to work in Experimental Gaming, but often the most interesting games do not fit comfortably into the game theoretic models unless care is taken to make explicit modifications to the assumptions leading to the definition of a game in the sense of von Neumann and Morgenstern.

In this paper, the simplest type of games are treated and the modifications of the von Neumann and Morgenstern assumptions are noted when the games are used for experimental purposes.

There are two types of representation of a game employed by von Neumann and Morgenstern. They are respectively known as the *extensive form* and the *normalized form*. The first can be displayed by means of a *game tree* and the second by means of a *payoff matrix*. A simple example serves to indicate the difference. Consider the very simple parlour game where each of two players has only one move. They must make this move simultaneously without any knowledge of each other's action. Each must select either a red or a black chip. If both players select the same color the first player wins $1, if they select different colors, the second player wins $1. In each instance the losing player must pay the $1 to his competitor.

The payoff matrices to players 1 and 2 are given below in their familiar form.

Payoff to Player 1

	Red	Black
Red	$1	−$1
Black	−$1	$1

Fig. 1

Payoff to Player 2

	Red	Black
Red	−$1	$1
Black	$1	−$1

Fig. 2

---

* Received February 1961.

† Research undertaken by the Cowles Foundation for Research in Economics under Task NR-047-006 with the Office of Naval Research.

Equivalently the two matrices can be combined into one as is indicated below. The first figure in each cell is the payoff to the first player and the second figure is the payoff to the second player.

	Red		Black	
Red	$1,	−$1	−$1,	$1
Black	−$1,	$1	$1,	−$1

FIG. 3

The second type of representation of this game can take two forms which are equivalent in the strict sense of game theory (but which may be psychologically viewed as being different).

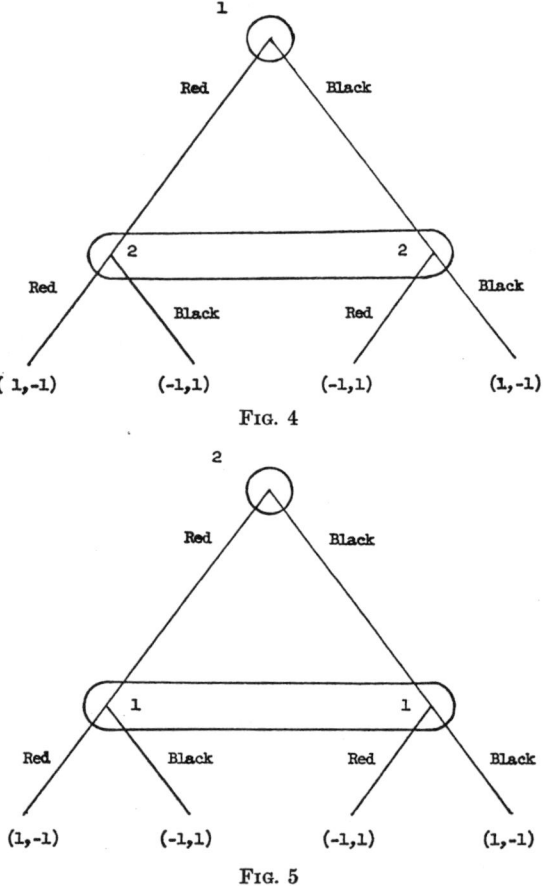

FIG. 4

FIG. 5

In Figures 4 and 5, the vertices in the game trees represent the points at which a player must make a choice. The number next to each vertex indicates which player must make the choice. Thus in Figure 4 the topmost vertex is a choice point for the first player and the other two are choice points for the second, while in Figure 5 it is vice versa. There are four terminal vertices at the end of which are pairs of numbers which represent the payoffs to the players. The first number in each case represents the payoff to the first player and the second the payoff to the second player. Each branch in the game tree represents an alternative. In this case, red or black. In Figure 4 two of the choice points of Player 2 are encircled together and the one choice point of Player 1 has a circle around it. In Figure 5 the two choice points of Player 1 are encircled together and the one choice point of Player 2 is encircled. These enclosures represent *information sets*. If a set of choice points are encircled together this signifies that although the reader acting as *Deus ex Machina* or the referee can distinguish between them, the player confronted with a choice is unable to do so. The game portrayed in Figure 4 has Player 1 move first, then Player 2 moves *in ignorance* of what Player 1 has done. If this is so then the game is equivalent to the game in which they both move simultaneously. The game portrayed in Figure 5 has Player 2 move first, after which Player 1 makes his move *in ignorance* of what Player 2 has done, hence this game is also equivalent to the game in which they move simultaneously.

The games portrayed in Figures 6 and 7 are very different games. In the first case the advantage is all to the second player and in the second case, the advantage is all to the first player. The information conditions have been changed as is exhibited in Figure 6 by the separate encirclements of the choice points for Player 2. He can now distinguish between them. This means that he knows what Player 1 has done prior to making his own move. This obviously gives him a complete advantage in this simple game. If the first player selects red, the second chooses black and wins. If the first player selects black, the second chooses red and wins. The same analysis with the roles of the players interchanged holds for the game illustrated in Figure 7. They are by no means equivalent games. The change in information conditions has had a considerable effect. The *normalized forms* of the two games illustrated in Figures 6 and 7 will be respectively a

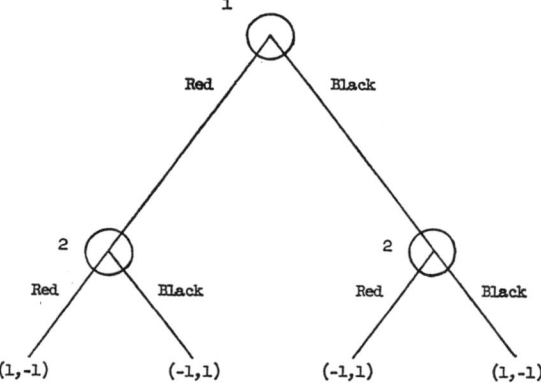

Fig. 6

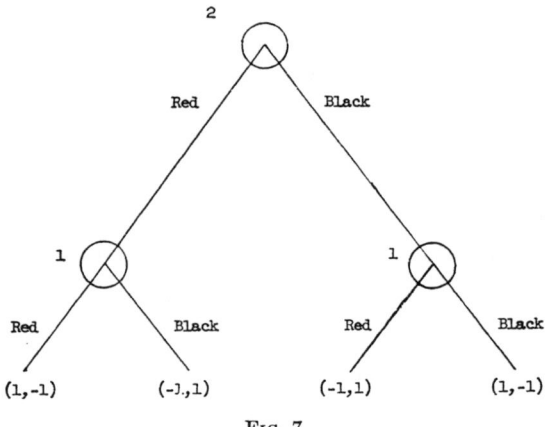

Fig. 7

2 x 4 matrix and a 4 x 2 matrix. The reader who wishes to verify this is referred to the excellent exposition by Luce and Raiffa.[1]

The type of game defined and dealt with by von Neumann and Morgenstern is a game of *finite length*. All the rules are fully defined to all players. All the payoffs are known to all players; and the worth of every payoff to each player is known. Furthermore the payoff comes at the end of the game.

Within the rules of the game is a complete description of what constitutes a possible choice under all circumstances. Furthermore, the state of information available to the players at any point during the game is implicitly contained within the rules. For example in chess the rules are such that at any point during the game both players are completely informed as to the disposition of all their pieces. Chess is a game with *perfect information* available at every move. This does not hold true for Poker. The players are not perfectly informed, they do not know the cards their competitors hold. Nevertheless the players in Poker are assumed to be totally cognizant of the rules.

One of the major confusions prevalent concerning Game Theory is the confusion caused by a failure to distinguish between a *lack of knowledge of the rules of the game* and the lack of knowledge of a player in a game where the rules specify that moves may have to be made *without perfect information*. In the first instance it would be as though one were to play Poker without knowing what defined a winning hand. In the second instance, the game of Poker where all players know the rules serves to provide an example.

In most instances of interest to the behavioral scientist a valid model of a social process is one in which the players are ignorant of the rules of the game. Thus the von Neumann and Morgenstern theories (zero-sum and non-zero sum) of how individuals should act are not usually germane. Modifications must be made if a theory of behavior in these pseudo-games or games with lack of information about the rules, is to be developed. Much (although by no means all) of the confusion and misunderstanding in the abortive attempts to apply the

---

[1] Luce, R. D., and H. Raiffa, *Games and Decisions*, (New York: Wiley, 1957), Ch. 3.

concepts of the Theory of Games to political bargaining and threat phenomena[2] springs from the misunderstanding of information conditions and the meaning of the rules of a game.

Most life processes have an uncertain point of termination. Even the inveterate gambler seated at Las Vegas playing roulette does not know when he will go broke or when he will die. With any luck the second stochastic process will get him before the first one does. Theoretically he has a chance to keep going forever. At least at any day when he is still alive and still in the chips there is usually a fair sized chance that he will be still alive and still in the chips on the morrow. It is precisely the possibility that at any point in an infinite game there is a possibility that there will be a tomorrow that makes it the natural vehicle for the study of threats and counter-threats. The players in a game without a calculable finite termination must always take into account the possibility that they will have to live with each other on the morrow. In life there are rarely once and for all settlings of debts; in finite game models there are. This is possibly why certain games of finite length provide a "mathematician's delight" in the form of many paradoxical solutions in the light of human behavior.

The simplest type of game is the one represented by a 2 x 2 matrix as is shown in Figures 1, 2 or 3. For experimental purposes it is convenient to have subjects play this type of game several times. In doing so it is important to remember that in fact the players are now playing in a "super-game" in which the individual games are sequentially arranged segments of the overall situation. Suppose we played the game illustrated in Figure 4 two times instead of once. The game tree for this new game is shown in Figure 8. It is still a finite game tree, but is much larger than before. In spite of the increase in size it is a well-defined game in the von Neumann and Morgenstern sense.

If, instead of telling the players that they are to play a simple game say, 5 or n times, where n is a given fixed number the players are instructed that they will play many times but that their last play will be determined randomly, then this is no longer a game in the Von Neumann and Morgenstern sense. If the players are told the distribution of probabilities with which the game will be stopped on any trial, then it is a well-defined dynamic, stochastic or infinite game (the name for this type of game has still not been completely standardized). There exists a body of literature on this type of game.[3]

If the players are not even told how the last play is to be determined, they are then forced to supply their own assumptions concerning the termination of play. This is another form of lack of knowledge concerning the rules of the game. Attempts have been made to cope with situations where the players do not know the probabilites of certain events which are needed to well-define the game. The writings on *Games Against Nature*[4] deal with this. In dynamic situations consideration must be given to learning about the rules when the players

---

[2] For example see the incorrect normalizations of games presented in: Schelling, T. C., *The Strategy of Conflict* (Cambridge: Harvard University Press, 1960).

[3] Dresher, M., A. W. Tucker and P. Wolfe (eds.) *Contributions to the Theory of Games*, Vol. III, (Princeton: Princeton University Press, 1958).

[4] Savage, L. J., *The Foundations of Statistics*, (New York: Wiley, 1954).

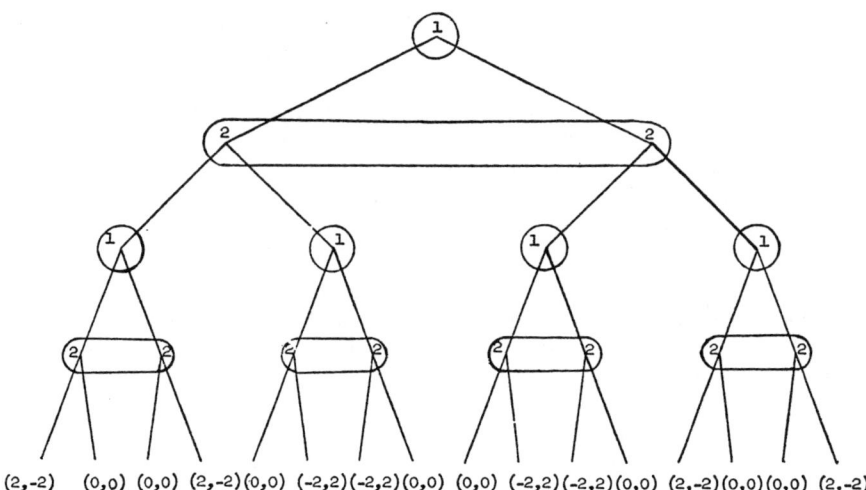

(2,-2) (0,0) (0,0) (2,-2)(0,0) (-2,2)(-2,2)(0,0) (0,0) (-2,2)(-2,2)(0,0) (2,-2)(0,0)(0,0) (2,-2)

FIG. 8

are not completely informed initially. This has given rise to investigations of game learning.[5]

2. Theories of Solutions to a Game

Section 1 dealt with the formalization of the description of a game. An economic analogy would be the formalization of the description of a market structure. Nothing was suggested about how the game will be or should be played. A theory for the solution to a game (as defined by von Neumann and Morgenstern or in a modified version) is either a normative theory or is based upon description of human behavior. The work in experimental gaming is a step in an attempt to validate theories which purport to be descriptive, or to produce descriptive theories.

Restricting the discussion to games of finite length with the rules known by all of the players, several different concepts of solution are noted. For ease we limit ourselves to two-person games. Let $P_1(s_1, s_2)$ stand for the payoff to the first player if he employs his strategy s_1 and the second player employs his strategy s_2; and similarly $P_2(s_1, s_2)$. At this point, the distinction must be made between *strictly competitive* and not strictly competitive games. The game displayed in Figure 3 is strictly competitive. An increase in the welfare of one player implies a decrease in the welfare of the other. In this instance we observe that:

$$P_1(s_1, s_2) = -P_2(s_1, s_2).$$

This relation does not hold for the game illustrated in Figure 9. Here there is room for cooperation which will yield rewards to both. It can be observed that

---

[5] Flood, M. M., "On Game-Learning Theory and Some Decision Making Experiments," R. M. Thrall, C. H. Coombs and R. L. Davis (eds.) in *Decision Processes*, (New York: Wiley, 1954); and more recently the work of S. Seigel and other experimental psychologists.

this is a non-constant sum game; i.e., the sum of the payoffs to both players in different outcomes is not constant. In the first game the sum is constant.

	1	2
1	5, 2	−10, −13
2	4, 1	−20, −23

Fig. 9

von Neumann and Morgenstern suggest that in a non-constant sum game the players should jointly maximize and then work out some arbitrated division of the proceeds between them. This presupposes that they are in a position to communicate with each other and are also in a position to make side-payments. They do not explicitly include the bargaining and haggling over side-payments as part of their description of the play of the game, but as something which takes place outside of it. The description of the behavior of the players in the game is given mathematically by the condition:

[1] $$\text{Max.}_{s_1} \text{Max.}_{s_2} (P_1(s_1, s_2) + P_2(s_1, s_2)).$$

This merely states that each player should select his strategy in such a manner that the sum of their payoffs is maximized. Applied to the game in Figure 9 this calls for each player to select his strategy 1.

John Nash has suggested a theory of non-cooperative play[6] which is a generalization of economic theories of equilibrium. His theory applies to situations where communication is limited between the players and they are not in a position to make side-payments. Nash shows that in any finite game which can be described by a set of payoff matrices (such as 1 and 2, or using a more compact notation, 3) there will exist at least one pair of strategies s_1^* and s_2^* such that the two conditions

[2] $$s_1^* = \text{Max}_{s_1} P_1(s_1, s_2^*)$$
$$\text{and} \quad s_2^* = \text{Max}_{s_2} P_2(s_1^*, s_2)$$

are simultaneously satisfied by the choice of s_1^* and s_2^* by the first and second players respectively. In words, if the first player believes that the second player will utilize s_2^* against him, his optimal reply (in the sense that it will maximize his own payoff) is s_1^* and vice versa.

It is possible that both players may have a very pessimistic view of the world and strive to play in a manner that minimizes the worst that can happen. In other words, on the assumption that the competitor is hostile each may assume that the other is going to minimize his competitor's payoff and each will strive

---

[6] Nash, J. F., Jr. "Non-Cooperative Games," *Annals of Mathematics*, LIV (September, 1951), pp. 286–295.

to maximize his payoff on that assumption. This can be expressed as:

[3]
$$\text{Max}_{s_1} \text{Min}_{s_2} P_1(s_1, s_2)$$
and
$$\text{Max}_{s_2} \text{Min}_{s_1} P_2(s_1, s_2).$$

Another possibility is that the competitors adopt the attitude that it is more important to maximize the difference in gain between them than it is to maximize individual gain. This is the type of thinking prevalent in tactical calculations of damage exchange rates. This is expressed as:

[4]
$$\text{Max}_{s_1} \text{Min}_{s_2} (P_1(s_1, s_2) - P_2(s_1, s_2)).$$

3. Some Experimental Games and Their Theoretical Solutions

The following six games were played by five pairs of students (Yale seniors in a class on Industrial Organization) in order to illustrate the interrelationship between behavior and structure in a market, as well as to gather data on their behavior.

Game 1

	1	2
1	6, 3	6, 7
2	10, 3	10, 7

FIG. 10

Game 2

	1	2
1	1, 3	2, 3
2	1, 1	2, 1

FIG. 11

Game 3

	1	2
1	2, 1	−1, −1
2	−1, −1	1, 2

FIG. 12

Game 4

	1	2
1	3, 3	−1, −1
2	−1, −1	2, 2

FIG. 13

Game 5

	1	2
1	3, 3	−2, 7
2	7, −2	−1, −1

Fig. 14

Game 6

	1	2
1	5, 2	−10, −13
2	4, 1	−20, −23

Fig. 15

In Figure 16 all four solution concepts given in Section 2 have been applied to the six games given above and the resultant strategy pairs which are the solutions are noted. For example the expression (1,1) stands for the strategy pair where each player selects his first strategy, i.e., $s_1 = 1$ and $s_2 = 1$. This gives the pair of payoffs in the upper left-hand corner of the payoff matrix.

	Solution [1]	Solution [2]	Solution [3]	Solution [4]
Game 1	(2,2)	(2,2)	(2,2)	(2,2)
Game 2	(1,2)	(1,1) (1,2) (2,1) (2,2)	(1,1) (1,2) (2,1) (2,2)	(2,1)
Game 3	(1,1) or (2,2)	(1,1) or (2,2)	* {2/5,3/5} and {3/5,2/5}	(1,2)
Game 4	(1,1)	(1,1) or (2,2)	* {3/7,4/7} and {3/7,4/7}	(1,1) (1,2) (2,1) (2,2)
Game 5	(1,1)	(2,2)	(2,2)	(2,2)
Game 6	(1,1)	(1,1)	(1,1)	(1,1) (1,2) (2,1) (2,2)

Fig. 16

* These both involve mixed strategies. The probabilities employed by each player are indicated in the curled brackets.

All four solution concepts when applied to Game 1 yield the same solution pair (2,2). A closer examination of the game shows the structural reason why this is so. The players are strategically independent. Their fates are not interlinked. This game illustrates the atomistic isolation between any two competitors in a purely competitive market. Regardless of their intentions, the structure is such that their behavior is the same in all instances. A non-constant sum game in which the fates of the players are not interlinked is referred to as an *inessential game*. It is inessential in the same way as is a strictly competitive game. There is nothing to be gained by discussion, negotiation or collusion. In essence, collusion has no meaning in this context.

In Games 3, 4 and 6 the jointly maxima and the non-cooperative solutions coincide illustrating the inherent aspects of implicit collusion built into the structures of these games. In Game 4 the strategy pair (1,1) is the only non-cooperative solution in the strict sense.[7]

The solution concepts may be regarded as *operators* which act upon the payoff functions (which reflect structure) to produce *behavior*. As is illustrated here, it is possible for many different operators to produce the same behavior, depending upon the structure. The operator embodies the intent of the player.

4. A Hypothesis and the Conditions of the Experiment

Indications from other work[8] suggest that the non-cooperative concept of solution will serve as the best predictor of modal behavior of the four suggested here, in a pseudo-game situation, i.e., not a well-defined game in the von Neumann and Morgenstern sense, but one with modifications. These modifications are that the games be played under conditions of incomplete information concerning the rules. These were manifested in two ways.

The players were only informed of their own payoff functions. They did not know the payoff functions of their competitors. Partners were chosen randomly and no communication was permitted except via a monitor who transmitted the information concerning the choice of each player after every play of the sub-game.

The players were initially not informed about the number of plays of the subgames; this was done to avoid certain aspects of so called "end effects," especially the possibilities for lengthy "backward induction." Nor were they given any probability distribution for the ending. The trial lengths were selected prior to the runs and were announced two to three trials before termination. The students were given approximately one minute per trial.

The solutions illustrated in Section 3 were for the single-period subgame. These games when played for a finite number of trials give rise to a supergame which is said to have *perfect recall*.[9] It has been shown that for such games it is legitimate to solve them by solving the series of sub-games individually,[10] thus for one of the games presented in Section 3, if it were played n times under conditions of complete information concerning the rules, the various solution concepts would call for a strategy pair in each subgame consistent with the strategy pairs predicted for a single play of the original 2 x 2 matrix game.

[7] Luce, R. D., and H. Raiffa, *Op. Cit.*, p. 107.

[8] The work referred to here is "Some Experimental Games" by Merrill Flood in *Management Science* Vol. 5 pp. 5–26, a Princeton Ph.D. Thesis by David Stern, (1959); "An Experimental Business Game" by Austin Hoggatt in *Behavioral Science*, Vol. 4, 1959; The business game of George Feeney and some preliminary investigations of quantity-variation Duopoly by L. Fouraker, M. Shubik and S. Siegel.

[9] Luce, R. D., and H. Raiffa, *Op. Cit.*, p. 161.

[10] *Ibid 6*, p. 162.

For the pseudo-games used in this experiment there exist no mathematically developed theories of solution, unless we wish to postulate that a theory of solution to a well-defined finite or infinite length game will apply them. By itself such a theory should not be enough as obviously learning must be taking place. The lengths of play were selected with consideration given to the difficulties in learning and the need for the players to learn in order for them to perform optimally in a game. This is discussed in Section 5 and the Appendix.

The data used to test the hypothesis that the non-cooperative solution theory is the best predictor among the four theories were the last five trials of each game.

The hypothesis is not fully defined until meaning is given to the word "best." Every solution concept will predict the mode correctly, have no resolution power or be incorrect. On the basis that: yes $>$? $>$ no, solutions can be ranked for all games. A vector domination can then be used for comparing solutions applied to several games. This is a rather weak condition; but is used here. For some purposes (especially if there are high stakes in the game) a loss function measure is needed.

	Strategy Pair	Pair 1	Pair 2	Pair 3	Pair 4	Pair 5	Prediction of Solution [1]	[2]	[3]	[4]
Game 1	(1,1)	—	—	—	—	—				
	(1,2)	—	—	—	—	—				
	(2,1)	—	—	—	—	—				
	(2,2)	5	5	5	5	5	×	×	×	×
Game 2	(1,1)	—	—	—	—	—		×	×	
	(1,2)	—	—	—	1	5	×	×	×	
	(2,1)	2	—	3	2	—		×	×	×
	(2,2)	3	5	2	2	—		×	×	
Game 3	(1,1)	—	—	—	5	5	×	×	6/25	
	(1,2)	—	—	—	—	—			4/25	×
	(2,1)	—	—	—	—	—			9/25	
	(2,2)	5	5	5	—	—	×	×	6/25	
Game 4	(1,1)	⑤	5	5	5	3	×	×	9/49	×
	(1,2)	—	—	—	—	1			12/49	×
	(2,1)	—	—	—	—	1			12/49	×
	(2,2)	—	—	—	—	—			16/49	×
Game 5	(1,1)	—	—	—	—	—	×			
	(1,2)	—	—	—	—	—				
	(2,1)	3	1	1	—	1				
	(2,2)	2	4	4	5	4		×	××	×
Game 6	(1,1)	5	4	5	5	5	×	×	×	×
	(1,2)	—	—	—	—	—				×
	(2,1)	—	①	—	—	—				×
	(2,2)	—	—	—	—	—				×

Fig. 17

The players were instructed to maximize their individual scores. As there was no monetary incentive involved, doubts can be raised as to the effectiveness of this instruction. In one instance in a final play, one subject played in a manner to inflict a loss on his competitor. Upon being questioned after he indicated that as it was the last trial of the last game he thought that he would teach his competitor a lesson. Beyond that, however, no end of the game pathologies were observed. The time series for the last five trials of all pairs in the fifth game are given in the Appendix.

Figure 17 gives the results for the last five trials. The frequencies encircled must be rejected for reasons indicated in the comments of the players given in the Appendix.

Examining Figure 16 we observe that all four theories are of equal power on Game 1 and are in complete agreement with the results. Given the nature of the game this is to be expected if the players behave rationally; hence the result can be regarded as a check of the rationality of the players.

In game 2 solutions 1 and 4 must be rejected, while solutions 2 and 3 have no power of resolution and hence are at least not inconsistent with the data. In this game there appears to be an attraction to the lower right hand corner of the matrices which could be due to geometric considerations. It would be desirable to rerun this game with the rows or columns interchanged.

In game 3, solutions 1 or 2 account for all the data. Solutions 3 and 4 must be rejected.

In game 4, solutions 1 and 2 are consistent with the data of four pairs (the data from pair #1 had to be rejected). The data from pair #5 shows some variability. The interview material in the Appendix indicates that the first player in this pair was attempting to damage his opponent even at cost to himself. This stresses the desirability of adequate incentives. Solutions 3 must be rejected and 4 has no resolution power.

In game 5 the players indicated that they had difficulty in learning. Solution 1 must be rejected; all the other solutions are acceptable.

In game 6 solutions 1, 2 and 3 are in complete agreement with the data; solution 4 has no power of resolution.

		Solutions			
		1	2	3	4
	1	Yes	Yes	Yes	Yes
	2	No	?	?	No
Games	3	Yes	Yes	?	No
	4	Yes	Yes	?	?
	5	No	Yes	Yes	Yes
	6	Yes	Yes	Yes	?

We can see that for at least one game, in the strict sense of the testing of the hypotheses that any of these theories of solution is consistent with the data, all must be rejected as frequencies appear when none are predicted. A learning (and teaching) modification needs to be introduced. Section 5 discusses the perceptions of the players for their opponents' payoff matrices; this gives some indication of the state of learning and teaching by the end of the game. For the weaker hypothesis concerning the prediction of mode, the table on p. 226 presents the data analysis. A "?" indicates that the solution has no resolution power.

5. Some Aspects of Learning

At the end of the games, the subjects were required to rank the entries in their opponents' matrices for each game. The table for the perceived matrices in Game 1 is given below. The entries are ranked a > b > c > d. The letter in the upper left hand corner is the value in the actual matrix. The five entries below are the perceived entries.

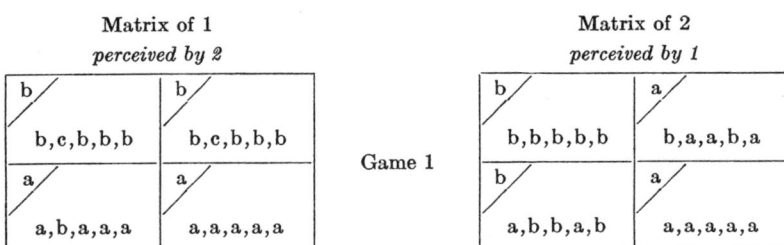

Game 1

We note that the players' perceptions of each others' matrices were accurate in eight out of ten cases. In the two instances in which the first player attributed an incorrect structure to his opponent's matrix (pairs 1 and 4), the misperception would make no difference to non-cooperative behavior. The players had nothing to gain by correcting their misperceptions.

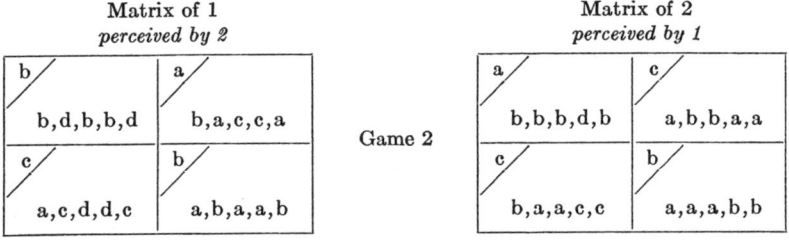

Game 2

In game 2 in only two instances out of ten the structure of the competitor's matrix was "guessed" by a player (the second players in pairs 2 and 5). The game was designed so that no information concerning an individual's payoffs was provided by knowing his move.

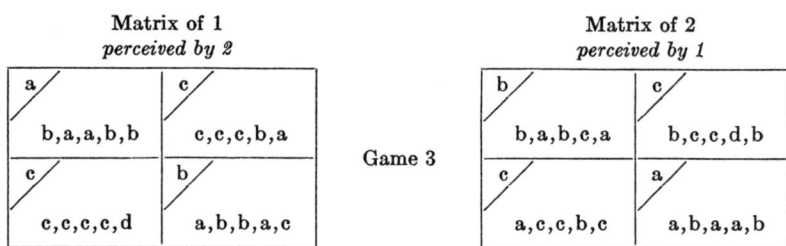

In game 3 three players correctly perceived their opponents' matrix. Four others misperceived the structure in a manner peculiarly in their favor! This game was selected to have two equilibrium points, one of which favored one player and the other, the other player. The misperceptions of four of the players were based on the belief that the strategy pair which resulted in an equilibrium favoring them also contained the most desirable outcome for their competitors.

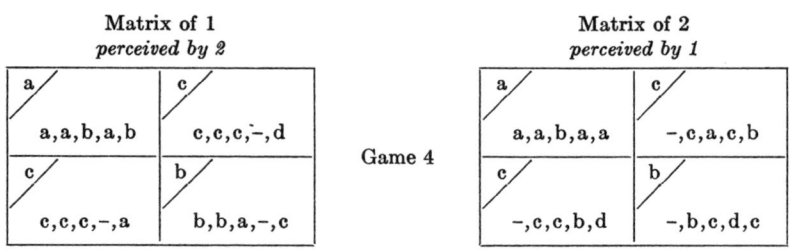

In Game 4, seven players correctly surmised that the upper left entry was the best for both themselves and their competitors. In two instances the players made no further exploration and the information concerning one entry in their opponent's matrix was sufficient for playing, no other estimate of the structure was given.

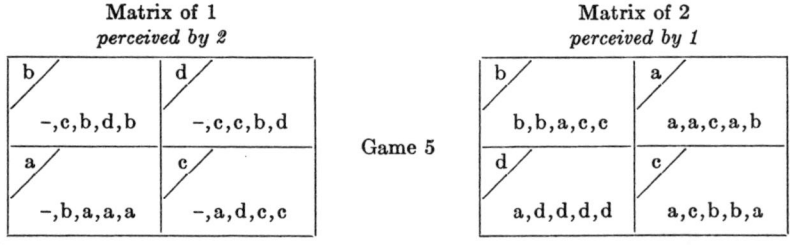

Game 5 was the classical "Prisoner's Dilemma" game. The sole equilibrium point is jointly minimal. Two players, (one from pairs 2 and 3) had an accurate perception of the competitor's matrix; six were able to locate the opponent's maximum, and five, the minimum.

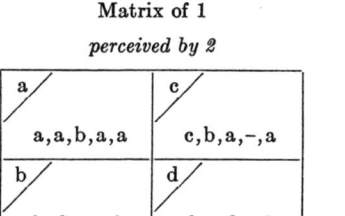 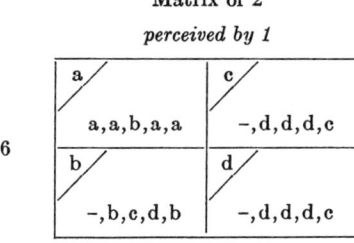

Game 6

In the sixth game, eight players out of ten correctly anticipated that the optimum for their competitor coincided with their own. A ninth player (pair 3, the second player) predicted a matrix whose structure would result in the same behavior in spite of his misperception.

As can be seen from some of the comments (given in the Appendix) there is not only an individual learning process, but a communication process taking place as well. These signals are closely connected to the as yet rather unsatisfactory concept of *threat* strategy.

6. Conclusions

There are great difficulties to be faced in attempting to use experimental games *in vitro* to learn about human processes *in vivo*. It is the belief of this writer that there may be many types of learning which are worth distinguishing from each other. The environment of the simple experimental game probably destroys many of the factors contributing to the acceptance of stable standards of behavior in societies or even in industries and trade associations. Nevertheless the games appear to provide a promising tool to help to separate out variables in the study of bargaining, threats and other aspects of competition and cooperation.

In particular the relationship among theories of games, learning and gaming need further clarification. Many of the proponents and opponents both of the theory of games and of gaming have failed to appreciate the need for extreme care in interpreting and relating the axioms behind the various theories of games to the experimental conditions.

Appendix

I. *Formats and Time Series*

The games were played for 10, 12, 15, 17, 22 and 10 trials each. Games 1 and 6 require the least learning; game 5 the most. The following two tables are the actual formats used for game 5. In order to save space, the sheet for Player 1 has been truncated and the time series for the last five moves of the five sets of players have been entered in the table for this game. The untampered format is given for Player 2.

Player 1   GAME 5

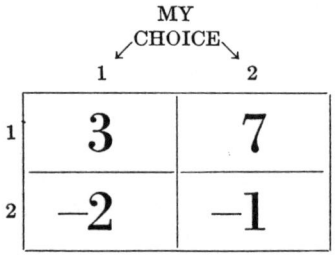

Trial Number	My Move					His Move					My Payoff
1	2	1	1	2	1	1	2	2	2	2	
2	2	2	2	2	2	1	2	2	2	2	
3	2	2	2	2	2	1	2	2	2	2	
4	2	2	2	2	2	2	2	2	2	2	
5	2	2	2	2	2	2	2	2	2	2	

Player 2   GAME 5

MY CHOICE

	1	2
1	3	7
2	−2	−1

Trial Number	My Move	His Move	My Payoff
1			
2			
3			
4			
5			
6			
7			
8			
9			
10			
11			
12			
13			
14			
15			
16			
17			
18			
19			
20			
21			
22			
23			
24			
25			

II. Comments of the Players

The players were required to comment on all games after they had played all six. These comments are given below. The comments marked with an * occasionally make life for the experimenter an unrewarding experience.

Pair 1, Player 1

Game 1: I started out hoping for the best and it turned out all right.

Game 2: I hoped I could get to the intent of 1. I felt the opposition had a rather simple matrix, with no great feeling between one or the other. I decided to stick with strategy 2 and let him settle on something.

Game 3: Seeing my opponent's dislike of strategy 1, I shifted to strategy 2 to take a steady profit of 1.

Game 4: The proctor told me to stick with 1, so I did and the game turned out very well indeed.

Game 5: Not quite able to figure this out. Stuck with two after unsuccessful attempts to learn something. Sought to minimize loss but wound up slightly ahead.

Game 6: Stuck with the best choice 1 and it turned out as I had hoped—No thought on opponent's strategy except that I must have been best.

Pair 1, Player 2

Game 1: My opponent's strategy could also have a reason other than the matrix I gave as my guess: a matrix which looks like the one of game 6. Analogous reasoning for game 6.

Game 2: This game seems to be biased in favor of player one.

Game 3: His strategy 2 seems to offer the same as my strategy 2: possible losses less than possible gains.

Game 4: We were in agreement: no need to change strategy, our best point seemed to be 1/1 for both of us.

Game 5: 2/2 is presumably his best point.

Pair 2, Player 1

Game 1: Both maximize at 2—no problem—his best move is 2 as is mine.

Game 2: Again we both maximize at 2—move indifferent.

Game 3: We both maximize on 2 although my best move is 1.

Game 4: Easy—both same matrix and we do the same.

Game 5: Very difficult—my best move is 1, his best is 2; we minimize our losses at 2 each.

*Game 6: Easy—same matrix, same choice, +1 each; I screwed him on the last move to make him lose much, me little.

Pair 2, Player 2

Game 1: In agreement from the start—I played my best and he played his.

Game 2: He played one presumably because it was his best but he had two:one in mind; when I played 1 he switched to 2 because he felt it was his best in that situation. He then found me switching and played 2:2 as better than 1:2.

Game 3: We seemed to jibe pretty well after the start. I think his payoffs were the reverse of mine.

Game 4: Hit it off at the start. 1:1 was his best or he would have switched at least once.

Game 5: Thoroughly confused.

Game 6: Hit it off at the start

Pair 3, Player 1

Game 1: 2 was my best move and 2 was also his best move—result immediate accord.

Game 2: We soon settled on 2 and 2 as best for us both, but he displayed a touch of irrationality (or just "playing around") at the end.

Game 3: 1 my best—he displayed irrationality again (it seems) after staying on 1 for 3 turns, he suddenly switched to 2 and wouldn't be budged—strange.

Game 4: We soon settled on a saddle point with 1,1—no trouble.

Game 5: He wouldn't budge from 2 (no matter how hard I tried to "force" him) so I had to content myself with minimizing my losses.

Game 6: We hit the saddle on the first try and never deviated.

Pair 3, Player 2

Game 1: Can't lose if I move 2, win same regardless of opponent's move, opponent can't lose if play 2, win same regardless of opponent's move. Me -2 always, opponent -2 always.

Game 2: Game beyond me. His safest 2, his best 2, 2.

Game 3: He wanted me to play 1 for 1, his 1 together is his best, second best is 2,2. I psyched him, although losing for a while to take second best, in end he came to me where 2,2 was my best.

Game 4: 1+ (on assumption either losses on any other combination would be same or proportional) was a matter on who could convice other to play their way. I won.

Game 5: Convinced him I'd never play 1. He took third best, me 2, him 1. For long run game both lost by not compromising.

Game 6: You both win a great deal by playing 1 no matter what opponent does. Both lose by playing 2 no matter what opponent does.

Pair 4, Player 1

Game 1: Choice 2 was equally profitable whether he chose 1 or 2. I stayed at 2 and maximized my return.

Game 2: My object was to try to elicit a response of 2 from opponent. He bounced equally with mine alternating 1,2. Seeing this I reversed his pattern getting a response of 2,2 on the last two moves.

Game 3: The choice 1,1 was my most profitable and 1,2 was my least loss. I stayed on 1 most of the time. He moved around trying to get me to change. I did move but settled down. It appeared that he settled down to minimize losses.

Game 4: We settled on 1,1 and no variation proving it was probably most mutually profitable point.

Game 5: I failed in trying to get him to give a response of 1. I got a few 1 in the beginning but when I went to 2 he went to 2 and would not be deterred.

Game 6: We settled on 1,1, both fearing any variation.

Pair 4, Player 2

Game 1: 2 is optimum move for me regardless of what opponent does. Since he made no attempt to see what I would do if he chose 1, he also must have enjoyed 2.

Game 2: My strategy here was to get player 2 to choose 1. I had no preference as to what I chose. He seemed to prefer choosing 2. I don't see much rhyme or reason to him.

Game 3: Player 2 only chose 2 twice, each time I think he expected me to play 1.

Game 4: Player 2 seemed perfectly happy with 1,1. We never tried anything else.

Game 5: Neither liked 2,2 or 1,1. This amounted to various probing until we figured out that the best either of us could do was 2,2.

Game 6: Again, 1,1 was the universal choice.

Pair 5, Player 1

Game 1: I believe his matrix is something like this; my payoff was 100.

4	8
4	8

Game 2: Matrix here is like this:

−1	4
2	−1

and my payoff is 21.

Game 3: Matrix 3:

2	1
−1	1

and my payoff is 9.

Game 4:

6	2
−2	1

My payoff was 21. Here I tried to make my opponent move from (1,1) because I figure that was his best position. I could have stayed with (1,1) if I had wished, with my payoff a steady 3.

Game 5:

−1	4
−2	5

Due to some of my opponent's mistakes, my nose stayed slightly above water in this fixed matrix. Payoff 3.

Game 6:

5	−1
3	−1

My payoff 50.

Pair 5, Player 2

Game 1: Both possibilites in choice 2 for my opponent were better than either possibility in choice one. The same was true for me, so we were locked into 2,2. We were cooperating for our mutual gain.

Game 2: Again we were cooperating after we found common ground. The first four tries were experimental and again we settled on one set of moves mutually beneficial.

Game 3: Cooperation again after solution was found. Neither of us was getting maximum profits but both were winning nevertheless.

Game 4: A battle all the way. He was continually trying to get me to play one so that he could play 2. 1,1 was suitable for me but he had a chance for great returns on 1,2 and so he was continually trying to get it.

Game 5: 2,2 was a slight loss to him. 1,1 was great gain so he was trying to get me to play 1 so that he could also.

Game 6: 1,1 was mutually agreeable although it may have not been maximum profit for him it was better than risking great loss.

[29]

"So long sucker"— a four-person game

M. HAUSNER, J. NASH, L. SHAPLEY, AND M. SHUBIK

This parlor game has little structure and depends almost completely on the bargaining ability and the persuasiveness of the players. In order to win, it is necessary to enter into a series of temporary unenforceable conditions. This, however, is usually not sufficient; at some point it may be to the advantage of a player to renege on his agreement. The four authors still occasionally talk to each other.

This game was invented in 1950 by Messrs. M. Hausner, J. Nash, L. Shapley, and M. Shubik. The aim was to produce an interesting, social game in which coalitions are both profitable and unstable. Technically, it is an essential four-person, no-side-payment game, in extensive form, with perfect information and no chance moves after the first. It has been played extensively in gatherings of different sorts, provoking a wide variety of reactions. The authors will welcome further reactions and comments.

Rules
1. A four-person game.[1]
2. Each player starts with 7 *chips*[1] (playing cards, or other

---

[1] For a longer game, more chips may be used. If the game is attempted

markers may be used instead), distinguishable by their color from the chips of any other player. As the game proceeds, players will gain possession of chips of other colors. The players must keep their holdings in view at all times.

3. The player to make the first move is decided by chance.
4. A *move* is made by playing a chip of any color out onto the playing area, or on top of any chip or pile of chips already in the playing area.
5. The *order of play*, except when a capture has just been made, or a player has just been defeated (Rules 6 and 9) is decided by the last player to have moved. He may give the move to any player (including himself) whose color is not represented in the pile just played on. But if all players are represented in that pile, then he must give the move to the player whose most-recently-played chip is furthest down in the pile.
6. A *capture* is accomplished by playing two chips of the same color consecutively on one pile. The player designated by that color must kill one chip, of his choice, out of the pile, and then take in the rest. He then gets the next move.
7. A *kill* of a chip is effected by placing it in the "dead box."
8. A *prisoner* is a chip of a color other than that of the player who holds it. A player may at any time during the game kill any prisoner in his possession, or *transfer* it to another player. Such transfers are unconditional, and cannot be retracted. A player may not transfer chips of his own color, nor kill them, except out of a captured pile (Rule 6).
9. *Defeat* of a player takes place when he is given the move, and is unable to play through having no chips in his possession. However his defeat is not final until every player holding prisoners has declared his refusal to come to the rescue by means of a transfer (Rule 8). Upon defeat, a player withdraws from the game, and the move *rebounds* to the player who gave him the move. (If the latter is thereby defeated, the move goes to the player who gave *him* the move, etc.)
10. The chips of a defeated player remain in play as prisoners, but

---

with more than four players, then the number of chips per player should be reduced.

are ignored in determining the order of play (Rule 5). If a pile is captured by the chips of a defeated player, the entire pile is killed, and the move rebounds as in Rule 9.
11. The *winner* is the player surviving after all others have been defeated. Note that a player can win even if he holds no chips and even if all chips of his color have been killed.
12. *Coalitions*, or agreements to cooperate, are permitted, and may take any form. However, the rules provide no penalty for failure to live up to an agreement. Open discussion is not restricted, but players are not allowed to confer away from the table during the game, or make agreements before the start of the game.

[30]

A NOTE ON A SIMULATED STOCK MARKET

MARTIN SHUBIK, *Yale University*

ABSTRACT

This article describes the design of a stock market exercise to be run with a business game (or a simulated business game). It reports on an exercise in which around 100 traders participated in a market, trading four stocks. The results obtained are reported: however, the work was primarily a feasibility study to see if very large group behavior could be studied in a controlled experiment. The experience gleaned from the experiment indicates that this is a fruitful approach.

HISTORY

In 1959 the General Electric Marketing Game designed by George Feeney[1] was run at the General Electric Management center at Crotonville. The marketing game provides as a background a simulated oligopolistic market with four firms of equal strength. The optimal size for an active team appears to be no more than three or four (although over ten individuals tried to run one team by majority vote, with a fair amount of chaos resulting). As there were well over sixty people available it became apparent that not all could participate actively in the game. It was decided that the surplus of available participants presented an opportunity both to illustrate to all the details of the functioning of a stock market and to investigate the feasibility of experimenting with a stock market game. As there was no previous experience whatsoever many errors were made in the setting up of the controls over this exercise. Furthermore the market was somewhat "thin." This experience and the subsequent one both indicate that the optimum size for a stock market is above 100 participants, possibly around 150.

The market or business game required about 20-25 minutes per period to run. A full day's run can cover between 16-20 periods.

On January 20, 1960 the same game was run at the Stanford Research Institute. There were between 80 to 100 participants. The market game ran all day and the stock market opened at 9:13 a.m. and ran until 4:35 p.m. covering 14 out of the 20 periods of play. Both members of the individual firms and the

---

*Research undertaken by the Cowles Commission for Research in Economics under Contract Nonr-3055(01) with the Office of Naval Research.

[1] Feeney, G. J., 'Simulating Marketing Strategy Problems', *Marketing Times*, 2(1), January 1959, pp. 8–24.

"trading or investing public" held shares. This already introduced a degree of complication which is not necessary unless one has an experiment designed to consider differences between corporate insiders and the general public. A far more important aspect of experimental control was made clear by both plays of the game. If the institutional problems in running the stock market game do not make it imperative to actually run a business game simultaneously to generate the inputs to the stockmarket, then a much more satisfactory procedure is to use a "canned game," i.e., the outputs from a previously played game or a set of constructed outputs from a fictional game chosen by the experimenters. Among the very important advantages associated with such a procedure is that the experimenter becomes immune to machine failures and to delays in the market game play. New information may be posted every 10 minutes and the stockmarket game may proceed at the speed of 5-6 periods per hour. Furthermore the experimenter will be able to control the non-zero sum aspects of the market in which the firms operate and hence be more careful about the briefing given to the investors or traders.

In the game which was run at the Stanford Institute the participants were sold shares at $1.25 per share which were at a premium of 59 cents over asset liquidation value at that time. They were briefed that the profits of the industry as a whole would grow sufficiently that the issue price (on the average) would be equalled or exceeded. Furthermore they were briefed that at the end of the exercise all shares would be redeemed in direct proportion to the asset values of the firms (the non-zero sum aspect of the game meant that the "house" stood to pay for any growth that had occured). The details of the specific stockmarket structure and the briefing are given in the Appendix. Four members of the San Francisco stock exchange were invited and came to give advice on the versimilitude of the exchange and to operate the four trading posts.

Short sales were permitted, however another and apparently unnecessary source of distortion was introduced when the players were instructed that those who were caught short at the end of trading would have to pay a penalty of twice the final liquidation price of the shares. Although this might realistically reflect the effect of a corner, experimentally it appears as an unnecessary complication. In the actual play there were very few short sales relative to long purchases. In any future run it seems to be reasonable to permit short sales and to require a cover at final market price.

In *toto* 428 shares were issued (some in units of less than 8 in spite of the initial briefing) and in the course of play 449 shares were traded in a total of 174 trades, which were broken down as follows:

Table 1

	Bid	Offer	Trade	Volume
Company 1	67	64	40	124
Company 2	50	61	39	130
Company 3	49	57	44	94
Company 4	64	68	51	101
			174	449

The stock market opened at the third period of play of the game. The first two periods were preprogrammed and designed to give both the members of the firms and the participants some feeling for the nature of the market. It had been intended to run the stock market up to the last period in the functioning of the business game however owing to machine failures and other administrative difficulties the stock market closed at the 16th period although the liquidation values were nevertheless based upon the 20th period. This introduced another source of "noise" in the game. Owing to time constraints people knew that the stock market had to close somewhat before 5 p.m.

PLAYERS AND GOALS

The players were upper middle and upper echelon managers from firms associated with and using Stanford Research Institute as well as some staff members of S.R.I. and faculty members of Stanford. They used their own money to buy the initial issue of the shares and appeared to understand the money-making possibilities available to them.

USES FOR THE GAME, HYPOTHESES AND ANALYSIS OF DATA

A simulated stock market appears to provide an environment for the experimental study of trading and expectations. Dividends, stock splits, information time lags, brokers fees and the other myriads of institutional details are not present. The value of the stock at the close of the exercise can be calculated in a manner that is a stock market analyst's dream. The financial reports are the same for all companies and cannot be used to conceal or confuse. All trading is done directly to the individual's own account.

The stock market in this simplified form serves as a stage for the study of expectations or anticipations. If the individual were able to judge the abilities of the different firms he would be in a position to adjust his portfolio accordingly.

Two hypotheses were made. The first concerns the relationship between stock market price and asset value.

Hypothesis 1

The market average will be always above the average asset value of the firms.

This is a relatively weak hypothesis for the start as the subscription price for the shares was 59 cents more than the then asset value. The hypothesis becomes more interesting when the average asset value of the firms is greater than $1.25.

The second hypothesis concerns the spread between stock price and asset value as related to the growth of the firms.

Hypothesis 2

(a) The ratio of initial to final average (first 4 and last 4 periods) "spread" between asset value and market price will be greatest for the most profitable firm and (b) least for the least profitable firm.

This hypothesis is in keeping with the "growth stock" psychology that has caused 60 and 80 times price earnings ratios to be not uncommon for "glamor issues."

There are many other hypotheses primarily involving dynamics and the time series aspects of expectations; however the two hypotheses noted were the only ones formulated prior to examining the data. A more careful running of a similar but better controlled stock market is called for before much faith could be attached to the results of a time series analysis.

The stock market price and asset information are presented in Table 2 below. The prices quoted are for the last trade of the period. When two prices are given such as 1.10 - 1.25 this indicates the bid offer spread in a period during which no trade took place. In one instance (Company 4, period 3) there was an offer but no bid. This is indicated by −1.25.

The final asset values (rounded to the nearest five cents) which were used to determine the liquidation worth of the shares were:

155	175	100	135

these may be compared with the final trades and the asset values at the end of the 16th period which were respectively:

160	170	100	150
127	144	97.9	138

The following five graphs display the information in Table 2. Whenever there was a bid-offer spread the midpoint was plotted.

Table 2

Quarter	Company 1		Company 2		Company 3		Company 4		Average	
	Assets	Price	Assets	Price	Assets	Price	Assets	Price	Assets	Price
3	69.02	.80-.85	69.99	.80	69.78	1.25	69.42	1.25	69.5	
						1.40				
4	69.85	.95	71.40	1.00	70.85	1.15	70.66	.95	70.8	101.3
5	71.57	1.05	71.76	1.05	71.17	1.10	71.23	1.00	71.6	106
		1.20				1.25				
6	73.26	1.20	73.01	1.00	72.31	1.10	72.53	1.10	72.8	110
7	75.37	1.25	75.66	1.20	74.35	1.10	74.68	1.05	75.0	115
8	77.73	1.25	78.15	1.45	77.46	1.10	78.28	1.20	77.9	125
9	82.42	1.25	82.77	1.50	78.69	1.10	82.21	1.25	81.5	127.5
10	89.15	1.40	89.87	1.50	80.64	1.05	86.61	1.20	86.6	128.7
11	96.22	1.75	97.41	1.65	83.40	1.00	88.93	1.10	91.5	137.5
12	103.21	1.75	106.00	2.15	86.29	1.05	97.08	.90	98.1	153
		1.90						1.30		
13	100.65	1.20	118.42	2.00	88.87	1.15	110.59	1.25	104.6	143
				2.25						
14	108.00	1.30	129.85	1.75	82.72	1.00	123.39	1.50	111.0	142
								1.75		
15	119.32	1.40	137.52	2.10	93.16	1.00	130.78	1.50	120.2	151
		1.50								
16	127.71	1.60	144.48	1.70	97.94	1.00	137.95	1.50	127.0	145
issue price	1.25		1.25		1.25		1.25			
asset value at issue	68		68		68		68			
final asset value	1.55		1.75		1.00		1.35			
stock last trade price	1.60		1.70		1.00		1.50			
assets value at period 16	1.27		1.44		.979		1.38			

A simple inspection of Figure 5 shows immediately that Hypothesis 1 is confirmed. Furthermore the average price at period 16 was higher than the final liquidation value 4 periods later! If we wish to consider that the market discounts the future it apparently did so by between 6 to 8 periods as can be seen by inspecting Figure 5.

It is of interest to note that from period 4 to 20 the assets of the firms approximately doubled or were growing at approximately 17% per annum.

In Table 3 the per period growth and the spread between the stock prices and asset values for the four firms is presented. The first column gives Assets(t) − Assets $(t−1)$ and the second the spread.

Inspection of Table 3 shows that the first part of Hypothesis 2 is confirmed here. The average spread actually widened for the major growth stock. The second part of the hypothesis was also confirmed. The least profitable firm by Period 16 showed the narrowing of average spread. There are obviously considerable serial effects and on any repeated run an appropriately sophisticated statistical analysis is called for.

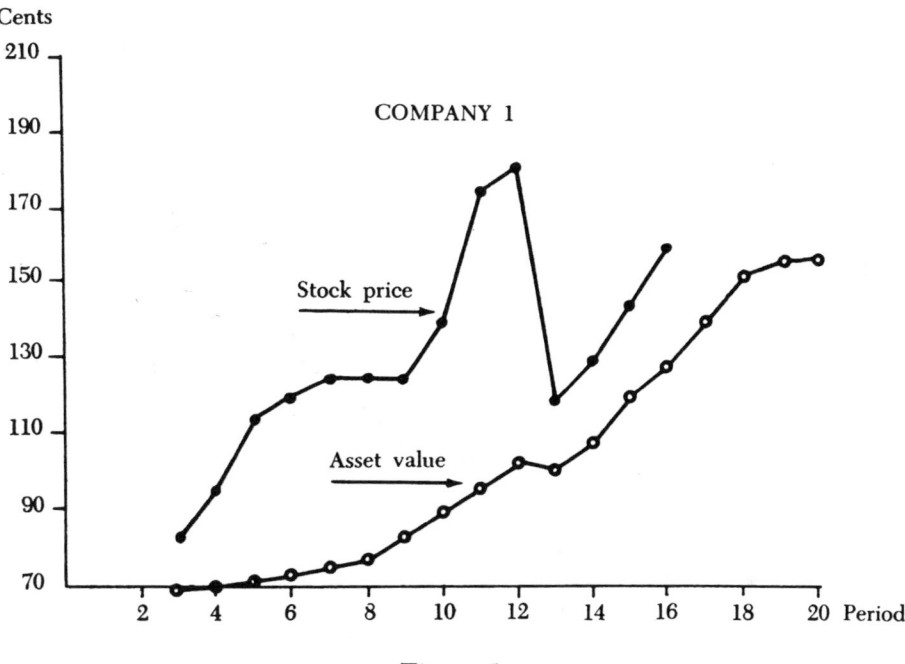

Figure 1

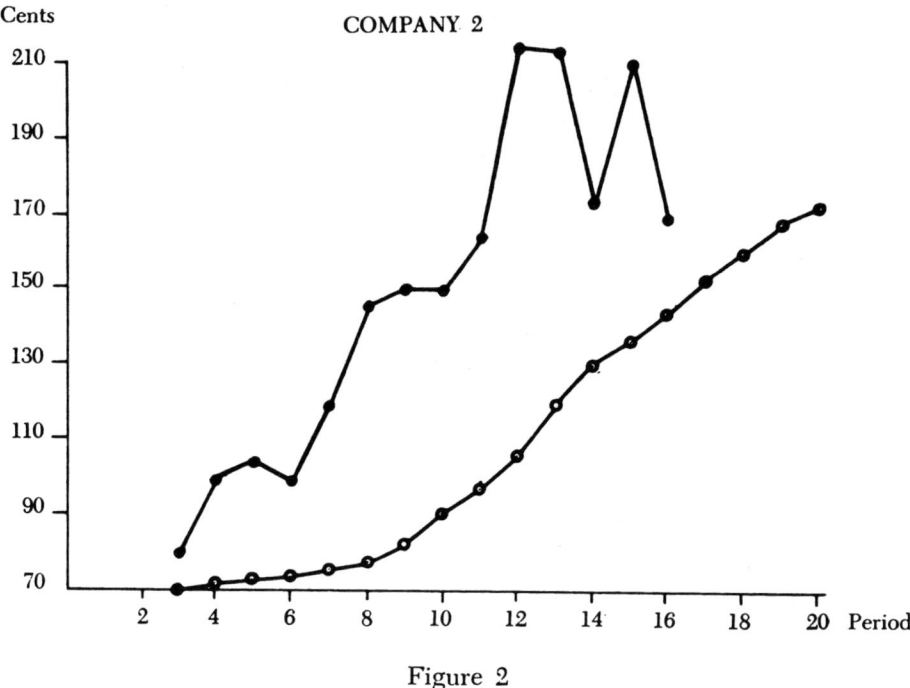

Figure 2

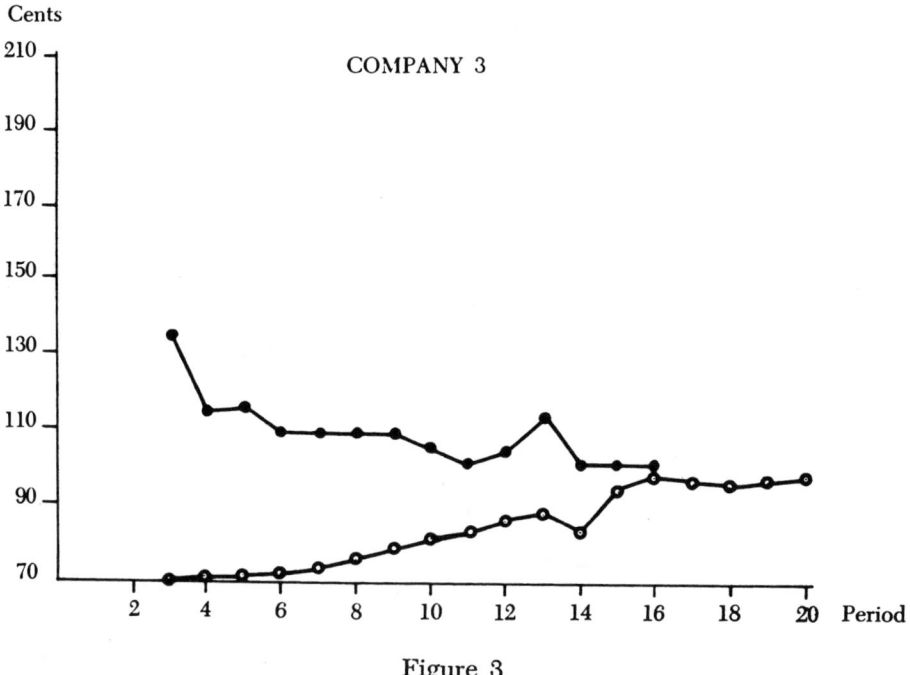

Figure 3

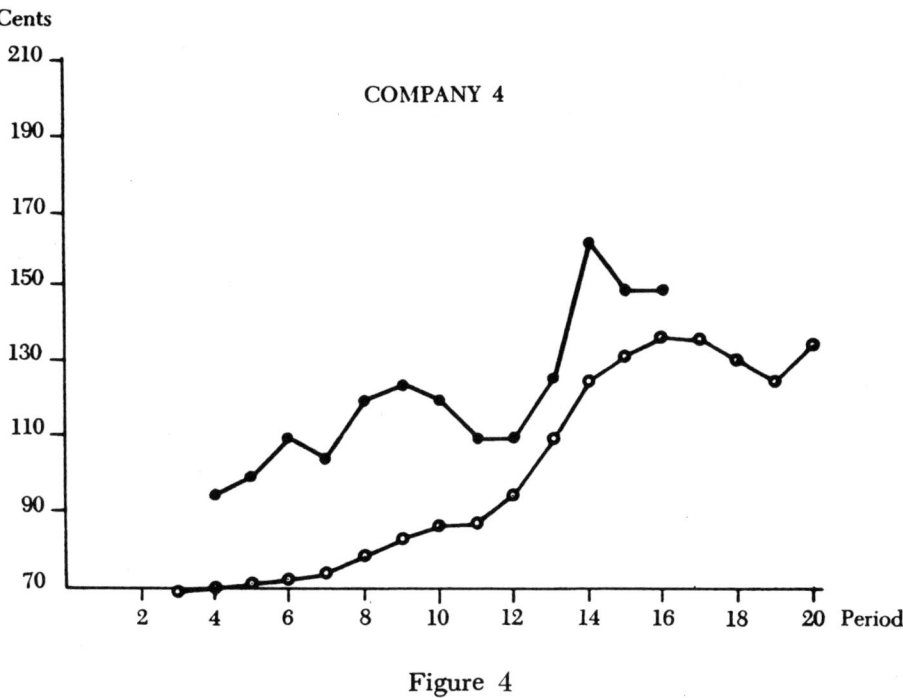

Figure 4

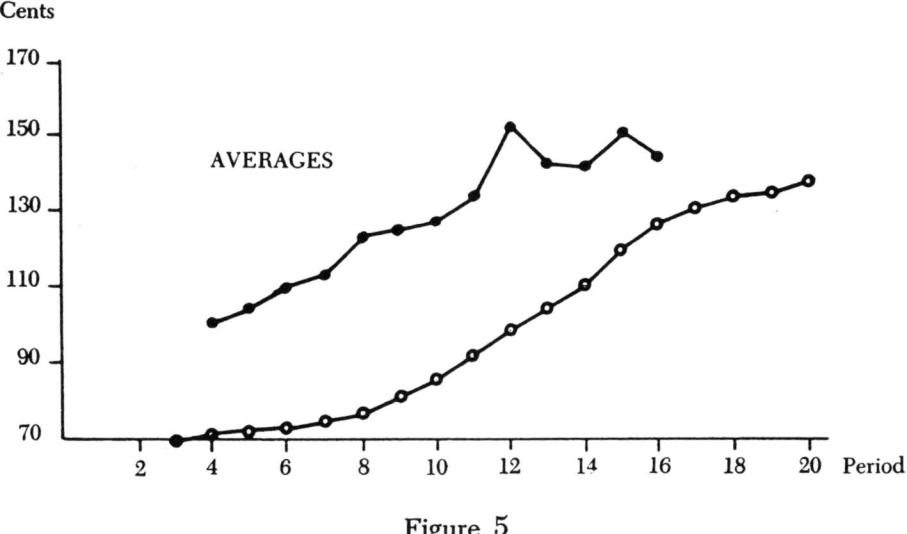

Figure 5

It should be noted that in Period 13 a panic developed in the trading of the stock of Company 1. It fell around 60 cents or by about 1/3 immediately after losses were announced following five periods of good growth. People actually charged from the bulletin and information boards and mobbed and fought to get to the trading post of Company 1 during this period. *Ex post facto* this was precisely the time to have bought the stock of Company 1.

Table 3

Growth of Assets* and Spread Between Stock Price and Asset Value

Quarter	Company 1		Company 2		Company 3		Company 4	
	Growth	Spread	Growth	Spread	Growth	Spread	Growth	Spread
3	1	13	2	10	-2	62	1	—
4	1	25	1	29	1	44	2	24
5	2	40	1	33	0	46	0	29
6	1	47	1	27	1	38	2	37
7	2	50	3	46	2	36	2	30
8	3	47	2	67	3	33	3	42
9	4	43	5	67	2	31	4	43
10	7	51	7	60	2	24	5	33
11	7	79	7	68	2	17	2	21
12	7	79	9	109	3	19	8	13
13	-2	19	9	94	3	26	14	14
14	7	22	12	45	-6	17	12	39
15	11	25	8	72	10	7	8	19
16	9	32	6	26	5	2	7	12
$\frac{\text{final spread}}{\text{initial}}$	$\frac{108}{125}$ = .86		$\frac{237}{99}$ = 2.39		$\frac{52}{190}$ = .27		$\frac{84}{120}$ = .70‡	
Assets at Period 16	128		144		98		138	

*Both sets of figures are rounded off to the nearest cent per share. As there were no dividends the first column gives: Earnings$(t-1)$ = Assets(t) − Assets $(t-1)$.

‡The missing first period spread was taken as the average of the other three at the very worst it could have been 56 cents (difference between asset value and offer).

CONCLUSIONS

The results from the analysis of the data are slim, but somewhat suggestive. The major purpose of this exercise was to explore the possibility of using a simulated stock market for gaming to study individual mass behavior (the trading floor was more a mass of individuals than a large group, except during the panic when it came close to being a mob).

It appears that the game is feasible and although not easy, neither very hard nor costly to run. The main problem is to be able to find more than 100 participants for a period of 4-6 hours. Around 8-10 people are needed to operate the game (4 traders, 1 or 2 for a clearing house, a floor president and 2 or 3 to post information and spell others).

The game not only has research value but after the debriefing on both occasions it was run there were comments concerning the added insights obtained into the role and functioning of a stock market.

It appears that not only should a prepared market be used as the input to the stock market (so that around 30 periods could be covered within 6 hours); but that it might be of interest to use the appropriately edited information together with verbal releases on four or five firms. Furthermore it might be of considerable interest to play this game with a group of participants such as the New York Society of Market Analysts.

APPENDIX

DESCRIPTION OF AND INSTRUCTIONS FOR A SIMULATED STOCK MARKET GAME

Relation To The Planning Simulation Exercise

The business game utilized in the planning simulation exercise serves as a data generation source for the stock market game.

The stock market game provides a model of an "ideal" stock market where all traders are completely informed about all firms whose stock are being traded; and in which all "frictions" such as brokerage fees, taxes, transfers costs, etc., are zero.

The addition of the stock market to the planning simulation exercise permits a much larger active participation than would otherwise be possible. It furthermore serves as both an exercise and experiment in market evaluation and expectations.

Description Of Simulated Stock Market Game

The Exchange

The Exchange is a large room with four trading posts each equipped with a large pad and easel upon which bids, offers and trades can be prominently displayed. In the room there is a series of graphs indicating the states of the four companies. These are up-dated every quarter. The yearly plan of each company is also displayed and up-dated every year.

Specialists

Four specialists act as the major focal points for trading. Each maintains a market in one stock. They are permitted to trade to their own accounts if they wish to do so. They may also subscribe as investors to the initial issue.

Shares and the Initial Issue of Shares

The shares issued by each company are in the form of bearers' certificates of seventy cents per value. The actual form of these certificates is shown below. The subscription price for the initial issue is $1.25 per share.

The initial offering is in lots of eight (8) shares of their own companies to officers in the four companies. To all other investors, the offering is in lots of eight (8) shares consisting of two (2) shares in each of the four companies.

Trading Including Short Sales

Trading is carried out only at the four specialists' posts. Short sales are permitted and when they take place a signed note is issued indicating the number of shares and who was short to whom. The type of the note is shown below:

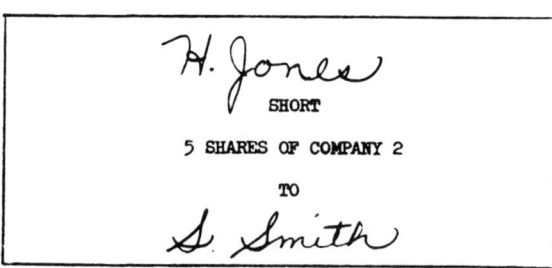

Trading is permitted in units of .05 of a dollar.

The specialists maintain records as indicated below. Bids, offers, trades, times, quantities and prices are noted. In the

example given, the first bid has been satisfied by the second and fourth offers. The third bid placed at 8:49 was withdrawn at 8:51.

A short sale offer is denoted by an asterisk. The third offer is for a short sale.

At the end of each quarter, the specialist will note the time in the TRADE column. In this example QUARTER 1 ended at 8:52. A quarter ends as soon as information concerning a new quarter commences to be posted.

COMPANY 1

	BID				OFFER				TRADE			
No.	Amt.	Price	Time	No.	Amt.	Price	Time	Nos. B	Nos. O	Amt.	Price	Time
1	5	.85	8:40	1	3	.90	8:30	1	2	2	.85	8:42
2	10	.70	8:48	2	2	.85	8:41	1	4	3	.85	8:50
3	3	.85	8:49 8:51	3	2	1.00*	8:43			QUARTER 1		8:52
4				4	3	.85	8:50					
5				5								
6				6								
7				7								
8				8								
9				9								
10				10								

Value of Shares at End of Exercise

At the end of the game the shares are evaluated in direct proportion to the final asset standing of each company. For example, if the initial assets of a company were $2,000,000 and the par value of the shares were 70 cents, then at the end, if the terminal assets were $4,000,000, the final redemption value of the stock would be $1.40.

Any short position outstanding at the end of the play will be required to cover at twice the redemption value of the shares involved. New shares will be issued to shorts at the end of play.

[31]
An Artificial Player for a Business Market Game

MARTIN SHUBIK, GERRIT WOLF, and SCOTT LOCKHART
Yale University

Much work in artificial intelligence is devoted to the construction of computer programs designed to simulate intelligent behavior. However, social intelligence and intellectual intelligence may not be the same things. The term "artificial intelligence" when applied to problem-solving, chess-playing, theorem-proving, or zero-sum games covers only a narrow band of what we usually wish to consider when we talk about intelligence. The type of artificial player needed to play in a nonconstant-sum game may be considerably different from the type of player needed to play chess. The first may do better with a "pleasing personality" and only a second-class intellect. The second needs a first-class intellect.

AUTHORS' NOTE: *This paper was supported by National Science Foundation Grant 2840.*

[27]

ARTIFICIAL PLAYERS AND NONCONSTANT-SUM GAMES

The concept of an artificial player is by no means as new as it may appear to be. The work in oligopoly theory over the last thirty to forty years can really be regarded as preliminary to the construction of such a player. Attempts by Stackelberg and Peacock (1952) and others to describe leader-follower situations and reaction curves can be viewed as attempts to specify the heuristics whereby one would be able to construct the appropriate artificial player.

More recently, under the title of the behavior theory of the firm (Cyert and March, 1960), attention has been paid to specifying heuristics describing the behavior of individuals within the firm. A heuristic is another word for a "rule of thumb." It is our contention that a reasonably competent artificial player can be constructed by connecting a reasonable group of heuristics and tuning the parameters within them.

An artificial player of worth should be expected to function with the same information conditions as a human player. It would be trivially easy to construct an artificial player given the complete structure of the game. Even this is not completely trivial, in the sense that a problem still remains even if all players know the structure of the game in detail. Nevertheless, both in actual economic life and in reasonably interesting gaming experiments, it is usually safe and useful to assume that the players do not know the complete structure of their environment.

The player we construct is assumed to know no more about his environment than does his human competitor. As we shall see in the discussion of the heuristics in the section immediately below, the intelligence of the player is not particularly great, and many of his acts have to be described more in sociopsychological than in intellectual terms. His reactions to his fellow players are probably more important than his ingenuity.

Success in the construction of an artificial player can be measured in several ways. One of the most important is the Turing Test (Turing, 1963), where the criterion used is that the human player is unable to distinguish whether or not he is playing a computer program or another human being. Another test used in this report is how well the artificial player does against the live players, regardless of whether or not they have perceived of him as human. A third criterion is, given that the human players are informed they are playing a program, to ascertain their reactions to playing the program, to find out if, in their opinion, it was possible to fathom its construction, and, furthermore, to find out if the program provided the live players with a challenge sufficient to make the competition neither boring nor trivial.

In the course of the past few years, Levitan and Shubik have constructed and analyzed a business oligopoly game (summarized in Shubik and Levitan, 1970).[1] This game can be played in several modes, but the most common mode has the players make decisions on advertising, price, production, and dividends for a one-product firm. The artificial player that has been designed makes these decisions.

The market structure of this particular game is derived from quadratic utility functions and quadratic payoff functions for the firms. The mathematical details of the market structure need not concern us here. However, it is noted that a detailed analysis of many solution concepts as applied to this market structure is available elsewhere (Shubik and Levitan, 1970).

THE ARTIFICIAL PLAYER

The artificial player makes four decisions each period: price, production, advertising, and dividends. Each decision is made by hierarchically processing a set of heuristic rules. This hierarchy, similar in structure for each decision, has three levels. Working backward in his processing from the first level of the actual decision, the player chooses between a boundary value and a learned value, choosing the first only if the boundary is exceeded. The learned value, the second level, is the product of two variables, an expectation and a personality tendency. Each of these is composed of several variables which are dependent on the last period's decisions, expectations, and personality tendencies. This processing forms the third level.

The artificial player's set of rules represents a decision maker of limited memory and processing ability. Within these limitations, learning occurs both by changing personality tendencies to cooperate or compete, and the memory used for prediction. This memory may be either strong or weak for the other player's decisions from the prior period— making the player a follower or a potential market setter, respectively. This player is like a small-business businessman who adapts from period to period. He only remembers what happened last period.

The artificial player is programmed to play up to ten other players. But because this study is only for a duopolistic market of one human player and the artificial player, player subscripts are omitted in what follows, for clarity. Where other's choice is referred to, in the general case this variable would be the average of all other players in the game. The only subscript used is that for time—t. Also, variables are mnemonically labeled with P preceding a label standing for price, A for advertising, B for production, and D for dividends. The set of rules is presented in hierarchical order for a particular decision, the final level first, with a description of the set following. Price is presented first.

$$P_t = \max(Pmin, Plearned_t) \qquad [1]$$

$$Pmin = \text{profit percent (unit cost + fixed cost/prod capacity)} \qquad [2a]$$

$$\text{Plearned}_t = (\text{Pexpected}_t)(\text{Ptendency}_t) \quad [2b]$$

$$\text{Pexpected}_t = \text{Psmooth} + \text{Ptrend}_t \quad [3a]$$

$$= (1-a-\beta)\,\text{Psmooth}_{t-1}$$

$$+ \text{Pother}_{t-1}\,(a+\beta) + \text{Ptrend}_{t-1}\,(1-\beta)$$

$$\text{Ptendency}_t = (1+\text{Pbias}_t)(1+\text{rv}_t[\text{Pvar}_t+\text{Psd}_t]) \quad [3b]$$

$$\text{Pbias}_t = (\text{Pbias}_{t-1})(\text{Pdecay})$$

rv_t = random variable with uniform distribution between zero and one

$$\text{Pvar}_t = (\text{Pvar}_{t-1})(\text{Pvardecay})$$

$$\text{Psd}_t = \frac{\text{Ppred}_t}{\text{Psmooth}_{t-1}}$$

(see equation 32a)

Working upward through the hierarchy, price personality tendency [3b] has four components, each having values between 0 and +1. Pbias$_t$ reflects a tendency to raise or lower price. This tendency decays exponentially over time through multiplication by a constant between 0 and 1. Pvar$_t$ represents a variability in this tendency. This variability decays similarly and enters in by being added on to a computed variability variable and multiplied by a random variable (normally distributed with values between -1 and +1). The effect of these variables is to produce a personality variable having values near 1.

The expected variable [3a] is the sum of two predictions. One prediction, $Psmooth_t$ is the convex combination of the prior Psmooth and other's price last period. The other prediction, $Ptrend_t$, which measures price trend, is a convex combination of the $Ptrend_t$ last period and the difference between other's price last time and $Psmooth_t$ last time. The α and β weights are between zero and one, and reflect a trade-off between dependence on prior predictions and the dependence on other's price.

Personality tendency and expected price are multiplied at the next level [2b], resulting in a learned price. Through this rule, the expected price relates to last period's price and is modified by personality, a percent upward or downward. The learned price is then compared with the minimum price. The minimum price [2a] is simply based on costs, capacity, and a minimum profit margin. This comparison [1] is the final level, in which the higher of the two is chosen. It prevents the player from choosing a price below which he could not pay his bills.

Advertising follows with a similar pattern:

$$A_t = \max(Amin, Alearned_t) \quad [11]$$

$$Amin = 0 \quad [12a]$$

$$Alearned_t = (Aexpected_t)(Atendency_t) \quad [12b]$$

$$Aexpected_t = a' \, Aexpected_{t-1} + (1-a') \, Aother_{t-1} \quad [13a]$$

$$Atendency = \left(\frac{[1+Abias_t]}{[1+v_t \, Avar_t]} \right) \quad [13b]$$

$$Abias_t = (Abias_{t-1})(Adecay)$$

$$\text{Avar}_t = (\text{Avar}_{t-1})(\text{Avardecay})$$

v_t = random variable with uniform distribution between zero and one

Advertising is simpler than price, in that it has only three personality components, no trend analysis, and a simple minimum of zero. The personality tendency [13b] has a tendency to raise or lower advertising and a variability factor, both decaying with time at a rate depending on the nearness of the decay constants to one. The Aexpected$_t$ variable [13a] is a convex combination of last period's expectation and other's advertising last period. Personality multiples expectation by a value near one, producing Alearned$_t$ [12b]. If this variable is greater than zero, it is selected as the advertising for that period [11].

Dividends has a familiar pattern.

$$D_t = \max(0, \text{Dlearned}_t) \qquad [21]$$

$$\text{Dlearned}_t = \text{Dexpected}_t - \text{Dtendency} \qquad [22]$$

$$\text{Dexpected}_t = \text{cash balance}_t - \text{loans}_t \qquad [23a]$$

$$\text{Dtendency} = (1+\rho^1+\rho^2)M(\text{constants}) \qquad [23b]$$

The production has a slightly different pattern and is computed after price and advertising

$$B_t = \min(\text{Production Capacity}, \text{Max } [B_t^1, B_t^2]) \qquad [31]$$

$$B_t^1 = (\text{sales}_t)(\text{demand}_t) - \text{inventory}_t \qquad [32a]$$

$$\text{demand}_t = \min(2,\ 1.5\ \frac{\text{Ppred}_t}{\text{Psmooth}_{t-1}} + 1)$$

$$\text{Ppred}_t = \theta\,\text{Ppred}_{t-1} + (1-\theta)$$

$$\text{Pother}_{t-1} - \text{Psmooth}_{t-1}$$

$$B_t^2 = 2(\text{sales}_t - \text{inventory}_t) \qquad [32b]$$

$$\text{sales}_t = \frac{\text{Sexpect}_t\ \text{Pexpected}_t\ \text{Alearned}_t}{\text{Plearned}_t\ \text{Aexpected}_t}$$

$$\text{Sexpect}_t = \text{Sactual}_{t-1}\,(2\text{-}a''\text{-}\beta'') + \text{Strend}_{t-1}\,(a''\beta'') + \beta''$$
$$\text{Ssmooth}_{t-1}$$

Simply, the player chooses the greater of two estimates. The one estimate, B_t^2 is the difference between inventory and predicted sales, multiplied by two, because only half of one's production can be sold each period. The other estimate, B_t^1, modifies predicted sales, by a predicted demand which may be such to reduce production. Predicted sales has a similar form to that of the previously discussed predicted price or advertising. If the greater of B_t^1 and B_t^2 is greater than the player's capacity, he produces his capacity. Otherwise, he chooses the larger of B_t^1 or B_t^2.

This completes a discussion of the player's strategies.

METHOD

The procedure for testing the artificial player included setting the artificial player's parameters and initial values, and pitting a varied sample of decision makers against the player, using an on-line version of an oligopoly game (Shubik and

Levitan, 1970). Simply, the game was set to be competitive in price but not in advertising. The human and artificial players are described as follows:

The price parameters and initial values are $Pbias_0 = .1$; $Pvar_0 = .1$; $Pdecay = .7$; $Pvardecay = .7$; $Ptrend_0 = 0$; $Psmooth_0 = \$240$; $Pother_0 = \$240$; $\alpha = .5$; $\beta = .5$. Initially, the player is competitive, but this decays slowly. His dependence is split between prediction and what other did last time. The advertising parameters are: $Abias_0 = .05$; $Avar_0 = .05$; $Adecay = .95$; $Avardecay = .98$; $Aexpected_0 = \$6,000,000$; $Aother_0 = \$6,000,000$; $\alpha' = .5$ and $\beta^1 = .5$. The decay is very slow and the bias is very small, which means advertising will not change much. Dividend requires values for ρ^1, ρ^2, and M. These are: $\rho^1 = .01$; $\rho^2 = .01$; and $M = 0$. The production parameters are: $Ssmooth_0 = 200,000$; $Strend_0 = 0$; $Sactual_0 = 200,000$; $'' = .5$; $'' = .5$; $Ppred t_0 = \$240$; and $\theta = .7$.

Two kinds of human decision makers played the game: fourteen students and eight faculty. All were from the Administrative Sciences Department or Economics Department. Formal training in economics was not measured. A player would come to the small groups laboratory and be given computer and minimal economic instructions. The time at the computer console to play twelve periods was about an hour to an hour and a half. After playing, each decision maker was debriefed.

How the players performed is reported in the next section.

RESULTS

The basic raw data for each of twelve periods and each of twenty-two players are price, advertising, dividends, and production decisions and outcomes of sales, gross profit, net profit, and inventory. It is not reasonable or necessary in this first look at the performance of players in this game to look

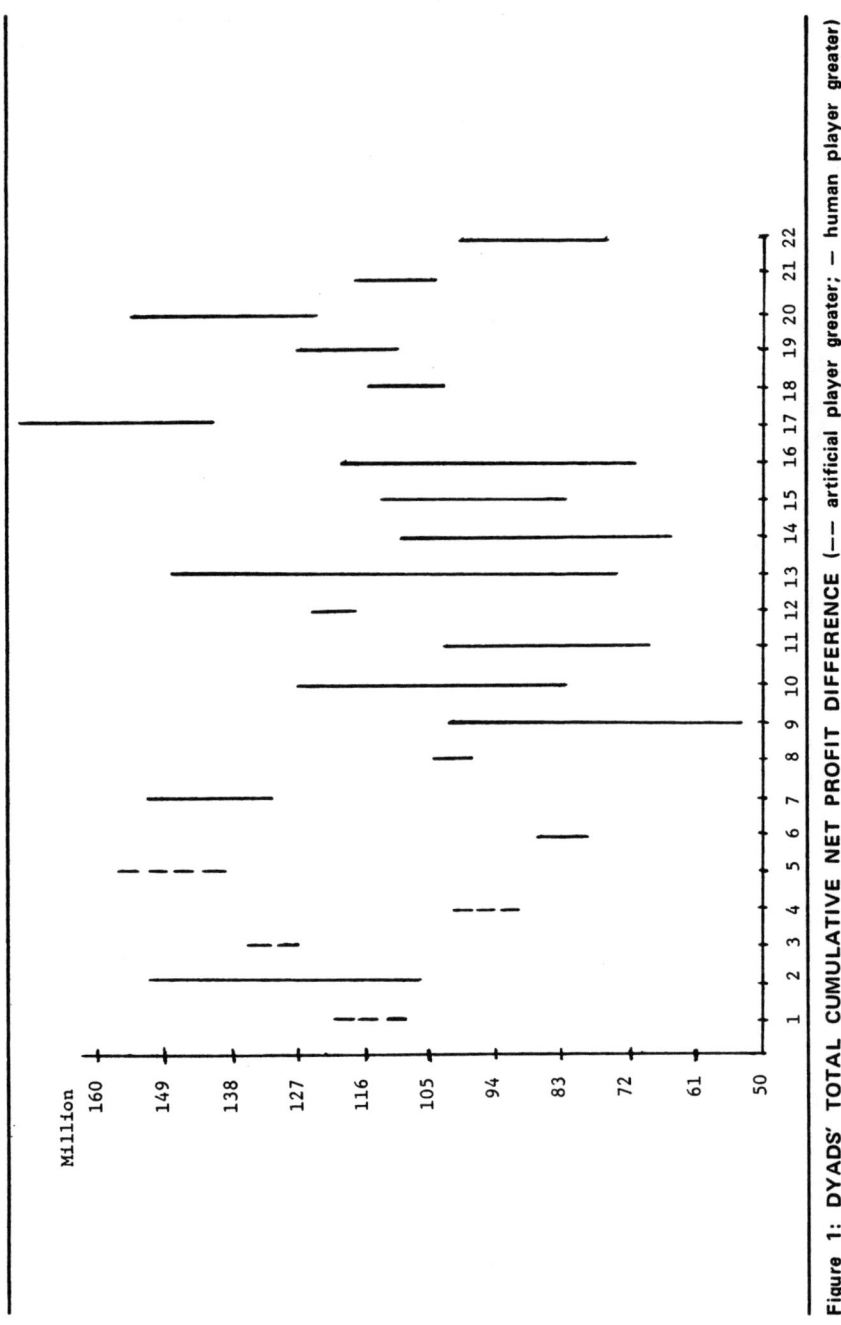

Figure 1: DYADS' TOTAL CUMULATIVE NET PROFIT DIFFERENCE (-- artificial player greater; — human player greater)

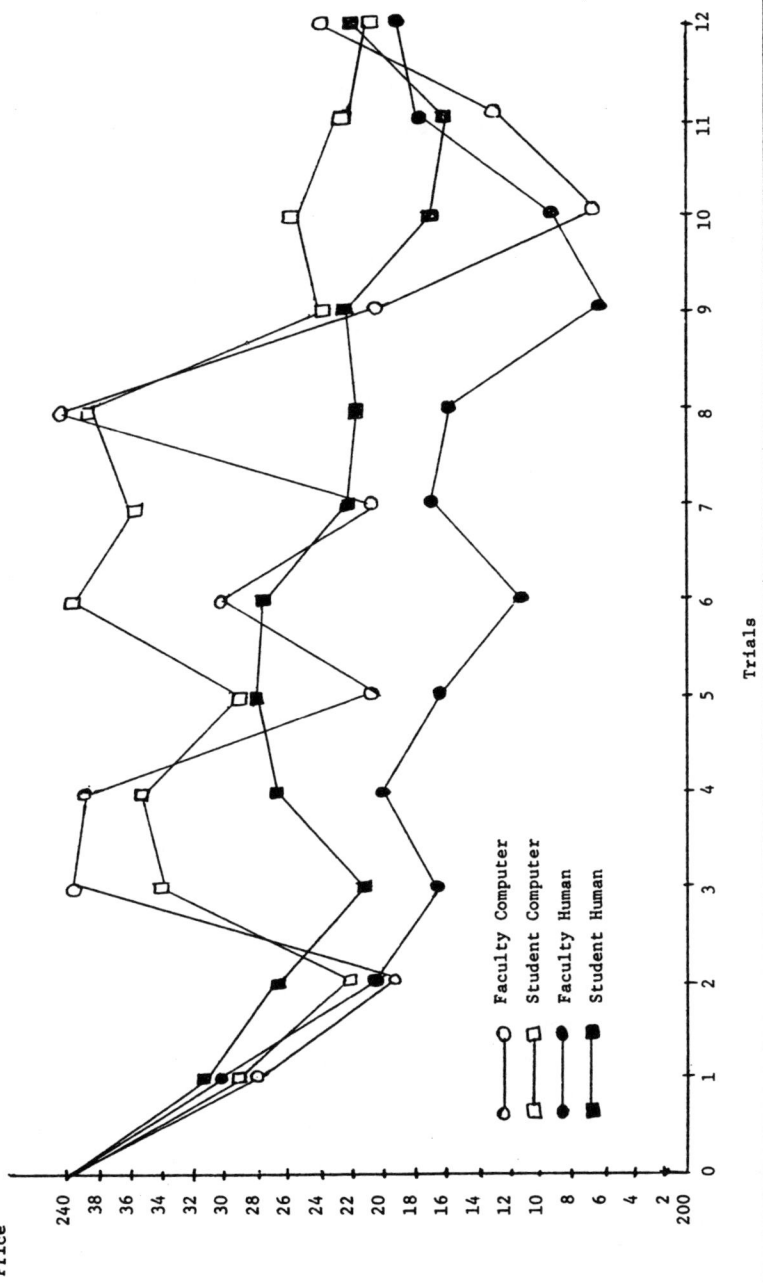

Figure 2: AVERAGE HUMAN AND ARTIFICIAL PLAYER PRICE RESULTS

at all the decisions or outcomes. The main decisions are price and advertising, and an interesting and reasonable outcome is net profit. The data for these variables are described and graphed followed by some statistical tests.

How credible was the artificial player in terms of net profit? Did he usually go bankrupt or send the human into bankruptcy, or are the results somewhere in between? Figure 1 describes final net profit results for each dyad. No one went bankrupt, and most of the human decision makers fared better than the artificial player. Four did not do as well. As an informal Turing Test, ask yourself if you can tell which player is artificial.

Besides comparing the players against each other, the players can be compared to several criteria which indicate the kinds of goals and motives of the players. The formal market structure of the game allows for prediction of net profit assuming a particular motivation. If the players were maximizing the joint profit to the two oligopolists, the cumulative net profit would be about $420 million. At the other extreme, a purely competitive motive of do-better-than-the-other would result in a cumulative net profit of $132 million. Between these motives, a player could try to do well for himself, irrespective of the player or firm in the market. This result would be $216 million. These three levels show that the actual performance (Figure 1) is closest to the competitive equilibrium performance.

Turning to the price decisions which contributed to the profit outcome, Figure 2 of mean prices by period and kind of player shows the following salient characteristics. Both artificial and human players cut price in the first two periods from $240 at the initial period to about $220. The artificial player in the next two periods raised price back to $240, while the humans only slightly raside price. The artificial player came down again in period five, and then moved up in periods 6, 7, and 8. The last four periods found him coming down to around $220, the price toward which the human players had been moving, with the faculty setting lower prices than the students.

To understand what is happening, one must look at the raw data. In the two periods, three-fourths of the humans each period cut price. The artificial player always cut price in the first period and only raised price in the second period for that quarter of the human players who had not cut in the first round. By the third round, the artificial player realized he had cut too much and raised his price in half of the games. But this cooperative response by him was overdone. The artificial player was left high, while the human players were deciding to become more competitive.

The cycle that was started continued. The humans were not as competitive in the beginning. The humans then became more competitive in response, and the artificial player became more cooperative. However, he overcompensated compared to the human and continued to overrespond on successive cycles. The final period prices were fairly similar at about $200.

Supportive of this description, a $2 \times 12 \times 2$ (faculty-student concomitant variable, and twelve time periods, and human-computer repeated measures) analysis of variance with unequal Ns (Winer, 1962: 319-337, 374-378) finds

(1) the between dyadic variance large (see first error term in tables);

(2) no difference between faculty and student dyads;

(3) an average difference over the twelve periods between humans and computer; and

(4) an interaction between the time and computer-human factors (see Table 1).

The advertising shows a similar, but much subdued, cyclic pattern. The between dyadic variability again was high, resulting in nonsignificant differences between faculty and

TABLE 1
ANOVA FOR EFFECT ON FACULTY OR STUDENT DYAD HUMAN OR COMPUTER PLAYER AND TIME AND PRICE CHOICES

Source	SS	df	MS	F
Between dyad	1,512,427.8	20	75,621.4	—
A (faculty-student)	3,913.4	1	3,913.4	—
Error	1,508,514.4	19	79,395.5	—
Within dyad	347,400.0	479	725.2	—
B (time)	13,307.4	11	1,209.8	—
AB	2,909.7	11	264.5	—
Error	273,190.0	207	1,319.8	—
C (human-computer)	5,910.0	1	5,910.0	28.91[a]
AC	113.3	1	113.3	—
Error	3,888.7	19	204.4	—
BC	5,676.6	11	516.1	2.64[b]
ABC	1,903.0	11	173.2	—
Error	40,506.4	207	195.68	—
Total		499		

a. p = .001
b. p = .01

students, between time periods, or between human and computer. Students tended to raise advertising higher than faculty (7.95 million > 6.90 million), but this was not significant. The only significant effect was the out-of-phase cyclic patterns of artificial player and human players (see Table 2).

Interpretation of these results follows in the discussion section.

DISCUSSION

The artificial player in this, his first runs, passed a face validity Turing Test, in the sense that the human decision makers in the debriefing reported involvement and challenge in playing the game. Also, the player won a few of the games and was not bankrupted in other interactions.

TABLE 2
ANOVA FOR EFFECT OF FACULTY OR STUDENT DYAD, HUMAN OR COMPUTER PLAYER AND TIME ON ADVERTISING CHOICES

Source	SS	df	MS	F
Between dyad	3,574.6	20	178.7	—
A (faculty-student)	116.2	1	116.2	—
Error	3,458.4	19	182.0	—
Within dyad	1,411.6	479	2.9	—
B (time)	32.5	11	2.95	—
AB	30.5	11	2.78	—
Error	870.2	207	4.20	—
C (human-computer)	1.0	1	1.0	—
AC	1.0	1	1.0	—
Error	152.8	19	8.04	—
BC	30.5	11	2.78	2.12[a]
ABC	22.7	11	2.06	1.57
Error	270.4	207	1.31	

a. $p = .01$

The follower nature of the artificial player was exploited in several cases. This happened by the human player feigning cooperation and suckering the artificial player to raise price, then undercutting for the next few trials. If this practice is done often enough, the artificial player can be bankrupted. The extreme of this procedure is for the human player to absurdly raise his price beyond the point of having no sales. The artificial player follows in the next period by pricing himself out of the market. In this next period and the following, the human player sells as much as he can produce by appropriately setting his price.

These problems are easily corrected by making the player asymmetric in the rate he becomes cooperative compared to the rate at which he competes (lowers price). Also, by adding a rule for checking if his price would price him out of the market, he prevents the extreme case of being suckered from occurring.

As a further check on the artificial player, with his present parameters, pitting him against himself resulted in a spiraling price war—ending in bankruptcy after several dozen trials. This confirms the follower nature of the player. He goes initially where his biases point him, and he stays there if the other player does not initiate a new direction. This result can also be seen by the nature of the player's rules as difference equations.

Further experimentation with the artificial player would include trying different biases, initial conditions, and dependences. More competitive biases against a human player may influence the human players to be more competitive. The same may be true for cooperative biases. Less dependence on other could lead to a reversal in dependence.

These results point to other kinds of artificial players. A much simpler player might lead to the same results. This simpler player would change his decisions from his last period decisions by a percentage with a random variation included. A more complex player would have a mechanism for getting out of lock-ins.

It remains to be seen if any artificial player can decrease the individual differences in strategies used by human players. The largest amount of variance occurred between dyads. It may be that only instructions and information will cut down on this variability. It appears that humans, basically, have either a competitive bias in this economic environment, or a live-and-let-live heuristic. The motive of cooperating or colluding with the other is practically nonexistent.

In conclusion, these initial runs show that experimenting with an artificial player is an interesting and useful line of research which provides a number of directions for exploring decision-making in multi-person environments.

NOTE

1. See Levitan and Shubik's publications *A Business Game for Teaching and Research Purposes*, Parts I (a general description of the game); IIA (theory and mathematical structure of the game); IV (mathematical structure and analysis of the nonsymmetric game CFDP 219); and VII (the nonsymmetric game, the generalized beat-the-average solution). The first two are available from the IBM Corporation in Yorktown Heights, N.Y. The last is available from the Cowles Foundation for Research in Economics at Yale University. For information on the third listed section above, please correspond with the authors.

REFERENCES

CYERT, M. and J. G. MARCH (1960) A Behavioral Theory of the Firm. Englewood Cliffs, N.J.: Prentice-Hall.
SHUBIK, M. and R. LEVITAN (1970) Market Structure and Behavior.
STACKELBERG, H. V. and A. T. PEACOCK (1952) The Theory of Market Economy. Oxford: Oxford Univ. Press.
TURING, A. M. (1963) "Computing machinery and intelligence," pp. 11-38 in E. E. Fiegenbaum and J. Feldman (eds.) Computer and Thought. New York: McGraw-Hill.
WINER, B. J. (1962) Statistical Principles in Experimental Design. New York: McGraw-Hill.

[32]

The Dollar Auction game: a paradox in noncooperative behavior and escalation[1]

MARTIN SHUBIK
Department of Administrative Sciences, Yale University

The Game

There is an extremely simple, highly amusing, and instructive parlor game which can be played at any party by arranging for the auction of a dollar. This game illustrates some of the difficulties with the noncooperative equilibrium concept and games in extensive form (von Neuman and Morgenstern, 1945).

The game is simplicity itself and is usually highly profitable to its promoter. The auctioneer auctions off a dollar bill to the highest bidder, with the understanding that *both* the highest bidder and the second highest bidder will pay. For example, if A has bid 10 cents and B has bid 15 cents, then the auctioneer will obtain 25 cents, pay a dollar to B, and A will be out 10 cents.

Suppose that bids must be made in multiples of 5 cents. Furthermore, suppose that the game ends if no one bids for a specific length of time. Ties are resolved in favor of the bidder closest to the auctioneer.

These rules completely specify the game except for a finite end rule; i.e., as specified, bidding could conceivably never cease. We could add an upper limit to the amount that anyone is permitted to bid. However, the analysis is confined to the (possibly infinite) game without a specific termination point, as no particularly interesting general phenomena appear if an upper bound is introduced.

In playing this game, a large crowd is desirable. Furthermore, experience has indicated that the best time is during a party when spirits are high and the propensity to calculate does not settle in until at least two bids have been made. For the purposes of the discussion and analysis, we limit ourselves to an auctioneer and two bidders, as the basic difficulties with this game can be illustrated at this level.

Let us assume that the auction has started, A has bid 5 cents and B has raised to 10 cents. By raising to 15 cents, A stands to gain 85 cents; by standing pat, he will certainly lose 5 cents. This argument holds (with modifications on gains and losses) at any stage. In particular, a turning point in the game occurs when the bidding stands with, say, A having a bid of 50 cents and B with a bid of 45 cents. At that point, it may appear to B that he should bid 55 cents and take his chances, rather than take a certain loss of 45 cents. If B bids 55 cents, then a critical zone has been passed for the auctioneer. No matter what happens to bidding, he will always make money, as the sum of the two top bids is now larger than a dollar.

The next critical zone appears in its most spectacular form when one of the bids is at a dollar. Suppose that B had bid one dollar, and A had previously bid 80 cents. At this point, A may elect to bid $1.05 rather than lose 80 cents with certainty. Beyond this point, both bidders

---

[1] This research was supported by National Science Foundation Grant GS-2840.

will be losing, but still may escalate their bids in order to cut down on losses.

Once two bids have been obtained from the crowd, the paradox of escalation is real. Experience with the game has shown that it is possible to "sell" a dollar bill for considerably more than a dollar. A total of payments between three and five dollars is not uncommon.

Some Formal Analysis

Considering the auction with an auctioneer and two bidders; this can be viewed as a three person constant-sum game. Let the auctioneer be Player 1 and the bidders, players 2 and 3. The characteristic function (von Neuman and Morgenstern, 1945) is:

$V(1) = -95$ cents, $V(2) = V(3) = 0$;
$V(1,2) = 0, V(1,3) = 0, V(2,3) = 95$ cents;
$V(1,2,3) = 0$.

The auctioneer cannot prevent a loss of 95 cents to himself if the two bidders form a coalition with one bidding 5 cents and the other refraining from bidding. Any coalition involving the auctioneer and only some bidders can obtain nothing. For any size of game, the only coalition that has a positive value is the one of all bidders.

When the auction is viewed as a cooperative game, it is evident that the auctioneer is at a disadvantage. When we switch to a noncooperative analysis, the locus of the disadvantage changes to the bidders.

There is a trivial and quite unsatisfactory noncooperative equilibrium point where the first bidder bids $1.00 as his opening bid and no one else bids. This yields a payoff of zero to all.

Another solution concept which points to a further difficulty with the equilibriums at the bid of $1.00 is that of:

$$\text{Max-Min} (P_A - P_B),$$

where P_A and P_B are the payoffs to bidders A and B respectively.

The "max-min the difference" solution can be considered in terms of a damage exchange rate. The bidders are concerned with their relative gains or losses rather than their absolute gains or losses.

Suppose that A had opened with a bid of $1.00. Then, for the cost of 5 cents, B can inflict damage of $1.00 on A by bidding $1.05. The damage exchange rate is 20 to 1. Unless there is an upper boundary to the bidding, there is no boundary to the escalation in the damage exchange rate.

On Threats and Communication

The key to the understanding of the processes at work in this game is in communication conditions. Generally in a crowd the individuals bid independently. They do not have lengthy discussions with each other. Furthermore, they do not sign agreements and specify strategies.

If it were possible to specify one's complete strategy, the first bidder would bid 5 cents and say, "If anyone else bids, I will immediately bid $1.00 if he bids less; or I will bid 5 cents more than he, if he bids $1.00 or more." If the other bidders believe him, then this strategy will block them from bidding and he will gain 95 cents.

If there is no formal mechanism for precommitment, we would need to specify the degree of belief of the other bidders in order to check upon the stability of the market.

In fact the bidders do not communicate directly more than their immediate bid, with no contingent statement whatsoever, except whatever might be signaled by facial expression, tone of voice, or other acts associated with bidding. In this sequential process a person is required to "put your money where your mouth is." The only communication is the bid, and the only signals are the history of bidding in the auction. There is no option to go back upon your word, as you do not have a word to go back upon.

Game Theory, Social-Psychology, Institutions and Escalation

This simple game is a paradigm for escalation. Once the contest has been joined, the odds are that the end will be a disaster to both. When this is played as a parlor game, this usually happens.

Can we generalize from this formal structure to interorganization fights or internation escalation? Only in a limited manner is the generalization useful. The internation negotiation has communication conditions considerably different from the parlor game. Signals and quasi-commitment are possible and common.

The game theory analysis of the game in extensive form shows us that the game theory model alone does not appear to be adequate. A general description of a typical play of the parlor game shows this. Why should anyone bid in the first place? Usually, it is because of fun or desire to participate in a parlor game rather than because of individualistic analysis.[2] Bidding proceeds fairly briskly until the point when the sum of the two top bids is greater than a dollar, after which a look of realization comes onto the faces of many participants. There is a pause and hesitation in the group when the bid goes through the one dollar barrier. From then on, there is a duel with bursts of speed until tension builds, bidding then slows and finally peters out.

The game's play appears to depend upon virtually only the social-psychology of the players, or other unstated factors of the environment in which it is played. It is far simpler than a real auction where the bidders need to evaluate the worth of items to themselves and others. It is even simpler than the Prisoner's Dilemma, where at least the concept of a 2×2 payoff matrix must be taught.

In bargaining between bureaucracies or nations, very often the negotiations are carried out by fiduciaries. Large time lags are present in the system. Furthermore, statements and explicit displays of intent concerning future behavior can be, and are, made. As much of the bargaining depends upon finding out one's own powers and wants as well as the powers and wants of the other side, the dynamics will be critically influenced by the perceptions and clarity of purpose of the negotiators.

There is no neat game theoretic solution to apply to the dynamics of the Dollar Auction, or to escalation between two nations *in abstracto*. The static game theory analysis is trivial, and although of some value, it is not enlightening concerning how to proceed from statics to dynamics.

The Dollar Auction is sufficiently simple that it may be a useful experimental game, as it contains an extremely simple aspect of escalation. Even were we to obtain clear results from such a study, it would be of only limited value in understanding escalation between nations. The latter requires a specific understanding of the mechanisms for the enforcement of agreement and the meaning of threat (Shubik, 1966). The game theoretic model for bargaining between nations must differ considerably from the Dollar Auction; and although a game theory analysis alone will probably never be adequate to explain such a process, it can serve to delimit the threat and enforcement possibilities.

---

[2] Technically, it is not difficult to modify the game in such a manner that two individuals are randomly selected as having bid 5 cents and 10 cents respectively, thus starting the process.

REFERENCES

Shubik, M. Towards a theory of threats. In A. Mensch (ed.), *Proceedings of a Conference Under the Aegis of the NATO Scientific Committee, Toulon, June–July 1964*. London: English Universities Press, 1966.

von Neuman, J., and O. Morgenstern. *The Theory of Games and Economic Behavior*. Princeton, N.J.: Princeton University Press, 1945.

ON THE SCOPE OF GAMING*†

MARTIN SHUBIK‡

Yale University

Gaming and simulation mean different things to different people. Currently there exist separate schools of individuals working on interrelated but basically different areas. Each has its own special goals and terminology. Yet there is a sufficient overlap among them that it is important to clarify the common and different interests and terminology.

The general topic of gaming is ripe for an examination to see to what extent there exists a basic methodology and theory of gaming. This paper addresses itself, in part, to this problem. Different types of games and different purposes are discussed. It is stressed that there is not one validation problem but many validation and specification problems which must be addressed if professional standards are to be attained.

1. Introduction

There are many forms of gaming, stretching from complex mathematical models to free-form verbal interchanges. Individuals whose world view and professional backgrounds are utterly different may all regard themselves as being involved in "gaming."

The subjects are different, their purposes are different and the criteria of validation differ, but the name is the same. In this paper, an attempt is made to sort out these major differences.

In a companion paper, definitions of the words gaming, game theory and simulation are given to provide a context both for the discussion here and there.

The prime purposes of this classification are:

(1) to call attention to the important prevalidation problems of *specification*, i.e., stating purpose and devising criteria by which to judge the attainment of one's goals;

(2) to indicate the possibility that *in spite* of the diversity there may be a common core of knowledge and professional skills of importance to all gamers; and

(3) to suggest that all specialists stand to benefit from an understanding of the diversity of gaming because frequently different types of gaming overlap and errors or important phenomena that may be completely ignored by one specialist may be obvious to another who sees the same game from a somewhat different viewpoint.

2. The Many Goals of Gaming: Teaching

Figure 1 shows the six main divisions of the goals of gaming, together with a finer breakdown of the categories of teaching and training. The breakdowns of the other categories are given subsequently.

In teaching and training, the audience for different games is extremely varied with respect to age, occupation, and reasons for using a game. A useful breakdown which correlates well (but not perfectly) with age is the type of educational operation:

Preschool,
Elementary School,
High School,

---

* Received April 1971.

† An early version of this paper appeared as a RAND Corporation publication, P-4608, March 1971. This is a heavily revised version with the additional research undertaken by the Cowles Foundation for Research in Economics under Contract with the Office of Naval Research.

‡ The author wishes to acknowledge the assistance of James Mayberry and Clayton Thomas.

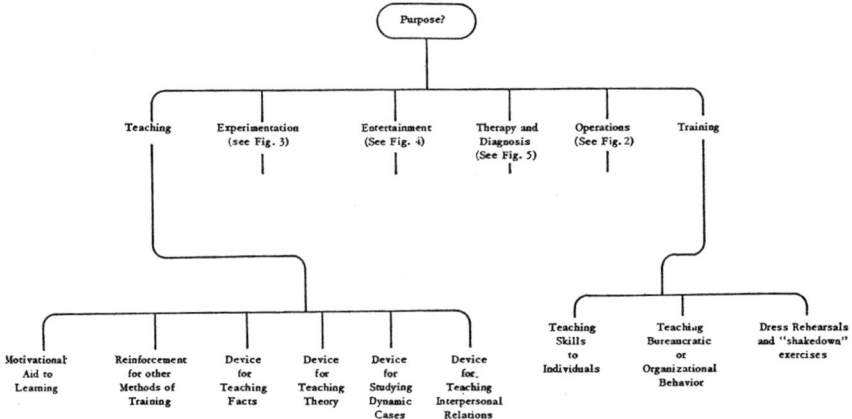

FIGURE 1

Undergraduate college,
Graduate, and
Adult educational programs.

An individual's occupation and his reasons for using a game are highly correlated. Without going into great detail, four reasons are suggested which broadly describe why most players are involved with teaching or training games:

They volunteer to play,
They are advised to play and follow the advice,
They are ordered to play by a superior, or
Bureaucratic or organizational rules require that they play.

Most games in most educational institutions are parts of courses or programs. There may be electives prior to registering for the program; however, once a student is in a program the organizational rules will require that he participate. In many colleges and universities in the United States there is a considerable amount of voluntary gaming.

Where the participants are members of large bureaucratic organizations such as the military, other parts of government service, or private corporations, they have, for the most part, been advised or ordered to participate. On occasion they may be volunteers. When this is the case, the type of volunteering is usually of the type where a department head is told to supply three out of his twenty men for a game. He may call for volunteers. It is not uncommon that the volunteers may be the three least busy or most junior men in the department.

Concerned citizens groups, curious students and "buffs" form the hard core of volunteer gamers. A crude estimate indicates that in 1970 there were between 15,000 to 25,000 war-gaming amateurs in the United States.[1] Currently there is a trend towards games stressing social interaction and the problems of society. This is manifested in the growth of a number of board games in the penumbra between education and entertainment; thus we have had a progression from MONOPOLY to SMOG. Even with war games there appears to have been an upswing in the last ten years of games calling for diplomacy, negotiations and grand strategy, such as DIPLOMACY and

---

[1] Based on the number of subscribers to the Avalon Hill publication, *The General*.

SUMMIT as contrasted with straight war games. From a technical game theoretic point of view there has been a shift from two-person zero-sum games or noncooperative individualistic enterprise games to nonconstant sum games where coalitions are of importance.

The overall trend in voluntary gaming in the last thirty years has been from an almost exclusive emphasis on military games to military-diplomatic games and to business games and now more recently to games concerning society.

2.1 *Different Roles in Gaming for Teaching*

Before questions concerning validation can be asked with respect to a single game, it is desirable to consider goals and criteria of success from several different points of view.

In particular any game should be considered in the context of its impact on individuals engaging in four activities related to it. They are:
the players,
the builders,
the controllers or directors, and
the sponsors.

Frequently an individual may play more than one role. Furthermore, the roles are often more finely differentiated than the breakdown noted above. For instance, the game direction may consist of a team which contains not only umpires or teachers who direct the game, but also experts who are called upon to judge the feasibility of certain acts while otherwise having no control role.

At the university level, especially with graduate students, more may be learned by the students in constructing games than in playing them. The locus of the learning experience is by no means centered with the players.

In gaming for teaching purposes, especially at the high school level or younger, the worth of a game is frequently no more than that of the teacher. An inspired teacher can direct a mediocre game with good results, and the best of teaching games can be of little use if it calls for considerable direction from an inadequate teacher.

The breakdown of roles noted above applies to gaming used for purposes other than teaching. It is referred to again later.

2.2 *Motivational Aid to Learning*

One of the major attractions of gaming has been as a motivational device. It appears to attract the attention of and involve the players deeply where other methods have far less impact. There is reasonable consensus on this point among those who have used games and a small amount of experimental evidence as shown by the work of Wing [29] and others [5]. Creators of educational games such as Layman Allen [21] stress the positive motivational features of educational games. However it is easy to slip from conjecture to unsubstantiated advocacy as is exemplified by the writings of Clark Abt [1]. Coleman has stressed the value of games in teaching disadvantaged children [9].

2.3 *Reinforcement for Other Methods of Teaching*

In the universities and schools, games are frequently used as part of a program along with more traditional methods of teaching. This is also true of the business schools and military academies. Gaming proponents claim that the mix of methods is most effective.

2.4 A Device for Teaching Facts

In virtually every type of gaming including the diplomatic-military games of the Studies, Analysis and Gaming Agency,[2] and business games such as the Carnegie Tech Game or INTOP, [28] gaming practitioners and players have claimed that gaming is an extremely useful way to learn and organize facts. A game usually provides a handy scheme for supplying associative links between facts, and as such it may aid both learning and remembering; although to date there is little hard evidence substantiating these claims.

2.5 A Device for Teaching Theory

At the advanced undergraduate and graduate level the building of games appears to be extremely useful in encouraging students to think in terms of models and abstractions. This improves their ability to theorize. In the social sciences especially, the importance of improving the ability of an individual to enable him to construct abstract representations of complex systems cannot be overemphasized. The discipline in constructing a playable game provides a deep appreciation of logical consistency and completeness, as well as stressing the connection between the model and its subject matter.

On the other hand, it is important to stress that before a game can be used with any success to teach theory it is rather desirable that the theory exists to be taught. In the exploitation of business games over the last decade this has not always been the case. A flagrant example of potential misuse has been in the modeling of advertising in business games. Even a brief glance at the literature on how advertising affects sales is sufficient to indicate that there is little substantiated theory in advertising, yet in many of the business games played both at universities and in business training programs advertising has been thown in as an *ad hoc* modification on demand with teaching results which could be damaging were it not for the basic skepticism of most of the players. It is critically important that players be warned against learning false or unsubstantiated principles.

2.6 A Device for Studying Dynamic Cases

Several business schools, especially the Harvard Business School, favor the use of the case method. A specific historical case may be taken up, a "scenario" written describing it, and the class is required to consider the problems it poses and the ways in which they were handled or might have been handled.

A game lends itself with great ease to providing a dynamic context to a case. Furthermore like the Czech experimental theater at Expo '67, it provides a natural means whereby alternative histories can develop.

A formal game, especially a large and complex one, has both the advantages and disadvantages of an institution. It may take on the inertia of an institution itself, as is exemplified by the Carnegie Tech game [7]. However, this may be an advantage as it is extremely difficult to explain or reproduce in the classroom the ambience of decision-making within a bureaucracy.

2.7 A Device for Teaching Interpersonal Relations

Many of the basic games for younger children and disadvantaged groups, as well as community action games to study urban redevelopment or other social problems, stress

---

[2] This agency is the successor to the Joint War Gaming Agency.

interpersonal relations both from the viewpoint of the individuals and their roles. In many of the uses of gaming "seeing the other individual's point of view" by role playing his position appears to be of value. Thus, for example, a slum child may begin to appreciate that a policeman's lot is not a happy one. Furthermore it might even be possible for a United States official to appreciate that to a North Vietnamese he does not necessarily appear as the epitome of sweetness, light, reason and democracy.

At the more direct level an appreciation of the need for bargaining, communication and compromise can be obtained from many of these games. A good example of such a game is DEMOCRACY [8]. Some of the insights gained here do not pertain only to personality factors but to a basic game theoretic phenomenon that in an n-person nonconstant sum game there is no neat unique way of defining socially rational behavior. There are many different criteria for social rationality, and (as evinced by the lack of a core) [22] it is frequently not possible to satisfy the demands of all groups even if each group can show that its demand is within the scope of its own power if it fails to cooperate with the remainder of society.

3. The Many Goals of Gaming: Training

Teaching blends into training, training into operational uses and so forth. Nevertheless it is useful to make the distinctions among different goals for gaming although they may blend together at the boundaries between them. In particular the major distinction between teaching and training concerns the emphasis placed on the *why* of the process. There are several quite effective small games which can be of use in improving an individual's performance in production and inventory scheduling without ever going into the depths of why certain methods work. An operator does not have to get a course in dynamic or in integer programming to become a better manager of production and inventory scheduling.

Many individuals can be taught to drive safely by means of analogue device trainers without having to learn much about Newtonian mechanics or how an automobile works. Training games for simple manual skills especially those requiring a fair amount of coordination are not particularly exciting, but they can be of tremendous use and can provide valuable simulated experience that would be costly in the extreme to obtain from the field.

In general when games are used for training, the only role occupied by the individual being trained is that of player. This contrasts with gaming for teaching where because the *why* is so important it is highly desirable in some instances to have students build or supervise as well as play games.

3.1 *Bureaucratic and Organizational Behavior*

In a complex society, licenses must be obtained, permits granted, rules checked, expectations examined, accounts audited, telephone calls made and routines for processing torrents of communication must be established. Training games offer the possibility not only of training individuals to acquire individual skills but also to learn bureaucratic routines.

3.2 *Dress Rehersals and Shakedown Exercises*

Rehearsals in the theater, field maneuvers and battle exercises are all examples of operations devoted to seeing that individuals know "their lines" and are able to cooperate in team action. They differ from the previous category only inasmuch as they are usually aimed at preparing for coordination in a temporary context such as a specific play or a projected offensive. The phrase "shakedown" appears to come from the naval usage "shakedown cruise" which is the original cruise of a ship devoted to getting the crew to coordinate and to check to see if the equipment works.

4. The Many Goals of Gaming: Operational Gaming

The different goals of operational gaming are indicated on Figure 2. In contrast with gaming for teaching, operational gaming is used almost exclusively by adults in military, governmental or corporate organizations.

There is an overlap between operational and training games in the domain of field exercises. It is difficult to say where the dress rehearsal and coordination aspects of an exercise cease and where planning, strategy testing, and exploration begin. In Figure 2 the category "shakedown" has been included under operational gaming as well as in Figure 1 under training.

By far the largest use of operational games to this day is military or diplomatic-military. Relative to these uses corporate operational gaming is insignificant and the use of operational gamings for social planning is in its infancy.

Because of the nature of the bureaucratic structure of decisionmaking a clear understanding of the roles and goals of the players, builders, controllers, and sponsors of operational gaming exercises is far more important to the professional who wants to know "what is going on" than is such detailed understanding of the use of gaming for teaching.

Operational gaming is "where the money is" currently and the goals of a consulting firm wanting to build a large game, a general wishing to advocate a weapons system and a colonel assigned to play in or operate the game can be sufficiently diverse that the mismatch makes an objective evaluation of such a game harder than reading the Rosetta stone.

4.1 Cross-Checking and Extra Validation for Other Methods

A game may be used as a back-up procedure to provide an extra insight into a process that has been investigated by other means. For example, a recommendation may be presented in report form. The basis for the recommendation may be expert opinion and/or empirical evidence. A gaming study of the same problem may turn up insights or raise questions overlooked by the approach. As operational games in general tend to be somewhat expensive in both time and money, the problem has to be of sufficient importance to merit the extra effort.

There is also the danger that a game may be employed to give a pseudoscientific window dressing to a recommendation.

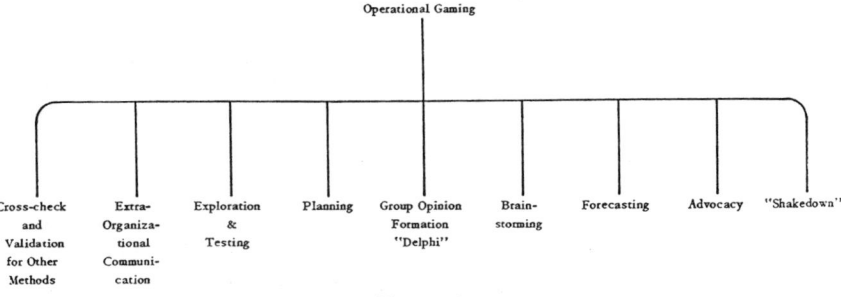

FIGURE 2

4.2 Extraorganizational Communication

There may be a game outside of the game being played. With operational games it is critical to understand both the stated and the unstated purposes of gaming by the individuals involved in the exercise. In particular gaming, along with short courses and seminars, is used to establish informal means of communication. In some instances the main objective may be to arrange to get two or three ranking individuals trapped together for two or three days on neutral ground.

Participants in diplomatic-military war games frequently comment on the value of being able to watch the decisionmaking styles of different high-ranked individuals.

The use of a game as a means for establishing informal communication will vary heavily with the style of play. If the game is held in an isolated locale over an intense period of play for three or four days or more, the effect may be quite striking. If, on the other hand, it is played in an intermittent manner over several weeks or months, then it is easy for most of the participants to minimize the disturbance to their set patterns.

4.3 Exploration, Testing, and Planning

The strict meaning of a strategy in the sense of game theory, while precise and worthy of note to a gamer, is not particularly useful to a planner. Planning involves the selection and aggregation of information. Even with the aid of high-speed digital computers the number of alternatives which can be explored is miniscule. Games such as those played by the SAGA[3] operation or the SIERRA series[4] of The Rand Corporation and many others have been used for planning, exploration and the testing out of a limited number of alternatives.

An intense amount of preparation goes into a game of this type. The preparation is in general far more extensive than the play. Two or three moves on each side may be taken, and in a debriefing session after the game there will be an attempt to summarize and note the consequences, alternatives or facts that had been overlooked prior to the commencement of play.

A planning game to be of use must utilize individuals sufficiently involved in the process that they can be privy to the actual problem and the major considerations. In military and governmental games these may range from colonels to five-star generals and cabinet officers.

There is some evidence that some high-ranking officials enjoy participating in gaming exercises; but there appears to be little evidence beyond the occasional testimonial as to what was accomplished. This last comment applies to gaming regarded as a "brainstorming" exercise as well.

4.4 Group Opinion Formation and "Delphi"

In the behavioral sciences and in the study of organizations, in evaluating many aspects of the present, and in forecasting the future, we have very little "hard" knowledge in the sense of the sciences in which experiments are performed and *replicated* frequently. In most professions much use is made of expert opinion. Up until recently little systematic thought was given to the study of how expert opinion is used and what the techniques are for optimizing the use of this scarce resource. Furthermore, little was known of the relative worth of using the opinion of more than one expert. When do diminishing returns set in? What sort of controls should there be over the interaction, and so forth?

---

[3] The type of game used here was originally suggested by Goldhamer and Speier [15].
[4] See, for example, Northrop [24].

An operational game may be regarded as a formal structure to elicit group planning—a process which involves both evaluation and prediction of the likelihood of contingencies.

Helmer [17] and Dalkey [10] have advocated the use of Delphi techniques, which consist of having a group of experts who are anonymous to each other respond to questionnaires, after which the results of their responses are processed and returned to them so that they can adjust their estimates in the light of the new information. Dalkey currently is engaged in large-scale experimentation [11] on the properties of the Delphi method.

One important feature that differentiates a formal operational game from Delphi is that there has been less emphasis on the aspects of motivation in relation to performance with the former than with the latter. To date there has been little effort to blend these two approaches. However the potential appears to be worthwhile.

4.5 *Forecasting*

In general a game is *not* a forecasting device. A good operational game may make use of good forecasting procedures but it is not in itself aimed at providing forecasts. This should not be confused with its use in discovering new alternatives and in helping to evaluate future possibilities. *Forecasting* and *contingency planning* are related but extremely different activities. In particular, a good forecaster may not be in the slightest interested in the importance or worth of his forecast. Accuracy may be a goal for the forecaster in and of itself, not because of its relevancy to the planning process.

A game may be a useful device for stressing the need for coordination of forecasting activities with planning and decisionmaking processes. In this sense the involvement of forecasters in the design and play of operational games may be of considerable use.

4.6 *Advocacy*

Last, but not least, we must note the use of operational games for advocacy. A competent game designer can build biases of almost any size into a game. Advocates for specific policies or weapons systems can load the dice so that the game has a great probability of producing the results they want to see. Games are fun. They are great propaganda devices. The exploitation of the AMA business game provides one such example [3]. Action groups of nonprofessionals can easily be hornswoggled by a latter-day snakeoil salesman peddling a game to cure all ills.

Smog, fog, the crime rate, central city decay, impotency, war, lack of understanding among nations, the evils of unemployment and the drug culture, the curse of the automobile and the lack of a good 5¢ cigar will all be cured if we only have a big enough data bank tied into a game room with large fancy maps.

Recently there has been a move for the building of a "World Game" by several extremely well-meaning individuals [13]. As a mild advocate of gaming this author believes that there are many good reasons to proceed with the use and building of large games for operational purposes, especially in areas dealing with social policy. However one must not confuse conversational feasibility with operational feasibility.

In some instances a game can be used as a euphemistic way for informing others of a change in policy by asking them to participate in an exercise whose outcome is a foregone conclusion. The Japanese war gaming prior to Pearl Harbor could be interpreted in this manner.[5]

---

[5] See Wohlstetter [30].

5. Experimental Gaming

Human beings fortunately are more difficult to experiment with than rats or guinea pigs. Even so, there is now a fast growing literature on experimental gaming in which human decisionmaking behavior is studied by observing the performance of individuals in formally structured games. In order to pursue this type of work fruitfully it is important that the experimenters have at least a basic elementary understanding of game theory and social psychology. In a companion article a background of game theory relevant to gaming has been presented [25].

Much experimentation has been done with simple 2 × 2 matrix games under relatively restricted conditions.

The experimental subjects have been, for the most part, undergraduates at various universities; some army personnel have been used, as have been some inmates of local jails and some middle- and upper-level corporate personnel.

These experiments are psychological-light-years distant from preschool educational games or from military-diplomatic free-form war games. The criteria for validation belong to more or less accepted statistical methods familiar to physical scientists, econometricians, and experimental psychologists.

Some experimentation has been performed with business games of middling or of considerable complexity [19] and with political, diplomatic, and war games [18]. In general, owing to the greater complexity and smaller degree of control on these games they have been harder to control, and hypotheses have been difficult to test. In some instances (Hoggatt [19], Shubik, Wolf and Lockhart [27]) players have been faced with artificial players as competitors.

5.1 *Validation of Hypotheses*

In general although the goals of the game designers are usually clear in experimental gaming, the goals of the players are by no means clear. There exists an enormous, and frequently poorly handled, problem in specifying, controlling, and measuring the goals and motivations of players in simple as well as in complex experimental games.

A separate article is needed to do justice to the literature on experiments with 2 × 2 matrix games, and another article is needed to discuss experimental work on the analysis of human factors in complex competitive systems. Nevertheless, without going into detail several disturbing features of work in gaming can be seen. Specifically much of the work with operational games presupposes that a considerable number of problems that belong to the domain of experimental gaming, i.e., basic research, have been solved; whereas, in fact, the expenditures and activities in experimental gaming are miniscule as compared with operational gaming.

Furthermore, although the word "validation" is popular and takes on a particularly

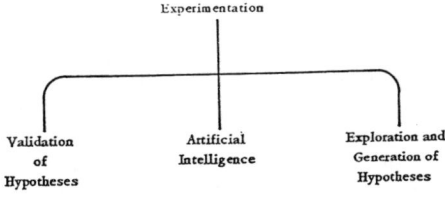

FIGURE 3

scientific flavor when applied to experimental games, if you do not know what you are trying to validate then all of the statistical apparatus you have may not help you (unless you are consciously searching your data to generate hypotheses). *Control* and *specification* are prevalidation procedures which even at this time are not yet carried out adequately on many of the experiments. The major contributing factors to the failure of control and specification are lack of cooperation among specialists (i.e., social psychologists who know no game theory, misunderstanding the competitive structure, or game theorists knowing no social-psychology, failing to allow for simple explanations of behavior) and lack of sufficiently automated laboratory facilities to enable the careful experimenters to obtain detailed observations and to run standard analyses at a reasonable cost.

5.2 *Artificial Intelligence*

Figure 3 has three branches. The first covers the type of experimentation that is more or less familiar in other disciplines. In the past decade there has been a considerable upsurge in the study of *artificial intelligence*, or in the study and the construction of computer programs which perform tasks that are usually regarded as requiring intelligence. No distinction has been made, in general, between the sort of intelligence required to solve difficult problems, such as playing chess, and to resolve interpersonal problems, such as those which arise in nonconstant sum games—bargaining, for instance.

Frequently the gamer is more interested in *social intelligence* than in individual intelligence. The problems in the construction of a good problem-solver or a socially intelligent player differ inasmuch as the criteria for the performance of the former are relatively easy to construct, whereas there are no such easy criteria that can be constructed to judge group or social rationality.

In particular, it appears that a good problem-solver, a program which can play chess well, for instance, requires efficient searching and calculating abilities and other features usually associated with intelligence and intellect. By the very nature of the game it need not have any "personality." A good chess-playing program has to be a "smart" or intelligent program, not a pleasant or nice one. This is not the case when we turn to nonconstant sum games. It is possible to build an artificial player for a business game [19] which plays in a manner comparable to human players. The rules or "heuristics" needed to construct such a player call more for an emphasis on his interpersonal relationships than on his ability to compute. A "nice," moderately cooperative and not particularly aggressive artificial player in a business game may elicit cooperation from his competitor and will do quite well.

The literature on artificial intelligence has very little on the subject of social intelligence. There has been and there is currently an extreme division of opinion on the nature of problem-solving, leaving aside the extension to social interaction. Simon, Minsky and Pappert [23], and many others are the proponents, whereas considerable criticism of the basis of artificial intelligence work has been offered by Bar Hillel and H. Dreyfus [12].

To enter into the debate on the pros and cons of artificial intelligence would take us too far astray from the work relevant to gaming; hence we confine our remarks only to those aspects of the subject relevant to those interested in gaming.

Along with the growth of interest in artificial intelligence has come a considerable growth in the design of protocols and ways to describe decisionmaking processes. Much work has been addressed to analogies between how one teaches a machine and how one teaches a child [23]. In particular those interested in experimentation with computer-aided instruction[6] need to be aware of the developments in artificial intelligence.

The experimental gamer is usually more interested in games which are more than problem-solving exercises. Many war games and games such as chess can be modeled as two-person zero-sum games, hence the main analytical problems they pose are in the domain of information processing and problem solving. Diplomatic-military, business, social development, and most other games do not fall under the zero-sum rubric. Social, political, or economic behavior all call for attention to interpersonal interaction. The construction of robots or artificial players in these games both provides opportunities to attempt to model socio-psychological processes in the building of the players and gives the experimenter greater control over his experiments, especially when he is able to replace a set of two-person experiments with a set of experiments consisting of a group of individual human players playing with the same artificial competitor.

5.3 *Exploration and Generation of Hypotheses*

Frequently, experimental games are used to explore decisionmaking processes and to generate hypotheses rather than to test specific hypotheses. Sometimes this is not the way things were planned, but this is how it works out. Prior to the experiment, several hypotheses may be suggested. After the experiment it appears that hypotheses can neither be accepted nor rejected, owing to insufficient definition or complications in the control of the experiment. Nevertheless, the running of the experiment clarifies the definition of the hypotheses, locates others, and locates the control difficulties.

The above reasoning is often used as an excuse or self-justification after an ill-conceived experiment has been run. However this is not always the case and pilot experiments play an extremely useful role when the topic being studied is both complex and ill defined.

6. Games for Entertainment

6.1 *The Theater*

It is important to remember the deep interconnection between gaming and theater. For example, many war exercises, fleet maneuvers, and "dry-runs" are identical in purpose with dress rehearsals. Huizinga [20], Callois [6], and many others have discussed the relationship between plays and games. It is not the purpose of this paper to explore the historical, anthropological and religious aspects of this interconnection. They form a fascinating subject in themselves. However, those who wish to use games for more mundane purposes should at least be aware of the interrelationship among the games, plays, theater in general, mass spectacles and ceremonial parades. The military parade itself is an extremely complex phenomenon being part entertainment, part training, partly a signaling process in a diplomatic dialogue and a device for influencing morale.

An important but open question is what are the basic features that differentiate good theater from good operational gaming? For example, how does the "realism" of the scenery affect both of these activities? The audiences are different, the role playing is different, and the stated purposes are different. Nevertheless, an analytical categorization of these differences is not an easy task.

6.2 *Gambling*

Three categories of individuals involved in gambling must be distinguished. There are those businessmen who run gambling ventures, professional gamblers who make a living from playing, and those who play for other reasons.

---

[6] See for example [2].

ON THE SCOPE OF GAMING P-31

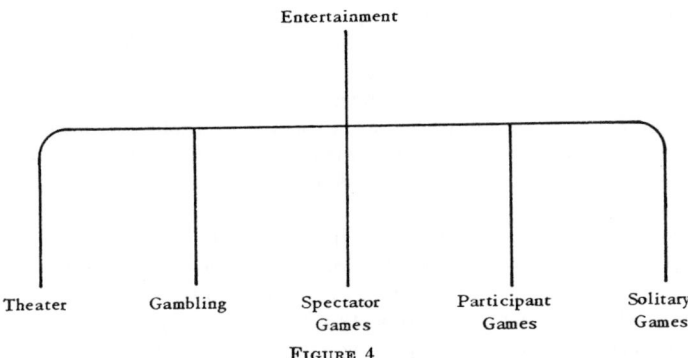

FIGURE 4

The individuals who run gambling establishments are in many senses not particularly distinguished from other businessmen except that possibly gambling as a business tends to be an enterprise with not very large components of risk as compared with a high innovation technology enterprise. The professional gambler such as the poker player (see for example H. O. Yardley [31], *The Education of a Poker Player*) does not seem to be far different from the professional arbitrageur. They both take risks but they are in the true sense of the word calculated risks and the individuals who devote professional attention to these occupations are usually skillful enough that they are able to earn a good (but in general not spectacular) living from their professional skills.

Although there is a large element of chance in a game such as poker in contrast with a game such as roulette it is primarily a game of skill and not of chance. The calculation of probabilities is one of the key aspects of good poker playing. There are obviously enormously important psychological factors in one's ability to judge the competence and style of the other players. Betting on horse races stands somewhere between a roulette game and a poker game in its skill component. The element of chance is extremely large. However, there are some useful calculations to be made concerning the odds being offered and the probable performance of the horses. The main factor, however, does come in the judgment of the horses and their performance on a specific track under the appropriate weather conditions.

In general, especially in large organizations, when someone states "we have taken a calculated risk," that frequently means that individuals have made a decision without doing the calculations necessary. In the case of the professional gambler, the very reverse is true. In general they have no bureaucratic structure around them and they are in a position where fast and explicit calculation of the odds is a central aspect of their very living.

In contrast with the fighter pilot, the poker player thinks in terms of odds explicitly. It is unlikely that the pilot calculates a probability of 0.15 of success by one avenue or 0.3 of success by another approach. There is undoubtedly an important difference in calculating explicit probabilities involving death and explicit probabilities involving money. Furthermore, the very nature of many gambling games of skill makes the calculation of probabilities a natural and explicit way of evaluating one's position. It is unlikely, however, that these features are in themselves sufficient to explain the fundamental difference in approach to thinking in terms of explicit probabilities evinced by professional gamblers as compared with, say, middle management or army colonels or, even more so, the citizen on the street. The literature on the social and personality aspects of the professional gambler appears to me to be surprisingly slight. This also holds true for the handful of special professions in which the risk-of-life component is sufficiently important to make the gambling aspect explicit, for example, test pilots and steeplejacks.

The interests of the individuals not professionally involved in gambling run the gamut from mild entertainment to deep addiction. Many individuals who lose $20, $30 or $100 at the tables in Las Vegas or Monte Carlo are paying an entertainment fee. For the most part they know that they are paying this fee and have decided that the entertainment is worth it. It is worth noting that the mere location and decor of main major gambling towns and main casinos lay stress on the theatrical aspects and the role-playing features of gambling. Las Vegas is designed so that the perfectly ordinary middle-class dweller of suburbia can lose his $100 or so in a socially acceptable manner in surroundings ranging from pseudo-luxury to pseudo-wickedness.

What are the risk-taking features of the ordinary individual who is not addicted to gambling, who plays small-stake roulette at a casino, or who buys the occasional ticket for the races? There exists a certain amount of literature in economics and psychology concerning gambling and the buying of insurance where the odds are in general extremely small for an event to occur. However, there is virtually no analytical literature on ordinary gambling behavior. Erving Goffman has several highly stimulating articles on con-games where the otherwise prudent and nonaddicted individual is taken for a sucker [14].

There are many individuals for whom gambling is an addiction. Dostoevski was a good example of one of these. There is a small psychopathological literature on gambling as is evinced by the somewhat unsatisfactory book of Dr. E. Bergler [4]. One of the difficulties in studying a subject such as pathological gambling is that it requires a multi-disciplinary approach. Psychiatrists will tend to see only the psychiatric aspects whereas, for example, those trained in a theory of games will undoubtedly lay heavy emphasis on the structural differences among various games.

From the viewpoint of those interested in operational games, especially games of a military or social variety, the study of addiction and extreme risk-taking would appear to be critical. The distance between the gambling addict and the drug addict may not be great. There also appears to be an important psychopathological risk-taking component in assassinations, in some forms of exploratory behavior, and in the actions of some extremist groups.

In summary, it appears to me that gambling behavior of virtually all types is a critical phenomenon in the understanding of many important features of risk-taking. Those who argue for operational games as a means for studying extremely original or surprising alternatives should also consider the need to explore the genesis of both "reasonable and pathological risk behavior."

6.3 Spectator Games

Many sports, such as football, baseball, hockey, basketball, cricket, etc. are primarily spectator sports. The vast majority of the participants are in the audience and derive vicarious pleasure from the play. There the analogy between the game and theater is possibly at its closest. There are the actors, and the great majority are spectators. The sports event is far more of a free-form play than is a theatrical performance. In the former, although the rules are given, the actual path of the play is not completely known in advance. In the latter, the complete path of the play has been specified except for the acting that has not been controlled by the direction. Spectator games may have a small advocacy and teaching component to them, inasmuch as they may inculcate an appreciation of teamwork and an ability to judge and understand the qualities of effective performance. However, for the most part they are pure entertainment. For a discussion of the vicarious pleasure and role identification aspects of spectator entertainment see Callois [6].

6.4 *Participant Games*

Bridge, poker, tennis, chess, football, charades, monopoly, and many board games, many of which can be played as spectator games, are most frequently played only by the active participants for their own amusement.

The distinction between participation in a poker game for amusement and for gambling purposes may easily vary as the size of the stakes. The importance of the payoffs to the players as an influence on the nature of the game cannot be overstressed. When an individual participates in a game whose stated purpose is operational or educational, but which nevertheless is formulated in such a way that the payoffs to him are not particularly clear, it becomes absolutely crucial to investigate the possibility that he has turned the exercise into a game for his entertainment.

It is a safe rule to apply when using games for teaching, experimentation, operations, or therapy to have as a null hypothesis that in fact the game was primarily theater or participant entertainment.

6.5 *Solitary Games*

Possibly one of the greatest sinks for the use of man-hours in gaming is the solitary game. Crossword puzzles, jigsaw puzzles, and solitaire are major examples of games "played to while away the time," although it can be argued that they may have an educational component. The origins of both the crossword puzzle and the jigsaw puzzle are relatively recent (within the last hundred years). Precisely what makes them so popular? Will they be supplanted by other solitary games? Could solitary games be designed that would be fun and more explicitly educational or experimental?

7. Therapy and Diagnosis

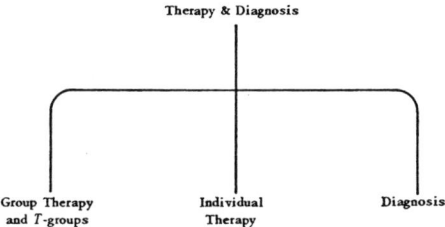

Games may well have both diagnostic and therapeutic value. Although these areas lie well beyond my own training and competence, it would be a glaring omission not to call them to the attention of the reader.

7.1 *Group Therapy and T-Groups*

In some ways group therapy sessions and T-groups might be regarded as "anti-games"; as such, the comparison between them and a formal operational game such as, say, a diplomatic-military game becomes of considerable interest. In the case of the latter, the individuals are encouraged to concentrate on certain aspects of role playing.

Very frequently an individual is required to simulate the decisionmaking process of someone else. In contrast with this activity, in group therapy individuals are encouraged to find out who they are. The stress appears to be in the other direction. Individuals will hopefully be able to examine where they have been role playing in a manner not consistent with their comprehension of self.

It appears to me that the paradigm of the game offers an extremely fruitful basis for joint work by psychiatrists, social-psychologists, and those interested in organizational decisionmaking.

7.2 *Diagnosis*

It is not difficult to design games that focus on relatively narrow band-widths of decisionmaking and of interactive behavior. Informal experimentation with several games such as "So-Long Sucker" [16] and "The Dollar Auction" [26] indicates that it is possible to obtain extremely strong participant reactions to relatively simple games. Experimentation with two-by-two matrix games of certain design has also indicated this. The use of small games for diagnosis might be relatively cheap and effective.

7.3 *Individual Therapy*

The use of games for therapeutic and corrective purposes is clearly closely related to, but somewhat different from, the use of games in teaching. Little appears to be known about the potentialities of this use.

8. Concluding Remarks

The scope of gaming is considerable. Many uses are utterly different from each other both in concept and purpose. Yet at the same time amid all of the diversity a certain common thread is present. The game is a paradigm for competitive and/or cooperative behavior within a structure of rules. The rules vary in formality in freeform gaming or in rigid rule gaming. They vary in portrayal of war situations, economics, social contract formation and so forth. But *all* games call for an explicit consideration of the role of the rules. A serious user of games is well advised to be broadly aware of the alternative uses and meanings of games as well as deeply specialized in his own type of gaming.

A key word concerning gaming that he hears frequently and from different sources is "validation." How can games be validated? Prior to asking this question it is imperative that we ask: "For what purpose and for whom?" A common vocabulary employed by different groups to mean different things is guaranteed to breed confusion. Around thirty considerably different purposes for gaming have been suggested. The criteria of validation for the success of a gaming endeavor are extremely different as we consider the different uses.

The size of the box office receipts is a good criterion for evaluating the success of a spectator sport from the viewpoint of the promoter. The number of people doing crossword puzzles is a good criterion for the owner of a newspaper. The criteria applied by a general, a zero-sum game theorist and a military hardware expert to force posture and allocation games are planets apart from the criteria that might be used to judge the success of a political-military exercise run by a mixed group of political scientists and top government officials. This in turn would be different from the judgments applied to evaluate the worth of a teaching game designed to give high school students an appreciation of international relations.

In light of the many different types of gaming, the different goals of the various interested parties and the problems of control, rather than talk about validation we are probably better off concentrating on four stages in the evaluation of gaming. They are:
(1) intention,
(2) specification,
(3) control, and
(4) validation.

The first refers to finding out generally what the goals of the concerned parties are. The second involves translating these goals into well-defined measures and in establishing that the measures can in fact be obtained from the game. The third refers to the actual control of the game necessary to guarantee that extraneous elements do not confound the obtaining of the measurements called for by the specification. The process of validation calls for interpreting the significance of the measurements in terms of the specification.

A methodology and a theory of gaming are only beginning to emerge. It is certainly premature to believe that there is such a thing as "the method for evaluating or validating all games." The "prevalidation" steps are still not always done adequately. The surprising feature of the growth in gaming is not that there is no single method of validation, but that so little attention has been paid to sorting out the different uses of gaming and to developing criteria and methods that apply to the special uses.

The promise from many of the different types of gaming appears to be considerable. The proof of the promise is by no means empty for some of the categories of gaming. There are some reasonable criteria available for judging the success of a social psychology experiment, the teaching value of some elementary games, and of some business games; the worth of some formal game theoretic and simulation models for weapons evaluation; the profitability of sports; the entertainment value of the theater and entertainment games and several other uses of gaming.

Our hard knowledge is extremely limited concerning how successful (and what are the criteria for success?) operational games are. What is really learned from political-diplomatic and military exercises? Who learns what from teaching games? The words *ad hoc* are frequently used in the pejorative sense. I would like to use them in a nonpejorative manner. It is my belief that the potentialities of gaming are considerable in many different fields of application. The *ad hoc* construction of specification, control and validation procedures with extreme attention paid to special purpose at hand could yield valuable insights and results from which the broader generalizations called for by a general theory of gaming might be constructed.

References

1. ABT, C. C., *Serious Games*, The Viking Press, New York, 1970.
2. ATKINSON, R. C., "Role of the Computer in Teaching Initial Reading," *Childhood Education* (1968).
3. BELLMAN, R., CLARK, C. E., MALCOLM, D. G., CRAFT, C. J. AND RICCIARDI, F. M., "On the Construction of a Multistage, Multiperson Business Game," *JORSA*, Vol. 5, No. 4 (August 1957).
4. BERGLER, E., *The Psychology of Gambling*, Hill and Wang, New York, 1957.
5. BOOCOCK, S. S. AND SCHILD, E. O., *Simulation Games in Learning*, Sage Publications, Beverly Hills, 1968.
6. CALLOIS, R., *Man Play and Games*, Thomas & Hudson, London, 1962.
7. COHEN, K. J., DILL, W. R., KUEHN, A. A. AND WINTERS, P. R., *The Carnegie Tech Management Game*, Richard D. Irwin, Homewood, Illinois, 1964.
8. COLEMAN, J., *Democracy*, The Johns Hopkins University Dept. of Social Relations and Academic Games Associates.

9. ——, "Social Processes and Social Simulation Games," in Boocock, S. S. and Schild, E. O., *Simulation Games in Learning*.
10. DALKEY, N. C., *The Delphi Method: An Experimental Study of Group Opinion*, The Rand Corporation, RM-5888-PR (June 1969).
11. —— AND ROARKE, D. L., *Experimental Assessment of Delphi Procedures with Group Value Judgments*, R-612-ARPA, The Rand Corporation (to appear).
12. DREYFUS, H., *Critique of Artificial Reason*, 1971 (to appear).
13. FULLER, B., *Presentations to Congress: The World Game*, Southern Illinois University, Carbondale, 1970.
14. GOFFMAN, E., "On Face Work," *Journal for the Study of Interpersonal Processes*, Vol. 18, No. 3 (August 1955), pp. 213–231.
15. GOLDHAMER, H. AND SPEIER, H., "Some Observations on Political Gaming," *World Politics*, Vol. 12 (1959), pp. 71–83.
16. HAUSNER, M., NASH, J. F., SHAPLEY, L. S. AND SHUBIK, M., "So Long Sucker," a four-person game in *Game Theory and Related Approaches to Social Behavior*, M. Shubik (ed.), John Wiley & Sons, New York, 1964.
17. HELMER, O., *A Use of Simulation for the Study of Future Values*, The Rand Corporation, P-3443 (1966).
18. HERMANN, C. F., "Validation Problems in Games and Simulations with Special Reference to Models of International Politics," *Behavioral Science*, Vol. 12 (May 1967), pp. 216–231.
19. HOGGATT, A. C., "Measuring the Cooperativeness of Behavior in Quantity Variation Duopoly Games," *Behavioral Science*, Vol. 12, No. 2 (March 1967).
20. HUIZINGA, J., *Homo Ludens*, Beacon Press, Boston, 1955 (translation).
21. LAYMAN ALLEN, *WFF'N PROOF*, New Haven.
22. LUCE, R. D. AND RAIFFA, H., *Games and Decisions: Introduction and Critical Survey*, John Wiley, New York, 1957.
23. MINSKY, M. AND PAPERT, S., "Artificial Intelligence Memo No. 200, Progress Report 1968–69," Massachusetts Institute of Technology (1970).
24. NORTHROP, G. M., *Use of Multiple On-Line, Time-Shared Computer Consoles in Simulation and Gaming*, The Rand Corporation P-3606 (1967).
25. SHUBIK, M., "On Gaming and Game Theory," *Management Science* (January 1972).
26. ——, "The Dollar Auction Game: A Paradox in Non-cooperative Behavior and Escalation," Yale University, Dept. of Administrative Sciences, Report No. 30 (1970).
27. ——, WOLF, J. AND LOCKHART, S., "An Artificial Player for a Business Market Game," *Simulation and Games* (1971) (to appear).
28. THORELLI, H. B. AND GRAVES, R. L., *International Operations Simulation*, The Free Press, Glencoe, 1964.
29. WING, R. L., *The Production and Evaluation of Three Computer-Based Economics Games for the Sixth Grade*, Board of Cooperative Educational Services, Westchester County, 1967.
30. WOHLSTETTER, R., *Pearl Harbor: Warning and Decision*, Stanford University Press, 1962.
31. YARDLEY, H. O., *The Education of a Poker Player*, Simon and Schuster, New Jersey, 1957.

[34]

Some experiences with an experimental oligopoly business game[1]
(with Gerrit Wolf and Herbert B. Eisenberg)

1. INTRODUCTION AND HISTORY

The work reported in this paper is based on a game which has existed in one form or another for some twelve or thirteen years. It was first conceived by one of the authors while at the Center for Advanced Study in the Behavioral Sciences. Three considerations made this type of game appear to be of value. They were:

1. the need to test oligopoly and game theory hypotheses;
2. the need to do so in a relatively rich environment (at least rich when compared with 2 x 2 or larger matrix games);
3. the need to automate or computerize as much of the experiment as possible.

The game which two of us constructed has about the same degree of complexity to the players as do the original AMA (1) and UCLA (2) games and many other of the current business or oligopoly games. It has 1–10 firms, each selling a (symmetrically) differentiated product in a single common market area. The firms can make pricing, advertising, production, dividend, and investment decisions each period. Sometimes they are called on to make fewer decisions. In particular, in the experiments reported on here, investment decisions were not made.

There have been four major series of games run. A batch of around 15 games were run more or less informally at the T. J. Watson Research Labs of IBM. Thirty-seven games were run at SDC under relatively good control conditions. Around 20 games have been run at Yale under varying conditions of control. Finally, there have been a series of runs with a larger multiproduct version of this game, known as the Financial, Allocation and Marketing Executive Game at IBM.

Versions of the game are now programmed for a timesharing system and the authors are currently engaged in a program of research investigating the behavior of individuals in monopoly games and in duopoly games when playing against an artificial player.

Our remarks are confined to the non-time sharing runs of the game, although it must be noted that the basic structure of the game and the body of analysis that has been developed is identical in all cases.

2. DESCRIPTION OF THE GAME

A

1. *Number of teams:* Less than or equal to ten.
2. *Number of players per team*: This is not in the program but is to be determined by the experimental, teaching, or operational design.
3. *Number of products per team:* In the current program there is one product per team.
4. *New products:* Not in this program.
5. *Intermediate markets:* Not in the program.
6. *Number of markets:* There is one market.
7. *Price:* Price is an independent variable under the control of each team.
8. *Production:* Production is an independent variable. Production costs are linear. There is an upper bound specified on the size of production during any quarter. Actual production and the production decision may be lagged. There are costs attached to changing the level of production.
9. *Distribution:* Distribution is not yet given as an independent decision variable.
10. *Development:* Development is not given as an indepedent variable (although, by a rewriting of the scenario, the advertising decision variable could play that role).
11. *Advertising:* Advertising appears as an independent decision variable. Competitive and cooperative effects of advertising are controlled by different parameters. The impact of advertising may be cumulatively lagged over several periods.
12. *Cycle:* A cycle in the overall economy can be utilized in plays of this program. (In general, we have run without a cycle.)
13. *Trend:* A trend can be specified.
14. *Inventories:* There are constant unit inventory carrying costs.
15. *Demand function:* The overall demand, leaving out the effect of advertising and assuming that the prices charged by all firms are identical, is linear. When the prices charged by the firms are not the same, there is a rationing scheme described in detail elsewhere (3) which computes the contingent demand schedules. Lost sales are also recorded. [61]
16. *Random variables:* The introduction of random variables is optional. One may be introduced to influence overall demand; others, to effect individual advertising.
17. *Supply:* A firm may offer for sale fewer items than it has in inventory. Thus an active inventory policy may be followed.
18. *Entry:* The entry routine calls for a new firm to enter the market whenever the entry-profitability requirement is exceeded. This calls for the return to investment to be above a certain level for a number of periods (specified by the user of the game). In the games reported on, entry was not considered.
19. *Exit:* A firm may voluntarily liquidate at any time, paying the proceeds of liquidation into a "bank account" where they will earn interest for the remainder of the game. A firm may be forced into bankruptcy and be

compelled to liquidate if it exhausts bank loans and runs out of short-term assets.

20. *Financing:* The only financing in the current pogram is by means of bank loans. These loans are made automatically as the cash needs of the firms call for them. There is an upper limit on the loan amount, and the repayment of loans has priority over the payment of dividends. (This rule can be easily modified.)

21. *Dividends:* During each quarter, the firms may pay dividends into a "bank" or outside account. These payments will earn interest, as distinguished from the cash held in the firm. Once paid out, however, dividends are no longer available as resources for the firm.

22. *Liquidation values:* The liquidation value of a firm may be specified as a percentage of its current book value.

23. *Discount rate:* There is a rate of interest specified for the economy. Dividends earn this rate, and this rate or more will be paid on bank loans.

24. *Overheads and depreciation:* The overheads are specified as parameters of the program. Depreciation depends in a linear manner upon the amount of capacity of the firms. (One may imagine that depreciation is spent upon maintaining capacity constantly; hence, the strategy of letting plant deteriorate is not considered.)

25. *Investment:* There is a simple investment program permitted in the procurement of new capacity. This can be purchased at a fixed cost per unit.

B

1. *Initial conditions:* Many parameters must be set to initialize the program. In starting a game, two broad considerations must be taken into account. The first concerns the 'realism' of the environment. Using the instructions given elsewhere (4), it is possible to select (at least crudely) parameters to give reasonable first order approximations to some of the salient features of different markets such as automobiles, refrigerators, etc. The second concerns the instructions given to the players. One unfortunate feature of gaming is that it is not possible to 'clear memory' of the players. As the game, even in a highly complex form, will tend to be a drastically poor environment in comparison to the usual world of the players, it is easy for the experimenter, teacher, or operational user of a game to supply misleading or biasing cues to the players in their initial briefing.

2. *Information conditions:* For the most part, the briefing on information conditions lies outside the program. The exception to this relates to the amount of information supplied concerning the specific actions and balance sheets of the competitors. For example, within the program, it is possible to select printouts which supply all firms with detailed information concerning the amount spent by each firm on advertising, to merely supply each firm with the average amount spent on advertising, or to supply each with no other information than that which he already knows, to wit, his own expenditure.

As a rule, many experimental and almost all business games that have been run are not games in the strict sense of game theory, inasmuch as the players are not generally cognizant of all the rules. They do not know the functional forms which provide the basic structure of the game (for example, the details of market structure). Furthermore, they do not know the values of parameters which delimit the sensitivities of response surfaces. Thus there are many different ways in which experiments can be performed with 'pseudo-games' varying the degree of incompleteness of the knowledge of the players concerning the rules, as well as varying the different levels of information in the sense of game theory.

3. *Dummy players:* The current program can be easily adapted to permit the playing of one or more positions to be done by a program which simulates behavior of a player according to certain simple theories of oligopolistic behavior. This has been done in the timesharing version.

4. *Analysis:* Although the program permits the playing of games with asymmetrical relations between the players, much of the current analysis, derivation of solutions, and experimentation for different theories of behavior is much easier if symmetry is assumed. Although the non-symmetric game has been analyzed, we discuss and report on only theory and experimentation with the symmetric game.

5. *Objective function and payoffs:* The presence of dividends, an interest rate, and a method for evaluating assets enables the experimenter to completely define the maximization of the expected discounted worth of assets plus dividends as a goal for the players. This is only one of [62] several goals that may be specified to the players. For example, maximize market share is another possibility. One of the problems encountered in gaming is that, although the players may be instructed to follow some particular goal, they may possibly be striving for another. This may easily happen when the stakes for which the players are ostensibly playing are of no interest. Thus, if they are not being rewarded in proportion to their performance in a manner that is of sufficient importance to them, the game may be transformed into one of status, where they strive to be 'first' according to some measure. It is necessary to check for this possibility, and our analysis program does so.

6. *End of play and termination rule:* As is well known from game theory analysis, it is possible to introduce many pathological effects by announcing a specified time of termination for a game played over many time periods. Most human affairs do not have a fixed point of termination; corporate existence is one of these. The announcement of a fixed date of termination introduces an 'end of the world' effect; individuals may be induced to demonstrate pathological behavior in reaction to their knowledge that at some point 'there will be no tomorrow'. In practice, the terminal pathologies do not appear to be as great as might be expected from a consideration of a formal, rationalistic theory of games; nevertheless, they can be sufficiently marked that they must be taken into account. There are many ways of doing so, such as allowing the end of the game to depend upon a random process.

The prime purposes for which this game was originally designed have been experimental, teaching, and operational economics. These are by no means the only purposes. A game of this type serves as a piece of laboratory equipment and a data organizing system which enables experimenters with different professional interests to perform several relatively small experiments simultaneously, using the same subjects and obtaining different, but consistent, sets of information which can be cross-correlated if the different specialists wish to consider the interaction between the various aspects of decision-making. Thus, for example, it is feasible to investigate simultaneously hypotheses concerning competitive, organizational, and learning behavior. An example of a three-way experiment would consist of a game with a steep response surface and a considerable lack of knowledge concerning the rules of the game. The economic hypothesis might be concerned with where average market performance will settle down, the psychological hypotheses could deal with the shape of the learning curves and by the appropriate design: say, in a six-team game, two different team sizes with a specified organization could be investigated.

The current game is easy to play and difficult to analyze. It is our belief that an approach to the understanding of multivariate behavioral systems can be made by constructing, investigating, and experimenting with games of a 'middling' size – more complicated than most of the strictly experimental games used in psychological and simple economic experiments, and less complicated than the larger business and war games. The policy of the authors towards the enlargement of this game is to add subroutines or gaming laboratory instructions only when they can be accompanied with at least some hypotheses specific to the variables being added, together with a scheme for analyzing the output. The game and current analysis, and other additional programs, are available at cost to those who wish to use them.

A sample player input sheet is given as Exhibit 1 in Appendix A. Two decision sheets appear to be adequate for most games. The non-timesharing games run by the authors and others have tended to be between 15 and 20 periods in length. They have been run on the basis of one decision period per day, with the teams turning in their decisions by noon and obtaining the outputs by the end of the day or early morning; or in a one-day session in a laboratory. The time taken to make the decisions for a quarter has varied from several hours to ten or fifteen minutes. The time-sharing games have tended to be 12 periods in length and require one or two hours to run.

A sample of the initial briefing and the outputs is provided in Appendix A.

3. SOME THEORY, MATHEMATICAL STRUCTURE, AND PARAMETERS

Elsewhere, for those interested in the mathematical structure of the game, we present the equations, the derivation of the joint maximum, non-cooperative equilibrium and beat-the-average solutions, together with a discussion of how to select parameters in order to produce the desired sensitivity of the payoffs to different strategies.

Here we merely note that we used the three solution concepts noted above to provide guidelines in investigating the data. They are static solutions and, hence, at the best can be regarded only as limiting steady state predictions when studying a game played through time.

In particular, joint maximum (as the name suggests) calls for more cooperation than is in general reasonable to expect when no direct communication is possible among the players. The non-cooperative equilibrium behavior is one that might result from reasonably self-centered, informed, purposeful individuals, each maximizing in their own interest, treating the other players as though they were part of the environment rather than active competitors. The beat-the-average solution is one in which players take their cues [63] from comparative performance. They strive to do better than the others rather than do well in any absolute sense.

4. SOME EXPERIMENTAL RESULTS

In total, 37 games were run at the Systems Development Corporation, 15 at the T. J. Watson Research Laboratories of IBM, and 7 at Yale University. In this discussion, however, we limit our treatment of data to the SDC games, as they are more easily comparable. Because of design flaws, these data are more useful to indicate problems in control and to suggest other studies than to provide conclusive evidence for simple economic hypotheses. In particular, the lesson that comes through loud and clear is that economic effects can easily be confounded by psychological and sociopsychological factors and that briefing, control of subject selection, and especially control of payoffs are of considerable importance in simple psycho-economic experiments of this variety.

We believe that, although the results are disappointing from the viewpoint of testing certain specific conjectures, the experiments and the difficulties with them merit reporting, if for no other reason than that they led the authors to perform two other experiments (5). Furthermore, the experiences with these games and the work done in arranging to run the series of experiments suggested the type of automated gaming laboratory system and data processing needed to perform this type of experiment effectively and cheaply (6).

In the remainder of this section we describe the experimental conditions, the hypotheses, and the results. They are further interpreted and discussed in Section 5. (Parameter settings and initial conditions, originally reported in an Appendix B, are available to interested readers.)

A total of 151 individuals, a majority of them male and employees of the Systems Development Corporation or undergraduates from a nearby college, participated. They were paid an hourly wage for their participation; thus their payments were not correlated with their performance. They were run in the gaming laboratory of SDC. The subjects were relatively naive and no further background data were collected.

Two major types of games were run. In series A, the players were actually monopolists; in other words, their moves did not influence each other's payoffs through the mathematical structure of the game. In 19 of the 21 monopoly games, the players did not know that they were monopolists. They saw the moves and payoffs of their 'competitors' after each period, along with the information on their own performance. In two games, a 3-person game and a 7-person game (denoted by

A3M and A7M), the players were told that they were monopolists – i.e., that their fates were not strategically intertwined.

The series C games had a competitive market structure, as can be seen from the parameters available from Appendix B.

In the series A, advertising had no effect even though it was a decision variable. In series C, it had an effect; however, under conditions of extreme competition the optimal advertising level would be zero.

The experiment factorially varied market structure and size. In addition, team assistance was varied in some conditions. For the most part a team consisted of a single individual in isolation. However, for 6 of the A2 and C2 games, one or two advisors were attached to the teams. We were unable to observe any clear effect of the advice on the quality of the decision making and, therefore, collapse the data, ignoring this experimental variation. Apart from noting it here, we make no further reference to this condition.

The market structure was either monopolistic or competitive. Markets of size two, three, five, seven, and ten firms were run. With this design, it was hoped to ascertain the extent to which the number of firms in a market affected the level of competition.

The following table shows how the players were distributed according to experimental conditions.

Market structure	Market size				
	2	3	5	7	10
Monopolistic					
People	16	18	10	21	20
Games	8	6	2	3	2
Competitive					
People	16	6	10	14	20
Games	8	2	2	2	2

As can be seen, the cells of the design have unequal Ns. Analyses took this into account. Also, individual acts instead of a play of the game were used as observations. This maximized the number of observations, but not all can be considered independent even in the monopolistic games. For example, in a market of size 5, ten people played two games, five in one game, five in another. Within a game, the observations are not independent. This poses problems both of analysis and design: the larger the market, the more people are needed. One exhausts subject supplies quickly the larger the market. Also, an uneconomically large number of observations would be needed to obtain equal Ns.

The instructions for the runs appear in Appendix A. It can be seen there that parties receive information about the variables and their relationship. However, the parameters of the model representing the market were withheld. One of the problems for each firm was to estimate the parameters and, in turn, to understand the market environment in which it was acting. [64]

Besides the four decision variables, output variables such as sales, inventory, lost sales, net profit, and gross profit were a natural product for each decision period. All firms played at least twenty periods, some even more, which usually took at least a day. Data were processed in batch, necessitating longer time intervals between decisions than would occur with an online computer processing decisions. Firms received balance sheets similar to those in Appendix A.

Price and advertising decisions were chosen for analysis because they would probably be the most revealing of behavior in these environments. Performance on these variables relative to optimal values (analytically solved for based on the model of the market) provided a way for determining how well people understood their environment and what kinds of motivation they had in relation to it.

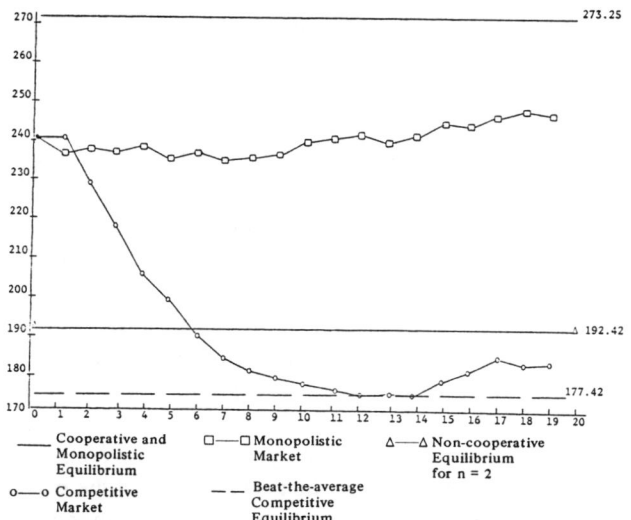

Note: These data exclude two A2 and four C2 games for which less than 20 periods of data were available. Unfortunately, during the running of the experiments, scheduling problems resulted in an uneven duration for the games. However, in no case were the players informed of the terminal play prior to playing.

Figure 1. Mean prices over time for monopolistic and competitive market situations.

Figure 1 reports the graph of mean price per economic quarter or decision period for each of the experimental market structure conditions. Everyone started at period zero with a price of $240. One sees the price in the monopolistic condition slightly increasing over time compared to a price decrease in the competitive condition. The monopolistic average is $240; it decreased to $235 by period eight and rose to $245 by period 20. The competitive average over all periods was $193, significantly different from the monopoly $240 ($F = 7.88$, $p < .01$). The competitive price barreled down to a low of $174 by period 13 and climbed back to $184 by period 20. These two dips in price, one in the monopolistic and one in the competitive market, but in different amounts, were significant also ($F = 5.81$, $p < .01$).

Advertising showed greater responsiveness to the experimental market variables, as seen in Figure 2. Everyone started at period zero with $6 million expenditure for

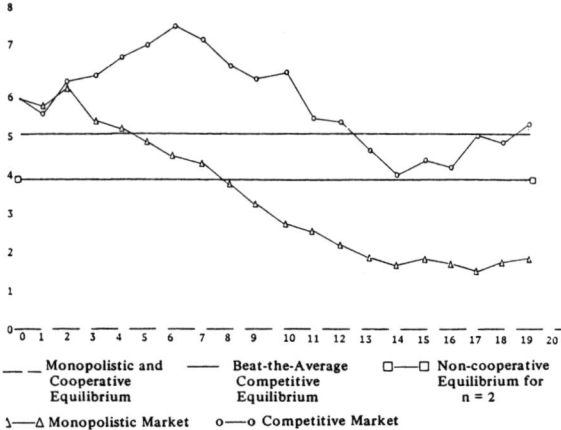

Figure 2. Mean advertising (millions) over time for monopolistic and competitive markets.

advertising. Competitive advertising in the first half dozen quarters went up by one-third and then declined for the rest of the periods to a low of $4 million and a final of $5 million, which is the competitive equilibrium. By contrast, non-competitive advertising regularly shrunk to a low of $2 million, which is approaching the optimal of zero advertising. These trends were significantly different ($F = 4.21$, $p < .01$). The average over time between the two structures ($6 million > $3.5 million) was also significant ($F = 27.7$, $p. = .01$).

Prior to running the experiment several hypotheses were considered.

Hypothesis 1: Within 15–20 plays, the average price of the players in any of the five markets would be within a 10% bandwidth of the monopoly price (where 100% is the difference between monopoly price and beat-the-average price, or 273.25–177.72).

Hypothesis 2: Within 15–20 plays, the average advertising would fall to within 10% of the optimum level for the monopolists (where 100% is the difference between the starting level of advertising and the optimum or 6,000,000–0).

In the first case, the range was defined independent of the starting point, and a better hypothesis might have made use of the initial price explicitly. However, it is important to note that both the distance between the monopoly price (or joint maximum) and the beat-the-average price as well as the starting point appear to be of importance. In this experiment, the advertising parameters were picked so that the A games were not influenced by advertising expenditures; hence, the [65] joint maximum – or for that matter any other type of solution – all call for the same expenditure, to wit, zero. This being the case, in our second hypothesis it was necessary to introduce starting point considerations explicitly.

Hypothesis 3: Within 15–20 periods, the overall average price in the C games would be in the bandwidth bounded below by the beat-the-average price and above by the non-cooperative equilibrium price for two players.

This is an extremely weak hypothesis and should be compared with Hypothesis 7.

Hypothesis 4: Within 15 to 20 periods, the overall average advertising in the C games would be in a bandwidth bounded above by the beat-the-average advertising expenditures and below by the non-cooperative equilibrium advertising for a two—player game.

The above four hypotheses concern aggregate behavior. The first two were emphatically not confirmed, as can be seen from Figures 1 and 2. There was in each case a tendency in the right direction, but it was by no means as fast as had been expected. More will be said about this in Section 5.

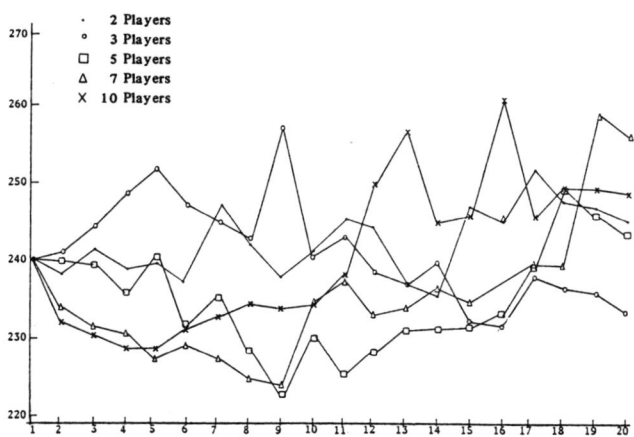

Figure 3. Mean monopolistic price for five markets over time.

The other two hypotheses are substantially confirmed, as can be seen from Figures 1 and 2. The weasel word 'substantially' is used because the hypotheses are not sufficiently carefully phrased. If interpreted literally, the last reading in Figure 2 can be regarded as disconfirming Hypothesis 4. However, from the economist's point of view, the interpretation is quite clear; they are substantially in the zone characterized by competitive behavior.

The other hypotheses concerned market size.

Hypothesis 5: The larger the size of the game for the A games, the quicker would the average price in that market achieve the 10% zone.

Hypothesis 6: The larger the size of the game for the A games, the quicker would the average advertising achieve the 10% zone.

Neither of these hypotheses is confirmed for the simple reason that convergence was far slower than expected (see Figures 3 and 4). However, it is of interest to note, as can be seen from the data in Appendix C and from Figure 4, that there appeared to be a tendency for advertising to decrease faster as market size rose. Only markets of size 7 violated the rank order. The basis for these two hypotheses was that the bigger the game, the more individual performances a player is able to see; hence, he has a large basis of information for judging his performance against others.

Hypothesis 7: For a C game of size n, within 15 to 20 plays, the overall average price would be within a bandwidth bounded below by the beat-the-average price and above by the non-cooperative equilibrium price for the n player game.

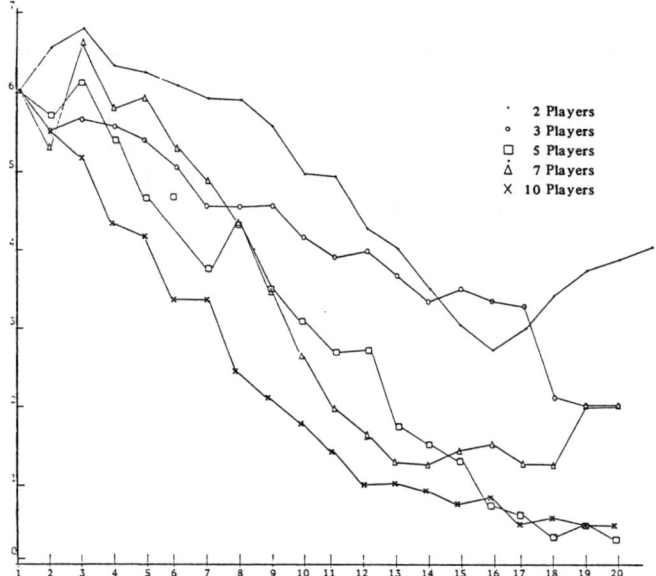

Figure 4. Mean monopolistic advertising for five markets over time.

Hypothesis 8: For a C game of size n, within 15 to 20 plays, the overall average advertising would be in a bandwidth bounded by the beat-the-average advertising expenditure from above and the non-cooperative equilibrium advertising expenditure for the n player game from below.

Unfortunately, neither of these hypotheses is confirmed by the data, as can be seen from glancing at the time series in Appendix C.

Figures 5 and 6 provide a little more insight into the importance of hypotheses concerning numbers. The solution in Figure 5 marked JM is not only the joint maximum for the C series, but is also all of the solutions for the A series (i.e., joint maximum, the non-cooperative equilibrium, and the beat-the-average solution all predict the same outcome in this case).

The curve marked NCE shows how non-cooperative equilibrium changes as the number of [66] players is increased. It can be shown mathematically that as the number of players increases the non-cooperative equilibrium converges to the beat-the-average solution.

Ex post facto, we have selected a single time period t = 15 to serve for our illustration. It was selected on the basis that we had expected 'things to settle down around the 15th to 20th period'.

The bars show the upper and lower bounds, the outcome for each game, and the average for each class of game during the 15th period of play.

Turning to Figure 6 we note that the non-cooperative equilibrium expenditures on advertising increase, with numbers approaching as their limit the beat-the-average expenditures. We may observe that the variance in advertising for any size of game appears to be larger than the variance in price. The jointly maximal amount of advertising is zero. This is the same as in the series of A games.

Political Economy, Oligopoly and Experimental Games 455

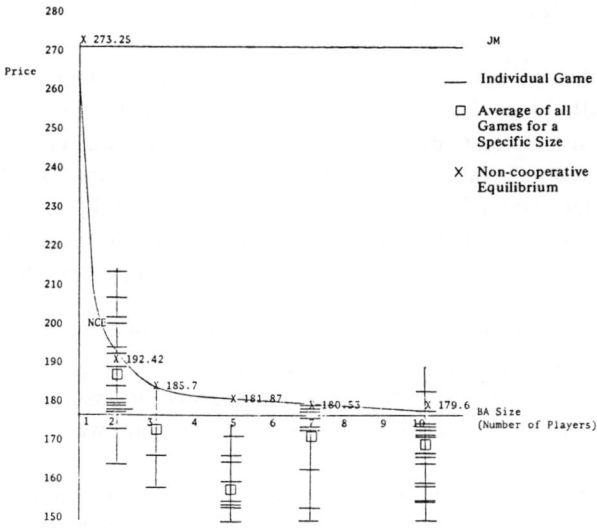

Figure 5. Joint maximum, non-competitive equilibrium, and beat average solutions, and the results of the 15th period of play for competitive markets of size 2, 3, 5, 7, and 10.

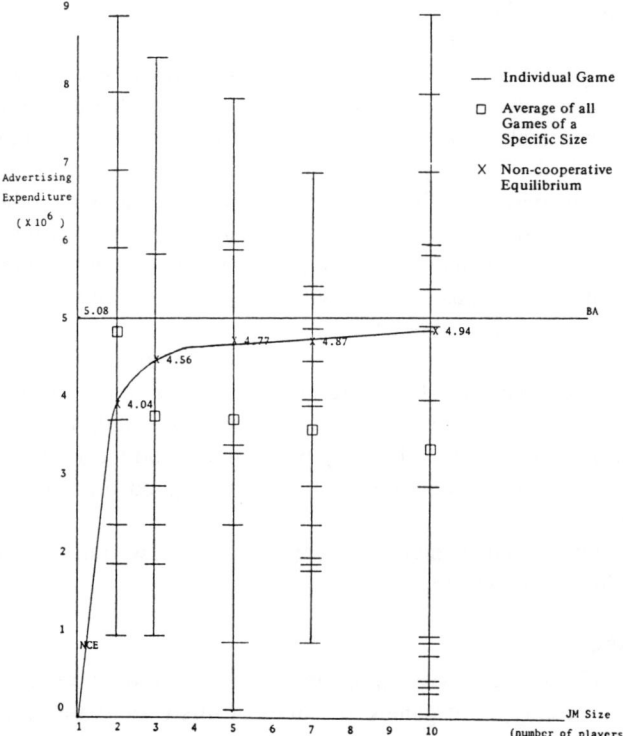

Figure 6. Advertising expenditure for markets of size 2, 3, 5, 7, and 10.

5. DISCUSSION

The hypotheses were framed in terms of static economic models, ignoring perceptual difficulties, learning processes, and dynamics in general. Furthermore, with one of us, at least, there was a somewhat sanguine view of 'how simple economics really is'. Ex post, this does not appear to be the case. There are strong indications that many of the players never really quite fully understood what was going on. This may account for the performance in the monopoly games (even when they were told in two of the games that they were monopolists, this did not produce significantly different behavior).

Even though the players may not have fully perceived the nature of the game they were in, the force of the economic variables in the C games was sufficiently powerful that in aggregate the economic theorizing associated with the non-cooperative equilibrium and beat-the-average solutions were of some predictive value.

The reason for including both of these theories and examining the bandwidth between their two predictions is as follows. The non-cooperative theory assumes that each individual is intent upon maximizing his profit and that he understands what he is doing. He, however, adopts a neutral attitude towards the rest of the world. He is neither hostile nor cooperative. He merely treats other competitors as part of the scenery and maximizes his payoff on the basis of his view of the environment. The beat-the-average solution is more directly interactive and competitive. Each individual looks to see how the rest are doing and attempts to do better. This type of behavior appears to fit three conditions: (1) where players are extremely competitive and strive for 'status' or to win rather than to maximize score; (2) when the actual stakes or rewards are low, or where the point score is not correlated with the actual payoffs; and (3) where the individuals are relatively ignorant and the best clues they have to play are to see how the others are doing.

When there are several players in a game and these players are not in direct face-to-face communication and the problem is somewhat complex, the presumption is that cooperation or collusion [67] will become more and more difficult as the numbers increase. Thus, although with two players we might not be surprised if cooperation were observed, with three or more it would be increasingly surprising. The non-cooperative equilibrium reflects a lack of cooperation, but not direct hostility or competition. If the players were being well paid, if their pay were related to their 'profits' or scores, and if they fully understood the game, we would expect that the non-cooperative equilibrium would be a good predictor for many-person games.

If pay is not correlated and/or perceptions are poor, we might expect the beat-the-average solution to do better.

Control of subjects
It appears to be desirable to adopt a somewhat more formal procedure for screening subjects for these types of games. If games of this variety are being run on a mass produced basis, then it may be worth keeping personality, IQ tests, or other records on each player. Leaving this aside, possibly the most important piece of information

would have been some type of test to see if the players understood the problem prior to playing. In a successful experiment influenced by the work here, this was done.

Simplicity

Although the game as it stands is relatively environment poor, there are still many experiments that could be done with this type of game in which the players could be required to make decisions on only one variable, such as price, production, or advertising, rather than on all three simultaneously. Experiments in triopoly with only one decision variable have been run successfully by several individuals.

The control over information in the games run here was such that the individuals saw the individual results of their competitors. A further simplification is to present the competitors with only aggregated summaries of their opponents' moves and fortunes, thus introducing a greater level of 'anonymity'. It is this possibility that stresses the distinction between games with two teams or with more than two.

Dynamics

No hypotheses about dynamics were made in this experiment. However, with these games, games played at IBM, Yale, and elsewhere, a distinct damped sinusoidal curve in the movement of prices was observed, with the first low appearing somewhere between the 8th and 15th periods (depending upon where the initial price is picked). The type of price behavior referred to is illustrated in aggregate in Figure 1.

In one game at IBM, the introduction of a sinusoidal change in overall demand appeared to be magnified in the price behavior of the teams. In another exploratory game, by selecting initial prices below the non-cooperative equilibrium, the fluctuation went in the other direction. We do not have a simple theory to explain the damped sinusoidal movement of price. In fact, in games of considerable length, somewhere around the 25th period or after we have observed an occasional 'flare-up' or increase in the oscillation, which may be related to boredom that appears to set in after having been in a steady state for a while.

There are several different factors, all of which contribute to making this type of sinusoidal behavior plausible; however, it would require several experiments and careful control to investigate the contribution of each. In particular, a simple expectations model, in which each maximizes on the assumption that the others will do as they did last time, will lead to a damped oscillation in this game. Production-inventory scheduling errors can lead to oscillations, which in this environment may or may not be damped. In particular, in some circumstances it can be shown that under price competition a simple non-cooperative equilibrium may not exist and there is a small inherent instability in the market which precludes any steady state and is manifested in the occasional large erratic swing (7). Finally, both individual and socio-psychological learning (signalling and teaching) could account for this type of behavior.

Artificial players

One way in which more experimental control can be obtained (and better automation of experimental procedures) is by using an artificial player or players in experimental games. This is especially useful in two-person games. In work reported elsewhere, an artificial player has been constructed for a time-sharing system on which this game can be played (8), an experiment has been run with an artificial player playing a single human in a duopoly situation. It is, of course, possible to run artificial players against each other, or to run a single human in an n-person market where the others are all artificial players. We have not yet done this, although we have the program that enables us to do so for $n \leq 6$ (this size restriction is caused by width of a paper roll).

The dimensions of investigation
This game was designed specifically to be simple enough to make analysis and the mathematization of various economic, psychological, and socio-psychological hypotheses tractable within the context of the game. Yet, at the same time, we believe that it is sufficiently rich and complex that many fundamentally different experiments can be done with it. In particular, in the SCD experiments reported here, our hypotheses were limited to overall aggregate behavior and to behavior broken down into aggregates of different sizes and types of game (for example, A2, C5). The breakdown available to us is:

Overall aggregate behavior,
Behavior in aggregate on a specific type [68] of game,
Behavior of one set of teams playing a specific game,
Behavior of individual teams,
Behavior of individual within teams.

We have not used the last three categories here.

Furthermore, it is important to break down behavior in a game of this type in terms of:

Accuracy of estimation of the physical environment,
Accuracy of prediction of the acts of other players,
Design of strategy, and
Internal efficiency and control.

These are all aspects of the management of a firm in an uncertain environment.

Finally, there is a need to understand the learning aspects that accompany the performance of the functions noted above.

Burial by data
It may be that there is a 'black body law' for gaming experiments. The value of the experiment may vary inversely as the fourth power of the output that is not immediately data processed. We were buried by data, and to some extent still are.

Any fool with a sufficient IQ, sufficient energy, and enough money can arrange to record on cards, tape, audio tape, visual tape, etc. everything that goes on in a gaming laboratory. It is seductively tempting to keep everything 'in order to analyze it at leisure later'. This is generally a snare and a delusion. It may be possible that many different individuals simultaneously wish to view the same experiment from different viewpoints. This is probably extremely desirable. However, the best way that this is achieved is by having each of them design his analysis package so that when the tons of paper start to pour out of the computer he knows what to do with them (or, better still, he does something that obviates the need for the cascade of paper).

If you really do not know what you are after, it is probably better to run a few pilot studies, labelled as such, rather than have experiments turn out that way by accident.

In the experiments noted above, we should have run a fair size pilot study at SCD prior to the main runs. A pilot study had been run at IBM previously. However, it appears to be a safe rule of thumb that, until gaming experimentation is far more organized than it is at present, any attempt to run an experiment at another locale calls for a pilot run at the new locale.

We still have most of the raw data and hope 'to be able to analyze it further at leisure later'.

The automated experiment

It is easy to show that there are literally thousands of small experiments that could be usefully done if it were possible to do them cheaply and quickly enough. The uninteresting noise to one individual may well be the main interest to another. Our experience with this and other runs of this game, as well as with other economic games, indicates that it is desirable to be in a position to automate as much as possible the mechanics, input, running, and analysis of gaming experiments. In particular, not only is it important to be able to do many small experiments easily and to examine one in the light of another, it is also highly desirable to have more exact replications of the same experiment than is currently the fashion.

A good experimental game should be looked upon more as a public good or a common piece of experimental apparatus rather than as an individual piece of work. It is too expensive for each individual to build the large volume of special tests and data processing needed for the effective test of any specific apparatus. It is too unwieldy to expect that many individuals will go to the effort of building the tests and programs from scratch to replicate someone else's experiment. If the apparatus in a *well-documented* form is available to many, then each can add routines of use to themselves and possibly to others who wish to replicate work or do closely related work utilizing the previous expenditure of effort.

Notes
1. The research on which this paper is based was supported by the National Science Foundation under Grant GS-2840, 1970.

References
1. Ricciardi, F. *Top Management Decision Simulation.* American Management Association, 1958.
2. Jackson, J. 'Learning from Experience in Business Decision Games'. *Calif. Mgmt. Rev.*, 1, 2, 1959; *The Executive Game.* Homewood, Ill: Irwin, 1966.
3. Levitan, R. 'Oligopoly Demand'. *IBM Corporation Research Report RC 1239*, Yorktown Heights, NY, July 1964.
4. Shubik, M. 'A Business Game for Teaching and Research Purposes', Part III. *Yale University, Cowles Foundation Discussion Paper 180*, November 1964.

5. Shubik, M., Wolf, G., and Lockhart, S. 'An Artificial Player for a Market Business Game'. *Simulation and Games*, 2, 1971, 27–43; Feder, P. and Wolf, G. 'Data Analysis of a Monopoly Business Game'. Yale University, Dept. of Administrative Sciences, Fall 1970.
6. Katz, R. and Wolf, G. 'A Computer Program for an Oligopoly Business Game'. Yale University, Dept. of Administrative Sciences, Fall 1970; Pomer, M., and Wolf, G., 'A User's Manual for an Oligopoly Game Data System'. Yale University, Dept. of Administrative Sciences, Fall 1970.
7. Shubik, M. *Strategy and Market Structure* (Ch. 5), New York: Wiley, 1959.
8. Shubik, M., Wolf, G., and Lockhart, S., op. cit.

APPENDIX A

Briefing and outputs

The major decisions which a player can be asked to make are:

price
production
advertising
dividends

and with a more long-run impact:

investment and
exit from the industry.

In most games, price, production, and advertising were the decisions. Sometimes dividends were paid.

Exhibit 1 shows the decision sheet for each team.

Exhibit 1: Decision record

DECISION RECORD

Company _____ Date _____

Quarter	Price	Advertising (000)	Production (000)	Dividends (000)	Investment (000)
16					
17					
18					
19					
20					
21					
22					
23					
24					
25					
26					
27					
28					
29					
30					

Exhibit 1

The output formats contain individual profit and loss and balance sheets for the firms and a report on industry statistics. The example presented here contains complete information on all industry statistics; the display of the amount of information on this output sheet is under the complete control of the manager of the game. It may be observed that the sum of the individual lost sales does not equal that of the lost sales to the industry. This is because they are not selling identical products, and there is a redistribution of unsatisfied demand to firms which have goods for sale. Only one-half of the current production in any period is available for sale during that period.

A sample instruction protocol which has been used by the present authors to brief players playing under conditions of almost complete knowledge of the rules is given as Exhibit 2. [70]

The following is an example of instructions given to players playing under conditions of complete knowledge of the rules of the game.

Exhibit 2: Instructions to players of business game #5

You may imagine, if you wish, that the product you are producing is some major appliance, say refrigerators. There are five (5) firms in the industry; you are to make quarterly decisions about production, quantity marketed, price, advertising, and dividends.

Goals for business policy Firms are expected to play in such a way as to get the largest possible current value of accumulated dividends and net assets. The dividends are assumed to earn compound interest for stockholders at 4% per quarter.

(In many games the dividend decision was not made.)

Dividends Firms are free to select any dividend policy except that dividends can only be paid to the extent that funds are available after bank loans are repaid.

Quantity marketed The quantity marketed by firms is bounded by starting inventory plus one-half of current production.

Ruin Firms must have $10,000,000 in cash to cover operating transactions. Firms start with a $50,000,000 line of bank credit. This will be drawn on to make up deficiencies where possible. If a firm cannot keep its cash amount above the ruin level, it will be liquidated at 60% of book value of net assets which will be credited to the dividend accounts on which interest will continue to accrue.

End of game At the end of the game, net assets will be liquidated at book value.

Start Initial cash is $20 million, and inventory is 100,000 units. Last production for each firm was 250,000 units, last advertising was $500,000.

Production capacity Quarterly production is limited to 500,000 units per firm. You are not able to purchase more capacity in this game.

Information conditions At the end of each quarter, you will obtain your balance sheet and detailed information on your market moves, as well as those of your competitors.

Costs are as follows:
1. Production costs
    $170 per unit produced plus
    $10 per unit change to production rate per quarter
2. Inventory
    $2 per average unit per quarter
3. Fixed costs
    $500,000 administrative overhead
4. Advertising
5. Depreciation
    5% per quarter
6. Interest
    4% per quarter on loans at beginning of quarter
7. Taxes
    50% on net profit. Rebate for losses.

Structure of the market The general form for the demand for your product given by:

$$\text{Demand}_i = \frac{1}{n}\left[.26 \times 10^7 - \frac{2}{3} \times 10^4 [p_i + 3[p_i - \bar{p}]]\right]$$

$$[1 + 10^{-4} \sum e_i][.8 + .2\frac{ne_i}{\sum e_i}]$$

where n is the current number of players
    p_i is the current price of player i
    $\bar{p}$ is the current average price

$$e_i = \frac{1}{2}a_i^t + \frac{1}{2}a_i^{t-1}$$

where a_i^t is the advertising expenditure of player i in the current period.

When there is excess demand (i.e., some firms cannot supply the demands made upon them) or when certain other conditions are not fulfilled, a rationing method re-allocates demand as a function of prices and advertising of those firms with inventories.

Initial period You may consider that you have taken over from a previous management who was operating your firm. Their performance and the performance of the rest of the industry is available to you in the form of last quarter's profit and loss statements, balance sheets and industry statistics.

APPENDIX B

Some statistics on the full sample of games
Details of the initial conditions and parameters for all of the games are available and can be supplied on request to the interested reader.

APPENDIX C

Table 1 SDC game: Mean prices by structure and market size over 20 quarters

	Firm size		N	1	2	3	4	5	6	7	8	9
Competitive	C_1	2	16	240.00	233.63	225.94	218.00	212.75	206.63	201.19	199.94	194.81
	C_2	3	6	240.00	237.17	220.00	208.50	190.00	190.67	182.67	180.67	177.67
	C_3	5	10	240.00	240.50	232.90	221.60	216.60	202.63	192.38	185.25	179.63
	C_4	7	14	240.00	240.07	224.64	210.71	198.07	190.14	180.29	177.29	176.69
	C_5	10	20	240.00	251.75	235.30	227.65	214.05	206.25	193.47	184.68	183.58
Average				240.00	240.62	227.76	217.29	206.29	199.26	190.00	185.56	180.83
Monopolistic	C_1●	2	14	240.00	239.36	242.29	239.93	240.93	238.43	248.00	243.14	238.71
	C_2○	3	18	240.00	242.28	245.50	249.67	253.11	248.56	246.06	242.22	258.67
	C_3□	5	10	240.00	241.50	240.50	237.00	241.80	233.10	236.50	229.80	225.50
	C_4△	7	21	240.00	235.00	233.57	232.10	228.86	230.29	229.24	226.71	226.00
	C_5x	10	20	240.00	232.95	232.05	230.25	230.80	232.70	233.90	235.10	235.70
Average				240.00	238.22	238.78	237.79	239.10	236.61	238.74	235.40	236.92
Average	→											
	B_1	2	30	240.00	236.49	234.11	228.96	226.84	222.53	224.59	221.54	216.76
	B_2	3	24	240.00	239.72	232.75	229.08	221.56	219.61	214.36	211.44	218.17
	B_3	5	20	240.00	241.00	236.70	229.30	229.20	217.86	214.44	207.52	202.56
	B_4	7	35	240.00	237.54	229.11	221.40	213.46	210.21	204.76	202.00	201.04
	B_5	10	40	240.00	242.35	233.67	228.95	222.42	219.47	213.69	209.89	209.64
Overall average				240.00	239.42	233.27	227.54	222.70	217.94	214.37	210.48	209.63

10	11	12	13	14	15	16	17	18	19	20	Average
195.88	193.25	191.06	190.81	190.63	190.13	188.71	197.71	196.71	193.10	193.38	202.71
179.67	176.83	174.83	174.83	176.50	175.83	178.83	175.67	175.00	174.17	172.67	188.11
171.88	172.50	173.25	166.38	169.75	163.50	177.00	168.75	175.00	175.67	175.67	190.04
175.69	175.31	172.77	172.62	172.46	174.00	175.08	174.38	173.54	179.33	182.83	188.26
181.05	173.79	172.00	167.22	170.22	171.22	176.39	188.89	213.78	197.44	194.78	197.18
180.83	178.34	176.78	174.37	175.91	174.94	179.20	181.08	186.81	183.94	183.86	*193.26*
243.21	246.64	243.43	239.29	236.79	248.33	246.08	253.58	248.75	248.17	246.83	243.59
242.06	245.00	240.94	239.28	241.89	234.00	233.87	240.00	239.00	238.13	236.00	242.81
231.40	227.10	229.90	232.20	232.40	232.80	235.30	241.00	251.20	247.30	245.20	236.57
236.48	239.00	235.57	236.39	238.76	237.17	247.14	241.05	241.81	259.36	256.07	237.55
235.90	239.60	251.00	258.75	246.90	247.65	262.00	247.25	251.70	251.30	250.55	242.30
237.81	239.46	240.17	241.16	239.35	240.10	244.88	244.58	246.49	248.85	246.93	*240.57*
219.54	219.95	217.25	215.05	213.71	219.23	217.40	225.65	222.73	220.63	220.10	223.15
210.86	210.92	207.89	207.06	209.19	204.92	206.35	207.83	207.00	206.15	204.33	215.46
201.64	199.80	201.57	199.29	201.07	198.15	206.15	204.88	213.10	211.48	210.43	213.31
206.08	207.15	204.17	204.45	205.61	205.86	211.11	207.72	207.67	219.35	219.45	212.91
208.48	206.69	211.50	212.99	208.56	209.44	219.19	218.07	232.74	224.37	222.66	219.74
209.32	208.90	208.48	207.77	207.63	207.52	212.04	212.83	216.65	216.40	215.40	

Table 2 ANOVA test of market structure, size and time on price decisions.

Source	SS	df	MS	F	$1-\infty$
Between subjects	*24,101,653.6*	*148*			
A (Structure)	1,209,684.6	1	1,290,684.6	7.882	.99
B (Size of market)	36,040.0	4	9,010.0	–	
AB	12,134.8	4	3,033.7	–	
error	22,762,794.2	139	163,761.1		
Within subjects	*6,959,914.6*	*2,639*			
C (Time)	236,420.2	19	12,443.2	4.799	.99
AC	286,406.0	19	15,074.0	5.814	.99
BC	42,504.2	76	559.3	–	
ABC	45,036.1	76	592.6	–	
error	6,349,548.1	2,449	2,592.7		

Table 3 SDC game: Mean advertising by structure and market size over 20 quarters

	Firm size	N	1	2	3	4	5	6	7	8	9	10
Competitive	C_1 2	8	6.00	6.38	7.94	8.19	8.94	8.81	9.19	8.00	8.13	7.38
	C_2 3	6	6.00	5.00	5.55	5.32	5.72	5.35	5.80	6.12	5.07	4.48
	C_3 5	10	6.00	6.17	6.76	6.82	7.80	8.24	8.19	7.62	6.44	6.51
	C_4 7	14	6.00	5.16	5.66	5.76	6.15	6.46	6.99	6.81	6.42	6.12
	C_5 10	20	6.00	5.76	5.51	6.69	6.16	7.93	8.87	8.92	8.06	7.51
Average			6.00	5.69	6.28	6.56	6.95	7.36	7.81	7.49	6.82	6.40
Monopolistic	A_1 2	14	6.00	6.54	6.79	6.31	6.23	6.09	5.86	5.86	5.54	4.89
	A_2 3	18	5.99	5.48	5.56	5.52	5.47	5.04	4.57	4.58	4.55	4.14
	A_3 5	10	6.00	5.73	6.15	5.38	4.62	4.65	3.73	4.32	3.49	3.03
	A_4 7	21	6.00	6.34	6.60	5.80	5.96	5.37	4.83	4.28	3.45	2.64
	A_5 10	20	6.00	5.53	5.14	4.38	4.12	3.28	3.25	2.42	1.92	1.51
Average			6.00	5.92	6.05	5.48	5.28	89	4.45	4.29	3.79	3.24
	B_1		6.00	6.46	7.36	7.25	7.58	7.45	7.52	6.93	6.83	6.13
	B_2		5.99	5.24	5.55	5.42	5.59	5.19	5.18	5.35	4.81	4.31
	B_3		6.00	5.95	6.45	6.10	6.21	6.44	5.96	5.97	4.96	4.77
	B_4		6.00	5.75	6.13	5.78	6.06	5.91	5.91	5.54	4.93	4.38
	B_5		6.00	5.65	5.32	5.54	5.14	5.61	6.06	5.67	4.99	4.51
Overall average			6.00	5.81	6.17	6.02	6.12	6.12	6.13	5.89	5.31	4.82

11	12	13	14	15	16	17	18	19	20	Average	
7.19	5.88	5.63	5.88	4.94	4.63	3.75	4.69	5.08	5.63	6.61	
3.81	3.35	4.47	4.70	3.82	3.87	4.70	4.82	4.58	3.72	4.81	
6.01	6.01	4.35	3.32	3.80	5.24	4.46	8.43	8.67	9.40	6.51	
6.84	5.99	5.52	4.77	3.72	4.42	4.92	4.47	3.17	3.80	5.46	
9.82	6.37	6.69	4.81	3.50	3.24	3.11	2.17	2.29	3.06	5.82	
6.74	5.52	5.33	4.69	3.95	4.28	4.19	4.92	4.76	5.12	5.84	A_1 Overall Average
4.20	3.82	3.49	3.14	2.67	3.01	3.40	3.75	3.85	4.04	4.77	
3.88	3.91	3.66	3.15	3.29	3.37	3.31	2.18	2.04	2.05	4.09	
2.45	2.47	1.76	1.37	1.12	0.77	0.64	0.37	0.48	0.30	2.94	
1.97	1.73	1.28	1.24	1.41	1.54	1.36	1.29	1.99	2.06	3.36	
1.41	1.06	1.14	0.93	0.79	1.10	0.56	0.62	0.47	0.53	2.31	
2.78	2.60	2.27	1.96	1.86	1.96	1.85	1.64	1.76	1.79	3.49	A_2 Overall Average
											B Overall Average
5.69	4.85	4.56	4.51	3.80	3.82	3.57	4.22	4.47	4.83	5.69	
3.85	3.63	4.06	3.90	3.55	3.62	4.00	3.50	3.31	2.88	4.45	
4.23	4.24	3.06	2.35	2.46	3.00	2.55	4.40	4.57	4.85	4.73	
4.38	4.40	3.86	3.40	3.00	2.56	2.98	3.14	2.88	2.58	4.41	
5.62	3.72	3.92	2.87	2.14	2.17	1.83	1.39	1.38	1.79	4.07	
4.76	4.06	3.80	3.33	2.90	3.19	3.02	3.28	3.26	3.46		

Table 4 ANOVA using advertising decisions as dependent variable.

Source	SS	df	MS	F	$1-\infty$
Between subjects	13,587.3	140			
A (Structure: Comp-Monop)	2,771.5	1	2,771.5	37.74	0.99
B (Size: 2,3,5,7,10)	615.2	4	153.8	2.09	0.90
AB	580.2	4	145.1	1.98	0.90
error	9,620.4	131	73.44		
Within subjects	23,599.4	2,324			
C (Time: 20 quarters)	3,191.4	19	168.0	19.5	0.99
AC	684.4	19	36.0	4.2	0.99
BC	530.1	76	7.0	–	
ABC	896.8	76	11.8	1.37	0.95
error	18,926.7	2,134	8.6		

[35]
Perception of Payoff Structure and Opponent's Behavior in Related Matrix Games

MARTIN SHUBIK
Department of Economics
Yale University

GERRIT WOLF
Department of Administrative Sciences and Psychology
Yale University

BYRON POON
(Deceased)

> Empirical results indicate to what extent judges recognize similarities between payoff matrices that vary in game theoretic information conditions. Judges failed to note similarities in patterns of payoffs between matrices but did recognize patterns of behavior of an opponent.

Assumptions made in the construction of models in the formal theory of games implicitly bypass many of the psychological aspects of perception and cognition. For example, according to the usual assumptions of game theory, a matrix game played twice is formally equivalent to a "super game" played once, as will be shown. Lack of psychological equivalence occurs because of assumptions about information conditions specifying memory for sequences of play.

AUTHORS' NOTE: *This research was supported entirely by National Science Foundation Grant GS-32396.*

We describe how information conditions can vary from repeated plays of a game. We then report empirical results that indicate to what extent subjects recognize similarities between payoff matrices that reflect the game theory assumptions about information conditions. We hypothesize that subjects would at least recognize similarities in the patterns of payoffs between matrices, even though they might not recognize the assumed information conditions implied by the matrices.

We found that subjects did not recognize similarities in patterns of payoffs between matrices but did recognize patterns of behavior on the part of an opponent. This recognition of behavior did not appear to be particularly well used for purposes of optimizing one's own score.

REPRESENTING A GAME AND INFORMATION CONDITIONS

In the literature on experimental gaming, a great amount has been written on the way in which individuals play in matrix games which are repeated over time. For example, there have been many studies devoted to behavior of players playing in an iterated Prisoner's Dilemma game. This game is represented as a 2 x 2 matrix as is shown in Table 1.

There are various experimental conditions under which the repeated matrix games are run. Usually, the players are informed of each other's moves after each play of the matrix game. They are, however, usually not paid off until the end of the series. In general, the players do not see each other and have no other means of communication beyond the actual behavior that is observed when they are informed of what the choice of the other player has been.

The matrix representation of the game is only one of the several representations utilized by game theorists. In particular, another important representation is known as the extensive form of a game. Two such representations of a simple game in extensive form are given in Figure 1. An examination of these representations will show that these extensive form diagrams are equivalent to a 2 x 2 matrix. The nodes labeled P_1 and P_2 represent the choice points for the players. The curve drawn around the

TABLE 1

3,3	0,4
4,1	2,2

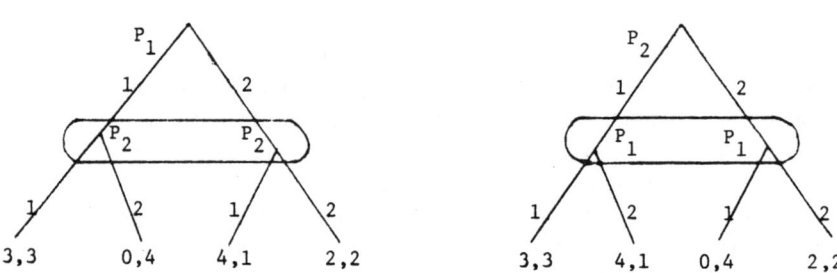

Figure 1.

two nodes at the split second level indicates that the player is called upon to move at these choice points but cannot distinguish between the nodes.

In other words, the player is assumed to be making his choice in ignorance of the play of the other player. The lines represent the moves available to each player. These lines are labeled 1 or 2 in both of the diagrams in Figure 1. The only difference between Figure 1a and 1b is that it is assumed in Figure 1a that Player 1 moves first. Player 2 will make his choice in ignorance of what Player 1 has done. In Figure 1b, it is assumed that Player 2 moves first and then Player 1 makes his decision in ignorance of Player 2's choice. These two games are regarded as strategically equivalent. It does not really matter who moves first, provided the other player is not informed of the move. In actual decision-making, this implies that there is no espionage or leakage of information.

Figures 2 and 3 show somewhat larger game trees or games in extensive form which are both related to the game shown in Figure 1, but under different information conditions.

Figure 2 shows the extensive form of a game which can be interpreted as a 4 x 4 matrix in strategic or matrix form. This alternatively can be considered as two plays of the 2 x 2 matrix on the assumption that the players are not informed how each has decided to play in the first play of the matrix, but they have to play with no information or communication whatsoever until the end of the second iteration.

The information sets indicate this as can be seen by looking at the two information sets at the third level. Player 1 knows whether he has selected a right or left branch; however, he does not know whether Player 2 has selected a left or right branch before he is called upon again to make his second move.

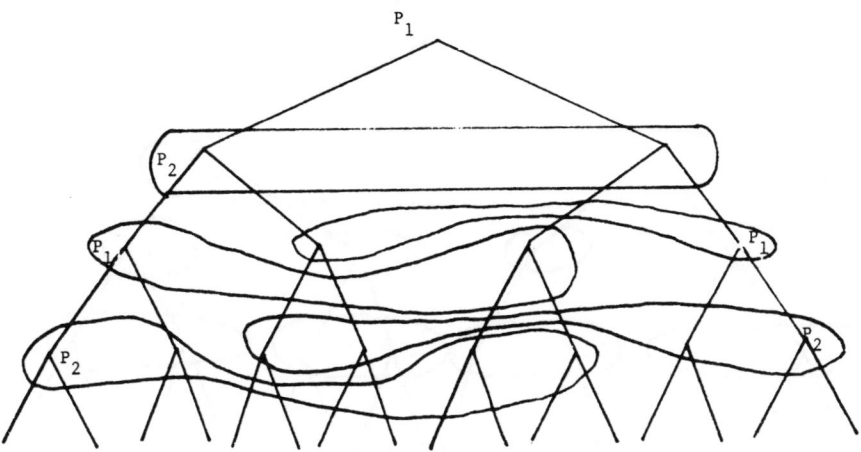

Figure 2.

Table 2 shows the 4 x 4 matrix related to Figure 2. A strategy by a player is now a pair of numbers indicating the individual's choice of his first move and of his second move. These have been "coded" as 1,1: 1,2: 2,1: and 2,2. They could have equally well been called 1, 2, 3, and 4.

Figure 3 is a representation in extensive form of a 2 x 2 matrix game played twice, where it is assumed that after the players have simultaneously played the game once, they are informed of each other's moves and then play the game simultaneously for a second time. This is shown by the four separate information sets for each player during the repeated game. These indicate that each is completely informed of what happened in the previous play.

The game shown in Figure 3 can be conceived of as a single 8 x 8 matrix game where each individual is presumed to use an overall strategy which he plans in advance to decide what to do in any contingency.

TABLE 2

1,1	6,6	3,7	3,7	0,8
1,2	7,4	5,5	4,5	2,6
2,1	7,4	4,5	5,5	2,6
2,2	8,2	6,3	6,3	4,4

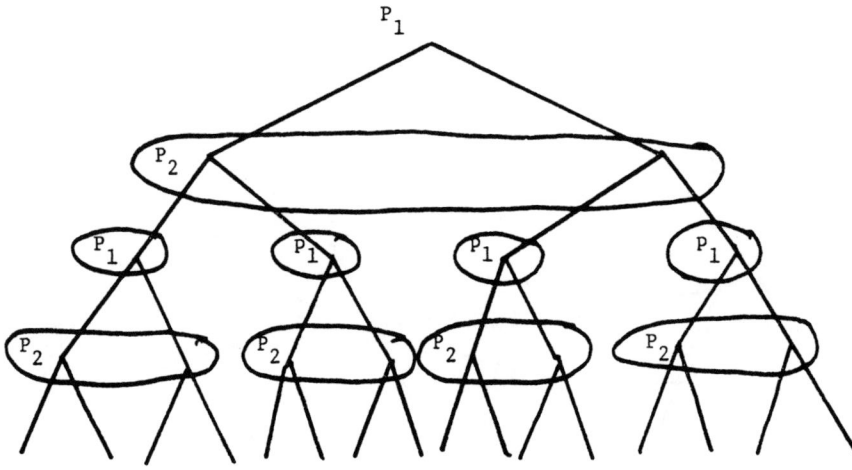

Figure 3.

THE EXPERIMENT AND ITS PURPOSE

The 2 x 2 matrix chosen for the game was an asymmetric Prisoner's Dilemma game as shown in Table 1. The 4 x 4 matrix game chosen had the same payoffs as shown in Table 2. An 8 x 8 matrix was also set up as the third game, Appendix 1. There were 12 trials in the 2 x 2 matrix game and 6 trials in the 4 x 4 and the 8 x 8 matrix games.

In order to randomize the sequence of playing the three matrix games, each subject was given an I.D. letter, A, B, ... , F, by which the order of the games presented to them was thus identified. Table 3 shows the I.D. letters and their corresponding order of matrix games presentation.

TABLE 3

	The Order of Playing the Matrix Games		
I.D. Letter	First	Second	Third
A	2 x 2	4 x 4	8 x 8
B	2 x 2	8 x 8	4 x 4
C	4 x 4	8 x 8	2 x 2
D	4 x 4	2 x 2	8 x 8
E	8 x 8	2 x 2	4 x 4
F	8 x 8	4 x 4	2 x 2

The subjects were told that they were playing against an artificial player in every game. They also knew the duration of each game through the spaces for the number of trials in each game. Furthermore, the subjects were asked to maximize their own scores.

When a subject arrived, and I.D. letter was assigned to him and he was given an instruction sheet which outlined the information about his opponent, the payoff cells for the subjects and his goal in the game. The subject was allowed to take as much time as he wanted in reading the instructions.

When the subject had finished reading the instructions, the first matrix game was presented to him. The subject was allowed to take as much time as needed to decide on his first move and subsequent moves. After the subject made a move, the referee wrote down next to the subject's move, the artificial player's move. After this information was given, the subject put down in one column his payoff for that trial and in another column, he wrote the cumulative payoff of the game up to that trial. When he finished one game, another game was presented to him.

After the subject had finished playing the three games, a questionnaire was given. He was not allowed to refer back to the game sheets when answering the questions. This questionnaire was denoted as the "Experimental Questionnaire." After finishing the questionnaire, the subject was given a second questionnaire in which the questions were the same, except an information sheet showing the three matrices and their respective number of trials was given. This was denoted as the "Control Questionnaire." This was the last stage of the experiment.

The referee acted as the agent for the artificial player. The artificial player's strategy was programmed and was fixed for all three matrices. Its strategy was to choose the "top" move on the first trial of each game. From the second trial on, its strategy was to follow the live player's move in the previous trial. In essence, the artificial player's move mirrored the live player's with one trial time lag.

Unfortunately, there was an error in the 8 x 8 matrix game. Nonetheless, the ability of the subject to spot the relationship between the 2 x 2 and the 4 x 4 should be bigger than the 8 x 8, since there exists a stronger, and easier to see, relationship. Thus, the key question of the experiment, which was the recognition of structure, would still be tested. This information was gathered by asking the following questions in both the Experimental Questionnaire and the Control Questionnaire.

Are the three matrices in the game related? If yes, in what way are they related?

There were six other questions in the Questionnaires. These were secondary questions related to performance-equivalence among the three matrices. These questions had multiple choice answers. The essence of these questions was of two forms. The first was:

> What was the equivalent move in the 8 x 8 (4 x 4) matrix game, if the subject chose Move x and Player A chose Move y in the first trial of the 2 x 2 matrix game; and the subject chose move x in the second trial of the 2 x 2 matrix game.

The second form of the questions was:

> Given a certain move in the 8 x 8 (4 x 4) matrix game, what were the equivalent moves for both players when the 2 x 2 matrix was played twice.

RESULTS

RECOGNITION OF MATRIX SIMILARITY

Are the three matrices in the game related? Yes = 56 and No = 42, for the experimental group and Yes = 64 and No = 34 for the control group.

Of the 56 players who responded yes in the experimental group, only 7 described the relatedness of the matrices in terms of the pattern of payoffs. Another 7 said the matrices were similar in that an optimal strategy existed, although they could not describe it. The remaining 42 players described the similarity not in terms of the pattern of payoffs in the matrix but, instead, in terms of the follower strategy of the artificial player.

The median number of correct answers to the 6 specific questions about the similarity of matrices was 7. The median number of errors was 32 and the median number saying they did not know the correct answer was 59.

In the control group, 22 said there were similarities in the payoff structures and 42 identified the strategy of the artificial player. The median number of correct answers to the specific questions was 13, the median error was 39 and the median "do not know" was 46.

ANALYSIS OF THE SUBJECTS' PAYOFF SCORES

For the 42 subjects who recognized the behavior of the artificial player, the matrix game can be considered as a one-person score-maximizing problem. Consequently, a hypothesis was looked at that the average payoff of these subjects would be greater than the average payoff of the subjects who did not perceive anything—those subjects who were categorized under the "No" subgroup. Furthermore, the payoff of the subjects who perceived the artificial player's strategy should be equivalent to the strategic optimal payoff in each matrix game. These two hypotheses were tested by looking at the average score of the last matrix game. Table 4 shows the results.

Table 4 shows that the average score of those subjects who detected the artificial player's strategy was higher than those who did not perceive anything in all three matrix games and than the average of all subjects. However, the average score of the subjects who had the behavioral perception was not as high as the strategic optimal payoff. Yet it was not as low as the noncooperative strategic payoff for the subjects was 8 x 8 matrix game. In this game, the average payoff for the subjects was only 0.3 cents higher than the noncooperative strategic payoff would predict.

Consequently, one may conjecture that, although there were 42 subjects who recognized the artificial player's strategy, they did not utilize the knowledge in maximizing score. This conjecture received more support from the fact that the scores chosen for the comparison were taken from the last game of the I.D. group in which each of the 42 subjects belonged. If any subject were to utilize the knowledge of the artificial player's strategy, he surely would have used it to maximize his score once he detected it. Therefore, any subject who did not utilize this knowledge in the last game would not have used it in earlier games either.

TABLE 4

	MATRIX DIMENSIONS		
	2 x 2	4 x 4	8 x 8
Strategic optimal payoff	37	32	43
Average payoff of subjects who recognized the artificial player's strategy	32.4	34	40.3
Average payoff of subjects who were in the "No" category	30.2	32.2	33.4
Average payoff of all subjects	31.2	32.0	39.0
Noncooperative strategic payoff	26	32	40

In addition, the subjects who did not perceive anything did not choose noncooperation as their strategy. This was a weak conjecture because their average payoff in the three matrix game was not uniformly higher or lower than what the noncooperative strategic payoff would predict. As a matter of fact, as is shown in Table 4, the average payoff was higher than the noncooperative strategic payoff in the 2 x 2 matrix game by 4.2 cents. In the 4 x 4 matrix game, the payoff of the two was almost the same. Yet the average payoff of the subjects who did not recognize the structural or the behavioral aspects of the experiment was lower than the noncooperative strategic payoff by 5.6 cents in the 8 x 8 matrix game.

CONCLUSIONS

The original purpose of this experiment was to explore the ability of individuals to recognize *structural* relationships. The questions asked concerned structure, *not* behavior; yet, out of 56 players who claimed that they saw a relationship among the matrices, 42 actually reported on a *behavioral relationship* among the artificial players in the different matrix games.

We suspect that, especially among naive subjects, there may be a general tendency to concentrate on patterns of behavior rather than on structure. This approach is consistent with the "chartist" approach to the stock market as contrasted with economic analysis and concern for structure. It is also consistent with the importance of behavorial processes as opposed to a concern with normative and authority structures in groups.

APPENDIX

8 x 8 Matrix Game Sheet

LIVE PLAYER (B)

		1	2	3	4	5	6	7	8
ARTIFICIAL PLAYER (A)	1	6, 6	3, 7	6, 7	6, 4	3, 8	3, 5	6, 5	3, 6
	2	7, 4	5, 5	7, 5	7, 2	5, 6	5, 3	7, 3	5, 4
	3	3, 6	0, 7	3, 7	3, 4	0, 8	0, 5	3, 5	0, 6
	4	7, 6	4, 7	7, 7	7, 4	4, 8	4, 5	7, 5	4, 6
	5	4, 4	2, 5	4, 5	4, 2	2, 6	2, 3	4, 3	2, 4
	6	8, 4	6, 5	8, 5	8, 2	6, 6	6, 3	8, 3	6, 4
	7	5, 6	2, 7	5, 7	5, 4	2, 8	2, 5	5, 5	2, 6
	8	6, 4	4, 5	6, 5	6, 2	4, 6	4, 3	6, 3	4, 4

TRIAL	PLAYER A's MOVE	PLAYER B's MOVE	PAYOFF	
			EACH TRIAL	TOTAL
1				
2				
3				
4				
5				
6				

NAME:

I.D. LETTER:

STARTING TIME:

FINISHING TIME:

TEAMS COMPARED TO INDIVIDUALS IN DUOPOLY GAMES WITH AN ARTIFICIAL PLAYER

GERRIT WOLF and MARTIN SHUBIK
Yale University

I. INTRODUCTION

A social, psychological, and economic analysis of two sets of duopoly games where one player is live and the other is an artifical player is presented here. The two sets of games differ according to whether the live player is a team or an individual.

The details of the structure of these games are given in other publications [5]. The complete structure of the heuristics for the artificial player is given in a separate publication [7].

The social psychology of duopoly and the economics of duopoly are explored in Sections II and III respectively. Hypotheses appear in Section IV. The games experimented with are discussed in Section V and the results and analysis are presented in Section VI. Interpretation and discussion in Section VII conclude the report.

II. THE SOCIAL PSYCHOLOGY OF DUOPOLY

The social psychology of play in a laboratory-simulated economic duopoly builds upon but goes far beyond play in a monopoly simulation. A duopoly game calls for social intelligence in addition to problem-solving intelligence required for simulated monopoly. The type of intelligence necessary for good performance in the monopoly game requires dealing with a number of choices and the uncertainty of outcomes, the ability to discern irrelevant factors and suboutcomes as well as sorting out the essence of the problem from the context, and achieving goals which vary in specificity and difficulty [10].

Social intelligence focuses on the responses to the effects of other's choices on one's own outcomes. This involves an interplayer relationship problem of separating the uncertainty of the outcome into that due to other's behavior and that due to randomness. It also includes a motivational problem of discerning one's own goal and other's goals. This problem arises because, unlike monopoly, duopoly provides several alternative goals from which to choose. Lastly, a player has to determine how the other player decides to accomplish his goals and then decide how to accomplish his own. Each one of these factors is explored below.

Interplayer Relationship

Relationships can be multidimensional in several senses. One sense is the number of decision variables open to a player, such as price and advertising in duopoly, which can affect both players' outcomes. The other sense is the dimensionality of the outcomes for a particular decision. This second sense is limited in the laboratory to one, say by specifying net profit as the dimension. In the real world, other dimensions such as market share or social responsibility might be relevant. (In face-to-face situations such as bargaining or conversation, outcomes are intrinsic to the two players' relationship, i.e., as compared to the outcomes extrinsic to the two players' relationship from the consumer in the anonymous market situation.)

Besides the issue of dimensionality, there is the question of describing the kind of relationship. This may usually be done in terms of degree of conflict, but questions of power and equity may also become relevant. These later types make sense in terms of the dimensions of market share or social responsibility. Laboratory game players usually assume, or are told, that conflict is the relevant dimension [2].

It is easy to blame another for one's misfortunes and oneself for success [3]. Whatever the actual cause of the outcome, players may tend to use the foregoing attribution in assessing the relationship. This means that any data able to be interpreted as part of a competitive relationship will be so interpreted rather than attributed to change or one's own strategy.

Motivation

Players seem to confuse market relationships with players' motivation or goal. As long as the relationship is not one of pure conflict, players can choose to cooperate or compete. Players in an economic simulation without direct verbal or written communication rarely think of cooperation. The simple logic is that games with low communication, in general, are competitive. A laboratory simulation is such a game and, therefore, one competes. Instructions to maximize net profit often become interpreted as the competitive goal to do better than the other player.

It does not take much to stimulate goals. Seeing the behavior of another player, whether in the same market or not, can provoke competitive behavior [9]. Students, who are used to comparing and competing with fellow students, have no trouble doing the same in a duopoly game.

Perception of the Other Player

Trying to guess the other's goals and, in turn, behavior, can be conceived as a problem in a sequence of "if he does that, I will do this; but he knows that I know, etc." While this is a possible approach to the problem of making a decision, players do not seem to use it. At another extreme, a player assumes the other is like himself and behaves accordingly.

Neither of these views takes learning into account. Assume players have finite memories and the ability to up-date impressions, if not test hypotheses about a) the strategy used by the players and b) the structure of the market. This view sees players as going through a mutual adjustment process.

The players appear to follow one of three learning strategies as part of the adjustment process. These three are the same as that for learning the structure of a monopoly [10]. The first learning strategy is that of incremental change. Behavior gradually changes by small amounts to an equilibrium level. The second learning strategy consists of search behavior at the beginning of learning with large changes in behavior in order to find out the boundaries of the payoff surface. The changes become smaller as learning progresses. The third learning strategy tests hypotheses in an all-or-none fashion. Changes in behavior would be large as different hypotheses are tested. The changes drop to zero when the correct hypothesis is discovered.

In duopoly, the mutual adjustment usually tends toward competitive behavior. Players are asymmetric in their tendencies to cooperate or compete. The asymmetry may reflect a lack of trust or just a behavioral competitive bias in games. Each tends to compete and in turn perceives the other as competitive. Those who are cooperative can find the cost great and are dragged down by a competitor or a slow cooperator [4]. As time goes on, competitive behavior may tend back toward cooperation. Players tire of competition, but cooperation is unstable and tiring also. Therefore, cyclic behavior could be expected.

Teams

A duopoly player may be a person or a team. One might expect similarities and differences between the behavior of an individual player and the behavior of a team player.

Three similarities are: a) experience and knowledge of the duopoly market and the other player; b) organization of the information processing task; c) confidence in one's ability and performance in duopoly decision-making. The first point is that the greater the knowledge by person or team, the better the performance. The second point is that individuals vary in cognitive complexity and groups analogously vary in the complexity of the authority structure among team members. The more differentiated this complexity, the better the performance. The third point is that individuals vary in self-esteem and groups analogously vary in cohesiveness. The greater the positive effect, the better the performance.

Individuals differ from teams in ways only partially related to the problem of duopoly decision-making. Some experimental results have been interpreted as showing that groups are willing to behave in a more risky manner than individuals. If we can assume that cooperation entails risk, this may imply that teams are more cooperative than individuals. Also, groups have been shown to be more effective problem solvers for complex tasks than individuals, although man-hours might be more for groups. Groups also have been shown to be more effective learners than individuals, although they take longer. The preceding results are summarized by Davis [1].

III. THE ECONOMIC BACKGROUND

Although the production and overall market aspects of a duopolistic market may be so similar to those of a monopolistic market that if the firms acted in collusion they could not be distinguished from the monopolistic market, the presence of two firms without enforced cooperation completely changes the nature of the problem of individual optimization.

In brief, short of complete explicit collusion, there is no single "rational" way to behave in a duopolistic market. Even limiting ourselves to static or steady-state solutions, the economist has no uniquely dominant theory for oligopoly. What may happen appears to depend upon the details of number, communication conditions, technology, and market institutions. There are, however, some comments which can be made about market structure which are reasonably general, and there are several different concepts about solution which serve to provide benchmarks against which we can measure performance.

On Solutions

The three solutions which serve to provide points from which to measure performance are joint maximum or the joint monopoly solution; the noncooperative equilibrium solution; and the beat-the-average solution (when there are only two competitors, this is equivalent to maximizing the difference in score). The details on these solutions and how to calculate them are given elsewhere [5]. The actual values predicted by the theories for the parameter values used in the experimental games are given in Table I.

TABLE I
ECONOMIC EQUILIBRIUM SOLUTIONS FOR THREE DECISION VARIABLES

STATIC

SOLUTION CONCEPTS	PRICE	ADVERTISING	PRODUCTION
Joint Maximum	272.49	0	156,666
Noncooperation	191.15	4,073,796	265,128
Beat the Average	176.36	5,172,589	284,848

DYNAMIC

	PRICE	ADVERTISING	PRODUCTION
Joint Maximum	273.24	0	155,671
Noncooperation	192.41	4,002,302	263,444
Beat the Average	177.72	5,081,811	283,039

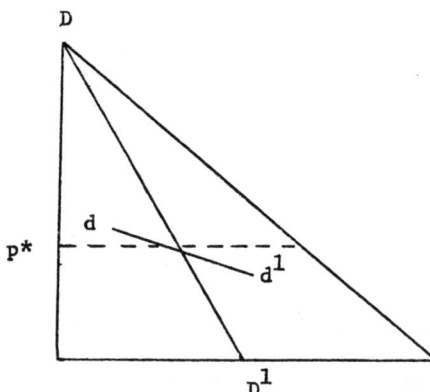

FIGURE 1—The relation of price and sales in a duopoly market.

A comparison with the report on the monopoly games [10] will show that the joint maximum here is directly comparable with the monopoly situation.

Figure 1 indicates the key additional aspect of the market structure in this game as contrasted with the monopoly game. This is the sensitivity of the sales of an individual as a function of both the price that the firm and its competitor charge. In Figure 1, the line DD' indicates what one will sell *on the assumption* that the other charges the same price. Suppose that one firm charges p*— then if the other changes its price given p* as fixed, its demand will vary along dd'.

IV. HYPOTHESES

The many details concerning oligopolistic demand structure have been investigated elsewhere [6].

H1: *Player's decisions and outcomes change from initial values toward equilibrium values.* If the market and the opponent are relatively competitive, the equilibrium values are competitive, and if the market and opponent are relatively cooperative, equilibrium values are cooperative. The competitive part of the hypothesis is tested in this paper, as well as the hypothesis that the amount of time spent to make a decision is expected to decline over time.

H2: It might be assumed that a reciprocity decision strategy combined with a "lock-in" preventive leads to good performance outcomes. Because the artificial player in the duopoly game used in this research can be described as using a "tit-for-tat" or follower strategy, we expect *the artificial player to score higher on the average than the live player*. In specific, the cooperative live player is taken advantage of in the initial stages of cooperation because of the artificial player's disinclination to be taken advantage of, and the competitive live player is taken advantage of by vigilance of the artificial player.

H3: *Teams perform better than individuals but take more time to do so.* This is because teams tend to behave less competitively in terms of price and advertising and have better control of production. This is a relatively effective strategy against a "tit-for-tat" strategy.

H4: *Top scoring live players behave less competitively and spend more time making decisions than do poor scoring live players.*

H5: Players exhibit learning through decreasing changes in decision behavior. This implies that the *differences between players decrease over time*. We hypothesize that rapid decreases indicate players are using a hypothesis-testing learning strategy. Moderate decreases indicate that players are using a search learning strategy. Small decreases or no decreases imply an incremental learning strategy. Increasing differences between players indicate different learning strategies by players or no learning.

H6: *Teams show smaller differences in final performance and decision behavior than individuals do.*

H7: *Top scorers show smaller variance in final performance and decision than poor scores do.*

V. METHODOLOGY

Experimental Conditions and Subjects

Four experimental conditions are reported in this paper.

Two of the conditions were composed by obtaining a sample of 26 subjects, each of whom played the game as an individual player, and a sample of 78 subjects who played the game as three-person team members. In summary, there were 26 individual players and 26 team players. Background data on each of these samples is described below.

The other two conditions consisted of top scorers and poor scorers. The top seven individual scorers and the top seven team scorers composed the sample of 14 top scorers. The bottom seven individual scorers and the bottom seven team scorers composed the sample of 14 poor scorers. Table II summarizes the aspects of the total sample.

All subjects were male and played the game once as individuals before playing the game for this study. The 26 individual players came from several graduate courses in microeconomics and game theory. All of these subjects were doing graduate work in management or economics and participated in the research as part of the course work. The 78 persons who formed teams were undergraduates with no particular major who were paid at an hourly rate of $2.50 to participate. The persons were attracted to the experiment through advertisements and then were scheduled for a particular time over the telephone.

The Game and the Artificial Player

Details about the game and the artificial player can be found elsewhere [7]. In brief, the game was parameterized for duopoly with reasonably high competitive levels for price and advertising, two of the decision variables available to the players. There were two other decision variables, production and dividends. The latter was not used in analyses in this paper. The exact parameter values for the game, including the market, accounting rules, and initial values are available from the authors.

The artificial player formed the second firm in the duopoly market. It was parameterized to perform competitively initially. Subsequent behavior was based on moving averages of a predicted performance of the live player. The artificial player converges to the behavior of the live player if the live player's behavior is stable. Otherwise, the artificial player pursues the live player as it follows the live player's behavior.

The number of periods played was twelve for the individuals and ten for the teams. Some of the teams played more than ten periods, but the analysis for this paper includes only the first ten periods.

Instructions and Team Structure

A sample instruction on the game and how it is played can be obtained from the authors. All players were instructed to maximize their own score.

The individual players were told that they were going to play twelve periods. The team players were given an hour to make as many decisions as they wanted.

The teams were instructed that the teams consisted of one highly experienced player and two moderately experienced players. The basis for this judgment came from a previous play of the game by each player. This information was intended to emphasize a status difference in the team.

Individual players, after the game, wrote papers on their experience. Team players filled out a number of questionnaires related to team feelings.

TABLE II

NUMBER OF OBSERVATIONS ACCORDING TO PLAYER TYPE AND PERFORMANCE QUALITY

Scorers	Player Type		
	Individual	Team	Total
Top	7	7	14
Moderate	12	12	24
Poor	7	7	14
TOTAL	26	26	52

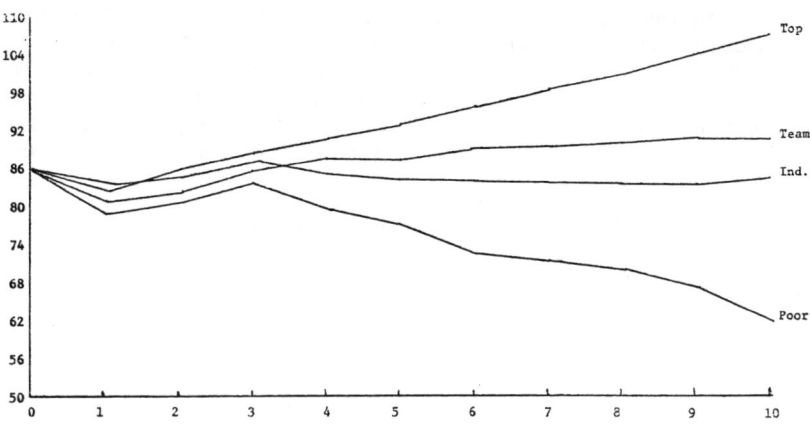

FIGURE 2—Mean score per period for four conditions.

VI. RESULTS

Hypotheses were tested for the outcome variable called score, decision variable of price, advertising, and production, and performance variable of intertrial interval. Score is the firm's net worth.

Score

H1: Players' score changes from an initial value of $85.1 million to a (theoretical) competitive value of $70.3 million, or a (theoretical) non-cooperative value of $95.6 million, but not the (theoretical) cooperative value of $130.[1]

Figure 2 graphs the mean score per period for teams, individuals, top scorers, and poor scorers. The period ten confidence interval for these four groups respectively is:

$$\begin{aligned} pr(87.1 \pm 12.5) &\geq .99 \\ pr(82.1 \pm 16.4) &\geq .99 \\ pr(105.1 \pm 7.6) &\geq .99 \\ pr(58.9 \pm 18.3) &\geq .99. \end{aligned} \quad (1)$$

[1] The theoretical values were derived by playing the game, against the artificial player, using a strategy of equilibrium price and advertising values for all decision periods and setting production to levels which minimized inventory costs. The cooperative equilibrium, in addition, required raising price over the equilibrium the first period only in order to bring the artificial player along and far enough. Also, advertising was not set to zero but cut in half each period.

We see that teams and individuals did not show significant differences from initial score while top and poor scorers did. Top scorers showed significant differences from all three theoretical equilibria, but in the direction of the cooperative equilibrium. Poor scorers showed significant differences from noncooperative and cooperative equilibria but not the competitive equilibrium.

H2: The artificial player scores higher than the live player.

Table III reports means and standard deviations for period ten for all four conditions for live and artificial players. In all four conditions, the mean score is higher for the artificial player than for the live player. These differences were significant.

H3: Teams score higher than individuals.

Table III shows that by period 10, teams do score higher than individuals, but the difference is not significant (t < 1.0).

H4: Top performing (live) players attain a statistically significant higher score than poor performers.

This hypothesis checks the legitimacy of forming two groups used in subsequent hypotheses. Table III shows that by period 10, the (14) top performers scored $105 million and the (14) poor performers scored $58 million. This difference is significant (t =

3.95, p = .01) and implies that the variability among the players was sufficient to construct these two groups of top and poor performers.

H5: Differences in score between players decrease over time.

The mean variance for three blocks of trials (1–3, 4–7, 8–10) are reported in Table IV. We see that the mean variances *increase* over time for teams, individuals, top performers, and poor performers. No matter what the condition, players were becoming

TABLE III

MEAN & S.D. SCORE, PRICE, ADVERTISING, PRODUCTION AND INTERTRIAL INTERVAL FOR PERIOD 10 FOR LIVE AND ARTIFICIAL PLAYERS

DEPENDENT VARIABLE	CONDITION	N	LIVE PLAYER $\bar{x}$ (S.D.)	ARTIFICAL PLAYER $\bar{x}$ (S.D.)
Score in Millions $	Team	26	87.1 (17.8)	97.2 (21.9)
	Individual	26	82.1 (23.4)	92.3 (19.4)
	Top	14	105.1 (7.8)	115.4 (12.7)
	Poor	14	58.9 (18.9)	72.4 (9.4)
Price	Team	26	198.9 (18.5)	188.1 (19.1)
	Individual	26	186.5 (24.2)	181.1 (20.7)
	Top	14	200.6 (15.2)	197.2 (17.3)
	Poor	14	186.8 (19.6)	171.1 (4.9)
Advertising in Millions $	Team	26	5.2 (1.5)	5.7 (1.6)
	Individual	26	4.1 (2.2)	4.5 (2.1)
	Top	14	4.6 (1.7)	4.7 (1.9)
	Poor	14	5.2 (2.1)	5.9 (1.6)
Production in 100,000 units	Team	26	2.11 (.64)	2.98 (.59)
	Individual	26	2.40 (1.25)	2.86 (.57)
	Top	14	2.20 (.73)	2.78 (.72)
	Poor	14	2.31 (.59)	3.16 (.29)
Time (Seconds)	Team	26	178.4 (62.4)	—
	Individual	26	223.4 (74.9)	—
	Top	14	177.4 (49.9)	—
	Poor	14	221.1 (89.5)	—

TABLE IV

MEAN VARIANCE FOR EACH OF THREE BLOCKS OF TIME FOR EACH OF FOUR EXPERIMENTAL CONDITIONS

DEPENDENT VARIABLES	CONDITION	PERIODS 1–3	PERIODS 4–7	PERIODS 8–10
Score in Millions $	Team	4.5	6.8	9.5
	Individual	3.5	12.6	14.8
	Top Performers	4.2	3.9	6.4
	Poor Performers	3.7	15.1	18.4
Price in $	Team	15.2	17.0	20.2
	Individual	17.9	16.5	14.4
	Top Performers	18.7	17.3	14.8
	Poor Performers	14.4	15.9	19.2
Advertising in Millions $	Team	1.6	2.1	2.2
	Individual	1.7	2.0	2.3
	Top Performers	1.9	1.8	1.9
	Poor Performers	1.3	2.3	2.5
Production in hundred-thousand units	Team	1.08	.78	.63
	Individual	1.49	1.19	1.09
	Top Performers	1.11	.95	.69
	Poor Performers	1.46	1.02	1.03
Intertrial Interval in Seconds	Team	153.9	57.3	60.3
	Individual	149.8	108.4	100.4
	Top Performers	147.8	71.4	58.8
	Poor Performers	155.9	94.2	101.9

dissimilar in score as time went on, partly because score is a cumulative variable. A decreasing variance would necessarily come about if all players were tending toward similar goals. This result indicates that these groups may not have similar goals.

H6: Teams show smaller variances in final score than individuals.

The data are in the predicted direction, but they are not significant (9.9 vs. 14.4).

H7: Top scorers show smaller variances in final score than poor scorers.

The mean variance for top scorers was 6.4 and the mean variance for poor scorers was 17.9. This difference is highly significant (t = 4.54, p = .01).

Price

H1: Players' price decision changes from initial value of $240 to a competitive decision of $176, or a noncooperative decision of $191 and not to a cooperative decision of $272. These decision values were determined theoretically from the structure of the market.

Figure 3 reports the mean price decision per period for teams, individuals, top performers, and poor performers. The confidence interval for period 10 for each of these groups respectively is:

$$\begin{aligned} \text{pr}(198.9 \pm 13.0) &\geq .99 \\ \text{pr}(186.5 \pm 16.9) &\geq .99 \\ \text{pr}(200.6 \pm 14.7) &\geq .99 \\ \text{pr}(186.8 \pm 19.0) &\geq .99. \end{aligned} \quad (2)$$

All four conditions show that the price decision decreased significantly from the initial value of $240. Teams and top performers fall near the noncooperative price of $191 and are significantly different from the cooperative and competitive price. Individual and poor performers attained price decisions not significantly different from the noncooperative and competitive equilibrium values.

H2: The artificial player set lower prices than live players. Table III reports means for period 10 that are in the predicted direction though not significant for all four conditions.

H3: Teams set a higher price than do individuals. Table III shows that by period 10, teams do set a higher price ($198.8) than do individuals ($186.5). This difference is significant at the .05 level (t = 2.04).

H4: Top scorers set a higher price than do

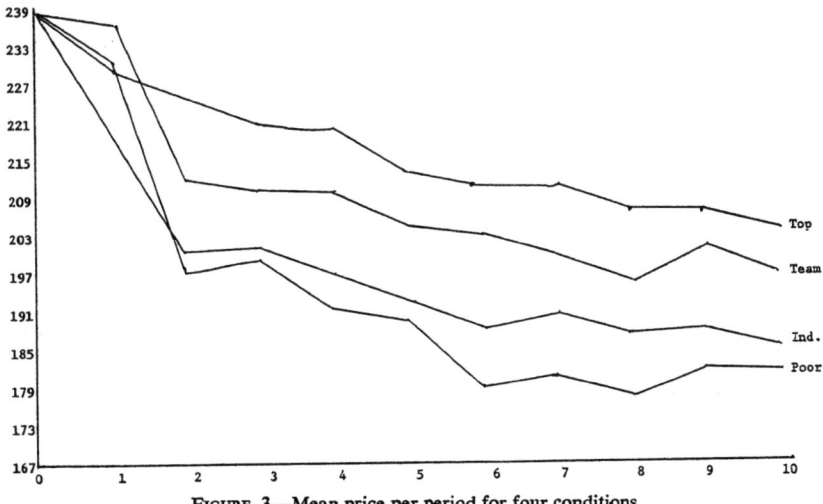

FIGURE 3—Mean price per period for four conditions.

poor scorers. Table I shows that by period 10, top scorers do set a higher price ($200.6) than do poor scorers ($186.8). This difference is significant at the .05 level (t = 2.01).

H5: Differences in price between players decrease over time. Table IV reports the mean variance for three blocks of trials for each group. Only the top performers show this trend.

H6: At the end of the game, teams show smaller differences in price than individuals do. Table IV shows the opposite result.

H7: At the end of the game, top scorers show smaller differences in price than poor scorers do.

Table IV shows the data are in the predicted direction.

Advertising

H1: Players' advertising expenditure changes from an initial value of $6 million to a (theoretically calculated) competitive value of $5.17 million, a noncooperative value of $4.07 million, or a cooperative value of $0.0.

Figure 4 graphs the mean advertising per period for teams, individuals, top performers and poor performers. The confidence interval for each of these groups respectively, for period 10, is:

$$\begin{aligned} pr(5.2 \pm 1.1) &\geq .99 \\ pr(4.1 \pm 1.5) &\geq .99 \\ pr(4.6 \pm 1.7) &\geq .99 \\ pr(5.2 \pm 2.0) &\geq .99. \end{aligned} \quad (3)$$

Only individual players showed significant difference from the initial value of $6.0 million. All were significantly different from the cooperative equilibrium, but none was significantly different from the noncooperative or competitive equilibrium.

H2: The artificial player set higher advertising than live players. The results were in the predicted direction for all four conditions (see Table III), but none was significant.

H3: Teams spend less on advertising than individuals spend. The results for period 10 show that individuals spend less than teams, but not by a statistically significant amount (t = 1.72).

H4: Top scorers spend less on advertising than poor scorers spend. Table I reports mean advertising in the predicted direction of this hypothesis, but the means were not significantly different.

H5: Differences in advertising expenditures between players decrease over time.

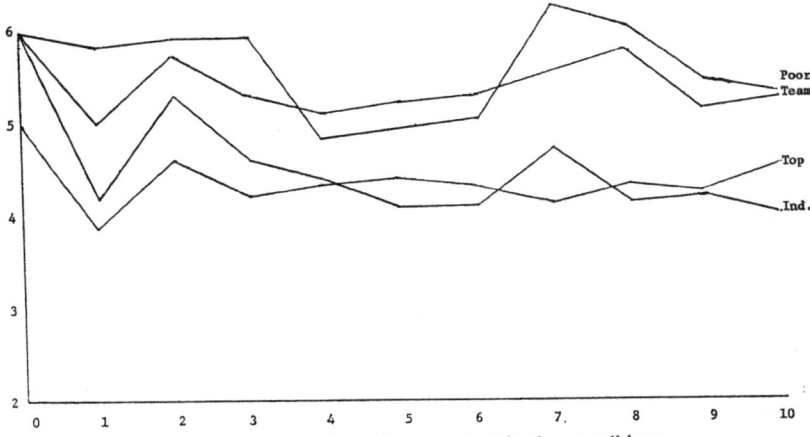

FIGURE 4—Mean advertising per period for four conditions.

Table IV shows means that offer support for this hypothesis.

H6: At the end of the game, teams show smaller differences among themselves than individuals show among themselves. While the mean variances are in the predicted direction, according to Table IV, they are not significantly different.

H7: At the end of the game, top scorers show smaller differences among themselves than poor scorers show among themselves. Table IV shows the mean variances are in the predicted direction but are not significant.

Production

H1: Players' production decision changes from initial values of 4.0 (hundred-thousand units) to a competitive level of 2.84 or a non-cooperative level of 2.65 and not to a cooperative level of 1.56.

Figure 5 graphs the mean production over ten periods for teams, individuals, top scorers, poor scorers. The ten-period confidence interval for these respective conditions is:

$$\begin{aligned}
pr(2.11 \pm .41) &\geq .99 \\
pr(2.40 \pm .88) &\geq .99 \\
pr(2.20 \pm .93) &\geq .99 \\
pr(2.31 \pm .85) &\geq .99.
\end{aligned} \quad (4)$$

These results show that in all conditions there is a change from initial levels downward. However, the variances are so large for the latter three that they show no differences from theoretical values. The team condition showed such a small variance that it did show significant differences from all theoretical levels.

H2: The artificial player produces more than live players produce.

The results are in the predicted direction for all four conditions and significant for two.

H3: Teams set a lower production than individuals do.

Table III shows that the results are in the predicted direction; however, a t-test did not reach acceptable significance levels.

H4: Top scorers set a lower production than do poor scorers.

Again, as Table III shows, the results are in the predicted direction.

H5: Differences in production levels between players decrease over time. All four conditions show the predicted trend, according to Table IV.

H6: At the end of the game, teams show smaller differences in production than individuals do. The results are in the predicted direction according to Table IV.

H7: At the end of the game, top scorers

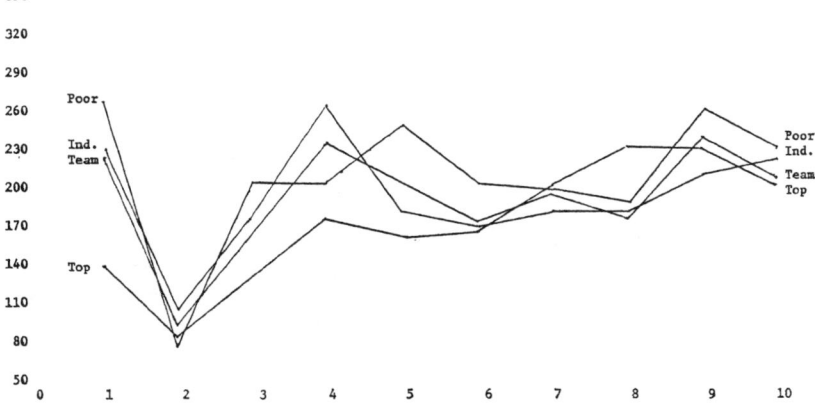

FIGURE 5—Mean production per period for four conditions.

show smaller differences in production than individuals do. The results are in the predicted direction according to Table IV.

Intertrial Interval

H1: The amount of time spent in making a decision decreases from initial (pre game) amount of time of 600 seconds to a minimal time to enter and receive data of 60 seconds.

Figure 6 graphs the mean intertrial interval over ten periods for teams, individuals, top scorers and poor scorers. The tenth period confidence interval for these respective conditions is (in seconds):

$$\begin{aligned} &\text{pr}(178.4 \pm 43.7) \geq .99 \\ &\text{pr}(223.4 \pm 52.4) \geq .99 \\ &\text{pr}(177.4 \pm 48.5) \geq .99 \\ &\text{pr}(221.1 \pm 86.8) \geq .99. \end{aligned} \quad (5)$$

Time does decrease in all conditions but is significantly different from the minimal time of 60 seconds.

H2: The artificial player, naturally because it is a computer program, takes a shorter amount of time to make a decision than does a live player. Including CPU and printing, the artificial player never takes longer than 30 seconds, which is significantly less than the human player. This result is reported for sake of completeness and does not have great import because the artificial player's time depends on the complexities of time-sharing.

H3: Teams take longer to decide than individuals. The results from Table I are the opposite, and these results approach significance ($t = 2.81$, $p = .06$). This result is surprising and is discussed in the next section.

H4: Top scorers take a shorter amount of time to decide than do poor scorers. The results from Table III are in the predicted direction but are not significant ($t = 1.54$).

H5: Differences in decision time decrease over time. All four conditions show this predicted trend, according to Table IV.

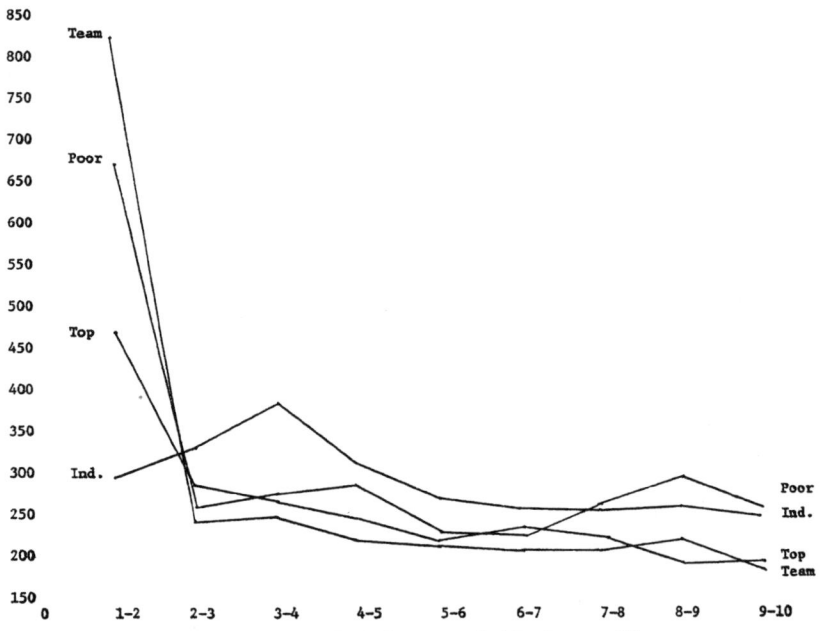

FIGURE 6—Mean decision time per period for four conditions.

H6: At the end of the game, individuals show smaller differences in decision time than teams. Table IV shows that the opposite prediction holds.

H7: At the end of the game, top scorers show smaller differences in decision time than poor scorers. Table IV shows this prediction receives support.

The results are discussed and summarized according to each of the seven hypotheses.

VII. DISCUSSION AND CONCLUSION

H1: *Players change behavior.*

To the extent that players' decisions changed from initial values, the direction was toward the noncooperative equilibrium. The competitive equilibrium could not be rejected, also, for low scoring players and for individual players. Advertising and production decisions showed less change than did the price decision. The large variance of score outcomes for the aggregate of all teams and individuals prevented these players from showing differences from the initial value. But, partitioning of these sets of players into top and poor performers did yield significant differences from the initial value for the score and price variables.

H2: *Artificial player performs better than live player.*

The artificial player's behavior was in the predicted direction for all hypotheses. It was significantly different from the live players for score and in particular conditions for price and production. The artificial player in all conditions scored higher, charged less, advertised more, produced more, and decided more quickly than the live player. The somewhat noncooperative price decisions as compared with the opponent's appear to provide the basic reason for its success.

H3: *Teams perform better than individuals.*

Teams scored higher, charged more, and produced less than individuals, with only the price variable significant, however. Contrary to prediction, teams advertised more, indicating competitiveness, and took *less* time to make a decision than individuals.

One explanation of this performance is that teams planned their strategy at the beginning of the game rather than developing it during the game. This would account for shorter time to make decisions. Figure 6 confirms that teams took a long time to make a decision at the beginning. The teams' planning found them more cooperative than individuals in terms of price but more competitive than individuals in terms of advertising. Because price was more important in this system, the "relatively" cooperative price led to a higher score for teams compared to individuals, even though the competitive artificial player won against any opponent.

These results do not necessarily support the superiority of teams but the importance of planning and initiating rather than merely reacting. The role of these variables for teams and individuals must be studied further.

H4: *Top scorers perform differently than poor scorers.*

The results showed that top scorers charge more, advertise less, produce less and take less time to make a decision than do poor scorers. Only the price decision was statistically significant. These results parallel the structure of the market where price is the most important variable, and the more cooperative the decisions by all parties the better the outcome. The artificial player responded cooperatively to a cooperative live player, albeit hesitantly. These results, therefore, show that top performing players performed well because of some understanding of the market and not because of an exploitation of some eccentricity in this situation.

H5: *Live players show at least incremental learning if not hypothesis testing for better players.*

If learning that would produce optimal performance was taking place, then the assumption was that the variance within a player's behavior should decrease over time. If this were happening for all players, one would then see decrease in variance between players. For top players price variance did decrease. Advertising showed no decreases, production showed decreases only for teams and top scorers, and time did show decreasing variance for all conditions.

This crude measure of learning detected very little learning. Top scorers did show some learning, and all other groups relatively little. Single teams or individuals may have exhibited learning but their behavior was obscured in the averaging of data. Future research should classify each player according to learning strategy and then look for performance differences between types of learning strategy. Even so it may turn out that for less than optimal performance there are many ways to achieve suboptimal performance. What determines these many ways lies somewhere in complexity and the understanding of complexity by the player in the player-environment system of this economic simulation.

H6: *Teams show less individual differences than do individuals by the end of the game.*

These data for this hypothesis were in the predicted direction for all variables except for price. The reason for the price result is unclear.

H7: *Top performers show less individual differences than poor performers by the end of the game.*

This result held for all variables.

In conclusion, we see that a duopoly market provides for considerable variability in behavior. If one added more players to the market, we would expect, based on previous results, that the economics of the situation would reduce this variability [8]. Some of the variability in a simulated duopoly can be accounted for by knowing whether the player is a team or an individual, a top performer or a poor performer. But these distinctions are only surrogates for potential genotypical variables of kind and amount of planning and degree of understanding of the complexity of the player-environment by the player.

Duopoly is probably the least favorable case for economic analysis. It clearly requires not only a detailed understanding of the socio-psychological dynamics, but undoubtedly depends upon technological and institutional details which determine the information and communication system and provide considerable structure to the strategic alternatives of the actors.

In spite of the obviously complex dynamics which make any single simple explanation of duopolistic behavior hardly worthwhile, the three economic (and game theoretic) solutions of joint maximum, noncooperative equilibrium, and competitive equilibrium serve as useful benchmarks and provide a means for considering different types of behavior. In particular, our results show that as crude predictions of overall price and profits the noncooperative or competitive equilibrium are far better than the joint maximum. The low level of communication in this duopolistic market, coupled with the slowness with which the artificial player can be coaxed into cooperating, makes it extremely difficult for most of the live players to work out a highly cooperative *quid pro quo* with the artificial player.

A problem which appears immediately for $n \geqslant 2$ is that the level of efficiency (in the sense of problem solving in maximizing in a monopolistic market), the degree of cooperation, and the ability to predict the behavior of the others all become confounded. It appears to be possible to sort out these various effects by varying communication conditions, behavior of the artificial player, parameters of the market, and number of competitors. We have some evidence concerning each of these [8], but we believe that neither we nor anyone else is anywhere near a satisfactory dynamic socio-psycho-economic institutional theory of duopoly.

For purposes of economic planning and antitrust policy, the benchmarks provided by economics and game theory together with institutional studies of markets probably provide enough guidance, but even an adequate description of how students play in a simple game calls for a richer and more developed set of theories and experimentation than have yet been postulated and tested.

REFERENCES

1. Davis, J. *Group Performance*. Reading, Massachusetts: Addison-Wesley, 1969.
2. Deutsch, M., "Socially Relevant Science: Reflections on some Studies of Interpersonal Conflict." *American Psychologist*, 24 (1969), 1076–1093.
3. Jones, E. E., D. Kanouse, H. H. Kelley, R. E. Nisbett, S. Valens, and B. Weiner. *Attribution: Perceiving the Causes of Behavior*. New York: McCaleb-Seiler, 1971.
4. Kelley, H. H. and A. J. Stahelski, "Social Interaction Basis of Cooperators' and Competitors' Beliefs about Others." *Journal of Personality and Social Psychology*, 16 (1970), 66–74.
5. Levitan, R. and M. Shubik, "A Business Game for Teaching and Research Purposes," Discussion Paper. Mimeo. Yorktown Heights, New York: IBM Watson Research Center, 1962, 1964, Parts I and II. New Haven, Connecticut: Yale University, 1964, 1967, Parts III–VII.
6. Shubik, M. *Strategy and Market Structure*. New York: Wiley, 1959.
7. ———, G. Wolf, and S. Lockhart, "An Artificial Player for a Business Market Game." *Simulation and Games*, 2 (1971), 27–43.
8. ———, G. Wolf, and H. Eisenberg, "Some Experiences with an Experimental Oligopoly Business Game." *General Systems*, 27 (1972), 61–75.
9. Wolf, G., "Effects of Comparative Information and Decision Complexity in a Monopoly Game." *Simulation and Games*, 4 (1973), 145–158.
10. ——— and M. Shubik, "Decision Making in a Single Firm Market," Technical Report, Department of Administrative Sciences, Yale University, 1974.

[37]
Cooperative Game Solutions

AUSTRALIAN, INDIAN, AND U.S. OPINIONS

MARTIN SHUBIK
Department of Economics
Yale University

> As part of several lectures the audiences in Australia, India, and the United States were asked how the gain from a cooperative agreement should be divided among three individuals. The responses are considered in the context of various cooperative solutions that have been suggested in the theory of games. More questions are raised than answered.

In 1973 at several Australian universities several different sets of individuals were asked to give their opinions on how a certain three-person nonconstant sum game should be played. The individuals were presented with a diagram as shown in Figure 1. This figure shows various solutions to a three-person nonconstant sum game with the following characteristic function:

$$v(1) = v(2) = v(3) = 0,$$
$$v(12) = 1, v(13) = 2, v(23) = 3,$$
$$v(123) = 4$$

AUTHOR'S NOTE: This work relates to Department of the Navy Contract N00014-77-C-0518 for the Project Center for the Study of Competitive and Conflict Systems issued by the Office of Naval Research under Contract Authority NR 047-006. However, the content does not necessarily reflect the position or the policy of the Department of the Navy or the Government, and no official endorsement should be inferred. The United States Government has at least a royalty-free, nonexclusive and irrevocable license throughout the world for government purposes to publish, translate, reproduce, deliver, perform, dispose of, and to authorize others so to do, all or any portions of this work.

JOURNAL OF CONFLICT RESOLUTION, Vol. 30 No. 1, March 1986 63-76
© 1986 Sage Publications, Inc.

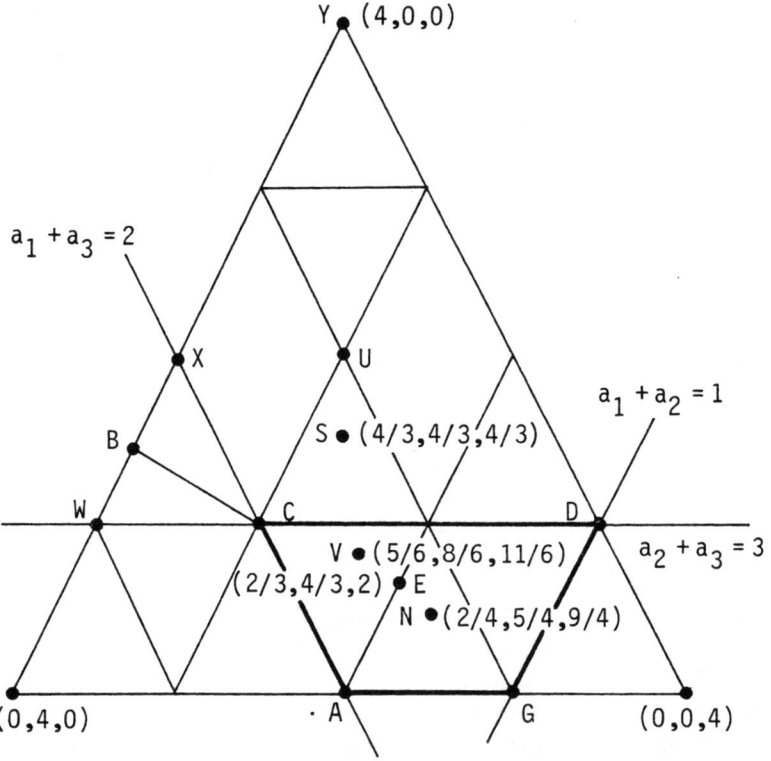

Figure 1

On the figure various suggested solutions have been noted. These solutions are discussed in the Solutions section.

In 1977 the same game was run with a group of students in a game theory class at Yale. The results from these investigations were reported in an article in 1978 (Shubik). In December 1978 and January 1979 it was possible to run one of the versions of this game together with two other highly related games for a much larger sample size. The results of these new investigations are reported here and compared with the previous ones. The new games were run in India.

BRIEFING AND RESPONDENTS

In Australia the game was run six times, four times with an abstract scenario and twice with an explicitly economic, or productivity scenario. Three of the four Australian audiences for the abstract scenario were composed of graduate students and faculty in economics. The fourth was graduate students and faculty in mathematics. The other two were social scientists and economists.

The Americans were given the same briefing as the Australian economic briefing. They were undergraduate students in game theory with backgrounds in economics or other social sciences at Yale.

The Indians were all given a productivity briefing differing from that given to the Australians. No diagram was supplied, nor were any outcomes suggested. Instead, a relatively simple verbal briefing as is indicated in the appendix was given to all.

The game was used seven times in India in conjunction with several other games that are noted later. The groups were as follows: 20 faculty and graduate students in economics at the University of Delhi; 49 third- and second-year students (with some faculty included) in mathematics and economics, Saint Steven's College, Delhi; 56 first- and third-year students in economics at Presidency College, Calcutta; 11 graduate economists and members of the staff of the Indian Statistical Institute, Calcutta; 59 students and faculty in economics at the University of Madras; 41 students and faculty in economics of the University of Madurai; and 12 faculty members of the Indian Institute of Management at Bangalore.

In all instances in Australia, the United States, and India, the game served as an introduction to a lecture on game theory. The lecture began with the briefing and the playing of the game, and players were informed that the game would serve as a useful basis for understanding the lecture.

SOLUTIONS

The core of an n-person game is a peculiarly economic solution. It has the property satisfied by any imputation or division of all proceeds within the core:

Individual rationality—no individual obtains less than he could get by himself

Group rationality—no group obtains less than they could get by themselves

Societal rationality or
Pareto optimality—all together waste nothing

In Figure 1 the vertices represent the three points at which one individual obtains all and the other two nothing. The side of the triangle opposite any vertex is where the individual favored at the vertex (individual 1 at vertex Y) obtains nothing. Thus, along the base of the triangle, individuals 2 and 3 vary their returns, leaving nothing for individual 1.

The core is the area bounded by ACDG. No individual or pair of individuals can obtain more by independent action than they can obtain when offered any imputation in the core. An imputation $a = (a_1, a_2, a_3)$ in the core satisfies the inequalities:

$$a_1 + a_2 \geq 1, a_1 \geq 0$$
$$a_1 + a_3 \geq 2, a_2 \geq 0$$
$$a_2 + a_3 \geq 3, a_3 \geq 0 \text{ and}$$
$$a_1 + a_2 + a_3 = 4$$

All three-person games with a core may be regarded as having arisen from an underlying economic structure involving trade and prices (Shapley and Shubik, 1976). Work in economics has shown that all trading economies, when portrayed as cooperative games, give rise to games with a core and the trade called forth by a price system distributes wealth in a manner that the outcome lies within the core (Shubik, 1959; Debreu and Scarf, 1972). Unfortunately, without further economic information there is no unique way that we can pick out the full economic structure of the game. In some instances the core may not exist; in other instances it is large.

Shapley (1953) has suggested a single-point solution to a cooperative game that always exists. This is the value. In essence it awards each individual the expected value of his marginal contributions to all coalitions on the "sociologically neutral" assumption that the entry of any individual into any coalition is perfectly random.

The value for this game is given by $V(5/6, 8/6, 11/6)$. The calculations for player 1 are indicated below.

All Orders		
123	1 enters first and adds nothing	0
132	1 enters first and adds nothing	0
213	1 enters second and adds $1-0$	1
231	1 enters third and adds $4-3$	1
312	1 enters second and adds $2-0$	2
321	1 enters third and adds $4-3$	1
		5

All orders have a probability of 1/6 of occurring; hence 1's expected return is 5/6.

A different one-point solution has been suggested by Schmeidler (1969). We may imagine moving in all the "walls" of the core as shown in Figure 1 by rewarding all coalitions some bonus ϵ_1. For example, suppose $\epsilon_1 = 1/2$ then

$$a_2 + a_3 = 3\frac{1}{2}$$

$$a_1 = \frac{1}{2}$$

$$a_1 + a_3 = 2\frac{1}{2}$$

$$a_1 + a_2 = 1\frac{1}{2}$$

The core of this game would be a line through the point N parallel to AG bounded by (1/2, 1, 2) and (1/2, 2, 1). We can repeat the rewarding of coalitions who form the walls of this new core with the proviso that we stop giving extra to coalitions if that would destroy the new core completely. Thus, for example,

$$a_2 + a_3 = 3\frac{1}{2}$$

$$a_1 = \frac{1}{2}$$

tells us that $a_1 + a_2 + a_3 = 4$, hence any further subsidy to these is not possible. But we can continue subsidies to the coalitions (1,3) and (1,2). Let that be ϵ_2. It is straightforward to check that the biggest we can let ϵ_2 become is $\epsilon_2 = 1/4$. For any larger ϵ_2 the core would disappear

$$a_1 + a_3 = 2\frac{3}{4}$$

$$a_1 + a_2 = 1\frac{3}{4}$$

together with $a_1 = 1/2$ determine the imputation that (2/4, 5/4, 9/4) satisfies. This is the point N in Figure 1; it is called the *nucleolus*.

The point S(4/3, 4/3, 4/3) is the point of pure symmetry, essentially ignoring questions of productivity.

An important distinction between the groups in India and elsewhere is that the former were not given the geometrical diagram with cues as to possible outcomes.

Table 1 shows the responses to the game.

The groups in India were large; the lecture room culture is different, and thus there was somewhat less control in eliciting responses from some of the audience.

There appear to be several items of interest that can be noted from the data in Table 1. They are shown in Tables 2 and 3.

Table 2 shows the percentage of respondents[1] who selected (4/3, 4/3, 4/3), another point not in the core, or a point in the core. It should be noted that the points not in the core, other than (4/3, 4/3, 4/3), were all close to its boundary. The Australians are divided into the abstract and economic briefing.

The samples from India were large, yet there is no indication of the value or nucleolus being selected. After the lectures there were indications that some members of the audience were intrigued with the value as a normative criterion, but without the prompting offered by Figure 1 to the Australian and U.S. respondents, the value and nucleolus were not suggested independently.

The only overlap of significance appears to be (2/3, 4/3, 2), and apart from the observation that the ratio 1:2:3 is the same as the ratio of the payoffs to the coalitions, no explanation or justification is offered here.

1. Those who failed to reply or made logical errors were eliminated from the calculation.

TABLE 1

	Australia Abstract Game					Econ. Briefing Australia				U.S.		India								Totals
	1	2	3	4	%	5	6	7	%	7	%	1	2	3	4	5	6	7	%	
S (4/3, 4/3, 4/3)	8	3	7	3	31.6	5	4		29	2	11.1	1			1	1°	3		2.8	7
E (2/3, 4/3, 2)^c	2	1	2	4	8.8	6	2		25.8	4	22.2	1	7		4	3	1	1	6.8	17
V (5/6, 8/6, 11/6)^c	1	2	1	2	7.0	2	3		16.1	6	33.3									
N (2/4, 5/4, 9/4)^c	2	3	1	3	10.5		1			3	16.7									
(1, 3/2, 3/2)^c	6		2		14.0							4	5	2	7	8	5	3	12.5	31
(1, 4/3, 5/3)^c		1	1											1	4				2.0	5
(0, 2, 2)^c	4			2	7.0										1$^+$				1	1
(7/6, 8/6, 9/6)^c	1					1	1		6.5			1	4	1	2	16	5	3	12.9	32
(1, 1, 2)^c		1	1			1														
(0, 3/2, 3/2)^x			2*			1							6		6	6	8		10.5	26
(1/2, 3/2, 2)^c													5	4	13	13	13	6	21.8	54
Other points in core chosen^c	1	2	1		7.0	1									1	1	1		1.2	3
Points not in core other than 5	1																			2
Core as a whole^c																				
No reply or				4	20		3		9.7	3	16.7	11	22†		2*	11	5	2	28.2	20

*The briefing of one of the first set of games in Australia contained a "should" and "would" version; that is, the respondents were asked how they believed that the proceeds should be divided, then they were asked how they thought it would be played without an outside arbitrator. This instance has been discussed by Shubik (1975) elsewhere. Two respondents confused the "should" and "would" scenarios and gave a two-person coalition outcome for the game.

†The "no reply or errors" includes several cases in which the respondents gave ordinal answers that, though formally correct, could not be easily coded (for example, $a_1 < a_2 < a_3$).

°One respondent added "if I were only taking equity into account, then instead of choosing what I have, I would choose (4/3, 4/3, 4/3)."

+One respondent assumed that "1 is the lowest form—one need not give an inducement. Obviously there are ethical connotations of starving—we can give him a subsistence wage to live on. Inducements are most essential for the working of the system."

^{c}In the core.

^{x}Not an imputation.

TABLE 2

	Australia 1	Australia 2	U.S.	India
(4/3, 4/3, 4/3)	28.8	29.0	–	3.9
Other not in core	1.4	–	–	1.6
In core	69.8	71	100	94.5

TABLE 3

		Australia 1	Australia 2	U.S.	India
Value	(5/6, 8/6, 11/6)	8.2	16.1	22.2	–
Nucleolus	(2/4, 5/4, 9/4)	12.3	3.2	33.3	–
	(2/3, 4/3, 2)	12.3	25.8	11.1	9.6
	(1, 3/2, 3/2)	11.0	–	16.7	17.4
	(1, 1, 2)	2.7	3.2	–	18.0
	(1/2, 3/2, 2)	–	–	–	14.6

GAMES WITH AND WITHOUT CORES

The game used in all of the instances above has a large core. The data gathered appear to offer some evidence for the proposition that there is a high probability that a point in the core will be selected. A natural question to ask is: How robust is the selection of a point in the core if we change its size? In order to obtain some insight into this question, the last three groups in India were asked to make their decision on how to divide the proceeds from three different three-person games. They are compared below.

The first is the game we have already described and investigated. The second is a game that has a one-point core that gives nothing to the first two players, and the third has no core whatsoever.

Game 1

$$v(1) = v(2) = v(3) = 0;$$

$$v(12) = 1, \ v(13) = 2, \ v(23) = 3;$$

$$v(123) = 4$$

Game 2

$$v(1) = v(2) = v(3) = 0;$$

$$v(12) = 0, \ v(13) = 4, \ v(23) = 4;$$

$$v(123) = 4$$

Game 3

$$v(1) = v(2) = v(3) = 0;$$
$$v(12) = 2\frac{1}{2}, \ v(13) = 3, \ v(23) = 3\frac{1}{2};$$
$$v(123) = 4$$

The second characteristic function portrays a game in which player 3 is critical. Both 1 and 2 are equally productive in combination with 3. This game has a single-point core yielding (0,0,4) that satisfies

$$a_1 \geq 0, \ a_2 \geq 0, \ a_3 \geq 0$$
$$a_1 + a_2 \geq 0, \ a_1 + a_3 \geq 4, \ a_2 + a_3 \geq 4$$
$$\text{and } a_1 + a_2 + a_3 = 4$$

The last game has no core whatsoever, as can be seen from the following considerations:

$$a_1 \geq 0, \ a_2 \geq 0, \ a_3 \geq 0$$
$$a_1 + a_2 \geq 2\frac{1}{2}$$
$$a_2 + a_3 \geq 3\frac{1}{2}$$
$$a_1 + a_3 \geq 3$$
$$\overline{2(a_1 + a_2 + a_3) \geq 9}$$

but $\quad a_1 + a_2 + a_3 = 4 \quad$ which is a contradiction.

The three games are distinguished by a "fat" core, a single-point core, and no core. All have single-point values and each has a single-point nucleolus. Table 4 shows all of these solutions.

The conjecture made prior to running Game 2 was that fewer than 10% of the respondents would select the core even though it is a single point. The reasoning was that it is grossly discriminatory to the first two

TABLE 4

	Core	Value	Nucleolus
Game 1	"fat core"	(5/6, 8/6, 11/6)	(2/4, 5/4, 9/4)
Game 2	(0, 0, 4)	(4/6, 4/6, 16/6)	(0, 0, 4)
Game 3	no core	(13/12, 16/12, 19/12)	(5/6, 8/6, 11/6)

players. After all, they are worth something to the production process. It is hard to believe that this point would not influence most respondents.

Tables 5 and 6 show results from Madras, Madurai, and Bangalore. Three points of interest can be noted. The percentage of nonrespondents in the audience rose in the three games. Limiting our statistics to the three institutions at which the audience responded to all three games, the percentages of nonrespondents[2] changed from 16.1% to 42% to 82.1%. Hence for the latter (and to some extent more complex) games the respondents had been somewhat self-selective.

Of the active replies,[3] as can be seen from Table 6, 46.2% were at (0,0,4), or in the core. Out of 65, 22, or 33.8%, were at an imputation that treats 1 and 2 symmetrically. The average for the 21 responses that were completely specified (i.e., leaving out ($\epsilon, \epsilon, 4 - \epsilon$)) was (.85, .85, 2.3); this compares with the value at (.66, .66, 2.74).

The fall-off from the core is striking, but not as large as expected.

A GAME WITHOUT A CORE

Only 20 of 112 individuals responded to the third game. The data are displayed in Table 7. It is of importance to note that in a small seminar consisting of faculty at the Indian Institute of Management (column 7) the size of respondents essentially did not shrink. The first game was answered by 10 of 12; the second and third by 9 of 12. This appears to be due to both the audience's small size and sophistication.

The specific imputations chosen by the respondents at IIM noted in Table 7 as "other" were (1,1.25,1.75) by three respondents, (1.2,1.3,1.5), (1.17,1.35,1.48), (1, 1.33,1.66), (1.1,1.3,1.60), (1.17,1.21,1.61) by one each.

Averaging over the replies from all 22 respondents in the three games, we obtain the results shown in Table 8. Although the average is between

2. These include some who made errors, but almost all were nonrespondents.
3. In Table 5 the percentages are based on the whole population of 112; here the no reply is eliminated, leaving 65.

TABLE 5

		India			
		5	6	7	%
	(4/3, 4/3, 4/3)	–	–	–	
Core	(0, 0, 4)	23	5	2	26.8
Value	(2/3, 2/3, 8/3)				
	(1, 1, 2)	8	2	4	12.5
	(1/2, 1/2, 3)	5		1	5.4
	(4/5, 4/5, 11/5)			1	.9
$0 < \epsilon < 2$	$(\epsilon, \epsilon, 4-\epsilon)$			1	.9
	(0, 2, 2)	3	1		3.6
	(1, 3/2, 3/2)	4			3.6
	Others	2	3		4.5
	No reply	14	30	3	42.0

TABLE 6

	India 1-4 G1	India 5-7 G1	India 5-7 G2
Percentage in core	96.2%	94.7%	46.2%

the nucleolus and the value, the sample size is somewhat small to attach much significance to this.

CONCLUSIONS

There is a surprising difference in the selection of the equal split (4/3,4/3,4/3) outcome between the Australians on the one hand and the U.S. and Indian respondents on the other. The Australians were significantly more biased toward the equal-split solution than the others.

Beyond the equal-split outcome the core was an extremely good predictor of the responses to Game 1. However, when cues were supplied (Australian and U.S. sample) there was considerable selection of the value and the nucleolus; when the cues were not supplied there was no selection of either the value or the nucleolus.

In the second game clearly the core at (0,0,4) flagrantly violates equity. Furthermore, if we considered players 1 and 2 as a syndicate or union, the resultant two-person game becomes symmetric:

TABLE 7

		India			
		5	6	7	%
	(4/3, 4/3, 4/3)	2			1.8
Nucleolus	(5/6, 8/6, 11/6)			1	.9
Value	(13/12, 16/12, 19/12)				
	(1/2, 3/2, 2)	5			4.6
	(1/2, 1, 5/2)		2		1.8
	(1, 3/2, 3/2)	3	1		3.7
	Other			8	7.4
	No reply	49	38	3	82.6

TABLE 8

	Player 1	Player 2	Player 3
Responses	22	22	22
Average	.8926	1.3517	1.7543
Variance	.0843	.0228	.1012
Nucleolus	.83	1.33	1.83
Value	1.08	1.25	1.58

$$v(\overline{12}) = 0, \quad v(\overline{3}) = 0$$

$$v(\overline{12}, 3) = 4$$

The core for this game stretches from (0,4) to (4,0). But in both the debriefing and student comments sessions, questions are raised as to why and 1 and 2 should not form a union. The abrupt change of the core from (x, 4 - x), $0 \leq x \leq 4$ to (0,0,4) appears to be too drastic between Game 2 and its two-person relation. The value moves from (2,2) to (2/3, 2/3, 8/3).

The results from the second game show that core selection drops off heavily in comparison with Game 1. Symmetry appears to be more important than the core property.

The third game raises more questions than have been adequately answered. The specific game here was chosen to have no core and no symmetry. An innate ordering of productivity was selected (3 more productive than 2 and 2 more than 1), and this was reflected in the fact that 22 of the 22 answers obtained rewarded 3 more than or as much as 2 and 2 more than or as much as 1. Further experimentation with games without symmetry and no core could help explore the worth of the nucleolus and value as predictors of average behavior.

Context appears to count. The antiseptic numbers in the game appear to be embedded in a context reflecting both cultural and professional perceptions and attitudes toward productivity, symmetry, fairness, and power.

Appendix
A PROBLEM IN COOPERATION

Three individuals have the opportunity to work together on a task. Call them 1, 2, and 3.

If they proceed independently each earns	0
If 1 and 2 work together they can earn	1
If 1 and 3 work together they can earn	2
If 2 and 3 work together they can earn	3
If 1, 2, and 3 all work together they can earn	4

Suppose that they decide that they should all work together to earn 4 but they come to you for advice as to how they should divide the 4 units between them. You are the judge.

Your task is to write down three numbers all greater than or equal to zero such that $\alpha_1 + \alpha_2 + \alpha_3 = 4$

$$\alpha_1 =$$
$$\alpha_2 =$$
$$\alpha_3 =$$

and to write down your explanation of why you chose these numbers.

REFERENCES

DEBREU, G. and H. SCARF (1972) "The limit core of an economy," in C. B. McGuire and R. Radner (eds.) Decision and Organization. Amsterdam: North Holland.
SCHMEIDLER, D. (1969) "The nucleolus of a characteristic function game." SIAM J. of Applied Mathematics 17: 1163-1170.
SHAPLEY, L. S. (1953) "A value for n-person games," pp. 307-317 in H. Kuhn and A. W. Tucker (eds.) Contributions to the Theory of Games II. Princeton, NJ: Princeton Univ. Press.

———and M. SHUBIK (1976) "Competitive outcomes in the cores of market games." Int. J. of Game Theory 4: 229-237.

SHUBIK, M. (1959) "Edgeworth market games," pp. 267-278 in A. W. Tucker and R. D. Luce (eds.) Contributions to the Theory of Games IV.

———(1975) Games for Society, Business and War. Amsterdam: Elsevier.

———(1978) "Opinions on how one should play a three person nonconstant sum game." Games and Simulation (October): 302-308.

von NEUMANN, J. and O. MORGENSTERN (1944) The Theory of Games and Economic Behavior. Princeton, NJ: Princeton Univ. Press.

PART IV

GAME THEORY AND OPERATIONS RESEARCH

Solutions of N-Person Games with Ordinal Utilities, Lloyd S. Shapley and Martin Shubik, Princeton University.

The standard definition of solution of a game, given by von Neumann and Morgenstern in *Theory of Games and Economic Behavior*, makes use of two substantial restrictions on the nature of utility: linearity and transferability. Because of this, some economists have criticized the application of game theory to economics. It is shown here that these restrictions are not essential. When the assumption of linearity is dropped, the phenomenon of players distributing risk among themselves is observed. This is analogous to situations in business where the more conservative members of an organization will consent to a policy that appears maximal to the rest only if they can be given extra financial guarantees. When transferability of utility is abandoned, side payments may be made in any commodity, whose utility for one player may bear no resemblance to its utility for another, and may be different for the same player in different situations.

The unrestricted definition of solution may be stated as follows: Let each player have a utility function mapping the possible outcomes of the game (including probability mixtures of direct outcomes) into a simply ordered set. Each outcome is then represented by a vector in the Cartesian product of these sets. Call a coalition S of players "effective" for a particular vector x if there is a joint strategy ξ_S available to the players in S which guarantees to each of them that the outcome y will be at least as desirable as x, regardless of the strategies adopted by the players outside S. A joint strategy ξ_S will in general include a scheme of side-payments among the members of S, in addition to specifying how they shall behave in the actual game. Let $R(x)$ denote the set of vectors z "dominated" by x in the following sense: the set of players who prefer x to z, or some subset of that set, constitutes an effective coalition for x. (Intuitively, in open negotiation, z would be rejected in favor of x.) Finally, let C denote the bounded set of outcomes in which every player receives at least as much as he could achieve by himself. Then a *solution* is defined to be any subset V of C fulfilling the two conditions: (1) every vector in $C - V$ is in $R(x)$ for some x in V; (2) no vector in V is in $R(x)$ for any x in V. These conditions may be compressed to the single equation $V = C - R(V)$. Because of its implicit form, this definition, like that of von Neumann and Morgenstern (to which it reduces when their assumptions are satisfied), provides no systematic method for finding solutions to particular games.

[39]
A METHOD FOR EVALUATING THE DISTRIBUTION OF POWER IN A COMMITTEE SYSTEM

L. S. SHAPLEY AND MARTIN SHUBIK
Princeton University

In the following paper we offer a method for the *a priori* evaluation of the division of power among the various bodies and members of a legislature or committee system. The method is based on a technique of the mathematical theory of games, applied to what are known there as "simple games" and "weighted majority games."[1] We apply it here to a number of illustrative cases, including the United States Congress, and discuss some of its formal properties.

The designing of the size and type of a legislative body is a process that may continue for many years, with frequent revisions and modifications aimed at reflecting changes in the social structure of the country; we may cite the role of the House of Lords in England as an example. The effect of a revision usually cannot be gauged in advance except in the roughest terms; it can easily happen that the mathematical structure of a voting system conceals a bias in power distribution unsuspected and unintended by the authors of the revision. How, for example, is one to predict the degree of protection which a proposed system affords to minority interests? Can a consistent criterion for "fair representation" be found?[2] It is difficult even to *describe* the net effect of a double representation system such as is found in the U. S. Congress (i.e., by states and by population), without attempting to deduce it *a priori*. The method of measuring "power" which we present in this paper is intended as a first step in the attack on these problems.

Our definition of the power of an individual member depends on the chance he has of being critical to the success of a winning coalition. It is easy to see, for example, that the chairman of a board consisting of an even number of members (including himself) has no power if he is allowed to vote only to break ties. Of course he may have prestige and moral influence and will even probably get to vote when someone is not present. However, in the narrow and abstract model of the board he is without power. If the board consists of an odd number of members, then he has exactly as much power as any ordinary member because his vote is "pivotal"—i.e., turns a possible defeat into a success—as often as the vote of any other member. Admittedly he may not cast his vote as often as the others, but much of the voting done by them is not necessary to ensure victory (though perhaps useful for publicity or other purposes). If a coalition has a majority, then extra votes do not change the outcome. For any vote, only a minimal winning coalition is necessary.

Put in crude economic terms, the above implies that if votes of senators

---

[1] See J. von Neumann and O. Morgenstern, *Theory of Games and Economic Behavior* (Princeton, 1944, 1947, 1953), pp. 420 ff.

[2] See K. J. Arrow, *Social Choice and Individual Values* (New York, 1951), p. 7.

were for sale, it might be worthwhile buying forty-nine of them, but the market value of the fiftieth (to the same customer) would be zero. It is possible to buy votes in most corporations by purchasing common stock. If their policies are entirely controlled by simple majority votes, then there is no more power to be gained after one share more than 50% has been acquired.[3]

Let us consider the following scheme: There is a group of individuals all willing to vote for some bill. They vote in order. As soon as a majority[4] has voted for it, it is declared passed, and the member who voted last is given credit for having passed it. Let us choose the voting order of the members randomly. Then we may compute the frequency with which an individual belongs to the group whose votes are used and, of more importance, we may compute how often he is *pivotal*. This latter number serves to give us our index. It measures the number of times that the action of the individual actually changes the state of affairs. A simple consequence of this formal scheme is that where all voters have the same number of votes, they will each be credited with $1/n$th of the power, there being n participants. If they have different numbers of votes (as in the case of stockholders of a corporation), the result is more complicated; more votes mean more power, as measured by our index, but not in direct proportion (see below).

Of course, the actual balloting procedure used will in all probability be quite different from the above. The "voting" of the formal scheme might better be thought of as declarations of support for the bill, and the randomly chosen order of voting as an indication of the relative degrees of support by the different members, with the most enthusiastic members "voting" first, etc. The *pivot* is then the last member whose support is needed in order for passage of the bill to be assured.

Analyzing a committee chairman's tie-breaking function in this light, we see that in an *odd* committee he is pivotal as often as an ordinary member, but in an *even* committee he is never pivotal. However, when the number of members is large, it may sometimes be better to modify the strict interpretation of the formal system, and say that the number of members in attendance is about as likely to be even as odd. The chairman's index would then be just half that of an ordinary member. Thus, in the U.S. Senate the power index of the presiding officer is—strictly—equal to $1/97$. Under the modified scheme it is $1/193$. (But it is zero under either interpretation when we are considering decisions requiring a two-thirds majority, since ties cannot occur on such votes.) Recent history shows that the "strict" model may sometimes be the more realistic: in the present Senate (1953–54) the tie-breaking power of the Vice President, stemming from the fact that 96 is an even number, has been a very significant factor. However, in the passage of ordinary legislation, where perfect attendance is unlikely even for important issues, the modified scheme is probably more appropriate.

[3] For a brief discussion of some of the factors in stock voting see H. G. Gothman and H. E. Dougall, *Corporate Financial Policy* (New York, 1948), pp. 56–61.

[4] More generally, a minimal winning coalition.

For Congress as a whole we have to consider three separate bodies which influence the fate of legislation. It takes majorities of Senate and House, with the President, or two-thirds majorities of Senate and House without the President, to enact a bill. We take all the members of the three bodies and consider them voting[5] for the bill in every possible order. In each order we observe the relative positions of the straight-majority pivotal men in the House and Senate, the President, and also the 2/3-majority pivotal men in House and Senate. One of these five individuals will be the pivot for the whole vote, depending on the order in which they appear. For example, if the President comes after the two straight-majority pivots, but before one or both of the 2/3-majority pivots, then he gets the credit for the passage of the bill. The frequency of this case, if we consider all possible orders (of the 533 individuals involved), turns out to be very nearly 1/6. This is the President's power index. (The calculation of this value and the following is quite complicated, and we shall not give it here.) The values for the House as a whole and for the Senate as a whole are both equal to 5/12, approximately. The individual members of each chamber share these amounts equally, with the exception of the presiding officers. Under our "modified" scheme they each get about 30% of the power of an ordinary member; under the "strict" scheme, about 60%. In brief, then, the power indices for the three bodies are in the proportion 5:5:2. The indices for a *single* congressman, a *single* senator, and the President are in the proportion 2:9:350.

In a multicameral system such as we have just investigated, it is obviously easier to defeat a measure than to pass it.[6] A coalition of senators, sufficiently numerous, can block passage of any bill. But they cannot push through a bill of their own without help from the other chamber. This suggests that our analysis so far has been incomplete—that we need an index of "blocking power" to supplement the index already defined. To this end, we could set up a formal scheme similar to the previous one, namely: arrange the individuals in all possible orders and imagine them casting *negative* votes. In each arrangement, determine the person whose vote finally defeats the measure and give him credit for the block. Then the "blocking power" index for each person would be the relative number of times that he was the "blocker."

Now it is a remarkable fact that the new index is exactly equal to the index of our original definition. We can even make a stronger assertion: *any scheme for imputing power among the members of a committee system either yields the power index defined above or leads to a logical inconsistency.* A proof, or even a precise formulation, of this assertion would involve us too deeply in mathematical symbolism for the purposes of the present paper.[7] But we can conclude

---

[5] In the formal sense described above.

[6] This statement can be put into numerical form without difficulty, to give a quantitative description of the "efficiency" of a legislature.

[7] The mathematical formulation and proof are given in L. S. Shapley, "A Value for N-Person Games," *Annals of Mathematics Study No. 28* (Princeton, 1953), pp. 307–17. Briefly stated, any alternative imputation scheme would conflict with either *symmetry*

that the scheme we have been using (arranging the individuals in all possible orders, etc.) is just a convenient conceptual device; the indices which emerge are not peculiar to that device but represent a basic element of the committee system itself.

We now summarize some of the general properties of the power index. In pure *bi*cameral systems using simple majority votes, each chamber gets 50% of of the power (as it turns out), regardless of the relative sizes. With more than two chambers, power varies inversely with size: the smallest body is most powerful, etc. But no chamber is completely powerless, and no chamber holds more than 50% of the power. To illustrate, take Congress without the provision for overriding the President's veto by means of two-thirds majorities. This is now a pure tricameral system with chamber sizes of 1, 97, and 435. The values come out to be slightly under 50% for the President, and approximately 25% each for the Senate and House, with the House slightly less than the Senate. The exact calculation of this case is quite difficult because of the large numbers involved. An easier example is obtained by taking the chamber sizes as 1, 3, and 5. Then the division of power is in the proportions 32:27:25. The calculation is reproduced at the end of this paper.

The power division in a multicameral system also depends on the type of majority required to pass a bill. Raising the majority in *one* chamber (say from one-half to two-thirds) increases the relative power of that chamber.[8] Raising the required majority in all chambers simultaneously weakens the smaller house or houses at the expense of the larger. In the extreme case, where unanimity is required in every house, each individual in the whole legislature has what amounts to a veto, and is just as powerful as any other individual. The power index of each chamber is therefore directly proportional to its size.

We may examine this effect further by considering a system consisting of a governor and a council. Both the governor and some specified fraction of the council have to approve a bill before it can pass. Suppose first that council approval has to be unanimous. Then (as we saw above) the governor has no more power than the typical councilman. The bicameral power division is in the ratio 1:N, if we take N to be the number of councilmen. If a simple majority rule is adopted, then the ratio becomes 1:1 between governor and council. That is, the governor has N times the power of a councilman. Now suppose that the approval of only one member of the council is required. This means that an individual councilman has very little chance of being pivotal. In fact the power division turns out to be N:1 in favor of the governor.[9] If

---

(equal power indices for members in equal positions under the rules) or *additivity* (power distribution in a committee system composed of two strictly independent parts the same as the power distributions obtained by evaluating the parts separately).

[8] As a general rule, if one component of a committee system (in which approval of all components is required) is made less "efficient"—i.e., more susceptible to blocking maneuvers—then its share of the total power will increase.

[9] In the general case the proportion is $N-M+1:M$, where M stands for the number of councilmen required for passage.

votes were for sale, we might now expect the governor's price to be N^2 times as high as the average councilman's.

Several other examples of power distribution may be given. The indices reveal the decisive nature of the veto power in the United Nations Security Council. The Council consists of eleven members, five of whom have vetoes. For a substantive resolution to pass, there must be seven affirmative votes and no vetoes. Our power evaluation gives 76/77 or 98.7% to the "Big Five" and 1/77 or 1.3% to the remaining six members. Individually, the members of the "Big Five" enjoy a better than 90 to 1 advantage over the others.

It is well known that usually only a small fraction of the stock is required to keep control of a corporation. The group in power is usually able to muster enough proxies to maintain its position. Even if this were not so, the power of stockholders is not directly proportional to their holding, but is usually biased in favor of a large interest. Consider one man holding 40% of a stock while the remaining 60% is scattered among 600 small shareholders, with 0.1% each. The power index of the large holder is 66.6%, whereas for the small holders it is less than 0.06% apiece. The 400:1 ratio in holdings produces a power advantage of better than 1000:1.[10]

The preceding was an example of a "weighted majority game." Another example is provided by a board with five members, one of whom casts two extra votes. If a simple majority (four out of seven votes) carries the day, then power is distributed 60% to the multivote member, 10% to each of the others. To see this, observe that there are five possible positions for the strong man, if we arrange the members in order at random. In three of these positions he is pivotal. Hence his index is equal to 3/5. (Similarly, in the preceding example, we may compute that the strong man is pivotal 400 times out of 601.)

* * *

The values in the examples given above do not take into account any of the sociological or political superstructure that almost invariably exists in a legislature or policy board. They were not intended to be a representation of present day "reality." It would be foolish to expect to be able to catch all the subtle shades and nuances of custom and procedure that are to be found in most real decision-making bodies. Nevertheless, the power index computations may be useful in the setting up of norms or standards, the departure from which will serve as a measure of, for example, political solidarity, or regional or sociological factionalism, in an assembly. To do this we need an empirical power index, to compare with the theoretical. One possibility is as follows: The voting record of an individual is taken. He is given no credit for being on the losing side of a vote. If he is on the winning side, when n others voted with him, then he

[10] If there are two or more large interests, the power distribution depends in a fairly complicated way on the sizes of the large interests. Generally speaking, however, the small holders are better off than in the previous case. If there are two big interests, equal in size, then the small holders actually have an advantage over the large holders, on a power per share basis. This suggests that such a situation is highly unstable.

is awarded the probability of his having been the pivot (or blocker, in the case of a defeated motion), which is $1/n+1$. His probabilities are then averaged over all votes. It can be shown that this measure gives more weight than the norm does to uncommitted members who hold the "balance of power" between extreme factions. For example, in a nine-man committee which contains two four-man factions which always oppose each other, the lone uncommitted member will always be on the winning side, and will have an observed index of 1/5, compared to the theoretical value of 1/9.

A difficulty in the application of the above measure is the problem of finding the correct weights to attach to the different issues. Obviously it would not be proper to take a uniform average over all votes, since there is bound to be a wide disparity in the importance of issues brought to a vote. Again, in a multicameral legislature (or in any more complicated system), many important issues may be decided without every member having had an opportunity to go on record with his stand. There are many other practical difficulties in the way of direct applications of the type mentioned. Yet the power index appears to offer useful information concerning the basic design of legislative assemblies and policy-making boards.

* * *

APPENDIX

The evaluation of the power distribution for a tricameral legislature with houses of 1, 3, and 5 members is given below:

There are 504 arrangements of five X's, three O's, and one ϕ, all equally likely if the nine items are ordered at random. In the following tabulation, the numbers indicate the number of permutations of predecessors ( ) and successors [ ] of the final pivot, marked with an asterisk. The dots indicate the pivots of the three separate houses.

$$
\left.\begin{array}{l}
\text{O Ȯ O X X } \dot{\phi} \text{ Ẋ X X} \\
\quad (60) \qquad * \quad [1] \\
\text{O Ȯ X X } \dot{\phi} \text{ Ẋ O X X} \\
\quad (30) \qquad * \quad [3]
\end{array}\right\} \text{150 pivots for X}
$$

$$
\left.\begin{array}{l}
\text{O X X Ẋ X X } \dot{\phi} \text{ Ȯ O} \\
\quad (42) \qquad\qquad * \quad [1] \\
\text{O X X Ẋ X } \dot{\phi} \text{ Ȯ O X} \\
\quad (30) \qquad\quad * \quad [2] \\
\text{O X X Ẋ } \dot{\phi} \text{ Ȯ O X X} \\
\quad (20) \qquad *\quad\; [3]
\end{array}\right\} \text{162 pivots for O}
$$

$$
\left.\begin{array}{l}
\text{O Ȯ O X X Ẋ X X } \dot{\phi} \\
\quad (56) \qquad\qquad\quad * \\
\text{O Ȯ O X X Ẋ X } \dot{\phi} \text{ X} \\
\quad (35) \qquad\qquad * \quad [1] \\
\text{O Ȯ O X X Ẋ } \dot{\phi} \text{ X X} \\
\quad (20) \qquad\quad * \quad [1] \\
\text{O Ȯ X X Ẋ X X } \dot{\phi} \text{ O} \\
\quad (21) \qquad\qquad * \quad [1] \\
\text{O Ȯ X X Ẋ X } \dot{\phi} \text{ O X} \\
\quad (15) \qquad\quad * \quad [2] \\
\text{O Ȯ X X Ẋ } \dot{\phi} \text{ O X X} \\
\quad (12) \quad * \qquad [3]
\end{array}\right\} \text{192 pivots for }\phi
$$

Power indices for the houses are 192/504, 162/504, and 150/504, and hence are in the proportion 32:27:25, with the smallest house the strongest. Powers of the individual members are as 32:9:9:9:5:5:5:5:5.

5.

Does the Fittest Necessarily Survive?

Martin Shubik
Princeton University

Below is an application of a simple model from the theory of non-cooperative games to the evaluation of the strength of an individual in a situation involving three countervailing powers. This model emphasizes the difference between the maximization problems of the natural sciences and the "cross-purposes maximization" situations which are found on the political and social scene. A principle of strength through weakness is observed.

Still present in the more naïve folklore of the study of politics is the concept formulated by Spencer of the "survival of the fittest." The origins of this term may be traced to simplified interpretations or misinterpretations of Darwinism in attempts to draw social and political analogies from the work in the biological sciences.

Coexistent with the above much misinterpreted hypothesis has been Thomas Hobbes's frequently quoted description of the state of nature as the ultimate in individualism. "No arts, no letters, no society, and which is worst of all, continual fear and danger of violent death, and the life of men solitary, poor, nasty, brutal and short."

The quotes of atomistic competition are many and many analogies with "the laws of the jungle," "dog-eat-dog," have been made. Behind all of these statements there has been some type of implicit assumption that the strong fare the best. Voltaire observed that: "It is said that God is on the side of the heaviest battalions."

It is evident that if Spencer's statement is to be more than a mere tautology, some operational meaning has to be attached to the word *fittest*. This involves stating the properties of "power" and devising a method whereby we can measure the power in the possession or control of an individual. This is obviously one of the major and most difficult tasks of political science. We will avoid most of this difficulty here by restricting the discussion to a very simple model in which the concept of power is easily defined. Yet even in this almost trivially simple example certain interesting aspects of individualistic maximization against countervailing powers are illustrated.

Given a Hobbesian state of nature, given a definition of individual power,

43

then let us ask what sort of individual is best suited to survive in a state in which every man acts for himself and by himself. In the following two examples, the former appears to support the observation of Voltaire, whereas in the latter the "heaviest battalion" seems to have lost the Deity's favor.

The two-person duel at one time helped to rid society of its excess of aristocrats as well as the occasional man of talent in fields other than dueling. We consider a duel that is fought with revolvers; each man fires one shot; they randomize to decide who shoots first. There is a quite natural manner in which we can define "power" in dueling. We may imagine that the duelists have been rated for their shooting ability. A rating of 0.8, for instance, means that eight out of ten bullets fired by a marksman will hit the target. In this example we will regard the duel as being fought for a prize which will be split among the survivors in proportion to their strength. Suppose that the duelists are rated 0.8 and 0.7, respectively; then, if they both survived the duel, one would get $8/15$ and the other $7/15$ of the prize. (This artifice of introducing a prize is not necessary; however, it will make the computation in the second example considerably easier and will not have any effect on the qualitative results.) Which of the two duelists has the best chance of survival? A simple computation indicated that the one who is the best shot has the best chance to survive. This somewhat unimpressive result is in the spirit of the quotations given at the start of this discussion. The actual chances for survival are given below:

A's chances for survival $\quad 1 - (½)(.7)(1 - .8) - (½)(.7) = .58$
B's chances for survival $\quad 1 - (½)(.8)(1 - .7) - (½)(.8) = .48$

This takes into account the equiprobabilities of each duelist's having to fire either first or last.

We now consider a somewhat more complicated duel involving three individuals. Their ratings are respectively, A, 0.8; B, 0.7; and C, 0.6. They fight a duel by each firing one shot at either of the other two. They randomize to determine the order in which they fire, and they stand equidistant from each other. There are six possible orders of firing: ABC, ACB, BAC, BCA, CAB, CBA, and they are all equiprobable. We will compute through in detail only the first case. Suppose that the order of firing is ABC. This happens to coincide with the order of their relative strengths. As A wishes to maximize his chances of surviving, he is forced to fire at B, hence B's chances of survival are: $1 - (.8) = .2$. If B is alive when it is his turn to shoot, he will fire at C because A having shot no longer represents a threat to him whereas C could still kill him with a high probability. The chances of C's surviving depend upon B's rated accuracy as a marksman and upon B's chances of being alive, hence his chances are: $1 - (.2)(.7) = .86$. If C survives and only A survives, then C will shoot at A. If both A and B are still alive at this stage, C will shoot at A because although neither of them represents a threat to him, he would prefer to see A dead because if alive he would be entitled to more of the prize than B. The chances of A's surviving depend upon C's rated accuracy and upon the chances of C's being alive when it is his turn to fire, hence his survival possibilities are: $1 - (.86)(.6) = .484$. We note that the chances of survival are 0.484 for A, 0.2 for B, and 0.86 for C. The worst shot has by far the highest probability of survival! Upon reflection, the cause of this paradox is easily seen. If all participants act completely individually without any type or form of *esprit de corps*, then the strong will be forced to eliminate the strong in order to maximize their chances to survive. Thus in a noncooperative world in which more than two battalions fought all comers, Voltaire's observation would not hold. The meek would apparently inherit Hobbes's earth. This conclusion does not depend upon the order of firing in the above example. If we work out all other possibilities and then average the results, we obtain the chances of survival for A are 0.260; for B, 0.488; for C, 0.820.

The political analogies to this type of situation are many. In elections very often the competition of two strong candidates gives the third and weaker candidate a chance to win. A weak country's political position may be made stronger by being involved in dealings with two stronger countries which are acting noncooperatively. At various times recently Iran's position could have been described as one involving strength due to weakness when caught between the interests of Great Britain and Russia.

The problem of the three-person duel was originally proposed as a mathematics problem by H. D. Larson.[1] His formulation was somewhat different from the one presented here. Essentially the same is to be found in Kinnaird, *Encyclopedia of Puzzles and Pastimes* (p. 246). Many variants and complications may be introduced in order to make the model more "realistic," however, the basic feature of the noncooperation will not be changed. We note that if the disparity in relative strengths is great enough, then it is possible to find situations in which the strongest player does actually have the best chance for survival. In a noncooperative environment it apparently does not pay to be slightly stronger than the others for this invites action against oneself.

These two duel models are examples of what are called noncooperative games. As such, these simple mathematical models lay stress upon certain essential features of human interaction. They are information patterns, individual powers, and the elements of cooperation or conflict in situations which involve cross-purpose maximizations, i.e., situations in which there is not an absolute identity of interests. In the models above, power was easily defined as shooting ability; motivation was specified as the desire both to survive and to obtain as great a share of the prize as possible; coalitions were ruled out; the information conditions were such that the power of all participants was known beforehand; the information pattern was specified inasmuch as the firing order was determined and it was assumed that each would know when his predecessor had fired. As some of these conditions are changed slightly, the quantitative results will vary. Depending upon the amount and the type of change, the qualitative results may also vary. For instance, a change in the information pattern may be brought about by considering a silent instead of a

---

[1] H. D. Larson, "A Dart Game," *American Mathematical Monthly*, December 1948, pp. 640–641.

noisy duel. These types of duels have been studied in military situations using models in which the participants approach each other at given rates and may fire at any time but are unable to hear each other's shots.

We ruled out coalitions because we were primarily interested in examining models that portrayed as much individualism as possible. It is of interest to note that in any situation involving more than two parties, it is impossible to define pure opposition. There is always some element of common interest to some group if the participants are strategically interlinked. In the three-person duel discussed, it would pay the two strong ones to eliminate first the weakest participant before falling out among themselves. It is a matter of empirical research to determine whether or not in certain biological, sociological, and political situations the strong eliminate the weak or themselves first. We have noted that for three or more participants in a strategically interlinked situation there will always be a community of interests between at least two. Even when there are only two individuals involved, it is difficult to define a situation in which there is only pure opposition and hence no opportunity for cooperation. The closest approximation to such a situation can be found in games played for amusement, such as matching pennies, chess, or checkers. In these games the winnings of one side equal precisely the losings of the other side. A game of this variety is called "zero-sum." In most political situations it is possible for both sides to gain by cooperation and to lose by individual action. A situation of this type is called "non-zero-sum." Many wars are of this nature, both the winner and the loser may be worse off than they would have been had they negotiated rather than fought. The duels could have been set up as cooperative games in which the participants negotiate to decide how to split the prize and use their rated shooting power for threat purposes rather than actually resort to fighting the duel. The present state of game theory is such that no completely satisfactory theory of the solution to cooperative games exists. Nevertheless, the methods used here serve both to clarify some of the basic concepts of individual and group action as well as to indicate results that are by no means intuitively obvious and yet appear to arise even from very simple models of situations involving the presence of more than one decision-making group with power.

[41]

The assignment game I: The core
(With Lloyd Shapley)

1. INTRODUCTION

Two-sided market models are important, as Cournot, Edgeworth, Böhm-Bawerk, and others have observed,[1] not only for the insights they may give into more general economic situations with many types of traders, consumers, and producers, but also for the simple reason that in real life many markets and most actual transactions are in fact bilateral – i.e., bring together a buyer and a seller of a single commodity. Modern game-theoretic concepts, when applied to even the most elementary economic models, have often yielded suggestive results, sometimes reinforcing and sometimes challenging the more traditional doctrines based on behavioristic theories of the individual.[2] The present study, of which this paper is the first part, will concern a class of simple, two-sided market 'games' whose distinctive feature is the indivisibility and the ability to satiate of the goods for sale, e.g. houses or automobiles, so that the primary object of the game is simply to find suitable 'assignments' of buyers to sellers.[3] [1]

We intend to explore the properties of such assignment games from several different solutional viewpoints. In this first part we shall concentrate on the *core* of the game – i.e., the set of outcomes that no coalition can improve upon.[4]

1.1 The underlying economic assumptions

The assumptions of our model, though restrictive in many respects, do permit considerable latitude in size and structure. There may be many or few traders on either side of the market, there may be product differentiation, and the traders themselves may be quite dissimilar in their likes and dislikes. Thus, a wide range of specific models, from a situation where, say, two or three firms are bidding for the same building site, to a large 'noisy' milieu like the private-party market in used cars, are encompassed within the same framework and can be given a more or less uniform treatment.

Let us list the main economic assumptions:

1. Utility is identified with money.
2. Side payments are permitted.
3. The objects of trade are indivisible.
4. Supply and demand functions are inflexible. [2]

This is not the place for extended comment on the first assumption.[5] We do, however, stress one point. In the open marketplace, where there are many individuals buying and selling for money (often acting as fiduciaries or trustees – i.e., using other people's money), the full monetization of utility comes closest to a practical realization. Here if anywhere the assumption is defensible.

The second assumption is largely one of convenience in modeling and analysis. By permitting free transfer of money among all participants we avoid the necessity of providing by special rules for the ordinary payments from a customer to his supplier. To be sure, we are also permitting payments to third parties, i.e., 'side payments' in ordinary parlance. But it is worth noting that the avoidance of such payments, in situations where they are in fact avoided, is generally more a matter of custom or ethics than of hard and fast rules. The *core* solution, as it turns out, excludes third-party payments. We shall see later that the *value* solution usually requires third-party payments, for reasons related to the discussion in Section 5 below.[6] The *stable-set* solutions, which are interpreted as 'standards of behavior' in the von Neumann-Morgenstern theory, have it both ways; some standards allow third-party payments while others do not.[7] [3]

The third assumption is rather unusual, since it is more often the opposite hypothesis (i.e., fungibility and perfect divisibility) that is imposed in the name of mathematical simplification. Indivisibility is usually considered an 'imperfection' in the market – a technical and conceptual difficulty. Indeed, our present approach leads us towards combinatorics and linear programming, and away from the differential calculus methods that pervade so much of traditional economics. There is no dearth of economic applications under this assumption, however; a classic example is Böhm-Bawerk's horse market (see Section 4).[8] Our model is significantly more general, however, since it permits differentiation among the items in trade.

Inflexibility, the fourth assumption, in a way includes the third. Having postulated indivisible commodities, we further assume that each producer has a supply of exactly one item, and each consumer a desire for exactly one item. One can achieve much the same effect in terms of perfectly divisible goods by assuming individual supply and demand functions a step-function type, as illustrated below for the case of undifferentiated products (see Figure 1).

Our assumptions are admittedly restrictive, but they are not entirely unrealistic and we are not seeking wide generality for our model in any case. While excluding many important but secondary considerations, we wish to focus attention on what we consider the basic motive force of competitive economics, namely the profitable interaction between separately maximizing individuals. [5]

2. DESCRIPTION OF THE GAME

2.1 A real estate market

We shall now formulate the market game in detail, motivating it in terms of a market in private homes. Another interpretation, less specific and hence more versatile, will be outlined in Section 2.3.

Let there be m homeowners in the market, and n prospective purchasers. We shall

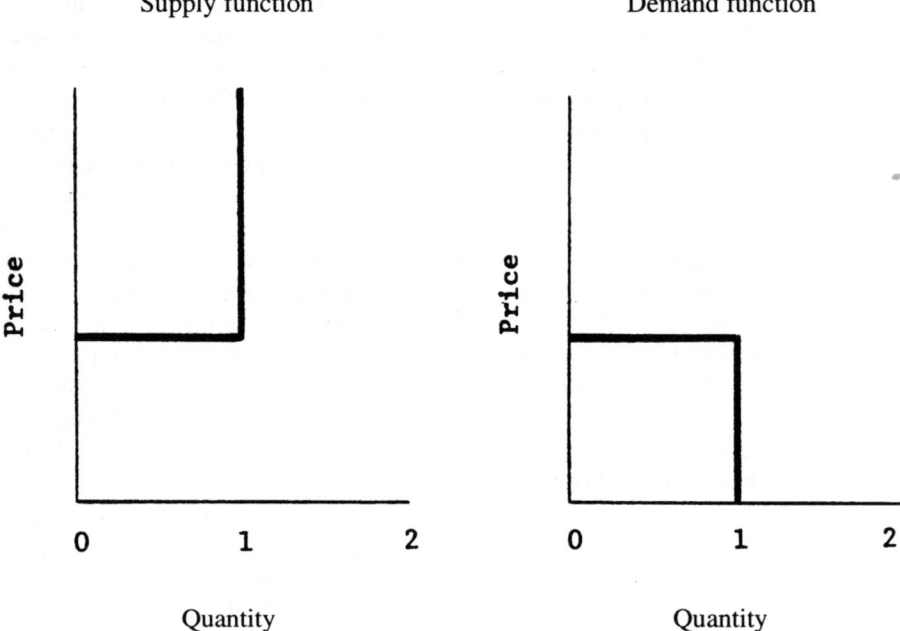

Figure 1

refer to them simply as sellers and buyers, respectively. The ith seller values his house at c_i dollars, while the jth buyer values the same house at h_{ij} dollars.[9] (These are meant to be actual utility valuations, not estimates of market price. But they might nevertheless be thought of as arising from the existence of other means of disposal or sources of supply, outside the model.) If $h_{ij} > c_i$ then a price favorable to both parties exists. We do not, however, assume that this inequality holds in all cases, or indeed in any case at all.

The possible moves in the game include the transfer of any house from its owner to any buyer, and the transfer of money from any player to any other. As we have pointed out before, we do not have to spell out the detailed scheduling of contacts, bids, offers, payments, etc., although such tactical features would be quite conspicuous in any account of the rules of the game in extensive form. All that we require, for our present purpose, can be summed up in the [6] one simple observation: if i sells his house to j at a price p_i, and if both avoid dealing with third parties, then i's final profit or gain is exactly

$$p_i - c_i, \qquad (2.1)$$

and j's is exactly

$$h_{ij} - p_i. \qquad (2.2)$$

We have no way at present of ascertaining the price p_i, but we can postpone the question. Since sales prices are not formally distinguished from other side payments, they will drop out of our initial calculations, only to reappear, rather unexpectedly, when the solution of the game begins to take shape.

2.2 The characteristic function

The characteristic function of the game, which states the worth v(S) of every coalition S, can now be determined. Let M denote the set of all sellers, and N the set of all buyers. To begin with, it is obvious that

$$v(S) = 0 \text{ if } |S| = 0 \text{ or } 1, \qquad (2.3)$$

since no player, without help from another, can effect a [7] profitable transaction. More generally, we see that all of the 'one-sided' coalitions are 'flat':

$$v(S) = 0 \text{ if } S \subseteq M \text{ or } S \subseteq N. \qquad (2.4)$$

In other words, only a 'mixed' coalition can ever hope to asssure a profit.

The simplest kind of mixed coalition consists of two players, one of each type. For this case we have[10]

$$v(\overline{ij}) = \max\,(0,\, h_{ij} - c_i) \text{ if } i \in M \text{ and } j \in N. \qquad (2.5)$$

For brevity, we denote this number by a_{ij}. Note that it does not depend on the sales price p_i that appears in (2.1) and (2.2), since that is merely an internal transfer as far as the coalition $\overline{ij}$ is concerned.

A moment's reflection reveals that these mixed pairs are the only essential coalitions in the game.[11] The best that a larger coalition can do is to split up into separate trading pairs and pool the profit. Hence the m × n matrix (a_{ij}) suffices to determine v completely. In fact, v may be characterized as the *smallest superadditive set function* on M ∪ N satisfying (2.3) *and* (2.5). [8]

In order actually to compute v for the larger mixed coalitions, we must pick out an optimal set of transactions, maximizing the coalition's total gain. Put symbolically, we have

$$v(S) = \max\,[a_{i_1j_1} + a_{i_2j_2} + a \ldots + a_{i_kj_k}], \qquad (2.6)$$

the maximum to be taken over all arrangements of 2k distinct players $i_1, \ldots, i_k$ in S ∩ N and $j_1, \ldots, j_k$ in S ∩ N, where k = min (|S ∩ M|, |S ∩ N|). The evaluation of an expression such as (2.6) is commonly called the 'optimal assignment problem' or simply 'assignment problem'; accordingly we refer to games of this form as *assignment games*.

For relatively small coalitions S (say, less than 10 players) the maximum in (2.6) can often be discovered by inspection; the reader may try his hand on the examples below. (In the first there are four optimal assignments, in the second there is just one; the respective values are 4 and 16.) [9]

	S ∩ N		
M	0	2	0
∩	2	0	2
S	0	2	0

	S ∩ N		
M	5	8	2
∩	7	9	6
S	2	3	0

For larger matrices, systematic methods are available for finding the optimal assignment.[12]

For the most part we shall be interested in evaluating (2.6) only for the all-player coalition S = M ∪ N. The number v(M ∪ N) is obviously very important, since it determines the maximum total monetary payoff, and hence the Pareto set and imputation space of the game. The optimal assignments for the other large, nonessential coalitions have little effect on the analysis of the game.

2.3 A game of partners

It will be observed that the characteristic function of the assignment game, when defined in terms of the numbers a_{ij}, is symmetric in its treatment of buyers and sellers, although the market model itself was not. The following alternative interpretation of the game, in sidestepping the preliminaries that went into the 'real estate' version, has the conceptual advantage of being symmetrical from the start.

We may imagine an economic milieu in which two types of agents, say 'producers' and 'consumers' (or 'men' and 'women'), are constrained for some reason to conduct their business under *exclusive*, bilateral contracts. Thus, after an initial period of jockeying for position, a number of partnerships are formed, and any transfers of goods or services [10] (but not money) are limited to exchanges between partners. In this model we do not care particularly about the nature of the goods or services; we only care that with every partnership $\overline{ij}$ there is associated a number $a_{ij} \geq 0$, denoting the potential for profit of that partnership if it forms. If we now regard the system as a cooperative game, it is apparent that its characteristic function is given by (2.4) and (2.6).

A contract between players i and j would of course specify how the gain a_{ij} is to be divided. A prudent, 'economic' man playing this game would be loath to enter a partnership for a stated share of the proceeds until he had satisfied himself that more favorable terms could not be obtained elsewhere. We can imagine that each player would set a price on his participation, and that no contracts would be signed until the prices on both sides of each partnership formed are in harmony, i.e., satisfy each partner and add up to what the partnership is worth. One may well ask: does a set of such 'harmonious' prices exist?

Mathematically, this question may be put as follows:
Do there exist numbers $q_i \geq 0$, $r_j \geq 0$ such that

$$\begin{cases} q_i + r_j = a_{ij} & \text{if i and j are ultimately partners,} \\ q_i + r_j \geq a_{ij} & \text{if they are not? [11]} \end{cases}$$

The answer is in the affirmative, as we shall presently demonstrate. This fact is of central importance throughout all the subsequent analysis of the assignment game.

2.4. Complementarity

Before proceeding to a study of the solutions of the assignment game, we shall digress briefly to describe an interesting property of the characteristic function.

The superadditivity of v in (2.6) is obvious, since combining two coalitions only enlarges the set of possible assignments. Less obvious is the way in which the *marginal value* of a player responds to changes in the make-up of the coalition to which he belongs. Intuitively, we should expect a player's value to a coalition to be small when there is an excess of players of his type, and large when there is an excess of players of the other type. In other words, similar players should appear somewhat as *substitutes* in coalitional matters, and dissimilar players as *complements*. The following theorem gives an expression to this intuitive idea.

A special notation will be useful: S^p will denote the coalition consisting of S and the added player p. (We shall use this notation only when p is not a member of S.) Then the *marginal value* of player p to the coalition S is given by [12]

$$m(p, S) = v(S^p) - v(S).$$

The following result is proved in [11].

THEOREM 1. *If p and q are of the same type (i.e., both in M or both in N) then*

$$m(p, S^q) \leq m(p, S), \quad \text{all } S \not\ni p, q.$$

If they are of the opposite type, then

$$m(p, S^q) \geq m(p, S), \quad \text{all } S \not\ni p, q.$$

It will be seen that this theorem is simply a statement about the sign of the "second difference" $v(S^{pq}) - v(S^p) - v(S^q) + v(S)$. When this quantity is positive, the players are complementary inputs to the coalition-forming process; when it is negative, they are substitutes. [13]

3. THE CORE

3.1. LP form of the model

It will be useful at this point to recast the assignment problem (2.6) into *linear programming* (LP) terminology.

Consider just the assignment problem for the coalition of all players – i.e., the problem of determining $v(M \cup N)$. Introduce mn nonnegative real variables x_{ij}, $i \in M$, $j \in N$, and impose on them the m + n constraints

$$\sum_{i \in M} x_{ij} \leq 1, \quad \sum_{j \in N} x_{ij} \leq 1. \tag{3.1}$$

(We may interpret x_{ij} in the real estate game as the fraction of the ith house sold to player j, or, in the partnership game, as the probability that partnership $\overline{ij}$ will form). The LP problem is then to maximize the following objective function:

$$z = \sum_{i \in M} \sum_{j \in N} a_{ij} x_{ij}. \tag{3.2}$$

It can be shown (see [3], p. 318) that the maximum value $z_{\max}$ is attained with all $x_{ij} = 0$ or 1. Thus, the fractions or probabilities artificially introduced disappear from the solution, and the (continuous) LP problem is effectively equivalent to the (discrete) assignment problem, so that we have [14]

$$z_{\max} = v(M \cup N).$$

As is well known, every LP problem can be transposed into a *dual* form, and the solutions of the two problems are intimately bound up in each other. In the present case, the dual has m + n nonnegative real variables, $u_1, ..., u_m, v_1, ..., v_n$, subject to the mn constraints,

$$u_i + v_j \geq a_{ij}, \, i \in M, \, j \in N, \tag{3.3}$$

and the objective is to *minimize* the sum:

$$w = \sum_{i \in M} u_i + \sum_{j \in N} v_j. \tag{3.4}$$

The fundamental duality theorem[13] tells us that $w_{\min} = z_{\max}$.

What meaning does the dual problem have in the present context? Let $(u, v) = (u_1, ..., u_m, v_1, ..., v_n)$ be a vector that minimizes (3.4), subject to (3.3). Then we have

$$\sum_{i \in M} u_i + \sum_{j \in N} v_j = w_{\min} = z_{\max} = v(M \cup N). \tag{3.5}$$

This means that (u, v) is an imputation[14] of the assignment game. Moreover, (3.3) tells us that for every pair $i \in M, j \in N$, [15]

$$u_i + v_j \geq a_{ij} = v(\overline{ij}).$$

It follows, by (2.6), that for any coalition S

$$\sum_{i \in S \cap M} u_i + \sum_{j \in S \cap M} v_j \geq v(S). \tag{3.6}$$

But (3.5) and (3.6) are exactly how the *core* of the game is defined: (3.5) ensures the feasibility of (u, v) and (3.6) ensures its non-improvability by any coalition. Conversely, any payoff vector in the core – i.e., satisfying (3.5) and (3.6) – clearly fulfills the conditions for a solution to the dual LP problem. Hence we conclude:

Political Economy, Oligopoly and Experimental Games 527

THEOREM 2. *The core of an assignment game is precisely the set of solutions of the LP dual of the corresponding assignment problem.*

3.2. Prices

Our venture into the field of linear programming has proved most fruitful. We have already garnered (1) a proof of the existence of the core, (2) a characterization of the points in the core, and (3) an assurance of effectively computational procedures for both the characteristic function and the core.[15] [16]

The reader schooled in LP will now be expecting an attempt to connect the dual solutions to a price mechanism, and we shall not disappoint him. In fact, the connection is very simple. In order to achieve the profit u_i promised by a given core vector (u, v), the owner of the ith house must sell it at the price

$$p_i = c_i + u_i. \qquad (3.7)$$

If all the owners attach such price tags to their houses, then the typical buyer j will be confronted by a choice among the m possible net gains:

$$h_{ij} - p_i, \quad i \in M. \qquad (3.8)$$

If these numbers are all negative, then j perforce stays out of the market, and ends the game with a profit of zero; otherwise he will seek to maximize (3.8). This is equivalent, in view of (3.7) and (2.5), to maximizing

$$a_{ij} - u_i, \quad i \in M.$$

We know by (3.3) that none of these numbers exceed v_j. On the other hand, the minimization of the dual objective form (3.4) ensures that at least one of them is equal to v_j. This means that buyer j's maximum profit is precisely v_j, and he is led to it by direct comparative shopping. [17]

For an example with specific numerical values, the reader is now invited to look ahead to Section 3.4.

What if several shoppers find that the same house is a 'best buy'? Can this happen? In the most common case, where the optimal assignment is unique and where the chosen vector (u, v) lies in the relative interior of the core (see Section 3.3), the answer is no. The price schedule (3.7) leads the players unambiguously to a nonconflicting allocation of goods that maximizes the welfare of the group as a whole and obtains for each individual the specified amount u_i or v_j.

In exceptional, 'degenerate' cases there may be ties in the buyers' preferences, and care must be taken to resolve the ties without assigning the same house to more than one buyer. This is always possible, as can be demonstrated by a simple perturbation argument.

3.3. Structure of the core

The core of the assignment game only rarely consists of just a single imputation.[16] Consequently, the price structure just described is seldom unique. In this section we shall attempt to account for the multiplicity of solutions and discuss what it means in the market context. [18]

In general, if numbers a_{ij} are chosen 'at random', the optimal assignment is unique.[17] This implies that a dual solution (u, v) exists in which the strict inequality:

$$u_i + v_j > a_{ij}$$

holds for all *pairs* $\overline{ij}$ that are not involved in the optimal assignment. Let $\overline{i'j'}$ be one of the min(m, n) pairs in the optimal assignment. Then a small amount can be shifted from $u_{i'}$ to $v_{j'}$ without spoiling any of the conditions for a dual solution. It follows that there are at least min(m, n) 'degrees of freedom' in the core. On the other hand, the dimensionality of the core is certainly never greater than min(m, n), because if the u-components of a core vector are given, then the v-components are completely determined, and vice versa.

A picture of the core is beginning to take shape: it is a closed, convex polyhedral set whose dimension is typically equal to min(m, n), but may be less in the presence [19] of degeneracies – i.e., special arithmetical relations among the a_{ij}. Note that the dimension of the imputation space, in which the core is situated, is $m + n - 1$, which is considerably larger than min(m, n).

We show that the core tends to be *elongated*, with its long axis oriented in the direction of market-wide price trends. There is a 'high-price corner' of the core, at which every seller gets his top profit and every buyer his bottom. There is also a 'low-price corner', where the reverse is true. These points are poles of a diameter; they prove to be at least as far apart as any other two points in the core. Thus, to a considerable extent, the fortunes of all players of the same type rise and fall together.

Let us try to make this intuitively plausible. For simplicity assume m = n. Suppose that we have a core vector at which all players show a positive profit. Then a small constant can be subtracted from *all* of the sales prices without upsetting the core conditions. Each buyer–seller pair makes the same profit as before, and no individual is forced to take a loss. This price-cutting can continue until some seller is priced out of the market, his profit going to zero. The process can of course be reversed, raising all prices by equal amounts until one of the buyers is driven out. Thus there is a natural 'degree of freedom' *within* the core that corresponds to market-wide movement.[18] [20]

This phenomenon is often encountered in the cooperative solutions to market games.[19] The heuristic principle may be stated as follows: *inter*group allocations are relatively indeterminate, *intra*group allocations are relatively precise.

THEOREM 3. *Over all imputations in the core, let* u_i^* *and* u_{*i} *denote the highest and lowest payoffs, respectively, to player* $i \in m$; *similarly define* v_j^* *and* v_{*j} *for player* $j \in N$. *Then the payoff vectors*

$$(u_*, v^*) \text{ and } (u^*, v_*)$$

are themselves in the core. Moreover, no two imputations in the core are further apart than these.

Figure 3, in Section 3.4 below, may help the reader to visualize this theorem. The heart of the proof is in the following lemma:

Lemma. Let (u', v') and (u'', v'') be any two imputations in the core. Define

$$\underline{u}_i = \min(u'_i, u''_i), \quad \underline{v}_j = \min(v'_j, v''_j),$$

$$\overline{u}_i = \max(u'_i, u''_i), \quad \overline{v}_j = \max(v'_j, v''_j). \quad [21]$$

Then the vectors $(\underline{u}, \overline{v})$ and $(\overline{u}, \underline{v})$ are in the core.

Proof. For any i, j we have

$$\underline{u}_i + \overline{v}_j = \min(u'_i + \overline{v}_j, u''_i + \overline{v}_j)$$
$$\geq \min(u'_i + v'_j, u''_i + v''_j)$$
$$\geq a_{ij},$$

using (3.3). Hence $(\underline{u}, \overline{v})$ is undominated. It remains to show that it is an imputation. It is obviously nonnegative; we must therefore show only that its components add up to $v(M \cup N)$. For convenience, label the players so that the pairs $\overline{11}, \overline{22}, \ldots, \overline{kk}$ describe an optimal assignment. (Here $k \leq \min(m, n)$.) Then:

$$\underline{u}_i = \min(u'_i, u''_i) = \min(a_{ii} - v'_i, a_{ii} - v''_i)$$
$$= a_{ii} - \max(v'_i, v''_i) = a_{ii} - \overline{v}_i$$

for $i \leq k$. Also, for $i, j > k$ (if any) we have $\underline{u}_i = \overline{v}_j = 0$. Hence

$$\sum_{i \in M} \underline{u}_i + \sum_{j \in N} \overline{v}_j = \sum_{i=1}^{k} a_{ii} = v(M \cup N).$$

This completes the proof of the lemma. [22]

To prove the theorem, simply take a finite collection of core vectors that includes all the extreme values going into the definitions of $u^*_i, u_{*i}, v^*_j, v_{*j}$, and apply the lemma repeatedly to construct additional core vectors, until (u^*, v_*) and (u_*, v^*) are reached. For the last statement of the theorem, note that any two points $(u', v'), (u'', v'')$ in the core necessarily satisfy the inequalities:

$$|u'_i - u''_i| \leq u^*_i - u_{*i}, \quad \text{all } i \in M,$$

and

$$|v'_j - v''_j| \leq v^*_j - v_{*j}, \quad \text{all } J \in N.$$

Thus the stated result holds not only for euclidean distance but for any distance function that depends only upon the absolute values of the coordinate differences.

3.4. A numerical example

Let there be three buyers and three sellers, as shown in Table 1. [23]

Table 1

Houses (i)	Sellers' basis (c_i)	Buyers' valuations (h_{i1})	(h_{i2})	(h_{i3})
1	$18 000	$23 000	$26 000	$20 000
2	$15 000	$22 000	$24 000	$21 000
3	$19 000	$21 000	$22 000	$17 000

These data lead at once to the following (a_{ij}) matrix, already considered in Section 2.2:

```
                      (buyers N)
                    1'    2'    3'       u:

                1   5    (8)    2         4
(sellers M)     2   7     9    (6)        5.5     (units of $1000)
                3  (2)    3     0         0

           v:       2     4    0.5
```

The unique optimal assignment is shown circled, the total gain being $16,000. One of the core vectors (u,v) is also shown. The reader can verify that the matrix ($u_i + v_j$) majorizes (a_{ij}), with equality only on the circled entries. The prices corresponding to this solution are $22,000, $20,500, and $19,000 respectively. Given these prices, the first buyer clearly prefers the third house, as his gain is $2000 rather than $1000 or $1500. Similarly, the second buyer prefers the $4000 gain at house 1 to the gain of $3500 or $3000 at house 2 or 3. Finally, house 2 is the only one that the third buyer would even consider, at the stated prices.

The nonuniqueness of this solution is evident, since none of the buyers' comparisons were closer than $500, and only one seller (the third) is at, or even close to, his cost. Thus, any one price could be raised by $500, or either of the first two prices lowered by $500, without [24] moving the outcome out of the core. Or all three prices could be increased together, in a market-wide price movement, until the 'weakest' buyer (the third) is driven out.

The geometry of the core is shown in Figure 2. Since optimality demands the unique set of assignments $\{\overline{12'}, \overline{23'}, \overline{31'}\}$, we have depicted just the rectangular, three-dimensional region in the (five-dimensional) imputation space in which the equations $u_1 + v_{2'} = 8$, $u_2 + v_{3'} = 6$, $u_3 + v_{1'} = 2$ are satisfied. (The imputations outside this region cannot be attained without the aid of 'third party' side payments.) The

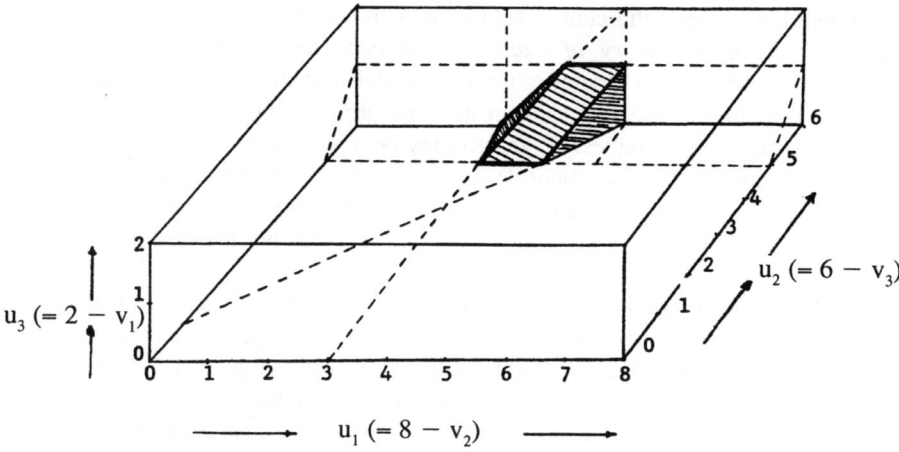

Figure 2

core is a five-sided polyhedron situated within this region, touching the boundaries $u_3 = 0$ and $v_{3'} = 0$. The solution point discussed above happens to lie in the bottom face, as shown by the open dot 'o' in Figure 3. The double-headed arrow indicates the direction of market-wide price changes. The low-price corner (u_*, v^*), where all sellers get their minimum gains, is at the lower left; the high-price corner (u^*, v_*) is at the upper right. The (euclidean) distance between them is $\sqrt{6}$ times $1000, or $2449; the distance measured in terms of the total effect on the sellers as a class is $4000. [25]

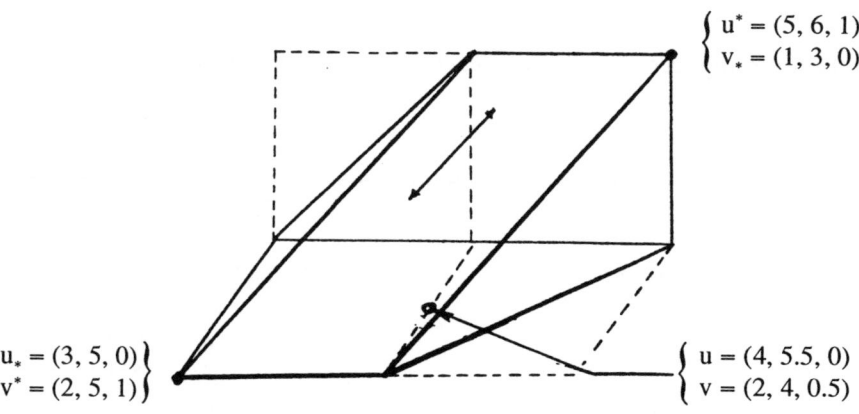

Figure 3 [26]

4. A SPECIAL CASE[20]

4.1 The horse market of Böhm-Bawerk

Discussions of price theory in economics have been enlivened by allusions to

'Böhm-Bawerk's horses', the central figures in a microcosmic market detailed in that author's *Positive Theory of Capital*, first published in 1891. Since this imaginary but numerically specific market model translates directly into an 18-person cooperative game, it makes an inviting target for the 'big game hunter' anxious to try out his techniques. (Eighteen may be a small number in economics, but it is a large number in game theory!) Apart from the test of techniques, there is the prospect of an instructive confrontation among the various game solutions – cores, stable sets, values, etc. – and the classical solution as put forward by Böhm-Bawerk.

The basic data of the market are shown in Table 2.[21] Eight individuals each have one horse for sale. Ten other individuals each wish to buy one horse. The horses themselves are all alike, but the traders have different 'subjective valuations', ranging between $10 and $30, of what it is worth to own a horse.[22] No restrictions are placed on communication, on transfers of money, or on transfers of horses. The basic game problem, informally stated, is to decide how the inherent profitability of the market, arising from the differences in subjective valuation, is to [27] be shared among the horse-traders. This is also, of course, the basic economic problem.

4.2. The characteristic function

The readers will recognize that the present example, and the class of markets it exemplifies, is a decided simplification of our general model in one important respect: the *absence of product differentiation*. The matrix of input data (h_{ij}) can accordingly be reduced to a vector (h_j). As a result, the assignment matrix (a_{ij}), now defined by

$$a_{ij} = \max (0, h_j - c_i), \qquad (4.1)$$

enjoys the following special property: *in each 2 × 2 submatrix with nonzero entries*,

Table 2

Sellers			Buyers		
A_1 values a horse at $10			B_1 values a horse at $30		
A_2	,,	,, $11	B_2	,,	,, $28
A_3	,,	,, $15	B_3	,,	,, $26
A_4	,,	,, $17	B_4	,,	,, $24
A_5	,,	,, $20	B_5	,,	,, $22
A_6	,,	,, $21.50	B_6	,,	,, £21
A_7	,,	,, $25	B_7	,,	,, $20
A_8	,,	,, $26	B_8	,,	,, $18
			B_9	,,	,, $17
			B_{10}	,,	,, $15

the sums of the diagonals are equal. This means that any set of assignments involving actual trades can be switched around, without loss or gain to anyone, since it only matters who buys and who sells, and not how they pair up. Hence the optimal assignment problem is rather trivial, and we shall be able to express the characteristic function in a very simple way.

Let S be an arbitrary coalition, and let its members be denoted by $A_{i_1}, ..., A_{i_l}$, $B_{j_1}, ..., B_{j_m}$, where the sellers A_i are arranged in order of increasing c_i and the buyers B_j in order of decreasing h_j. Then it is easy to see that [29]

$$v(S) = a_{i_1 j_1} + a_{i_2 j_2} + ... + a_{i_k j_k}, \qquad (4.2)$$

where $k = \min(l, m)$. In other words, for computational purposes we can assume that the 'strongest' buyer in any coalition buys from the 'strongest' seller, and so on down the line, until either the players of one type are exhausted or a pair is reached that cannot trade at a profit.

Applying this to the data in Table 2, we find that

$$v(M \cup N) = \$20 + \$17 + \$11 + \$7 + \$2 + 0 + 0 + 0 = \$57. \qquad (4.3)$$

This is the total amount of profit inherent in the market, and it is achieved whenever the first five buyers acquire the horses of the first five sellers.

4.3. The core

The core, too, is easy to determine, since the absence of product differentiation leads to a uniform market price. To see this let (u, v) be a typical imputation in the core. Suppose the amount of money received by some active seller i, which is $u_i + c_i$, happens to be less than the amount of money *paid* by some active buyer j, which is $h_j - v_j$. Then an improvement for $\overline{ij}$ would obviously be possible, since

$$u_i + v_j < h_j - c_i = a_{ij},$$

and (u, v) would not be in the core after all. [30]

In short, each active seller must receive at least as much as each active buyer pays out. But no money enters the system; hence in any given core imputation *all transactions take place at the same price.*

Thus, the core can be described by means of a single parameter. A vector (u, v) is in the core if and only if, for some p in a specified range, we have

$$\begin{cases} u_i = \max(0, p - c_i), & \text{all } i \in M, \text{ and} \\ v_j = \max(0, h_j - p), & \text{all } j \in N. \end{cases} \qquad (4.4)$$

The range over which p can vary is given by the requirement that $\Sigma u_i + \Sigma v_j = v(M \cup N)$. This can only be satisfied if the same number of players of each type are active – more precisely, if the number of i such that $p - c_i > 0$ does not exceed the number of j such that $h_j - p \geq 0$, and vice versa. Geometrically the core is a straight line segment, and in its interior exactly the same number of traders of each type make a positive profit.

In the numerical example (Table 2), a price in the interval $21.00 < p < £21.50$ permits exactly *five* players of each type to trade at a profit. If the price were to exceed the upper limit, the sixth seller would put up his horse for sale and an imbalance would be created that would tend to drive the price down again. Similarly, a price less than $21 would find six (or more) buyers competing for [31] five (or fewer horses). The endpoints of the core are therefore

(11.5, 10.5, 6.5, 4.5, 1.5, 0, 0, 0; 8.5, 6.5, 4.5, 2.5, 1.5, 0, 0, 0, 0, 0)

and

(11, 10, 6, 4, 1, 0, 0, 0; 9, 7, 5, 3, 1, 0, 0, 0, 0, 0).

These are of course the extremal imputations (u^*, v_*) and (u_*, v^*) of Theorem 3 in Section 3.3.

4.4. Discussion

The core that we have determined in the preceding section is *precisely the classical solution* as given by Böhm-Bawerk [1]. A uniform market-wide price is established at a level which, though not precisely determined, is constrained within fairly narrow limits. If the consumers and suppliers in the market are sufficiently numerous, and if the tastes of the former and the costs of the latter are sufficiently variegated, then something close to a determinate outcome can be predicted. It is the classical balance of supply and demand. (See Section 5.)

The one-dimensionality of the core represents the most extreme form of the enlongation described in Theorem 3; this is a consequence of the lack of differentiation among [32] the objects of trade.[23] If a small 'random' perturbation were applied to the model, giving the buyers slightly different subjective valuations for different horses, then although the dimension of the core might jump from 1 to 5 it would nonetheless remain very slender.[24] If the buyers should become more dissimilar in their tastes, however, the core would grow fatter. The ultimate in this direction would be a game in which each buyer has only one horse that he would accept at any price. Since the bargaining for each horse would then be independent of the others, the game would decompose into smaller games, and the core would take the form of a rectangular parallelepiped (i.e., a product of intervals). [33]

5. THE CORE AND THE COMPETITIVE EQUILIBRIUM

5.1. Convergence of the core

An extensive literature[25] now exists on the relation between the core of a market game and the notion of a competitive price equilibrium. Three basic features of this relationship are (1) that every competitive outcome is in the core,[26] (2) that the core will generally contain other outcomes, and indeed may exist in situations where no competitive outcome exists, and (3) that under suitable conditions involving the number, size, or similarity of the economic agents, the core approaches or even coincides with the set of competitive outcomes.

The reader familiar with these ideas will already have recognized that in the assignment game every core outcome is competitive and vice versa.[27] (See the discussion of prices [34] in Section 3.2.) There is no doubt, then, as to the convergence of the two solution concepts; even if there are only a handful of players they coincide exactly. The reason for this lies in the special nature of the supply and demand functions (see Figure 1), and the underlying preferences and indivisibilities that they reflect. The 'lumpiness' of the commodity causes essentially the same nonuniqueness in the competitive prices as it does in the allocations that all coalitions will tolerate.

There is, however, another sort of convergence of interest, already mentioned in Section 4.4. We shall describe it in general terms, having no actual theorems to offer. If the number of traders is increased, on both sides of the market, in such a way that their valuations for the products brought to market become more and more diverse (but remain bounded in a suitable sense), then the core will tend to shrink in size. The presence of a *large variety* of traders or products can in effect smooth out the supply and demand functions, and lead to more tightly determined competitive prices. In the Böhm-Bawerk model, where there is no product differentiation, this sharpening of the solution is a simple matter of narrowing the range of prices that will draw an equal number of buyers and sellers into the active market. In the more general model, however, the increasing dimensionality of the solution and the space in [35] which it is defined makes a precise discussion of the shrinkage phenomenon more difficult.[28]

5.2. Some weaknesses

From both the economic and game-theoretic viewpoint, certain weaknesses in the core (= competitive) solution soon become apparent. For one thing, the core gives no consideration to the ability of individuals or coalitions to *obstruct* outcomes, though this is often a powerful lever in bargaining.[29] For example, the coalition M of all sellers can certainly block any trade,[30] yet the core gives them no credit for this. Indeed, the inequality condition (3.6) for $S = M$ (or for $S = N$ for that matter) has no 'bite' at all, and does not eliminate any individually rational outcomes from the solution. The core is based on what a coalition can *do*, not what it can *prevent*. The core is therefore not a really satisfactory basis for a bargaining theory.

There is a more subtle bargaining tactic, not of an obstructive nature, that also fails to be reflective in the core or competitive equilibrium. It can be illustrated [36] in terms of the horse market of Table 2. Consider the 'weak' players in the game – i.e., sellers A_6, A_7, A_8 and buyers B_6 through B_{10}. Although they are not technically dummies, they are quite unable to prevent the others from obtaining the full $57 profit, or to interfere in any other way with the imputations in the core. Nevertheless, the marginal weak buyer B_6 (for example) has an important role: it is his valuation of $21 that establishes the price floor in the core. Were he to quit the scene, the floor would drop to $20, and the sellers as a group would very likely suffer. By his very presence – i.e., his willingness to pay any amount up to $21 for a horse – he performs a service to some of the players (and a disservice to others). What is it worth?

A similar capability exists for the marginal weak seller A_6, who sets the ceiling of $21.50 in the solution. In lesser measure, the other weak players also have some bargaining power of this kind. The seventh buyer B_7, for example, might argue that his participation in the bidding tends to limit the sixth buyer's 'leverage', since if B_7 should quit the scene, then B_6 might be able to command a larger bribe from the active sellers for remaining on hand to protect the $21 floor.

The marginal *strong* players, A_5 and B_5, also help to determine the price range. Since their expected profits in any case are rather small, their threats to behave [37] 'irrationally' could also carry some weight. In our example, the departure of B_5 would decrease the total social profit from $57 to $56 (see Table 2), but it would also cause the price range to shift downwards from $21–$21.50 to $20–$21. Would the other four buyers be willing to pay, say fifty cents each to bring this about?[31]

To complete the picture, we should note that the top or 'strongest' players of each type are in an extremely poor bargaining position. They are vulnerable to all the threats we have been describing, and they have no very credible counterthreats: they have too much to lose. A man who is desperately in need of a horse (like Shakespeare's Richard III!) is obviously at a disadvantage in haggling over the price, if his need is known. His situation practically invites collusion among the horse-merchants. But the classical solution (core) gives no hint of this.

We are not arguing against the classical solution as such.[32] It may not be possible to realize the bargaining potentials described above *within a given institutional form*. Sanctions against collusion or compensation, physical barriers to communication, or simply the cost and time of [38] bargaining, may be sufficient to vitiate the bargaining tactics we have been describing. The correct solution concept to apply in a given situation depends on the larger 'social' context from which the mathematical model was abstracted. It also depends, of course, on the purpose of the application – on the type of questions one wishes the theory to answer. A corollary is that when we are conducting a general analysis of the abstract model, as here, then it behoves us to explore and correlate a number of different solution concepts. This we hope to do in subsequent papers. [39]

NOTES

1. See references [1], [2], [5] at the end of this paper.
2. See for example [4], [9], [13], [14], [16].
3. This model was first treated by one of the authors in 1955 (see [8]); the present account has been extracted from the manuscript of a book in preparation. A short, nongame-theoretic account will be found in Gale [6], pp. 160–62.
4. For treatments of other solution concepts in certain special cases, see [9], [10], [13], also [17], pp. 555–86.
5. See for example [12], pp. 807–8.
6. See also [10] and [13].
7. See [9] and [17] loc. cit.
8. For a modern treatment of the problem of indivisibilities in a competitive market, see [7]. [4]
9. For an example, see Table 1, in Section 3.4.
10. We write $\overline{ij}$ rather than the more awkward '$\{i, j\}$'.
11. A coalition is 'essential' if it has more than one member and exhibits strict superadditivity for every split-up into smaller coalitions, i.e., $v(S) > v(S_1) + v(S_2)$ for all S_1, S_2 with $S_1 \neq \emptyset$, $S_2 \neq \emptyset$, $S_1 \cup S_2$

= S, and S1 ∩ S2 = ∅.
12. See for example [3], Chapter 15.
13. See for example [3], p. 129. The existence of feasible solutions to both primal and dual is apparent from an inspection of (3.1) and (3.3).
14. That is, a nonnegative, Pareto optimal payoff vector.
15. It is interesting that the core can be found without solving the multitude of assignment problems arising from the proper submatrices of $(a_{ij}|i \in M, j \in N)$. Because of this, the core in a large assignment game may be easier to compute than the characteristic function itself. The explanation of the paradox lies in the relatively small number of coalitions that are essential.
16. The first matrix in Section 2.2 provides an example: the core of the associated game contains just the imputation (0, 2, 0; 0, 2, 0). The other matrix in that section is more typical, however.
17. If the original data of the real estate version of the game are chosen 'at random' – more precisely, from continuous probability distributions that give zero weight to any particular numerical value – then the value 0 may be favoured over other values in the assignment matrix, and nonuniqueness in the optical assignment may occur with positive probability. But this merely reflects indeterminacy in the choice of *inactive* buyer-seller pairs, involving players who are actually 'priced out of the market'. The u_i or v_j for such players are identically zero in the core, and the dimensions of the core is reduced accordingly.
18. This will be illustrated in Section 3.4. If $m \neq n$ a similar discussion applies, except that it may no longer be possible to drive any of the players of the less numerous type down to zero.
19. For stable sets compare [9]; for cores compare the 'equal treatment' principle as expressed, for example, in Theorem 2 of [4]; for kernels compare [14].
20. Much of the following is drawn from [10].
21. See [1], p. 203. We have arbitrarily replaced pounds with dollars.
22. To own a second horse, however, is worth nothing more.
23. A technical explanation may be found in the large number of optimal assignments that are possible. Multiplicity in the primal solutions of an LP problem is commonly accompanied by reduced dimensionality of the dual solutions.
24. It can be shown generally that the core varies continuously as a function of the characteristic-function values.
25. Much of it, to be sure, has appeared since [8], our first treatment of the core of the assignment game. The basic paper of Debreu and Scarf [4] remains an excellent introduction; for a recent contribution to the subject (with a good bibliography) see [15].
26. A final distribution of goods and money, or the corresponding vector of utilities, is called 'competitive' if it can result from a behavioristic process in which all decisions concerning production, trade, and consumption are made noncollusively and in strict accordance with a set of prices that have been specified *ex machina* in such a way that the supply and demand of each commodity will be in balance.
27. More precisely, in the market models exemplified by houses or horses that we have been using to motivate the abstract assignment game. In the less specific 'partnership' model of Section 2.3, the notion of competitive outcome must be reconstrued.

    Incidentally, the indivisibilities and product differentiation in our model put it outside the scope of most of the general existence and convergence theorems in the literature.
28. Even in the Böhm-Bawerk case, if the price range does not narrow fast enough, the (euclidean) distance between the endpoints of the core may actually increase as the price interval goes to zero.
29. The core is sometimes described as 'the set of outcomes that no coalition can block'; this unfortunate and misleading description stems from a counterintuitive use of the term 'block' in the mathematical terminology used by some writers. A better rendering is 'the set of outcomes that no coalition can improve on'.
30. As shown by the fact that $v(N) = 0$.
31. If they were so willing, then Pareto optimality might reassert itself, in the guise of a counter-offer by the sellers, and induce B_5 to return to the market to his further profit.
32. An instructive critique of the classical solution will be found in [17] loc cit.; see also [10] and Section 3 of [13].

REFERENCES

[1] Böhm-Bawerk, E. von, *Positive Theory of Capital* (translated by William Smart), G.E. Steckert, New York, 1923 (original publication 1891).
[2] Cournot, A.A., *Researches into the Mathematical Principles of the Theory of Wealth* (translated by

N.T. Bacon), Macmillan and Co., New York, 1897 (original publication 1838).
[3] Dantzig, G.B., *Linear Programming and Extensions*, Princeton University Press, Princeton, 1963.
[4] Debreu, G. and Scarf, H., 'A limit theorem on the core of an economy', *Int. Econ. Rev.* 4 (1963), 235-46.
[5] Edgeworth, F.Y., *Mathematical Physics*, Kegan Paul, London, 1881.
[6] Gale, D., *The Theory of Linear Economic Models*, McGraw Hill, New York, 1960.
[7] Henry, C. 'Indivisibilités dans une économie d'exchanges', *Econometrica* 38 (1970), 542-58.
[8] Shapley, L.S., *Markets as Cooperative Games*, The Rand Corporation, P-629, March 1955.
[9] Shapley, L.S., 'The solutions of a symmetric market game', *Annals of Mathematics Study* 40 (1959), 145-62.
[10] Shapley, L.S., *Values of Large Games V: An 18-Person Market Game*, The Rand Corporation, RM-2860, November 1961.
[11] Shapley, L.S., 'Complements and substitutes in the optimal assignment problem', *Nav. Res. Log. Q.* 9 (1962), 45-8.
[12] Shapley, L.S. and Shubik, M., 'Quasi-cores in a monetary economy with nonconvex preferences', *Econometrica* 34 (1966), 805-27.
[13] Shapley, L.S. and Shubik, M., 'Pure competition, coalitional power, and fair division', *Int. Econ. Rev.* 10 (1969), 337-62.
[14] Shapley, L.S. and Shubik, M., 'The Kernels and bargaining sets of market games', (forthcoming).
[15] Shitowitz, B., *Oligopoly in Markets with a Continuum of Traders*, The Hebrew University, Department of Mathematics, RM-63, August 1970.
[16] Shubik, M., 'Edgeworth market games', *Annals of Mathematics Study* 40 (1959), 267-78.
[17] Von Neumann, J. and Morgenstern, O., *Theory of Games and Economic Behavior*, Princeton University Press, Princeton, 1944.

WHAT IS AN APPLICATION AND WHEN IS THEORY A WASTE OF TIME?*

M. SHUBIK

Yale School of Organization and Management, Box 1A, New Haven, Connecticut 06520

1. Preamble

It is both a pleasure and an honor to be called upon to address the plenary session.

It is hard to believe that the first issue of *Management Science* came out in October 1954 with the first article by Harold Francis Smiddy—Vice President of General Electric and head of the Management Consulting Services. I learned much from him and I would also like to acknowledge West Churchman, Tjalling Koopmans, George Dantzig, Bill Cooper, Merrill Flood, Tom Cowan, Bob Thrall, Murray Geisler, Charles Hitch, Harlan Mills and Herbert Simon for their influence on my thoughts concerning management science.

It is over 32 years since the appearance of *Management Science*. Prior to coming to this meeting I skimmed all of the issues starting in 1954. This occasion and the *Management Science* retrospective has caused me to realize that, although some of us keep going with the hope that our best work still lies ahead, the promotion that is guaranteed to all academics is to the rank of extinguished scholar.

Before one can theorize usefully, it pays to know a few facts. The knowledge of how to manage an institution does not imply an ability to abstract, generalize, theorize or for that matter even know the facts.

The time was ripe for the development of a management science in the 1950s. Growth took place in several locations, as was evinced at both General Electric and in RAND's relationship with the Air Force. In each instance the activity called for a group of scientifically-oriented sheep to study corporate and military wolves with the encouragement of the latter—and amazingly it worked!—possibly because the sponsors paid the bills but did not listen too closely to what their advisors said.

The brief written preview of my talk is, to a great extent, an accurate summary of my central theme. Much of what I have to note is an embellishment of the observation that:

> Theory, for theory's sake, can easily degenerate into an uninteresting art form. Yet, practice without theory can quickly become a dull and dangerous occupation. Unfortunately, the world is a complicated place and complicated solutions and processes are often required to make complex organizations run. The ability to live with uncertainty and the insight into both one's professional powers and limitations is the sign of a mature management science.

The themes of my discussion were motivated by

(1) Deep professional interests in the uses and limitations of mathematical economics,

(2) Memories of several pleasant years with General Electric,

(3) My role as a consultant, and

(4) My own use of theory (or lack of use) in my own decisions and my view of the uses of theory by others in practice.

---

* Accepted by Donald G. Morrison, acting as Special Departmental Editor.
Plenary address, The Institute of Management Sciences Meeting, New Orleans, May 4–6, 1987.

It seems to me that it is courteous of a speaker, especially a plenary session speaker, to give the essence of one's talk in a sentence or two so that the audience can sleep, or do other things such as work on your own papers. Thus I offer four brief observations which I trust will stay within your plenary-talk alloted attention span:

(1) I want to help stamp out the phrase "rational behavior" and replace it with: "context rational behavior." I call your attention to the 1962 *Management Science* article by I. J. Good on "How Rational Should a Manager Be?" (Good 1962).

(2) I want to stress that the phrase "there is no substitute for knowing your business" is not merely an old warhorse folktale or anti-theoretical statement, but that it will probably be a central theoretical finding as we begin to understand how our thoughts, goals and actions are interrelated. It has been observed (I believe by Frank Lloyd Wright) that experts do not think about what to do—they know.

(3) Formalizing or routinizing for others one's learning from experience, especially habitual or common experience, is *far* harder than we appreciate. For example, try to describe what you use your little finger for—possibly the best way to learn what your little finger does is to lose it.

(4) Simulation sometimes (in relatively rigid systems) can help us do contingent planning. But in formal systems building, in the assumptions and the model we often exclude the potential for finding an adequate solution, because in reality:

(a) A substitute activity may come from a conceptually ruled out highly different source.

(b) An activity or item may be useless except for an extremely rare event (a champagne swizzle stick or an expert in bankruptcy liquidations are both rarely required in everyday operations).

(c) The activity or item may be totally useless at this time, even though at some point in distant history it might have had an important function (the human appendix and many parts of a bureaucratic organization are examples).

Having offered a few capsule comments I wish to return to the specific theme of management science.

1.1. On Management Science and Management Scientists

In the premier issue of *Management Science* (October 1954), the first article was "The Evolution of the Science of Managing in America" by Harold F. Smiddy and Lionel Naum. Harold Smiddy was an entrepreneur, visionary and fanatic. As such he was receptive to both scientists and charlatans. He knew that managers needed to listen. At General Electric he promoted "parallel paths" of promotion, reward and prestige for line and staff, for managers and scientists with the zeal of a Jesuit.

He recognized the need for structural administrative change to provide incentives to all—he probably had great influence on General Electric but not necessarily in the way he had planned. This may be illustrated by Pappert's principle. Marvin Minsky in his new book, *Society of the Mind*, (1986, p. 202) quotes Pappert's Principle:

"Some of the most crucial steps in mental growth are based not simply on acquiring new skills, but on acquiring new administrative ways to use what one already knows."

Pappert is concerned with mental growth of the child. Smiddy was concerned with the growth of the corporation. I suggest the two are more closely linked than we may usually suspect.

The Smiddy-Naum article is worth reading now. It contains an excellent historical summary, noting Charles Babbage's (1832) *Economy of Machinery and Manufacture*; F. W. Taylor (1911), *Principles of Scientific Management*; *The Gilbreth's Motion Study* (1911); and the American Society of Mechanical Engineers with 10-year reports (Alford 1912, 1922, 1932); as well as the role of Louis Brandeis, Bernard Baruch, Edward Filene, Mary Parker Follett and many others.

WHAT IS AN APPLICATION?

Smiddy imposed on the managers at General Electric a 6-weeks sojourn at Croton-on-Hudson—General Electric's internal business school. It was a blend of boot camp, camaraderie almost without rank, highly technical lectures by fine, but often narrow, professionals mixed in with pitches, dog and pony shows and vaguely muddled right wing pseudo-philosophy topped off with a large supply of many profound books which were not read.

I have taken the time to comment on Harold Smiddy and will comment on the changes in *Management Science* because from the start the complex interrelationships among
   practice
   theory and
   organization;
conscious individual optimization, team work and process were all recognized as components of management. I must add that underlying all of these is: how to learn and how to teach. These are central organizational problems for the individual and the organization.

Before I discuss what is an application, a few observations of the changes in *Management Science* over the years are called for. First I note that its title is: *Management Science* not *Entrepreneurial Science*. Those of us who are still optimists believe that there is both a science and an art to management and that some progress has been made. None but those with deep and mystical faith believe that we have much to say scientifically about entrepreneurship.

Until February 1965 there was one publication for all. Preachers, philosophers, practitioners and theorists wrote for and read one publication. In 1965 a series for Theory and a different series for Application were introduced.

In September 1975 the double series was abandoned with the creation of a new journal *Mathematics of Operations Research*. This gave recognition to the fact that the fields surrounding management science had developed in such a way that there was a relatively technical set of mathematical papers of interest primarily to those trained in applied mathematics. These papers, though of indirect value, are of little direct concern to either managers or most practitioners.

In 1983 the Management Science Roundtable was started in recognition of the importance to maintain, encourage and enlarge broad debate on basic issues. The success in development of management science and operations research has been marked by the growth of cumulative bodies of technical knowledge. But future success depends critically both upon the growth of technique and theory and upon being able to maintain a broad enough view of management science as a whole.

2. What Is an Application?

   Cooks, cook;
   Mathematical theorists produce theorems;
   Consultants hold hands—and good consultants
   like good psychiatrists—hold mirrors.
   Managers manage and good managers listen.
   Entrepreneurs—well no one including themselves know what they do.

Most individuals are
(1) too set in their ways,
(2) too busy, or
(3) uninterested

to be able to establish understanding with those outside their field. Those who are wise and perceptive may be able to settle for a higher order of human communication than "understanding" that is "trust." But once you have made the investment in your routines you are somewhat like T. S. Eliot's Rum Tum Tugger

> He do do
> What he will do
> And there's no doing
> Anything about it.

Given that much of my own work has involved Game Theory and that, if anything, the meaning of application is less well defined in game theory than say in linear programming, I will stick to T. S. Eliot's view of Rum Tum Tugger and confine my examples to the area in which I have spent much of my professional effort.

I begin with a stark example of practice without theory leading easily to a dangerous mindset. Many tactical situations are well modeled as two-person zero-sum games. A reasonable case can be made out for the maximin solution as a sensible though conservative way to play. In military terms one looks at *capabilities* and *intentions*. But in a zero-sum game when A attempts to maximize for himself this is equivalent to minimizing the gain of B. Thus B can assume that A's intentions are the worst from his point of view and he will use his capabilities accordingly.

This type of thinking when generalized to strategic nuclear warfare can lead to a dangerous and possibly even paranoid view of defense. The zero-sum analogy does not carry through easily to strategic war.

Is persuasion, explanation and illustration concerning the distinctions between a zero and nonzero sum game an application? I suggest that it is.

As a one time simulator, long range planner and anticipator it took me many years to understand that all of these activities are *short-term activities*—part of a process involving constant updating—rather than the production of a plan to be believed for any length of time. A good consultant, staffman and scholar must appreciate the difference between the mere exhibition of expertise and the influencing of process. A successful planning operation influences ways of thought and process. The actual plans are short-term manifestations of the activity. Academics and those without line responsibility may take pride in being Cassandras. The pride is misplaced. A Cassandra is remembered for being right and ineffective.

The major opening work in game theory appeared in the 1940s (Princeton University Press required a subsidy to publish it). Within a few years von Neumann and Morgenstern's book (1944) was hailed as one of the great intellectual achievements of the twentieth century as was noted by Merrill Flood (1956) in his discussion of the objective of TIMS in 1955. Soon after, especially among the economists, questions were raised as to whether game theory had lived up to its promise. This question was raised, in my opinion, because depending upon who you are what constitutes an application differs.

Mathematical programming, simulation, game theory and computers in general are changing the world considerably. But not merely or even primarily in the easily identifiable ways.

This is an imperfect world. New ideas, concepts and processes are often misapplied before they are well used. It has been said that consultants recognize only three tenses in the language of consulting

> The past horrendous,
> The present imperfect, and best of all
> The future perfect.

False hopes frequently fuel the first contract—reality starts to settle in subsequently. The discourse on: "The Researcher and the Manager: A Dialectic of Implementation" published in *Management Science* by Churchman and Schainblatt (1965) amply illustrates the difficulties in implementation.

WHAT IS AN APPLICATION?

Beged-Dov in *Management Science* (1966) provides a partial answer as to when managers may well communicate effectively with operations researchers—when the latter becomes managers. His article is entitled "Scientists and Engineers—The Emerging Managers of Business."

But back to game theory for examples concerning application. I suggest that there are at least *three* game theories:
(1) High church game theory
(2) Low church game theory and
(3) Conversational game theory

1. *High church game theory* is the domain of mathematics, axiom systems, formal systems and solution concepts. Much of it can verge on "art for art's sake." There are subindustries of mathematics such as the exhaustive exploration of all four-person games in characteristic function form; or quasi-philosophical occupations such as the search for a single perfect noncooperative equilibrium point as a normative solution to a game in strategic form.

Who are the sponsors, producers and consumers of this work? What are the motivations? What defines application? Probably the most honest answer is that to the "high church game theorist" the work is fun and the ability to have the time to do it is provided by academia and a few foundations.

At most an application is an application of the mathematical techniques to the social sciences. Much of my own work over the years falls into this category. Beyond von Neumann and Morgenstern (1944), Nash (1950) and Shapley (1953) have been the key conceptualizers of solutions.

Von Neumann provided the maximin solution for zero-sum games. Von Neumann and Morgenstern offered the characteristic function as the description of a game in cooperative or coalitional form and suggested the stable set solution. Nash generalized the original insight of Cournot to the *noncooperative equilibrium* solution. Shapley and Gillies explored the *core* of a cooperative game as a solution concept; and Shapley constructed the *value* as a one-point solution to a cooperative game.

There are other solutions which have been considered and I omit discussing the work of Schmeidler (1969), Selten (1975) and others, not because of my lack of appreciation but for lack of time, as this is not an exhaustive survey of game theory.

Each solution concept picks up a different viewpoint of multiperson decision making:

(1) The maximin solution of two-person zero-sum games concentrates on *pure opposition*;
(2) Nash equilibrium concentrates on individual strategic power;
(3) The core reflects the countervailing power of groups; and
(4) The value portrays a combinatoric marginal value productivity of all individuals.

Much of my work has consisted of four applications to the economic theory of the price system, to voting and to the assignment of joint costs and incentives.

(1) The core of Shapley and Gillies, in the context of economic exchange, is intimately related to the *contact curve* discussed by Edgeworth (1881) and under an appropriate characterization of the countervailing power of many leads to the emergence of a price system. Although I was able to link the work of Edgeworth with the core and the core with the price system (Shubik 1959) more mathematics than I have was required to fully analyze this linkage. The work of Debreu and Scarf (1963), Hildenbrand (1974) and many others has subsequently done this.

(2) A different way of approaching the emergence of an efficient price system is via the effect of decentralized decision making of many competitors. By means of a class of models entitled *strategic market games* utilizing the Nash noncooperative equilibrium solution, several of us (Shubik 1973, Dubey and Shubik 1978, Shapley and Shubik

1977 and others) have been able to connect the emergence of efficient prices with various price formation and auction mechanisms and the structure of monetary and financial institutions.

(3) Using Shapley's *value solution* Shapley and Shubik (1954) were able to suggest a *power index* for evaluating the relative power of voting blocks, building on the observation that the relationship between the number of shares held and power is not linear—at over 50% of the vote you have complete control.

(4) Again, using Shapley's value solution in an article in *Management Science* I was able to suggest application to the assignment of joint costs in ways consistent with the preservation of various incentives (Shubik 1962).

In all of these instances the work was based upon the body of mathematical game theory and was brought one step closer to operating concerns but without direct sponsorship or immediate application.

2. *Low church game theory* in contrast involves sponsorship to work on a specific application producing, if only for illustrative purposes, actual calculations and possibly a parametric sensitivity analysis. There is a body of literature involving military applications of two-person zero-sum games, concerning weapons evaluation, dueling, search and evasion games and also some tactical advertising and marketing models (for conflict bibliographies, see Shubik 1983). I believe that these applications have been of some, but nevertheless relatively modest, worth, but nowhere near the applied value of linear programming.

I suggest, however, that in terms of application and value to management at the highest levels *Conversational Game Theory* which consists of advice, suggestions and counsel as to how to think strategically is of considerable worth. It shows how to understand the presence of paradoxes in so-called rational behavior when there are two or more players.

The concepts of
the zero-sum game
the nonzero-sum game
the prisoner's dilemma

are now part of common strategic thought. Simple games such as chicken, the battle of the sexes (see Luce and Raiffe 1957), the dollar auction (Shubik 1971), and "so long, sucker" (Hausner, Nash, Shapley and Shubik 1964) are not as broadly known but are all simple enough and striking enough that they can be used to illustrate basic problems in conflict, coordination and cooperation. For instance, so long sucker was designed so that in order to win it is necessary to form coalitions. But that is not sufficient. At some point an individual must doublecross his partner. In the dollar auction, the highest bidder obtains a dollar, but both he and the second highest bidder are required to pay. If A has bid a dollar and B has bid 95 cents, is it "rational" for B to raise his bid to \$1.05 in order to try to cut his losses?

Without delving into the mysteries or mathematics of the Shapley value, important business lessons can be learned.

Harry and Joe formed a corporation each with 50% of the vote. Joe says to Harry: let's attract a talented third partner by each giving him 1% of the corporation. Does Harry understand that the simple majority weighted voting game with weights (49, 49, 2) is a symmetric game? Paradoxes in costing can also be illustrated (Shubik 1962, Young 1985).

The broad literature on nuclear gamemanship—on Mutually Assured Destruction —is rich with game theoretic paradoxes. If nothing else, the game theoretic formulation and way of thought has forced attention to the *strategic audit* (see Shubik 1983) of competitive problems. In particular the attention paid to understanding what are the true payoffs is at the very basis of strategic wisdom.

WHAT IS AN APPLICATION?

Is nuclear war a zero-sum game? Clearly not. If not then what are the payoffs, how are they measured?

How do we measure damage caused by cancer deaths related to smoking? Do we use
(1) Marginal deaths,
(2) Lost man years of productivity, or
(3) Pain months in dying?

If marginal deaths are used, cancer will dominate automobile accidents. If lost man years of productivity are considered, then, when age is taken into account, automobile accidents may dominate cancer. But are either of these indices the appropriate measure to describe the concern of the public in allocating funds? Perhaps the way death should be considered is in terms of the fear of its unpleasantness (or at least this should be part of the measure). It is one thing to consider months of pain in a terminal hospital and another being shot to death by a jealous spouse.

How do we measure convoy effectiveness?
(1) By tons of goods delivered or
(2) The ratio of submarines to convoy ships sunk or what?

This example made the rounds of operations researchers from World War II. The idea of a high damage exchange rate between submarines and convoy ships sunk may look attractive at first glance, but this criterion can be optimized by sending convoys consisting only of antisubmarine vessels.

Without high church game theory the concepts, illustrations and stories of conversational game theory would hardly exist and certainly would not have a coherent intellectual basis. Without conversational and high church game theory, sponsorship for low church game theory would hardly exist.

Application is not just calculation and specific problem solving. It is also concept clarification, education and changing modes of thought.

3. A Look into the Future

Many years ago Herbert Simon coined a marvelously intuitive phrase. It was "satisficing man".

Several decades and thousands of simulations later it is still a slogan looking for a subject. Simon would be among the first to admit that the study of human individual and organizational decision making is analogous to attacking the Hydra. For every head cut off nine more spring into being.

Even earlier than "satisficing man" words such as "holistic" and "gestalt" have been used to describe the smile on the vanished Cheshire cat of Alice-in-Wonderland—the mystical whole of an organization that we cannot quite put our finger on.

Marvin Minsky (1986, p. 27) has the following warning.

> We're often told that certain wholes "are more than the whole of their parts." We hear this expression with reverent words like "holistic" and "gestalt" whose academic tones suggest that they refer to clear and definite ideas. But I suspect the actual function of such terms is to anesthetize a sense of ignorance.

My brother, a distinguished pathologist, has observed that one of the purposes of the uses of Latin names in medicine is to provide an authoritative label to a phenomenon that is not well understood. This, in turn, may generate misplaced confidence by the layman.

In contrast with the Will-O-Wisp of organization, of our inability to verbalize much of our everyday experience such as how does "a glass of cold water really feel on a hot day after heavy work", many theorists have operated with the clear clean picture of

Completely informed, costlessly calculating *rational man* who knows what he wants; can express it; Bayesian updates rather than learns; has no personality and neither possesses virtues nor faults.

His progeny is *game theory man* who dwells in the rationalistic world with one extra twist—he recognizes the power, strategies and goals of others.

I want to stress that in much of our teaching and professional emphasis we have *completely reversed* the true role of models of rational man in contrast with what we are:

As finite process machines working on horrendously ill-defined and probably complex problems the human being complete with personality and emotions is far more complicated and subtle than "rational man". The emotions such as rage or fear are higher order processes not necessarily base features that saints should strive to get rid of to improve their decision making. There may be a limit to striving for self-control unless we wish to limit ourselves to more machine-like behavior to fit the Bed of Procustes of the simulations we can handle or the formal models we can analyze.

This point is elegantly made in a poem by Theodore Melnechuk which I steal directly[1] from Chapter 5 of Minsky's new book.

PUNCH AND JUDY, TO THEIR AUDIENCE

Our puppet strings are hard to see,
So we perceive ourselves as free,
Convinced that no mere objects could
Behave in terms of bad and good.

To you, we manikins seem less
than live, because our consciousness
is that of dummies, made to sit
on laps of gods and mouth their wit;

Are you, our transcendental gods,
likewise dangled from your rods,
and need, to show spontaneous charm,
some higher god's inserted arm?

We seem to form a nested set,
with each the next one's marionette
who, if you asked him, would insist
that he's the last ventriloquist.

The nested set may remind us of the Russian dolls within dolls, or, a little more fancily, Benoit Mandelbrot's "fractiles." The individual agents at the appropriate atomic level may well be relatively simple, but their organization and orchestration is not.

The future of management science in particular and the behavioral sciences in general lies in our understanding of:

Context rational behavior,
Bounded rationality,
Satisficing,
Artificial Intelligence and
Theories of Organization.

Computational methods, algorithms, computer applications and special mathematics can and must continue to develop. But, as has already been seen, the appropriate

---

[1] With permission of Theodore Melnechuk.

WHAT IS AN APPLICATION?

techniques are encouraged by perceived needs, be they the Berlin airlift; oil refinery problems, transportation or inventories and production.

When the problems are less well perceived such as in marketing, bidding, auctions, incentive systems, learning and organizational design the communication among managers, operators, theorists and formalizers becomes more tenuous. There is a danger that the mathematics is lavished on a playland of models constructed in academia whose purpose and end application is defined in terms of publication in the learned journals and promotion in the academies.

Basically reviewing the last 40 to 50 years it is easy to be an optimist. Move the clock back 50 years and ask about the state of the art in

High speed digital computers,
Personal computers,
Linear programming,
Input output studies,
Dynamic programming,
Inventory and production models,
Simulation and
Game theory,

to name a few. The body of knowledge is enormous; it is cumulative and growing substantially. Yet some of the big problems of 30-40-50 years ago seem bigger today—such as good models of organization, Artificial Intelligence and Bounded Rationality—but this is probably a perceptual illusion because today we understand far more about the difficulties and the problems that we hardly perceived 30 or 40 years ago. Wisdom has accompanied our understanding of our lack of knowledge. Possibly this is cold comfort, but it is some comfort.

I am an optimist for another reason. As responsible members of society and management scientists we have to be concerned with the major problems of society as well as the local problems of the corporation, the academy or the profession. These include

Terrorism,
Overall economic growth,
Pollution,
The threat of nuclear war,

to name a few. As someone originally schooled to have a virtually irrational belief in rational behavior and also with a pessimistic military and zero-game theory bias towards a "capabilities" or "maximin" view of some processes rather than an "intensions" view I am pleasantly surprised as to how *good matters are* not how bad they are.

A little bit of simple rational analysis—say as a consultant to the terrorists and food poisoners—could quickly suggest how much worse matters could be. Fortunately terrorist groups do not appear to employ management science and operations research departments.

World per capita growth prospects may not appear to be too rosy to some of us but the predictions of doom and gloom of the Club of Rome illustrate the limitations of simple simulations and extrapolations.

Pollution of many sorts is and may well be a major problem area in the coming century along with crime and grime. Yet descriptions of New York, London or Paris in the gay 1890's do not appear better than today. It is possible that it may already be too late to turn back the greenhouse effect, but I doubt it.

We have been without nuclear war for over 40 years. I find this extremely comforting given the number of powers involved and the state of the world. Perhaps the view of Samuel Hoffenstein (1947) written in his collection of poems, *Pencil in the Air*, expresses it more accurately than I can.

Fear not the atom in its fission
The cradle will outwit the hearse.
Man on this earth has a mission
To survive and keep on getting worse.

How do individuals assess and act on risk? How do formal organizations assess and act on risk? How do societies assess and act on risk?

These have been major questions for the last 10–20 years and will remain so for many years to come. They are central both to some of the major problems of society and to the development of scientific knowledge.

In a recent seminar at Yale on risk and organization the importance of understanding *fiduciary risk* emerged, i.e. risk taken by A to the account of others. But even more striking was the finding of the general passivity and slowness of eventual reaction of the United States government and its agencies to virtually all major problems involving risk. This includes terrorism, nuclear power, nuclear proliferation and various forms of pollution.

3.1. *If I Were Tsar*: *Confession of a Secret Moralist*

In some occupations, you have a conjecture, you prove a theorem and solve your problem. You have a reasonably well-defined inventory problem and you take care of it.

The development of a management science is part of an *ongoing process* not an isolated set of problems. The old office joke is real.

Are you helping to solve the problem?

Or are you part of the problem?

Static equilibrium is a snare and a delusion. The solutions (whatever that means) are in the process. The institutions of our society are the carriers of process.

Our problem today is not to bring back the RAND Corporation or General Electric or the Office of Naval Research of the 1950s. They were not stable. Whatever stability they had was dynamic. Our problem is to create the equivalent institutions for the next 20 years.

In spite of the writings and our introspection we do not have the blueprints for a new RAND that equals the old or how to provide charismatic leadership for new research. A new think tank is not a new intellectual hamburger franchise.

I close with three small proposals which I believe all point in the right direction though I do not imply that they are adequate alone.

(1) *The "Big Bucks" Gaming Institute.* Wall Street and Las Vegas are currently respectively a very big and a moderately large use-your-own-money gaming institution. When I was considerably younger and somewhat less skeptical I proposed a *Tax Gaming Institute*. Offer a prize of $100–200,000 tax free to the team of budding young lawyers who in the game find the biggest loopholes in a proposed tax law. In those days, holding to a too narrow view of process I thought that locating and removing loopholes was what the legislative process required rather than concealing and maintaining them so that there might be hidden gold for friends and political allies.

Leaving this specific example aside I believe that we have hardly touched the potential of the large game employed on large problems with substantial prizes to the participants.

(2) *Court Jesters, Astrologers, Grand Viziers and Senior Scientists.* I have a conjecture that the success at RAND, GE and elsewhere in the 1950s is in part explained by the possibility that they were to some extent luxury items which were paid for by the ruling dukes and kings. They were in part listened to and may have had great influence in molding the view of the crown princes, but were not regarded as a threat to their rulers.

WHAT IS AN APPLICATION?

The operations researchers and management consultants could wander all over GE in the 1950s. They solved some practical problems, but, even more important, a dialogue was established with many managers and mutual education took place.

American industry is rich enough to afford at least small inhouse groups of court astronomers or astrologers. They are not that expensive and they help make the juices flow. A very distinguished computer scientist, when he was a graduate student, spent a summer at a cancer research laboratory. At the end of the summer while being driven to the airport by the director he did nothing but complain about the lack of scientific attitude at the lab. The director replied "it has been wonderful having you, John. You have been very stimulating. Next year we will invite a poet to write poems about cancer." Not only should there be luxury groups they should be required to give top briefings two or three times a year.

(3) *Cross-Fertilization or Mutual Incomprehension Sessions.* Academia needs visits from major managers not to give "top-of-the-mountain" talks but, even at the risk of much mutual incomprehension, to establish a common language and at least some shared perceptions as to what the problems are. Mutual respect and an attempt to link practitioners with theorists in trust is a key element in setting up the conditions for process as a solution.

4. Conclusion

Life cannot wait until the sciences may have explained the universe scientifically. Practice without introspection or question of context may succeed in a narrow sense. Solving mathematical puzzles without worry concerning context can provide, for some, a satisfactory existence. But the science and art of management calls for more. An application is when the context is understood, the theory is relevant and the decision process is influenced. Theory may become a waste of time for all but the theorists when there is no concern for relevance or application beyond the self-perpetuation of the club.

References

BEGED-DOV, A. G., "Scientists and Engineers—The Emerging Managers of Business," *Management Sci.*, 12 (1966), B580–B591.
CHURCHMAN, C. W. AND A. H. SCHAINBLATT, "The Researcher and the Manager: A Dialectic of Implementation," *Management Sci.*, (1965), B69–B87.
DEBREU, G. AND H. S. SCARF, "A Limit Theorem on the Core of an Economy," *Internat. Economic Rev.*, 4 (1963), 235–246.
DUBEY, P. AND M. SHUBIK, "The Noncooperative Equilibria of a Closed Trading Economy with Market Supply and Bidding Strategies," *J. Economic Theory*, 17 (1978), 1–20.
EDGEWORTH, F. Y., *Mathematical Psychics*, Kegan Paul, London, 1881.
FLOOD, M., "The Objectives of TIMS," *Management Sci.*, 3 (1956), 178–184.
GOOD, I. J., "How Rational Should a Manager Be?," *Management Sci.*, 8 (1962), 383–393.
HAUSNER, M., J. F. NASH, L. S. SHAPLEY AND M. SHUBIK, "So Long Sucker, A Four Person Game," in M. Shubik (Ed.), *Game Theory and Related Approaches to Social Behavior*, Wiley, New York, 1964, 359–361.
HILDENBRAND, W., *Core and Equilibria of a Large Economy*, Princeton University Press, Princeton, NJ, 1974.
HOFFENSTEIN, S., *Pencil in the Air*, Doubleday, New York, 1947.
LUCE, R. D. AND H. RAIFFA, *Games and Decisions*, Wiley, New York, 1957.
MINSKY, M., *The Society of Mind*, Simon and Schuster, New York, 1986.
NASH, J. F., JR., "Equilibrium Points in n-Person Games," *Proc. Nat. Acad. Sci. USA*, 36 (1950), 48–49.
SCHMEIDLER, D., "The Nucleolus of a Characteristic Function Game," *SIAM J. Appl. Math.*, 17 (1969), 1163–1170.
SELTEN, R., "Reexamination of the Perfectness Concept for Equilibrium Points in Extensive Games," *Internat. J. Game Theory*, 4 (1975), 25–55.

SHAPLEY, L. S., "A Value for n-Person Games," in *Contributions to the Theory of Games*. Vol. 2, H. Kuhn and A. W. Tucker (Eds.), Princeton University Press, Princeton, NJ, 1953, 307–317.

—— AND M. SHUBIK, "A Method for Evaluating the Distribution of Power in a Committee System," *Amer. Political Science Rev.*, 48 (1954), 787–792.

—— AND ——, "Trade Using One Commodity as a Means of Payment," *J. Political Economy*, 85 (1977), 937–968.

SHUBIK, M., "Edgeworth Market Games," in *Contributions to the Theory of Games*. Vol. 4, A. W. Tucker and R. D. Luce (Eds.), Princeton University Press, Princeton, NJ, 1959, 267–278.

——, "Incentives, Decentralized Control, the Assignment of Joint Costs and Internal Pricing," *Management Sci.*, 8 (1962), 325–343.

——, "The Dollar Auction Game," *J. Conflict Resolution*, 15 (1971), 109–111.

——, "Commodity Money, Credit, and Bankruptcy in a General Equilibrium Model," *Western Economic J.*, 11 (1973), 24–38.

——, *The Mathematics of Conflict*, North-Holland, Amsterdam, 1983a.

——, "The Strategic Audit: A Game Theoretic Approach to Corporate Competitive Strategy," *Managerial and Decision Economics*, 3 (1983b).

SMIDDY, H. F. AND L. NAUM, "The Evolution of the Science of Managing in America," *Management Sci.*, 1 (1954), 1–31.

VON NEUMANN, J. AND O. MORGENSTERN, *The Theory of Games and Economic Behavior*, Princeton University Press, Princeton, NJ, 1944.

YOUNG, H. P. (ED.), *Cost Allocation: Methods, Principles, Applications*, North-Holland, Amsterdam, 1985.

Name index

Abegglen, J.C. 65, 91
Abelson, R.P. 240
Abt, C.C. 429
Adelman, M. 159
Alexander, S. 57–8
Alford 540
Allen, R.G.D. 59
Anderson, R.M. 240
Ansoff, H.I. 233
Argyris, C. 65, 84, 108
Arrow, K.J. 113, 131, 149–50, 157–8, 163, 170, 189, 197, 204, 214, 510
Aumann, R.J. 168, 199, 205, 224, 226, 237
Axelrod, R. 241

Babbage, C. 540
Bagehot, W. 171
Bain, J.S. 79, 86, 160, 165, 230, 242
Balderston, F.E. 48, 53, 161
Barnard, C.I. 65, 84
Barraclough, S.L. 125–7
Baruch, B. 540
Basmann, R.L. 52
Bass, F.M. 150
Baumol, W.J. 155, 167, 230, 234
Bayes 240
Beckmann, M.J. 328, 335
Beged-Dov, A.G. 543
Bellman, R. 62, 166, 279
Bergler, E. 439
Berkowitz, M.L. 99
Berle, A.A. 86
Bertrand, J. 14, 36, 199, 201, 229, 236, 247, 289–90, 311, 313, 329, 333, 341
Black, R.D.C. 148–9, 170
Boehm Bawerk *see* Von Böhm-Bawerk
Bonini, C.P. 162
Boulding, K. 59, 156, 167
Brandeis, L. 540
Brems, H. 10, 19, 160
Brummet, R.L. 105, 108
Bryton, B. 48
Burck, G. 26
Burger, E. 354
Bush, R.R. 60, 66

Caesar, Julius 85
Callois, R. 437, 440
Carroll, L. 170

Case, J.H. 231
Cassady, R. 180
Chamberlin, E.H. 29, 36, 79, 99, 158, 162, 198, 201, 229–30, 289, 312–13, 324
Cherry, C. 21
Churchill, B.C. 87
Churchman, C.W. 70, 539, 542
Clark, J.M. 100
Clarkson 47
Clarkson, G.P.E. 162
Cohen, K.J. 48, 52, 161–2
Coleman, J. 429
Collingwood, S.D. 170
Marquis de Condorcet 170
Conway, R.W. 50
Coombs, C.H. 99, 376
Cooper, B. 539
Cournot, A.A. 14, 36, 79, 99, 158–9, 167, 197–9, 201, 206–8, 228–9, 247, 262, 289, 304, 312, 329, 333, 520, 543
Cowan, T. 539
Cowles, A. 31
Cramer, H. 26
Crecine, J.P. 161
Cross, J. 168, 180–81
Cyert, R.M. 48, 51, 150, 153, 155, 162, 233, 408

Dalkey, N. 19, 434
Dantzig, G. 60, 112–13, 154, 539
Davis, J. 480
Davis, R.L. 99, 376
Debreu, G. 113, 131, 150, 153, 157, 168–70, 185, 197, 204, 495, 537, 543
Dent, J.K. 90
Dewing, A.S. 89–90
Dierker, H. 239
Dirham, J.B. 159
Dodgson, C.L. 170
Domike, A.L. 125–7
Dorfman, R. 154, 166
Dostoevski 439
Dougall, H.E. 511
Downs, A. 170
Dresher, M. 375
Dreyfus, H. 436
Dubey, P. 236, 241, 351, 360, 367, 543
Dunlop, J. 5

Edgeworth, F.Y. 36, 163, 168, 197–9, 201, 203–5, 207, 215, 229, 236, 247, 261, 265, 289–90, 302–4, 311–13, 324, 328, 330, 333, 520, 543
Edwards, W. 61
Eells, R. 60, 86
Eilenberg, S. 360
Eisenberg, M. 358
Eliot, T.S. 542
Emett 45
Estes, W. 66

Farquharson, R. 170
Farrell, M.J. 187
Feeney, G. 99, 380, 394
Feigenbaum, E. 48, 162
Feldman, J. 162
Fellner 229
Fellner, W. 78, 86
Filene, E. 540
Firestone, O.J. 165
Flood, M. 99, 376, 380, 539, 542
Follett, M.P. 540
Forrester, J.W. 48, 52
Fouraker, L. 46, 60, 70, 99, 162, 380
Freud, S. 58, 75
Friedman, J. 162, 351

Gabszewicz, J.J. 235, 351
Gale, D. 157, 166, 216, 536
Gaskins, D.W. 351
Gass, S. 60
Geanakoplos, J. 241
Geisler, M. 539
Gillies 543
Gillies, D.B. 204
Goffman, E. 439
Goldhamer, H. 433
Gomory, R.E. 166, 187
Good, I.J. 540
Goode, H.H. 60
Gothman, H.G. 511
Graham, B. 161
Green, J. 232, 240
Grodal, B. 239
Grossman, S.J. 232

Hahn, F.H. 157, 171, 214
Harsanyi, J. 140, 168, 189, 238
Hart, A.G. 21
Hart, O.D. 231–2, 236
Hausner, M. 391, 544
Helmer, O. 434
Henderson, I.M. 156
Hensaw, R.C. 162

Hicks, J.R. 156
Hildenbrand, W. 543
Hirschleifer 225
Hitch, C. 539
Hobbes, T. 516, 518
Hoffenstein, S. 547
Hoffman, A. 276
Hoggatt, A. 48, 53, 99, 161–2, 380, 435
Hotelling, H. 229–30
Houthakker 153
Huizinga, J. 437
Hurwicz, L. 113, 157

Iklé, F.C. 181
Isoda, K. 354

Jackson, J. 60, 162
Johnson, B.M. 50

Kahn, A.E. 159
Kakutani 367
Kamien, M.I. 351
Kaplan, A.D.H. 87
Kaysen, C. 159
Keynes, J.M. 57, 166
Khaldun, I. 65
Kinnaird 518
Knight, F.H. 14, 21, 27–8
Koopmans, T. 539
Kramer, G.H. 170
Kreps, D.N. 232, 239
Kuhn 166
Kuhn, H.W. 63, 105

Laderman, J. 81
Lancaster, K.J. 154
Lang, T. 105, 109
Larson, H.D. 518
Lasswell, H.D. 59
Lemke, C.E. 166
Leontief, W.W. 166
Levi, A. 240
Levitan, R. 162–3, 242, 308, 311, 313, 316, 340, 348, 409, 415, 423
Lewin, B. 125
Littauer, S.B. 81
Lockhart, S. 435
Lorange, P. 233
Luce, R.D. 59, 81, 129, 137, 374, 380, 544

Machina, M.J. 240
Machlup, F. 86
Machol, R.E. 60
Mack, R.P. 52
Mandelbrot, B. 546

Mansfield, E. 154
March, J.G. 48, 51, 60, 65, 84, 150, 155, 162, 233, 408
Marget, A.W. 28
Markowitz, H.M. 165
Marris, R. 155, 160, 165
Marschak, J. 21, 27, 65, 68, 82, 84, 110
Marshall, A. 57, 148, 154
Mas-Colell, A. 236
Mason, E.S. 159
Maxwell, W.L. 50
Mayberry, J. 261, 264, 272, 275, 427
McCarthy, J. 47, 68
McCracken, D. 69
McFarland, W.B. 105, 109
McKenzie, L.W. 157
Means, G.C. 86
Melnechuk, T. 546
Miller, G. 68
Mills, H. 539
Minsky, M. 47, 60, 83, 97, 436, 540, 545–6
Mintz, S.W. 131
Montgomery, D. 360
Moore, O.K. 99
Morgenstern, O. 3–4, 6, 10, 14, 18–19, 28, 39, 62–3, 106–7, 150, 167, 227, 252, 261–2, 371, 374–7, 380, 424–5, 509–10, 542–3
Morrison, D.G. 539
Mosteller, F. 60, 66, 153
Musgrave, R.A. 163

Nash, J.F. 7, 20, 70, 99, 168, 201, 228, 235, 261, 264, 272, 275, 377, 391, 543–4
Naum, L. 540
Naylor, T.H. 233
Negishi, T. 158
Nelson, R. 209, 234
Newton 252–3, 320
Newton, I. 159
Nikaido, H. 157, 354
Nogee, P. 153
Northrop, G.M. 433
Novshek, W. 236, 351
Nti, K. 230, 351

Orcutt, G.H. 47, 161

Panzer, J.C. 230
Pappert 540
Pappert, S. 436
Peacock, A.T. 408
Phillips, T.R. 65
Pigou, A.C. 163
Plott, C.R. 170, 234

Ponssard, J.P. 225
Pontryagin, L.S. 166
Porter, M.E. 233
Postlewaite, A. 236

Quandt, R.E. 156, 234

Radner, R. 65, 84, 110, 157, 206, 240
Raiffa, H. 59, 81, 129, 137, 374, 380, 544
Rapaport, A. 167
Ratoosh, P. 70
Ricardo, D. 127
Riker, W.H. 170
Roberts, J. 353
Robinson, J. 158–9, 229
Rogawski, J.D. 236
Rosenkranz, F. 234
Rosenthal, F. 65
Ross, S. 232

Samuelson, P.A. 4, 152–3, 156, 164, 166
Savage, L.J. 29, 59, 61, 375
Scarf, H. 157, 166, 168, 185, 205–6, 495, 537, 543
Schainblatt, A.H. 542
Schelling, T.C. 167, 375
Scherer, F.M. 231
Schiff, M. 105, 109
Schmalensee, R. 230
Schmeidler, D. 236, 496, 543
Schumpeter, J.A. 156, 160, 240
Schwartz, N.L. 351
Seigel, S. 376
Selten 239
Selten, R. 543
Selznick, P. 64, 84, 90
Shackle 240
Shackle, G.L. 25, 61–2
Shakespeare, W. 536
Shannon, C. 22, 60, 67
Shapley, L.S. 3, 63, 70, 105, 108, 115, 132, 136–7, 156, 158, 164, 168, 184, 186, 198–9, 201, 206, 208, 211, 216, 235–6, 247, 261, 271, 276, 304, 308, 311, 313, 316, 391, 495, 512, 543–4
Shitovitz, B. 237
Shubik, M. 21, 40, 43, 48, 62, 78, 81, 92, 98–9, 103, 113, 129, 132, 137, 155–6, 161–4, 168, 170, 184–6, 198–9, 202, 205–6, 208, 211, 216, 228, 230, 233–7, 239–42, 261–4, 272, 274–5, 289–90, 304, 308, 311, 313, 316, 328, 332, 335, 340, 348, 351, 380, 391, 409, 414, 423, 426, 435, 493, 495, 498, 543–4
Siegel, S. 46, 60, 66, 70, 97, 99, 162, 380

Simon 47, 436
Simon, H. 58, 60, 65, 83–4, 153, 155, 233–4, 539, 545
Singleton, R. 276
Smiddy, H.F. 539–41
Smith, A. 206
Smith, V.L. 162, 234
Sobel, M.J. 367
Sokal, R.R. 154
Solow 166
Sonnenschein, H. 236, 351, 353
Speier, H. 433
Spence, A.M. 351
Spencer 516
Stackelberg see Von Stackelberg
Stern, D. 99, 162, 380
Stigler, G.J. 77
Stone, J.S. 99
Sun, T. 65

Taylor, F.W. 540
Telser, L.G. 153
Thomas, C. 427
Thompson, G.L. 103
Thrall, R.M. 99, 376, 539
Tinbergen, J. 101
Tolstoy, L.N. 59
Triffin, R. 158, 230
Tucker, A.W. 63, 105, 375
Turing, A.M. 409

Vial, J.P. 235, 351
Vickrey, W. 156, 170, 261

Viner 14
Viner, J. 58
Voltaire 516–18
Von Böhm-Bawerk, E. 198, 520–21, 531–2, 534, 537
Von Neumann, J. 3–4, 6, 10, 18–19, 28, 39, 62–3, 106–7, 150, 154, 159, 166–7, 227, 252, 261, 371, 374–7, 380, 424–5, 509–10, 542–3
Von Stackelberg, H. 36, 79, 159, 229–30, 247, 408

Wagner, H.M. 161, 167
Wald 18
Wald, A. 59
Walras, L. 204–5, 207
Weiss, L. 81
Williams, S.H. 170
Williamson, O.E. 232
Willig, R.D. 230
Wilson, R. 170, 232, 239
Winer, B.J. 419
Wing, R.L. 429
Winter, S. 209, 234
Wohlstetter, R. 434
Wolf, J. 435
Wolfe, E.R. 131
Wolfe, P. 112–13, 375
Wright, F.L. 540

Yance, J.V. 48, 52
Yardley, H.O. 438
Young, H.P. 544

Economists of the Twentieth Century

Monetarism and Macroeconomic
Policy
Thomas Mayer

Studies in Fiscal Federalism
Wallace E. Oates

The World Economy in Perspective
Essays in International Trade and European Integration
Herbert Giersch

Towards a New Economics
Critical Essays on Ecology, Distribution and Other Themes
Kenneth E. Boulding

Studies in Positive and Normative Economics
Martin J. Bailey

The Collected Essays of Richard E. Quandt (2 volumes)
Richard E. Quandt

International Trade Theory and Policy
Selected Essays of W. Max Corden
W. Max Corden

Organization and Technology in Capitalist Development
William Lazonick

Studies in Human Capital
Collected Essays of Jacob Mincer, Volume 1
Jacob Mincer

Studies in Labor Supply
Collected Essays of Jacob Mincer, Volume 2
Jacob Mincer

Macroeconomics and Economic Policy
The Selected Essays of Assar Lindbeck
Volume I
Assar Lindbeck

The Welfare State
The Selected Essays of Assar Lindbeck
Volume II
Assar Lindbeck

Classical Economics, Public Expenditure and Growth
Walter Eltis

Money, Interest Rates and Inflation
Frederic S. Mishkin

The Public Choice Approach to Politics
Dennis C. Mueller

The Liberal Economic Order
Volume I Essays on International Economics
Volume II Money, Cycles and Related Themes
Gottfried Haberler
Edited by Anthony Y.C. Koo

Economic Growth and Business Cycles
Prices and the Process of Cyclical Development
Paolo Sylos Labini

International Adjustment, Money and Trade
Theory and Measurement for Economic Policy
Volume I
Herbert G. Grubel

International Capital and Service Flows
Theory and Measurement for Economic Policy
Volume II
Herbert G. Grubel

Unintended Effects of Government Policies
Theory and Measurement for Economic Policy
Volume III
Herbert G. Grubel

The Economics of Competitive Enterprise
Selected Essays of P.W.S. Andrews
*Edited by Frederic S. Lee
and Peter E. Earl*

The Repressed Economy
Causes, Consequences, Reform
Deepak Lal

Economic Theory and Market Socialism
Selected Essays of Oskar Lange
Edited by Tadeusz Kowalik

Trade, Development and Political Economy
Selected Essays of Ronald Findlay
Ronald Findlay

General Equilibrium Theory
The Collected Essays of Takashi Negishi
Volume I
Takashi Negishi

The History of Economics
The Collected Essays of Takashi Negishi
Volume II
Takashi Negishi

Studies in Econometric Theory
The Collected Essays of Takeshi Amemiya
Takeshi Amemiya

Exchange Rates and the Monetary System
Selected Essays of Peter B. Kenen
Peter B. Kenen

Econometric Methods and Applications
(2 volumes)
G.S. Maddala

National Accounting and Economic Theory
The Collected Papers of Dan Usher, Volume I
Dan Usher

Welfare Economics and Public Finance
The Collected Papers of Dan Usher, Volume II
Dan Usher

Economic Theory and Capitalist Society
The Selected Essays of Shigeto Tsuru, Volume I
Shigeto Tsuru

Methodology, Money and the Firm
The Collected Essays of D.P. O'Brien
(2 volumes)
D.P. O'Brien

Economic Theory and Financial Policy
The Selected Essays of Jacques J. Polak
(2 volumes)
Jacques J. Polak

Sturdy Econometrics
Edward E. Leamer

The Emergence of Economic Ideas
Essays in the History of Economics
Nathan Rosenberg

Productivity Change, Public Goods and Transaction Costs
Essays at the Boundaries of Microeconomics
Yoram Barzel

Reflections on Economic Development
The Selected Essays of Michael P. Todaro
Michael P. Todaro

The Economic Development of Modern Japan
The Selected Essays of Shigeto Tsuru
Volume II
Shigeto Tsuru

Money, Credit and Policy
Allan H. Meltzer

Macroeconomics and Monetary Theory
The Selected Essays of Meghnad Desai
Volume I
Meghnad Desai

Poverty, Famine and Economic Development
The Selected Essays of Meghnad Desai
Volume II
Meghnad Desai

Explaining the Economic Performance of Nations
Essays in Time and Space
Angus Maddison

Economic Doctrine and Method
Selected Papers of R.W. Clower
Robert W. Clower

Economic Theory and Reality
Selected Essays on their Disparities and Reconciliation
Tibor Scitovsky

Doing Economic Research
Essays on the Applied Methodology of Economics
Thomas Mayer

Institutions and Development Strategies
The Selected Essays of Irma Adelman
Volume I
Irma Adelman

Dynamics and Income Distribution
The Selected Essays of Irma Adelman
Volume II
Irma Adelman

The Economics of Growth and Development
Selected Essays of A.P. Thirlwall
A.P. Thirlwall

Theoretical and Applied Econometrics
The Selected Papers of Phoebus J. Dhrymes
Phoebus J. Dhrymes

Innovation, Technology and the Economy
The Selected Essays of Edwin Mansfield
(2 volumes)
Edwin Mansfield

Economic Theory and Policy in Context
The Selected Essays of R.D. Collison Black
R.D. Collison Black

Location Economics
Theoretical Underpinnings and Applications
Melvin L. Greenhut

Spatial Microeconomics
Theoretical Underpinnings and Applications
Melvin L. Greenhut

Capitalism, Socialism and Post-Keynesianism
Selected Essays of G.C. Harcourt
G.C. Harcourt

Time Series Analysis and Macroeconometric Modelling
The Collected Papers of Kenneth F. Wallis
Kenneth F. Wallis

Foundations of Modern Econometrics
The Selected Essays of Ragnar Frisch
(2 volumes)
Edited by Olav Bjerkholt

Growth, the Environment and the Distribution of Incomes
Essays by a Sceptical Optimist
Wilfred Beckerman

The Economics of Environmental Regulation
Wallace E. Oates

Econometrics, Macroeconomics and Economic Policy
Selected Papers of Carl F. Christ
Carl F. Christ

Strategic Approaches to the International Economy
Selected Essays of Koichi Hamada
Koichi Hamada

Economic Analysis and Political Ideology
The Selected Essays of Karl Brunner
Volume One
Edited by Thomas Lys

Growth Theory and Technical Change
The Selected Essays of Ryuzo Sato
Volume One
Ryuzo Sato

Industrialization, Inequality and Economic Growth
Jeffrey G. Williamson

Economic Theory and Public Decisions
Selected Essays of Robert Dorfman
Robert Dorfman

The Logic of Action One
Method, Money, and the Austrian School
Murray N. Rothbard

The Logic of Action Two
Applications and Criticism from the Austrian School
Murray N. Rothbard

Bayesian Analysis in Econometrics and Statistics
The Zellner View and Papers
Arnold Zellner

On the Foundations of Monopolistic Competition and Economic Geography
The Selected Essays of B. Curtis Eaton and Richard G. Lipsey
B. Curtis Eaton and Richard G. Lipsey

Microeconomics, Growth and Political Economy
The Selected Essays of Richard G. Lipsey
Volume One
Richard G. Lipsey

Macroeconomic Theory and Policy
The Selected Essays of Richard G. Lipsey
Volume Two
Richard G. Lipsey

Employment, Labor Unions and Wages
The Collected Essays of Orley Ashenfelter
Volume One
Edited by Kevin F. Hallock

Education, Training and Discrimination
The Collected Essays of Orley Ashenfelter
Volume Two
Edited by Kevin F. Hallock

Economic Institutions and the Demand and Supply of Labour
The Collected Essays of Orley Ashenfelter
Volume Three
Edited by Kevin F. Hallock

Monetary Theory and Monetary Policy
The Selected Essays of Karl Brunner
Volume Two
Edited by Thomas Lys

Macroeconomic Issues from a Keynesian Perspective
Selected Essays of A.P. Thirlwall
Volume Two
A.P. Thirlwall

Money and Macroeconomics
The Selected Essays of David Laidler
David Laidler

The Economics and Politics of Money
The Selected Essays of Alan Walters
Edited by Kent Matthews

Economics and Social Justice
Essays on Power, Labor and Institutional
Change
David M. Gordon
*Edited by Thomas E. Weisskopf and
Samuel Bowles*

Practicing Econometrics
Essays in Method and Application
Zvi Griliches

Economics Against the Grain
Volume One
Microeconomics, Industrial Organization and
Related Themes
Julian L. Simon

Economics Against the Grain
Volume Two
Population Economics, Natural Resources and
Related Themes
Julian L. Simon

Advances in Econometric Theory
The Selected Works of Halbert White
Halbert White

The Economics of Imperfect Knowledge
Collected Papers of G.B. Richardson
G.B. Richardson

Economic Performance and the Theory of
the Firm
The Selected Papers of David J. Teece
Volume One
David J. Teece

Strategy, Technology and Public Policy
The Selected Papers of David J. Teece
Volume Two
David J. Teece

The Keynesian Revolution, Then and Now
The Selected Essays of Robert Eisner
Volume One
Robert Eisner

Investment, National Income and
Economic Policy
The Selected Essays of Robert Eisner
Volume Two
Robert Eisner

International Trade Opening and the
Formation of the Global Economy
Selected Essays of P. J. Lloyd
P. J. Lloyd

Production, Stability and Dynamic Symmetry
The Selected Essays of Ryuzo Sato
Volume Two
Ryuzo Sato

Variants in Economic Theory
Selected Works of Hal R. Varian
Hal R. Varian

Political Economy, Oligopoly and Experimental Games
The Selected Essays of Martin Shubik
Volume One
Martin Shubik

Money and Financial Institutions
A GAME THEORETIC APPROACH
The Selected Essays of Martin Shubik
Volume Two
Martin Shubik

DATE DUE

DEC 05 2003		
OCT 11 2007		
		Printed in USA

HIGHSMITH #45230